ON FRAZER'S RIVER.

of twelve miles an hour could be maintained on the navigable reaches, and of roads, the distance might be accomplished in the following time :

	Days.	Hours.	Mins.
at the rate of 12 miles an hour	5	2	20
at the rate of 7 miles an hour	3	1	0
above	0	13	0
	8	16	20

een hours and twenty minutes.

ior, being within two days journey of Toronto, if proper land roads were made the navigable reaches between the former place and Frazer's River, the total Langley might be accomplished in twelve, or, making every allowance for ac-, that is, during the season of navigation.

REPORT

ON THE

EXPLORATION OF THE COUNTRY

BETWEEN

Lake Superior and the Red River Settlement,

AND

BETWEEN THE LATTER PLACE AND THE ASSINIBOINE AND SASKATCHEWAN.

BY S. J. DAWSON, ESQUIRE, C. E.

First Published in 1859 by order of The Legislative
Assembly, Toronto
First Greenwood Reprinting, 1968
Library of Congress Catalog Card Number: 68-55884

PRINTED IN THE UNITED STATES OF AMERICA

RETURN

To an Address from the Legislative Assembly to His Excellency the Governor General, dated the 23rd ult., praying His Excellency to cause to be laid before the House "copies of the Reports and Plans (subsequent to those already laid before the "House) of the Exploration of the country west of Lake Superior, conducted by S. "J. Dawson, Esq., C. E., and party, during the last two years."

By Command,

C. ALLEYN,
Secretary.

Secretary's Office,
Toronto, 16th March, 1859.

REPORTS.

Red River Settlement,
4th July, 1858.

Sir,—I beg leave to acknowledge the receipt of your letters of the 16th and 20th April, containing instructions for the guidance of the Expedition during the present season. These instructions it will be our endeavour to carry out to the satisfaction of the Government.

I have now the honor to inform you, that as soon as the exploratory surveys, in progress at the date of my last report, were completed, I set out on the excursion I had proposed by way of the Manitouba and Winnipegoos Lakes to the Saskatchewan River, and returned by the Assiniboine, reaching this place on the 29th ult.

On my arrival I found the men brought here by Professor Hind waiting for me,—the Professor himself having gone west to the Souris River, and in order that as little time as possible might be lost, I immediately dispatched two of my assistants with eleven men and two canoes to commence the work indicated in your instructions, retaining one canoe and five men until such time as I could collect provisions enough to prevent the necessity of sending immediately to Lake Superior, where Professor Hind had left the supplies, bringing here only enough for his own party. We have now obtained a sufficient supply and are about to set out.

As the time at my disposal will not admit of my reporting so fully as I could wish on the various surveys which have been accomplished since the date of my last report, I shall for the present confine myself to a brief account of our last expedition, and a description of the extensive region through which we travelled.

On the 10th of May, having provided ourselves with such supplies as the settlement could afford, we crossed over to Manitouba Lake. There we embarked in canoes, and had a very tedious passage against strong head winds to the northwest end of Winnipegoos Lake; from whence we crossed by the Mossy Portage to Lac Bourbon.

Leaving my assistants to measure the distance and ascertain the difference of level between the lakes just named, I descended the Saskatchewan to the Grand Rapid and examined it. Returning again to the Mossy Portage I divided the party —sending my Chief Assistant, Mr. Wells, back by the western coast of Winnipegoos Lake, Lac Dauphin and the little Saskatchewan, as detailed in his report whicht I send herewith, while with the other division of the party I ascended Swan River, crossed from thence to Fort Pelly and came down by the Assiniboine.

In order to be the better comprehended in describing the general appearance of the country, I enclose a rough sketch,* hastily compiled from our notes, to which I would respectfully refer you.

A range of high lands, it will be seen, extends south-eastward from the Pasquia Mountain on the Saskatchewan, in latitude 35° 30 North, to the United States boundary line. This range has in all probability, at some period, formed the south-western embankment of a great inland sea, which covered the valley of Red River, and comprised within its mass Lakes Winnipeg, Winnipegoos, Manitouba and the numerous smaller lakes which are spread over the great alluvial flat in which they lie. The country, bounded on one side by this range, and on the other by Lake Winnipeg, and the high lands to the eastward of Red River is an almost unbroken level, sloping very slightly to the Red River and Lake Winnipeg.

Part of this extensive tract is open prairie land, but by far the greater portion is densely wooded. A line drawn north 75° west from the confluence of Red River with Lake Winnipeg to Lac Dauphin would pass through about an equal extent of wood land and prairie. From thence northward, a forest but rarely broken by prairie openings extends to the Saskatchewan. To the south the country becomes more open, until, on nearing the Assiniboine, the woods entirely disappear, and an apparently boundless

* The sketch above referred to is embraced in the general map now being published.

prairie spreads out on every side. The streams, however, are all bordered more or less with wood; a heavy growth of oak, elm, basswood, &c., extends in many places for a mile or two from the banks of the Assiniboine.

Proceeding by the road from the Red River to Manitouba Lake, the country for the first twelve miles or so presents the appearance of an unbroken level with clumps of trees rising here and there like islands in an otherwise boundless ocean. Further on, the wood becomes more frequent, and sometimes the prospect seems bounded by forests; on approaching these, however, other prairies open up, and other woods appear, and in this way wood-land and prairie alternate all the way to Manitouba Lake; although the ground seems level it is not precisely so, but slightly rolling or undulating. The elevations are of every width, from half a mile upwards, and run in a direction from north-west to south-east; between them in most places the ground is more or less marshy and covered with low bushes and willows, or presenting ponds growing bulrushes and rank grass. The road is, however, sufficiently dry to be travelled by wheeled vehicles at all times during the summer season. Sometimes little stony ridges occur, marking what at one time has been the shore of a shallow lake.

At the terminus of the road on Manitouba there is a small settlement, and the settlers are of opinion that their land is superior even to the soil at Red River, while it is not, like it, subject to be overflowed. My own opinion is, that, as regards the soil, it is precisely of the same character.

The north-eastern shore of Manitouba Lake, the coast by which we passed, is low, and of a character so uniform that the same description will apply throughout. By the action of the water, or ice, or both combined, a high beach of shingle has been thrown up, consisting of water-worn fragments of limestone mixed with occasional boulders of granite. On the top of this ridge there is generally a dense growth of wood, while, between it and the main land, an open marsh, varying in width from half a mile to two miles, extends along the whole coast, broken only by occasional points of higher land, which run down to the Lake. When we passed, the marsh was covered with withered bulrushes and long grass, which, although of last year's growth, still evinced the rankness of the vegetation peculiar to this region. The stems of some of the bulrushes on being measured were found to be an inch and three quarters in diameter. From the marsh, the main land, a rich alluvial soil, gradually rises to a moderate elevation, and is not subject to be inundated. The country bordering on the lower end of Winnipegoos Lake and the Singuissippi River, the stream which connects it with Manitouba, is of the same description; but about the middle of Winnipegoos Lake the land becomes slightly higher and the marshes disappear. The limestone rock then shows itself for a short distance, rising in horizontal strata to an elevation of 30 feet or so above the Lake. At the Mossy Portage a comparatively barren ridge separates the waters of Winnipegoos Lake from those of Lac Bourbon.

From the latter Lake to the Grand Rapid of the Saskatchewan the country has not a very inviting appearance. In many places the bare limestone rock appears on the surface, in others a thin coating of vegetable mould over it, scarcely supports a stunted growth of cypress, spruce and aspen. Some of the islands however, and there are many of them, appear to be fertile, especially at a little rapid just above Cross Lake. There the wood is of a large growth, and, although it was so early in the season (the 30th of May) when we passed that ice was still visible on the shores of Lac Bourbon, the foliage at these Islands was almost fully developed.

The Grand Rapid is about three miles in length, varying in width from 1800 feet at the head to about ¾ of a mile at the lower end. On the south side a perpendicular cliff of limestone rock rises abruptly from the water's edge and extends along the whole rapid. On the north side the banks rise precipitously, but present a face of rock only in certain places. For the first mile or so the water, confined in a channel so narrow for a river of such volume, rushes down with great impetuosity. The current then gradually moderates, and two miles farther on the Saskatchewan is lost in Lake Winnipeg. The total descent at the Grand Rapid may be safely estimated at upwards of 60 feet. Canoes and batteaux can easily be run down, and even towed up a part of the way. But, of course, in its present state, this rapid, with such a descent, must be regarded as forming an absolute break in the navigation, that is, to vessels of a considerable size.

Between the Grand Rapid and Lac Bourbon there are two little rapids which present obstructions of a less serious nature, but which could not yet be navigated in their present state by vessels of large size. From Lac Bourbon upwards, the navigation of the Saskatchewan is unimpeded for a long distance. On the 4th of June, having examined the Grand Rapid and ascertained the difference of level between Winnipegoos Lake and Lac Bourbon, I divided the party, as already explained, and crossed over to Swan River.

The country bordering on the western extremity of Winnipegoos Lake is, in general, of a fair elevation, and the land appears to be remarkably fertile; between Red Deer River and Swan River a level country extends to the base of the Porcupine Hills. It is well wooded, and upon the whole I should think this tract well adapted for settlement. Mineral springs occur in various places near the mouth of Swan River. One of these we visited, and found some people engaged in the manufacture of salt. At this place, in a bare flat of about 20 acres in extent, but slightly elevated above the level of the Lake, numerous springs bubble up, all of them emitting more or less gas. Some are exceedingly briny, while others taste exactly like the St. Leon water of Lower Canada, and on being drunk produce the same effect.

From Winnipegoos Lake to Swan Lake the distance is about six miles. The stream which connects them, here appropriately enough called Shoal River, varies in width from 150 to 300 feet. It is shallow and has a very swift course.

About Swan Lake the country is highly interesting. Numerous islands appear in the Lake: to the north an apparently level and well wooded country extends to the base of the Porcupine Range, while to the south the blue outline of the Duck Mountain is seen on the verge of the horizon.

Ascending from Swan Lake for two miles or so the banks of Swan River are rather low. In the succeeding ten miles they gradually become higher, until they attain a hight of nearly 100 feet above the river. The current is here remarkably swift, and the channel much embarrassed by round boulders of granite mixed with fragments of limestone, which latter is the rock proper to the country, although it does not crop out so far as we could see on any part of Swan River. Land-slips occur in many places where the banks are high, exposing an alluvial soil of great depth resting on drift clay or shale, of a slightly bituminous appearance.

About 30 miles above Swan Lake the prairie region fairly commences. There the river winds about in a fine valley, the banks of which rise to the height of 80 or 100 feet. Beyond these an apparently unbroken level extends on one side for a distance of 15 or 20 miles to the Porcupine Hills, and for an equal distance on the other, to the high table-land called the Duck Mountain. From this south-westward to Thunder Mountain the country is the finest I have ever seen in a state of nature. The prospect is bounded by the blue outline of the hills just named, while, in the plain, alternate wood and prairie present an appearance more pleasing than if either entirely prevailed.

On the 10th of June, the time at which we passed, the trees were in full foliage, and the prairie openings presented a vast expanse of green sward.

On approaching Thunder Mountain, which seems to be a connecting link between the Porcupine range and the Duck Mountain, the country becomes more uneven. Some of the ridges on the shoulder of the Thunder Mountain even show sand, but there are wide valleys between them.

On leaving Swan River to cross to Fort Pelly the land rises rapidly to a plateau elevated about 250 feet above the level of the stream. The road then follows for some distance a tributary of Swan River, which runs in a beautiful valley, with alternate slopes of wood land and prairie. Numbers of horses were quietly feeding on the rich pasture of this valley when we passed, and what with the clumps of trees on the rising grounds, and the stream winding among green meadows, it seemed as if it wanted but the presence of human habitations to give it the appearance of a highly cultivated country. The Hudson's Bay Company keep a guard here to take care of the numerous horses attached to their establishment of Fort Pelly.

Arrived at Fort Pelly we spent the greater part of a day, the 16th of June, there to refit our canoe and prepare for the journey down the Assiniboine. And here I should mention, that we were much indebted to Mr. McDonald, the gentleman in charge of the establishment, who kindly furnished us with horses and carts to convey our canoes and articles across from Swan River to the Assiniboine, and was otherwise most attentive and obliging.

Leaving Fort Pelly early on the morning of the 17th of June, we proceeded on our journey. For 18 miles or so downward (from Fort Pelly,) the Assiniboine is very narrow, crooked, and much embarrassed by shoals and rapids. It is then joined by a stream appropriately named the White Mud River, which flows from the westward, and seems to be the main branch. This river drains a considerable portion of the great alluvial prairies which travellers pass on their way to Carlton House, and which have excited such general admiration on account of their great fertility.

From the White Mud River to Fort Ellice, a distance of about a hundred miles, the Assiniboine winds about in a deep valley, varying from a mile to two miles or so in width. At the White Mud River the banks of this valley rise only to a moderate elevation. Near Fort Ellice they attain a height of about 250 feet. On ascending these heights a view is obtained of a rolling prairie, stretching away on either side of the Assiniboine as far as the eye can reach. It would seem as if the whole of this vast region were a sort of level plateau, and that the greater height of the banks at Fort Ellice indicated the descent which the river had made in its course.

With regard to the quality of the soil; on going inland a little we found it to be of an alluvial character, differing in no respect from the soil in the prairie lands at Red River.

The smallest brook that flows from the prairie has cut itself a valley almost as deep as that of the Assiniboine itself; and from the latter stream a fine view is often obtained of glens stretching far inland, with winding banks, covered in some cases with green herbage, and in others with forests which ascend to the level of the plain above.

The course of the Assiniboine is remarkably crooked. Occasionally it crosses the valley as much as three times in a direct distance of one mile. The margin of the stream is in general wooded; sometimes the woods extend across the whole valley; in other cases the green banks slope down from the prairie level to the water's edge. Where the river runs close by some steep promontory, it occasionally happens that half the hill has slipped down, disclosing a face of yellow loam or drift clay, resting on crumbling slate or shale, which again is curiously interstratified with other substances as soft as itself, some of which shew the presence of iron ore. The immediate banks of the river are of soft alluvial earth, and are constantly tumbling in. As might be supposed, the water is muddy, and yet it is not unpleasant to the taste. Besides the White Mud River, two considerable tributaries join the Assiniboine from the west, above Fort Ellice. These are the Broken-arm and the Qu'appelle Rivers. The latter stream drains a great extent of alluvial prairie land; and at the Touchwood Hills, near its sources, it is said that coal is to be found in abundance.

From Fort Ellice to the Rapid River the country is much of the same character that I have described it as being above that place; but on passing the Rapid River a change is perceptible: the high banks of the valley disappear, and the prairie slopes more gently to the river. A little above the Souris River a still greater change occurs; the alluvial banks giving place to sand-hills, which run in ridges from north-west to south-east. Through these the river cuts its way in an extremely tortuous course, sometimes running south-east, in a direction parallel to the ridges; then cutting across a ridge, and suddenly turning in an opposite course. These ridges, where the river has cut through them, expose sand resting on stiff blue clay. In several places the limestone rock is seen beneath the clay in horizontal strata, full of organic remains, with sandstone resting on the top of it.

This comparatively barren tract may be about 40 or 50 miles in width. It is evidently a continuation of the high lands at the Duck Mountain, and here forms the south-western embankment of the great alluvial valley of Red River. But even this tract, which is the poorest to be met with in the country, is not all barren; for those who have travelled inland say that the valleys between the sand-hills are very fertile, and the whole tract would at least afford excellent pasturage.

Leaving the sand-hills the Assiniboine winds by the Grand Portage, where the venerable Archdeacon Corcoran has formed a settlement of half breeds and Indians. The soil here is of the same character as at Red River, and superior to it in so far that it is never subject to be overflowed. But with regard to this part of the country, it has been so often described, and it is so near the well known settlements at Red River, that I need not detain you by alluding further to it. I shall therefore only refer to the tracts which appear to me to be most valuable for settlement in the region I have visited, and then describe the manner in which I conceive they would be most easily rendered accessible.

By far the greater portion of the lands bordering on the Manitouba and Winnipegoos Lakes is unquestionably well adapted for settlement. Experience already shews that wheat yields an abundant return on Manitouba, and at the little Saskatchewan. At the latter place even Indian corn is said to be a sure crop. This being the case, it is reasonable to conclude that wheat would thrive also at Sanguissippi Lake, and at Lac Dauphin, and along the western coast of Winnipegoos Lake. The valley of Swan River, in point of fertility of soil, is perhaps unsurpassed in any country; and, as regards climate, it cannot, I think, be inferior to the valley of Red River, inasmuch as, if further north, it is also much further west, and removed from the influence of the cold winds of Lake Winnipeg, which sometimes have a prejudicial effect on the crops at Red River.

The country bordering on Red Deer River is said to be very fertile; and the fact that maple is to be found there in considerable quantities would confirm the belief that the climate cannot be very unfavorable. Throughout this region wood is in sufficient abundance to ensure a supply of fuel for a long time to come, or until such time as the coal mines may be developed.

It is said that coal is to be found in various places on the Porcupine Hills, and on the Duck Mountain. I found some specimens of lignite on going up Swan River, which fully confirm the fact, but whether it is to be found in available quantities can only be ascertained through time.

The great alluvial valley drained by the Assiniboine and its tributaries above the Souris River, will no doubt become, at some period, one of the finest wheat-growing countries in the world. No one in this part of the country even pretends that, in point of soil or climate, it is unfavorable to the growth of agricultural produce.

In regard to the means of communication that could be most easily made available, the country bordering on the lakes so often mentioned might be reached by steamers or other craft. There is nothing to prevent a steamer of light draft running from Red River Settlement to the end of Winnipegoos Lake.

On reference to Mr. Wells' Report, it will be seen that the Little Saskatchewan River is navigable from Lake Winnipeg to Manitouba Lake. The latter lake is not deep, but there are not many shoals to be met with, and the bottom is in general as level as the surrounding country. I sounded wherever I went, and found an uniform depth of from 15 to 18 feet, after passing a few hundred feet from the shore. The Winnipegoos Lake is on a higher level by about 5 feet than the Manitouba, and the sounding line showed that it was by so much deeper except at the upper end, where it attains the depth of from 36 to 50 feet.

The Sanguissippi River, which connects the Winnipegoos and the Manitouba Lakes, has a general depth of from 6 to 8 feet, except at one place, near the middle of its course, where there is a very swift run, with a depth scarcely amounting to 5 ft.

The distance from Winnipegoos Lake to Lac Bourbon, by the Mossy Portage, is 4 miles and 18 chains, and the difference of level 4 feet; the Winnipegoos being that much higher than Lac Bourbon. The Saskatchewan was, however, very low, at the time of our visit, and it is probable, that when it is at its usual height the difference of level between the two lakes is but very slight.

From Lac Bourbon upwards for 400 miles, the Saskatchewan presents an unbroken reach of navigable water. Above that, accounts differ as to whether it is navigable or not, some asserting that it is too much embarrassed by rapids and shoals, and others that it might be navigable for boats of light draught and great power to the base of the Rocky Mountains. From what I have learned, I incline to the latter opinion, but the truth can only be ascertained by examination. At all events, from the Red River Settlement to Carlton House, following Lake Winnipeg, the Little Saskatchewan, Lakes Winnepegoos and Manitouba, Lac Bourbon and the Great Saskatchewan River, there is a navigable reach of 800 miles, broken only by the Mossy Portage. This one carrying place cannot be reckoned a great impediment in such a distance; steamers might be placed on the waters on either side, and a land road made across it, and this, I conceive, would be all that could be desired until settlement should have advanced so far as to render a more perfect means of communication necessary.

From what I have said, it will be seen that the Lakes, and the country bordering on the Great Saskatchewan, are easy of access.

In regard to the facilities for communication in the valley of the Assiniboine, wheeled carriages can already be driven over the whole territory, by the lines of route indicated on the plan, and it is only by such conveyances that settlers will, in the first instance, be able to supply their wants. The Assiniboine is only navigable for considerable vessels as far up as the Grand Portage; above that, indeed, canoes can be towed up, and batteaux can descend, except at extreme low water, but it has a very tortuous and rapid course, more especially among the sand hills, and it is in many places shallow, so that it could only be made navigable for vessels of any size at a great outlay. Everything considered, therefore, I am of opinion that transport could be more easily effected by land. The country is admirably adapted for railroads, but of course in an unsettled region these are out of the question.

On reference to the map, it will be seen that Lac Dauphin and a part of Winnipegoos Lake approach within 70 or 75 miles of the Assiniboine. These places being accessible to steamers, land roads might be made across the country, which would afford all the accommodation required by a settlement in its first stage; indeed so obvious is this way of reaching the Assiniboine, that the Hudson's Bay Company supply their establishment at Fort Pelly by way of these lakes and Swan River, carting their goods from the latter place across the country to the Assiniboine.

In speaking of navigable lines that might be made available, I should mention that at the Grand Portage, there is said to be an old water-course, by which the Assiniboine in all probability has at some period discharged its waters into the Manitouba Lake. Now, as the Assiniboine is navigable, or in a state that it might easily be made so, as far as the Grand Portage, it may yet be found advantageous to open a water communication between it and the Manitouba Lake. The advantages which this route would possess over that by Lake Winnipeg and the Little Saskatchewan, are, that it would be considerably shorter, and would be open somewhat earlier in the Spring.

In considering the project of colonizing a country so remote as this, and of which so little is, as yet, generally known, the question will naturally arise as to whether the native population would be likely to offer any opposition to settlement, or whether the country, if occupied, would be easily governed. Believing that any information which will enable the Government to judge of these important matters will be acceptable, I shall, before concluding, describe the present state of the country through which we passed, confining myself to what has come under my own observation.

At Manitouba, Lac Dauphin, and "Partridge Crop," there are small settlements of Indians and people of mixed origin. At the first and last named settlements, the people are rather industrious, and raise wheat, Indian corn, and a variety of articles. At Lac Dauphin the settlers simply grow potatoes and Indian corn. They are gradually acquiring habits of industry, but they can live so easily by fishing or hunting, as Mr. Wells says in his report, that they are slow to adopt the more laborious pursuits of civilized life. Ducks, geese, and aquatic fowl of all sorts frequent the waters in that quarter, in great numbers, and the lakes and rivers literally swarm with fish.

Near the Grand Rapid of the Saskatchewan, we saw about 15 families of Indians, from Swan River, who migrate there annually to catch sturgeon. They seemed to me to be a very quiet and orderly people, and I thought I could detect in their countenances that they were not wholly of Indian origin; and on enquiring as to this point, some of them were proud to boast of their descent from the Canadian Fur-traders who had occupied this country many years ago. At the upper end of Winnipegoos Lake, we only saw one family, (those already mentioned as being engaged in making salt,) and from thence in a journey of five hundred miles by the valleys of Swan River and the Assiniboine, we saw not a living being, except the few people in charge of the Hudson's Bay Company's establishments of Fort Pelly and Fort Ellice, until we got to the settlements in this neighborhood. At Fort Ellice we were told that the hunters were gone further west; but from all we could see or learn, there was no avoiding the conclusion that the population which once wandered over the vast plains of the Assiniboine had decreased to an unexampled extent. This I attribute partly to the indiscriminate slaughter of the Buffalo to supply the pemican required for the trade of the country, the introduction of horses having rendered these clumsy animals an easy prey to the hunter. And the Indian, so that he can supply his immediate wants in a profusion however boundless, never troubles himself about the future. Numerous, therefore, as the buffalo still are, there can be little doubt but that they will soon be destroyed.

Large supplies of pemican used to be obtained at Fort Pelly and Fort Ellice, but these establisments can furnish little now. At many places we observed the plains furrowed with old tracks of the buffalo, and in some cases the banks of the Assiniboine were strewn with their bones, but there were none of these animals, we were informed, to be seen within several days' journey of any place we were at. This shews how fast they are being destroyed, and it leaves but the reflection, that as their staple article of food diminishes, the Indians must decrease in numbers, unless they can be induced to adopt the habits of civilised life. That with proper management they may be so, is fully shewn by the success which has attended the labors of the zealous Missionaries at the Grand Portage, Red River, "Partridge Crop" and elsewhere. As to the people of the Red River Settlement, they are as orderly and quiet a community, I may safely say, as can be met with any where, and I believe there is nothing they desire more than to see the country opened up.

In leaving this part of the country I have much pleasure in saying that both from the people of this settlement, and the officers of the Hon. Hudson's Bay Company, we have met with every kindness and civility. It may appear invidious to mention names: and yet I cannot omit those of Chief Factor McTavish at Fort Garry, who furnished us with many articles which we could not otherwise have obtained; and of Chief Trader Murray at Pembina, who always had his hospitable quarters open for any of the party that passed his way.

I have the honor to be, Sir,
Your most obedient servant,
S. J. DAWSON.
C. E. in charge Red River Expedition.

The Hon. Provincial Secretary,
Toronto, C. W.

P. S.—Both Mr. Wells and I made a cursory survey, taking the courses and correcting the distances by numerous observations as we proceeded, and I have left a letter for Professor Hind informing him of our operations.

S. J. D.

RED RIVER SETTLEMENT,
30th June, 1858.

SIR,—After leaving the Mossy Portage on the morning of the 4th of June, I proceeded agreeably to your instructions to make a cursory survey of the west shore of Lake Winnipegoos, the River Dauphin, and the Lake of that name, the Little Saskatchewan, and the Southern part of Lake Winnipeg, and would beg leave to submit the following report of my operations :

The western shore of Lake Winnipegoos, in common with the other Lakes through which I passed, is much better adapted for settlement than the eastern one, inasmuch as the land is higher, and the climate, if any thing, a little better. In crossing Lake Winnipegoos from east to west, a distance of only about twelve miles, I found vegetation somewhat further advanced than on the side I had just left; the soil is also better, inasmuch as that it is higher. Timber, such as maple, elm, oak and poplar, covers the country to the water's edge. I visited several places where sugar had been made, and saw specimens of that article equal to any that I have ever seen in Eastern Canada.

The Duck Mountain, which occupies almost the entire background, commences to rise not far from the Lake shore, keeping a gentle ascent for 15 or 20 miles back, where it attains its greatest elevation, a height of 600 or 700 feet above the Lake. I learned from the people who reside in Duck Bay, that the entire face of the Mountain is a succession of gentle slopes and flat table lands, and that the summit itself is an extensive plateau of alluvial soil covered with a fine growth of timber.

There are three salt springs near the southern end of Lake Winnipegoos, one of which I visited, where there are works established for the manufacture of salt.

There are some 40 or 50 half-breed Indians who reside here, and at the Duck Bay, and though assured by them that all kinds of grain succeed well, yet they cultivate only a few potatoes, as fish and game are so plentiful and of such good quality, that they may be said to live almost without exertion.

From the salt springs I passed through the Dauphin River, almost 16 miles, to Dauphin Lake. The Dauphin River is a fine stream about 40 yards broad, and having 5 feet of water in the shallowest places. Its banks are of a strong grey clay, covered with black mould, and timbered with oak, elm and poplar. It has two considerable tributaries rising in the Duck and Riding Mountains, which appear to drain a country well adapted for settlement. There are several places on the Dauphin River where the Indians grow potatoes, Indian corn and melons. The wild grape, wild hop and wild vetch, are also common on the banks of the river.

Lac Dauphin is about 30 miles long from north-west to south-east, and six miles broad. Its western shore is bounded by the Riding Mountain, similar in all respects to the Duck Mountain already mentioned. Its southern shore is bounded by a prairie interspersed with wooded knolls, which I was informed extended without interruption south-east to the Assiniboine and Red Rivers.

I saw fixed rock only in two places in this part of the country, namely, at Snake Island, near the lower end of Lake Winnipegoos, and at the lower part of the Dauphin River. That on the Snake Island, a whitish limestone, is full of organic remains; the other is similar to the Manitouba limestone and nearly without organic remains.

From Lac Dauphin I returned by the north-east end of Lake Mamitouba and its discharge, the Little Saskatchewan River.

The Little Saskatchewan, as its name implies, has a very strong current, which I found, on measuring to be 2½ and 3 miles per hour, *but it is entirely free from rapids.* Its average breadth is 250 yards, with from 8 to 12 feet of water. It appears well adapted for steamboat navigation.

The country in the vicinity of the Little Saskatchewan presents every inducement for settlement, as is proved by the flourishing state of the present settlement at Fairford, or, as it is more generally called, "Partridge Crop,"—a mission established under the Bishop of Rupert's Land, about six years ago, by the Rev. Mr. Cowley.

The present establishment is some six miles higher up the river than the first one, which, being subject to inundation in times of high water, was abandoned. There are several well built houses, a chapel, school and mill, at this place, with a population of about 200 souls, Indians and half-breeds.

The Rev. Mr. Stag, the missionary now in charge, informed me that the school which is attached to the mission was usually attended by from 50 to 60 children, half of whom are Indian children; indeed, the Indians belonging to this place appear to be fast acquiring the tastes and habits of civilization, being more clean and better dressed than any I have seen in the country. Mr. Stag also informed me that, notwithstanding the ease with which the ground was cultivated, and the large returns of grain, he required to use all his influence to induce the Indians to cultivate the land, as their wants are so easily supplied by fishing and hunting.

From the Little Saskatchewan I returned through Lake Winnipeg, arriving at this place on the 26th of June.

The journey through Lake Winnipeg presented no feature of essential difference from the other lakes, except that the western shore is low, with occasional limestone cliffs, and the eastern shore high, with granite rock.

I am, Sir,
Your obedient servant,
A. W. WELLS.

S. J. Dawson, Esq.,
Red River Settlement.

FORT WILLIAM, LAKE SUPERIOR,
21st August, 1858.

SIR,—In my report of the 4th ultimo from Red River, I had the satisfaction to acknowledge the receipt of your letters of the 16th and 20th of April, since which time I am without any further communication from the Government.

I have now the honor to inform you that the party under my directions are at present actively occupied in the exploration of the country between Rainy Lake and Lake Superior; one of my assistants, well appointed with men and supplies, being at Lac des Mille Lacs, and another, equally well provided, at Dog Lake. For my own part, since reaching this place, I have been occupied in sending the necessary supplies to the different points where they will be required, and having accomplished this, I am now about to set out for Lac des Mille Lacs.

In regard to the progress which has been made in the brief period that has elapsed since we left Red River, as explained in my last report, immediately on my return from the Saskatchewan I despatched two of my assistants with most of the men to commence the operations indicated in the instructions which I had just then received, while I remained for a few days to collect what supplies the settlement could afford.

The party so despatched made all haste to the Lake of the Woods, from whence, proceeding to Rainy River, they ascended that stream more slowly, making traverses back into the country at intervals of a few miles so as to ascertain the extent of arable land on the British side. They then examined the country at the mouth of the River Seine, the stream by which Lac des Mille Lacs discharges its waters into Rainy Lake, and thence went to Nequaquon Lake, which, previous to their leaving Red River, I had appointed as the place of rendezvous, and there I joined them on the second day after their arrival.

At Nequaquon Lake the route from the Kaministaquia joins the route from the Grand Portage, and from this place it was my intention to have despatched a party at once to Lac des Mille Lacs. Finding, however, that we had made too great an inroad on the stock of provisions which we had been able to procure at Red River to render it quite safe to send them into such a wilderness, I hurried on to the Grand Portage, leaving a sufficient party to follow more leisurely and take the levels and measurements as they came. At the Grand Portage I purchased the requisite supplies at a small fur-trading establishment, and immediately despatched my chief assistant, Mr. Wells, to Lac des Mille Lacs and Rainy Lake.

As soon as the remainder of the party had completed the survey of the route by the Grand Portage, they came down here, when I at once despatched them to Dog Lake. They will now be employed in running lines and making traverses between the Kaministaquia, Thunder Bay, and Dog Lake, so as to ascertain the route most practicable for a road. They will also examine the country between Pointe de Meuron and Gunflint Lake.

From what I have said it will be seen that our surveys, so far as they have gone since we left Red River, enable me only to speak of the land available for settlement in the Rainy River country, and of the respective merits of the canoe routes by the Kaministaquia and Pigeon River.

With regard to the former of these subjects, the land immediately bordering on Rainy River, on the British side, is of an alluvial description, and almost as uniformly level as the prairies at Red River. For a mile or so inland from the main stream the ground is dry, and a dense growth of large timber, consisting of poplar, elm, oak, basswood and occasional white pines indicates a productive soil. For a mile or two beyond this, however, swampy ground predominates, while beyond that again the land gradually rises to a range of hills of no great eminence, which, as far as we could observe, seemed to run parallel to the river, at a distance of from four to eight miles back. The distance from Rainy Lake to the Lake of the Woods, following the windings of the stream, is about eighty miles, and throughout the whole of this extent the land fronting on the the river is fit for settlement, without I may say a single break; indeed, I have never seen any thing to equal it in my experience, except at Swan River, and on the Assiniboine. Farther inland, although the ground becomes in many places swampy and broken, there must be a fair proportion also fit for settlement, but to ascertain anything like the precise area that is so, would require a closer examination than we have as yet been able to bestow.

Passing from Rainy River to the Lake of the Woods, there is, around that extensive sheet of water, a coast line within the British territory, of more than three hundred miles. Taking the whole of this distance there is unquestionably a great deal of barren land, but there is also much that is fit for settlement; in some places high rocks line the shore, and indicate too surely the character of the country inland, but in others, there are gentle slopes where the growth of timber would argue a soil of great fertility, and, added to this, there are numerous islands in the Lake which, although many of them are comparatively barren, are in some instances well adapted for settlement, both from their situation and the extent of arable land they contain.

With regard to the extent of land fit for settlement about Rainy Lake, the surveys which we are now carrying on will afford us frequent opportunities of seeing that country, and as soon as I have made a further examination and obtained more precise information, I shall lose no time in reporting to the Government. In the meantime, I may safely say that a considerable extent of land available for settlement will be found in that country.

In reference to the comparative advantages of the two canoe routes which leave Lake Superior, the one at Fort William and the other at the Grand Portage Bay, and meet at Nequaquon Lake, I am only yet in a position to say that, regarded as canoe routes, that by Fort William and the Kaministaquia, although a little the longest, is incomparably the best.

The route by the Grand Portage and Pigeon River, passes through a country of greater elevation, and withal more rugged and mountainous than that by the Kaministaquia and the Lac des Mille Lacs.

The highest water level on the Grand Portage route is at an elevation of no less than 1053 feet above Lake Superior, while around the Lakes at this great altitude the mountains tower up to the height of at least five hundred feet.

The highest water level on the Kaministaquia route is that of the little pond at the west end of the Prairie Portage, which is 893 feet above Lake Superior, while at that elevation on this route the country is comparatively level.

The number of portages on the Grand Portage route, between Lake Superior and Lake Nequaqon, is 31, and their aggregate length 15 miles and 52 chains.

On the Kaministaquia route between the same points, the number of portages is but 20, and their aggregate length only 10 miles and 77 chains.

The Grand Portage route has, moreover, the disadvantage of touching on the head waters of four different rivers, viz., Pigeon River, Arrow River, Sageinaga River, which discharges its waters into the Upper Sturgeon Lake, and the stream which flows from White Wood Lake to Lake Nequaquon. For 90 miles between Fowl Portage and White Wood Lake, the route lies along the head waters of these rivers, and in that distance the streams between the lakes are so shallow and small as to be scarcely navigable for birch canoes, so that, in the event of a more perfect water communication becoming necessary, the means of supplying it could not be looked for in a region so elevated.

That as a canoe route the Kaministaquia and Mille Lacs way is the best is pretty conclusively proved by the fact, that it is the one which was eventually adopted by the North-west Company, and always followed by the Hudson's Bay Company. However, I shall be better prepared to report on this subject when the surveys now in progress are completed.

With respect to the carrying on of the surveys, they will, I am confident, advance rapidly under the present arrangements. There is, however, a wide extent of country to examine; and when the most advantageous route is discovered, instead of mere cursory surveys such as we have hitherto been making, the measurements and levels will have to be taken with great precaution, so that estimates can be founded on them afterwards. I think, therefore, that to complete the whole properly, the surveys should be continued during the winter; and, if the Government should decide on doing so, a great saving will be effected in sending supplies by canoe to certain points along the route before the close of the navigation, so that the surveying parties may continue uninterruptedly at work. I therefore send down the bearer, Mr. Charles de Salaberry, who will bring up such supplies as will be required; that is, in the event of the Government deciding to continue the surveys during the winter.

Before concluding this report, I should mention that our operations excite the greatest interest among the Indians, especially at Fort Frances and the Lake of the Woods, where they are very numerous. On coming up I found the principal chiefs waiting for me at various places; but the most important interview I had with them was at Fort Frances. The chiefs at that place guide and rule, more or less, the whole Saulteaux tribe; and feeling the importance of keeping on good terms with them, I went to a Grand Council to which they invited me on the moment of my arrival. After their preliminary ceremonies had been gone through with, the principal chief delivered a long harangue, with which I need not trouble you further than to say, that the point he aimed at was to ascertain what object the Government had in view in causing the country to be explored. I replied that I could not say what course might be ultimately adopted by the Government; that they need not fear, however, but that their interests would be consulted; and that we were merely examining the country, to our doing which we trusted they would offer no opposition. I then made them some presents which I had taken with me for the purpose, and said I would wait until the next day to hear what further they had to say.

On the following morning they called on me, and said they would not put me to the trouble of attending another council; that they had consulted among themselves, and come to the conclusion to allow us, in the meantime, to explore the country as we pleased; but that they trusted no settlers would be sent in without their being consulted; and concluded by requesting me to inform the Great Chief of the Canadians that they and their people would assemble at Fort Frances early in June next year, when, if he would send a person duly authorised to confer with them, he would not find the Indians unreasonable, as they wished to be friends with the Canadians.

I promised to convey their message to the Government, and then parted with them on the most friendly terms. Mr. Chatelain, the Hudson's Bay Company's Officer in charge at Fort Frances, kindly acted as interpreter, and he informed me that, about ten days previous to our arrival, no less than five hundred Indians had been waiting for me; but that they had been compelled to disperse on account of the scarcity of provisions. When we arrived I only saw the chiefs, and about twenty others.

The bearer, Mr. de Salaberry, has been very active, and of great assistance to me, especially in pushing on the men when *en route* with the canoes; and I therefore recommend him with much pleasure to your favorable notice.

I have the honor to be, Sir,

Your most obedient servant,

S. J. DAWSON, C. E.

To the Honorable Provincial Secretary,
Toronto, C. W.

Fort William, Lake Superior,
23rd September, 1858.

Sir,—I have the honor to inform you that for the last few weeks I have been engaged in exploring and making a cursory survey of the River Seine, the stream by which Lac des Mille Lacs discharges its waters into Rainy Lake. From this excursion I have only just returned; and as the mail boat is hourly expected, there is barely time to enter into a very minute description of the country, or prepare a map, without which a detailed statement would not be sufficiently intelligible. I shall therefore for the present confine myself to a very brief account of our recent operations, and the advantages which the River Seine presents as a line of communication.

Where that stream issues from Lac des Mille Lacs it is a fine river, over a hundred feet in width. For a considerable part of its course it winds about, with a general westward direction, through a flat, thickly wooded with cypress trees and poplar of a large size; while beyond this, on either side, low hills swell up gradually, covered for the most part with a dense growth of poplar interspersed here and there with tall pines, which rise singly or in groves above the surrounding forest.

At times the valley contracts, and where it does so the river presents cascades past which a portage has to be made, or little rapids which can be run with a canoe; but between these there is generally a considerable extent of navigable water. This description will apply to the country for about forty miles below Lac des Mille Lacs. The lower part of the valley presents a succession of lakes, varying from a mile to fifteen miles in length, until near Rainy Lake, into which the river, much increased in volume, discharges itself in a series of cascades, making a plunge of over 112 feet in the distance of five miles and a half. The lakes just referred to are bounded, for the most part, by low hills, generally wooded, but in some cases rocky, with an occasional valley between them presenting a less barren appearance.

The rock throughout is, I may say, entirely of the primitive formation,—that is, where we had an opportunity of observing it.—In one of the lakes an island rises like the peak of a half submerged mountain, exhibiting a description of rock somewhat different from that of the surrounding country; and here we found some specimens indicating the presence of copper.

Viewed in its general aspect, the valley of the Seine is far less rugged in its character than the country bordering on either of the other two routes which have as yet been followed to Rainy Lake. On the borders of the lakes of which I have just spoken, it is certainly somewhat rough, but not so much so, by any means, as to be impracticable for roads; and, should it ever be determined on to construct a railway from Lake Superior to Rainy Lake, from what I have as yet ascertained of the general features of the country, I am convinced that it must pass through the valley of the Seine.

Regarded as a navigable route, this river is susceptible of being made far superior to either of the other lines at present used.

The main objection to these latter is that they follow no continuous river system but pass over dividing ridges from the waters of one stream to those of another, and consequently, however much improved, can only afford at best a broken sort of communication. The Seine, on the contrary, with the Mille Lacs from whence it issues, and the Savanne River, its tributary, presents great facilities for making the water communication continuous, by means of lock and dam, from the Height of Land to Rainy Lake.

Regarding it in the meantime, however, merely as a canoe route, it is but very little inferior, even in its present state, to either of the other routes; the only formidable barrier to canoe navigation being the series of falls which I have already mentioned as occurring in the immediate vicinity of Rainy Lake. At low water these falls are passed by twelve short portages, but at high water it would perhaps be necessary to make a portage of five miles and a half to avoid them. It is doubtless on account of the obstacles which occur at these falls that this river has not been followed as the highway for canoes, for between them and Lac des Mille Lacs the carrying places, although there are as many as fifteen of them, do not equal the French Portage alone in their aggregate length.

On reference to the lithographed map accompanying the blue book containing the reports of the different members of the expedition, it will be seen that the discharge of Lac des Mille Lacs is marked as supposed to enter a deep bay on the north side of Rainy Lake. We have ascertained, however, that it does not enter that bay, but the first one to the eastward of it, from whence a reach of navigable water, which may properly be considered as an arm of Rainy Lake, extends, on the magnetic course North 70° East, for twenty miles further than the map indicates; and into the head of this arm the River Seine discharges itself.

Until we have had leisure to prepare plans, however, I need not refer farther to this river than to say that, at a very moderate outlay, it might be made an excellent route for canoes, or boats such as the Hudson's Bay Company use. A road, say six miles in length, might be made past the impediments near Rainy Lake, and by the construction of dams the portages above that place might be reduced to one-third of their present number; and this, it is important to consider, would be a step towards works of a more comprehensive nature.

With regard to the operations being carried on at present, I dispatched my principal assistant, Mr. Wells, from Mille Lacs, to make a cursory survey of a tributary of the River Seine, which takes its rise a little to the northward of the Prairie Portage, and—after a westerly course of over sixty-five miles, without a break, as the Indians report, in the navigation—joins the River Seine about twelve miles below Lac des Mille Lacs; from thence he will descend to Rainy Lake, ascend the River Maligne to the Upper Sturgeon Lake, and then make a cursory survey of the Sageinaga River and Lake.

The small party employed in this quarter have already completed two exploratory lines, one from Pointe de Meuron, and one from Current River to Dog Lake. The mountain range bordering on Lake Superior is the main barrier to a line of road. The line just run from Pointe de Meuron is very rough; that from Current River is better; but I am in hopes of finding still better ground between the bottom of Thunder Bay and Dog Lake, and in order to ascertain this point, I have sent a party to run an exploratory line through. When this work is completed, we shall examine the country between the Kaministiquia and Gun Flint Lake.

In my report of the 21st ultimo, I had the honor to suggest the expediency of establishing a party to continue the surveys during winter. To be more explicit, however, I would propose leaving one of my assistants and six men at Lac des Mille Lacs, to make a trigonometrical survey of the River Seine, the Savanne River, and Dog River and Lake, taking the levels throughout with the greatest precaution. Densely wooded as the entire region is, an instrumental survey can be much better accomplished in winter, when the lakes and rivers are frozen over, than it can in summer. I would also leave one of my assistants in charge of the instruments and stores at this place, with instructions to keep a meteorological register, and, when opportunity offered, to prosecute still further the surveys between this and Dog Lake on the one hand, and Sageinaga Lake on the other. This arranged, I should go down late in the fall, with my principal assistant, Mr. Wells, to make up plans, &c., of the exploratory surveys now accomplished, where we would have conveniences for work of that kind, which cannot be obtained here, and return to this place in the month of March, or immediately on the opening of the navigation.

The surveys now extend over four and a half degrees of latitude, and about thirteen degrees of longitude, so that we have a considerable amount of work to put together.

Mr. De Salaberry, who took down my last despatches, by some mischance lost a trip of the boat, so that he cannot now return so soon as I had reason to anticipate, and, as the season for canoe navigation is drawing to a close, I have in the meantime sent some provisions to the Savanne River, and Mille Lacs, so that we may be fully prepared, in the event of the surveys being continued during the winter. Should they not be so, the provisions will be in a convenient situation for next springs use —all the additional expense involved being the cost of maintaining a man where they are kept, to take care of them.—

I have the honor to be, Sir,

Your most obedient,

S. J. DAWSON,

C. E. in charge of Red River Expedition.

To the Honorable
The Provincial Secretary,
Toronto, C. W.

GENERAL REPORT

ON THE

PROGRESS

OF THE

RED RIVER EXPEDITION.

GENERAL REPORT.

TORONTO, 22nd February, 1859.

SIR,—I have the honor to submit to your notice, for the information of the Government, a report in reference to the exploratory surveys which have been carried on under my direction between Lake Superior and the Saskatchewan River, together with the following maps, illustrative of the topography of the region explored:

1. A general map on a scale of 10 miles to 1 inch.

2. A plan shewing the newly explored River Seine, and the various canoe routes between Lake Superior and Rainy Lake, on a scale of 2 miles to 1 inch.

3. A plan, in profile, shewing the relative altitude and length of the routes by Pigeon River and the Kamanistiquia.

4. A map of a portion of British North America, shewing the route which it is believed could be most easily made available through Canada and British Columbia.

5. A sketch shewing the surveys accomplished by Mr. J. F. Gaudet at the Savanne River, &c., up to 23rd December, 1858.

6. A sketch shewing the surveys made by Mr. L. A. Russell between Thunder Bay and Dog Lake, up to 10th January last.

On returning from the Saskatchewan and Assiniboine, I brought with me such specimens of fossils, shale, coal, &c., as I thought would best illustrate the geological formation of the country. These I submitted to Sir Wm. Logan, the Provincial Geologist, and to E. Billings, Esq., Palæontologist, his principal assistant.

Mr. Billings has kindly favored me with an article, which I have much pleasure in inserting, and to which, coming as it does from such high authority, I would invite particular attention. In respect to some of the fossils, it will be observed he has obtained the valuable opinions of Professor Dawson of McGill College, and Messrs. Meek and Hayden of Washington, who are considered the highest authorities in America on all points relating to the secondary and tertiary formations of the central portion of the continent.

Mr. Russell, of Ottawa, who has made the subject of colonization his particular study, and who has had a great deal of experience in opening up new countries, has favored me with his opinions in respect to the advantage of settling in a prairie region, as compared with a country entirely wooded; and his remarks, I am convinced, will be read with much interest.

As I write, despatches have come in from the party engaged in exploring the country between Lake Superior and Rainy Lake, and I have much satisfaction in saying that the surveys have been progressing as rapidly as could have been expected, as will be seen on reference to the annexed reports from Messrs. Gaudet & Russell.

I have the honor to be, Sir,
Your most obedient servant,

S. J. DAWSON,
C. E. in charge of Red River Expedition.

Hon. Charles Alleyn, M.P.P.
Provincial Secretary, &c.
Toronto, C. W.

REMARKS ON THE PROGRESS OF THE EXPEDITION.

The first exploratory excursions of the various parties composing the Red River Expedition having been dwelt upon at ample length in the Blue Book containing their correspondence and reports, I shall, for the sake of brevity, confine myself, in this report, to an account of the operations of the party under my immediate direction when the expedition consisted of several distinct divisions, and the progress which has been made since I have had the full charge.

Before proceeding further, however, I may mention that I was prevented from reaching Red River, at the same time as the other parties, by an attack of typhus which compelled me to remain for some time at a remote missionary establishment on the River Winnipeg. So infectious was the disease, and of a type so virulent, that four out of eight individuals who were in my canoe were attacked and suffered severely from it; I and one of the men found shelter with the Rev. Robert McDonald, who, strangers as we were to him, did all for us that kindness could prompt, or intelligence suggest, and on our getting a little better, perceiving my anxiety to rejoin my party, he prepared his canoe and accompanied me all the way to Red River, the man who had been with me having left a few days previously for home.

Mr. McDonald is a clergyman of the Episcopal Church, in connexion with the Church Missionary Society, and I may here mention an incident of the journey, which, as it is illusrtative of a practice common with all the missionaries when travelling in that remote region, will serve, in some measure, to shew the beneficial influence which their presence is producing among the native population.

Every evening as we proceeded down the Winnipeg, as soon as the necessary preparations had been made for passing the night, the whole party, including his people and mine, in number seventeen, and, with three exceptions, all either pure Indians or partly of Indian origin, was assembled, when prayers were read and appropriate hymns sung. The Indians all joined, and as the night closed in, it had a strange effect in that unbroken wilderness, to hear the anthem rising above the din of the rushing torrent, and see the children of the forest bent in prayer, where so lately they had been accustomed to invocations of another kind in the practice of their strange observances; and one could not but reflect on the great moral change which was thus—slowly, perhaps, but surely—being effected by the efforts of those disinterested men who, leaving home and the attractions of civilisation behind them, are content to seek out the savage, and instruct him in his native wilderness.

Mr. McDonald's station at Islington, on the Winnipeg, is a most important one; the Indians from Lac Seul, on the English River, and from Rat Portage, and the Lake of the Woods, sometimes assemble there, more especially when they go to collect the wild rice which is abundant in the neighbourhood.

They are a somewhat stubborn race, however, and he has, as yet, only a congregation of fifty or sixty, but his establishment is not without its influence on the far greater number who have not embraced Christianity. It is impossible that even these untutored savages can contemplate with indifference the efforts of a man who seeks them out in distress, ministers to them in illness, and does all in his power to relieve their necessities; and this I had an opportunity of observing, for, when I was there, the Chiefs came to see him as they passed, and it was not difficult to perceive the reverence and respect with which they all regarded him.

On reaching Red River Settlement, which I did on the 9th December, 1857, I found that no adequate provision had been made for carrying on the necessary operations during the winter, the supply of pemican and flour being only equal to the support of the smallest number of men that could be advantageously employed for about three months, but, with the exception of a blanket each in the possession of the officers, we had no camping equipage of any kind whatever. In this emergency I applied to Mr. McDermot, a merchant in the settlement, who obtained us credit at the Hudson's Bay Company's establishments and elsewhere, and was thus enabled to fit out a working party in the course of a fortnight, and commence the exploration of the country between Red River and the Lake of the Woods.

Subsequently, on considering to what we would be exposed should our credit fail, and our resources become entirely exhausted, Mr. Napier, who was in command of another branch of the expedition, and I determined on sending an express to Canada. Mr. C. de Salaberry was chosen for this service and despatched on the 18th of December. He returned on the 3rd of March following and, after his arrival, our power to obtain what we required was limited only by the capacity of the settlement to furnish it.

The explorations accomplished during the winter, and early in the spring, embrace the region between the Lake of the Woods and Red River; the Roseau River, of which a cursory survey was made; the Red River, which was subjected to an instrumental survey between Pembina and Fort Garry, and between the Stone Fort and Lake Winnipeg; a like survey having been made, also, of the coast of Lake Winnipeg between the mouth of Red River and Fort Alexander.

My report on the country between the Lake of the Woods and Red River, with a copy of the map accompanying it, is published in the "Blue Book" of last year.

The region embraced in the other surveys above enumerated will be more particularly described in this report.

When these surveys were progressing in the spring, I endeavoured to collect a supply of provisions for our then contemplated excursion to the westward, and this was a matter of some difficulty in consequence of most of the available supplies having been bought up by the Hudson's Bay Company, the unusual scarcity in the settlement, and the desire of the settlers to keep a supply on hand in case of their crops failing. In one place we succeeded in getting a bag of pemican, and in another a little flour, added to which we bought and salted some beef and pork, so that by the 10th of May, when we were prepared to set out, although the supply was by no means so great as could have been wished, the country abounded in game and we were not apprehensive as to the result.

In my report of the 4th July, 1858, I have given a brief account of our exploration to the westward of Red River. Before setting out it struck me that a better knowledge of the country, in reference to its geographical features and the facilities for navigation, could be obtained by travelling in canoes than in the usual manner on horseback, as, in the latter case, we could only follow the beaten lines of road where the country has been so often described by travellers, and the result fully justified my anticipations in this respect, for, while we had an opportunity of observing the rivers and lakes, we obtained from the men who were with us much valuable information in regard to the country. They were natives, and had travelled in every direction, either on their hunting excursions or in the service of the Hudson's Bay Company; added to which we made excursions from the rivers and lakes as often as opportunity permitted; on ascending Swan River we sometimes travelled as much as fifteen or twenty miles in one day through the adjacent forests and prairies.

Returning from this excursion we reached Red River on the 29th of June and found the men brought up by Professor Hind waiting for us. They came without supplies, contrary to what I anticipated, so that no provision had been made for them and we had, in consequence, again to scour the settlement in order to get enough of provisions to supply the party on the long journey to Lake Superior. There are, however, some circumstances connected with our being placed in this dilemma which render it necessary for me to make it the subject of a separate communication. In the meantime I confine myself to saying that we eventually succeeded in obtaining a supply and left the Red River Settlement: one division of the party on the second day after our arrival, and the other on the 10th of July.

The explorations subsequent to this period are detailed in my reports of 21st August, 23rd September and 28th October. They embrace an examination of the country bordering on the Rainy River and the various routes between Rainy Lake and Lake Superior, and will be further noticed in describing the section of country to which they have reference.

GENERAL DESCRIPTION OF THE COUNTRY BETWEEN LAKE SUPERIOR AND RED RIVER.

The region through which the explorations have extended embraces two sections of country widely different in physical character. The first, extending from Lake Superior to Lake Winnipeg, is of the primitive or crystalline formation. In its general aspect it is a hilly and broken country, intersected by rapid rivers and wide-spread lakes. The mountains, however, do not rise to any great elevation, except on the immediate borders of Lake Superior, and there are several fine alluvial valleys, the most extensive of which is that of Rainy River, which has been so often referred to in previous reports. The lakes and rivers present long reaches of navigable water, the principal of which, extending from Fort Frances to the western extremity of Lac Plat is 158 miles in length. Dense forests cover the whole of this region, and the most valuable kinds of wood are seen in various places and in considerable quantities. Elm is to be found on Rainy River, and white pine of a fair size and good quality, abounds on the borders of the streams which rush down the steep declivity of the eastern slope to Lake Superior; but it is still more abundant on the western slope, on the waters which flow towards Rainy Lake. On the Sageinaga River, and on the Seine and Maligne, there are extensive forests of red and white pine. Occasional white pines appear too in the beautiful valley of Rainy River and on the islands in the Lake of the Woods, but on proceeding westward they become more rare, and on nearing Lake Winnipeg disappear altogether.

When the pine forests in the neighborhood of Rainy Lake are considered in connection with the fertile region to the westward of Red River, where there is but little wood fit for economic purposes, and regarded in reference to what may be the future wants of that extensive district, they assume an importance not to be overlooked in estimating the resources of this part of the country.

Of the Lakes in this section, the Lake of the Woods is the most extensive. From Lac Plat, which may be regarded as its western extremity to White Fish Lake, which is a somewhat similar extension in an opposite direction, the distance is not far short of 100 miles, and from the mouth of Rainy River, at the entrance of the Lake to its outlet at Rat Portage, in lat. 49° 47′ North, and long. 94° 44′ West, the distance is about 70 miles, so that altogether it may occupy an area of sixteen hundred square miles. This extensive sheet of water, like all the other lakes on the line of route, is interspersed with Islands, and on some of these the Indians have grown maize from time immemorial and have never known it fail, so that the climate cannot be unfavorable for agricultural operations. To what extent the country on the main land may be fit for settlement can only be ascertained on further exploration; for, except on the borders of Lac Plat, it has not been at all examined. It would be difficult to conceive anything more beautiful of its kind than the scenery of this lake. Islands rise in continuous clusters, and in every variety of form. Sometimes in passing through them the prospect seems entirely shut in; soon again it opens out and through long vistas a glance is obtained of an ocean-like expanse, where the water meets the horizon. Not a trace of civilization is anywhere observable, but the Indians are numerous, and, indeed, this lake seems to be their favorite resort in summer; the wild rice on its borders, and the fish which abound in its waters, affording them an easy means of subsistence, not to mention the maize which they grow on the islands. They are a fine looking race, and if removed from the humanizing influences of civilization, they are also strangers to the vices which it brings. The men are generally tall and well formed, and some of the women remarkably comely, but they are not very cleanly in their habits, and there can be nothing more suggestive of indolence than their mode of life, which, however, has one feature to recommend it in the entire exemption from care with which it seems to be attended. Gliding in their light canoes from island to island, basking in the sunshine on some pebbly strand, and merely exerting themselves to an extent sufficient to supply their immediate wants, the future affects them not, and they appear to be supremely happy; but the winter brings its troubles, and they have then to betake themselves to the forests in single families, where, having only game to depend upon, they are sometimes sadly straitened.

From the Lake of the Woods to Lake Winnipeg, the Winnipeg River presents a succession of lake-like expanses and foaming cataracts, making altogether a descent of about 360 feet in a distance of 160 miles. The shores are in general rocky, but there is nothing to warrant the assumption that there may not be good land at some distance from the river; for on making enquiries of the Indians who were familiar with the country, we learned that the region inland was comparatively level, or rising only in undulations which scarcely deserve the name of hills. To explore it, however, would be a mere waste of time, as it is not on the line of route which, instead of following the Winnipeg, must strike across the country from the Lake of the Woods to Red River.

Some of the falls on the Winnipeg are remarkably beautiful, and acquire additional interest from the wildness of the surrounding scenery. The Silver Falls, Pointe du Bois and Chute à Jaquot are all very imposing, more especially the latter, which is one of the few places where the whole volume of the Winnipeg can be seen. In general, however, the scenery is comparatively tame. High rocky bluffs, woody islets and lakes, compose a picture pleasing enough in itself, but wearisome from its monotony.

THE RED RIVER COUNTRY AND THE REGION TO THE WESTWARD THEREOF.

This is a region differing as widely as may be in its physical character from that which has just been described. Although the distance across from the Lake of the Woods to Red River is but ninety miles, the country undergoes a complete change. Instead of the lakes with their woody islets, the clear running streams and foaming rapids, and the swelling hills covered with forests of pine, an undeviating flat spreads out every where, vast prairies open up where the eye seeks in vain for some prominent point to rest upon, and the rivers, richly bordered with trees of another kind, flow with a sluggish course through the great alluvial plain. However, if the scenic beauty which characterizes the region so near it to the eastward is wanting, this country is incomparably superior in all that can minister to the wants of man. Civilization has already set its impress upon the soil, and human habitations appear at intervals for a hundred miles along the Red River and the Assiniboine.

The tongue of land immediately to the eastward of Red River, within the boundary line, and between it and the Lake of the Woods and the River Winipeg, is remarkable, inasmuch as it divides the wooded from the prairie region, partaking to some extent of the character of both. Its eastern border, on the Lake of the Woods and the Winnipeg, is of the crystalline formation, of an uneven surface, and densely wooded. Its western, on the Red River, presents wide prairie openings, and for a distance of about thirty miles back is of an alluvial soil. Immediately to the westward of the Lake of the Woods, and but slightly elevated above it, there is a marshy plateau, scantily wooded, from which the Roseau River flows westward to Red River, the White Mouth River northward to the Winnipeg, and several inconsiderable streams eastward to Lac Plat, and the Lake of the Woods itself. Westward of this plateau the land descends evenly to the prairie bordering on Red River, and to the northward it declines very gently to Lake Winnipeg; another river, the Broken Head, taking its rise on the slope between White Mouth River and Red River, about six miles to the eastward of which latter it flows into Lake Winnipeg in a reedy marsh. In the marshy plateau just referred to lies Lac des Roseau, issuing from whence the Roseau, or Reed Grass River, flows westward almost at right angles to the general direction of the streams. On leaving Lac des Roseau it has at first a sluggish course through a reedy marsh, it then rushes rapidly through a wooded country, making a descent of about 250 feet in the distance of thirty miles to the prairie bordering on Red River, through which it winds its way with a comparatively gentle current.

The obstructions which occur in its course through the wooded region, and the great descent which it there makes, preclude the probability, not to say possibility, of its ever becoming available as a channel of communication, except for very small and light canoes, between Red River and the Lake of the Woods; that is, unless it should be thought advisable at some future period when the country becomes settled, to make it the course of a canal, taking a supply of water from the Lake of the Woods, a thing which is by no means impracticable, the Roseau Lake being almost on the level of that Lake and no high land intervening between them.

Besides the Roseau several inconsiderable streams, among which the Rat River and the Seine—mere brooks—are the principal, join the Red River from the eastward.

The Red River itself has a course nearly due north from the boundary line at Pembina to Lake Winnipeg, a distance in a direct line of 104 miles. It has an average width of about 300 feet, and is navigable to vessels of light draught, as it is, also, far to the south beyond the boundary line. Near Lake Winnipeg the banks are low and marshy, but from the Indian Settlement upwards they vary from twenty to thirty feet in height. Beyond these the prairie is almost level, having but a very slight inclination to the River. Frequently there are two or more terraces before the prairie level is attained, and these are subject to be overflowed in seasons of extreme high water. Twice within the memory of the present generation, once in 1826, and again in 1852, the water has spread over the upper terrace to the extent of several miles from the river but to no great depth, sufficient, however, to drive the settlers on the immediate banks of the river from their homes; but they are now building their houses on more elevated ground than they formerly did, and it is to be hoped they may not be subjected to a recurrence of the evils which they then experienced.

At Fort Garry, in latitude 49° 54′ North, and longitude 97° 21′ West, the Red River is joined from the westward by the Assiniboine, its principal tributary; several smaller streams also join it from the same direction, as delineated on the accompanying map, but they drain but a limited area and require no particular description.

The Assiniboine River, and the Manitouba and Winnipegoos Lakes having been referred to at considerable length in my report of the 4th July, 1857, I shall in this avoid as far as possible going over the same ground, merely endeavoring to convey a general idea of the country and supply what may have been omitted. The lakes acquire great importance from the easy access which they afford to the neighbouring territory. They extend in a direction from south-east to north-west about 300 miles. The stream by which they discharge their united waters into Lake Winnipeg is called the Little Saskatchewan, in contra-distinction to the Saskatchewan proper. There is another stream taking its rise on the Duck Mountain and flowing into the Assiniboine, called, also, the Little Saskatchewan, but more commonly known as the Rapid River, by which name I have designated it. I mention this to avoid confusion of names.

The country westward of Red River so far as we have explored it, presents three divisions, which, although possessing many features in common, are yet of a character in some respects distinctly different.

Of these the great alluvial flat extending from the 49th parallel to the Saskatchewan, bounded to the eastward and north-eastward by Lake Winnipeg and the wooded region between Red River and the Lake of the Woods, and on the south-west by the high lands which extend from the boundary line to the Pasquia Mountain, on the Saskatchewan, may be regarded as the first. It has a length of 340 miles, and an average width of 60 or 70, and may embrace an area of 20,400 square miles. About one-third of this extent is chiefly open prairie land, and the remaining two-thirds mostly wooded. From Pembina to Lake Winnepig the prairie land vastly predominates, but from thence north-westward to the Saskatchewan the forests gradually become more dense until they cover the entire face of the country. The whole of this region is very level, and, if exception is made of the lands immediately bordering on Lake Winnipeg and the Saskatchewan, the soil is of an alluvial description, and so rich that, as experience has shewn, wheat may be grown for twenty successive years without exhausting it. A considerable portion of the area is occupied by swamps and lakes, but the swamps, so far as I had an opportunity of observing them, are mere marshes, with a bottom of alluvial soil, similar to that of the dry prairie, and so firm that horses and cattle can wade through them in almost any direction. They seem to owe their existence solely to the extreme flatness of the country, and as they are at a much higher level than the streams, which all run in deep channels, they might be very easily drained; indeed, with a proper system of drainage the whole of this great alluvial flat might be brought under cultivation, except, of course, where it is periodically overflowed, and the extent to which it is subject to be so is quite insignificant as compared to the whole area. Of the lakes, the Manitouba, Winnipegoos and Shoal Lake are the principal, and these may occupy an area of about 2700 square miles.

The streams which flow through the prairie are all bordered more or less with forests, in which oak and elm of a fair size are to be met with, although not in very great quantities. In the wooded section, of which, however, less is known, poplar predominates, but on the borders of the lakes and streams, larch, spruce, birch and oak are to be found, of a size and quality available for economic purposes.

The second natural division embraces the hilly region which, as has been stated, forms the south-western boundary or embankment of the great alluvial flat which has just been described; it extends from the 49th parallel to the Saskatchewan, a distance in a north-westerly direction of 360 miles, and may have an average width of 40 miles.

This region is of a character more varied, and, perhaps, on that account more interesting than any other part of the country. High rolling banks and elevated plateaux, covered with dense forests, alternate with wide spread valleys of unsurpassed fertility. Numerous streams taking their rise among the hills, run with a rapid course towards the Assiniboine on the one side, and to the Manitouba and Winnipegoos Lakes on the other. Of these the principal are the Dauphin River, Spruce River, Duck River, Swan River, Red Deer River, and the Wauketsequapawoo or Floating Ice River, which flow into the Winnepegoos Lake; and Shell River, Birdtail Creek, Arrow River, Rapid River, and Oak River, which run to the Assiniboine. The hills are known as the Riding Mountain, Duck Mountain, Porcupine Hill, Thunder Mountain, &c. Of these the Porcupine Hill and Thunder Mountain alone deserve the name of hills; the Duck and Riding Mountains being nothing more than elevated plateaux of great extent, penetrated by deep glens. As seen from the Winnipegoos Lake, the Duck Mountain presents a perfectly even outline, rising to the height of perhaps 500 or 600 feet above the lake. Porcupine hill, or range rather, may have an altitude of 1500 feet above the surrounding country, while Thunder Mountain is but a higher swell in the undulating ridge which connects this range with the Duck Mountain. When first seen on ascending the valley of Swan River, Thunder Mountain has a very striking resemblance to the Montreal Mountain, and it acquires interest from the fact that the Indians report coal on its eastern declivity.

Throughout the whole of this region wood is in sufficient abundance to supply the wants of settlers for generations to come. The high lands are in general densely wooded, and the valleys present about an equal extent of woodland and prairie. Salt springs occur in various places on the shores of Winnipegoos Lake, and in the lower part of Swan River. At these springs the Hudson's Bay Company manufacture salt for their own use; and an enterprising individual of the name of Monkman has established works, where he makes enough for the supply of the Red River Settlement.

With respect to the nature of the soil, and appearance of the country, I cannot do better than transcribe an extract from a journal which I kept while ascending Swan River, the valley of which, from all I could learn, very much resembles that of Red Deer River, and the other streams which have been mentioned as flowing into Winnipegoos Lake.

7th June.

Crossed over this morning to visit the salt springs (on the western shore of Winnipegoos Lake, about 6 miles from the mouth of Swan River). On landing we ascended a steep bank, which has the appearance of an artificial embankment, within which, and but slightly elevated above the level of the lake, a bare flat, without tree, shrub, or grass of any kind, occupies an area of 20 or 30 acres. The surface of this flat is encrusted for the most part with a saline substance; and throughout its entire extent there are numerous little mounds, from the top of which issue springs, all of them emitting more or less gas. We found here a family of half breeds engaged in the manufacture of salt, and we learned from them that the mounds sometimes subside, when the ground begins to swell in another place, and another spring bursts forth. The water is, in taste, like that of the St. Léon water of Lower Canada, and on being drunk, produces the same effect. Similar springs, we were informed, occur on Shoal River, between the Winnipegoos and Swan Lakes. Leaving the salt springs, we ascended Shoal River, which is the name here given to Swan River, and crossed over Swan Lake, at the head of which we encamped for the night. The scenery of this lonely region is really beautiful of its kind. As we passed through Swan Lake, the sun was setting behind a range of hills which rose over a low wooded country to the west. To the south the blue outline of the Duck Mountain was just discernable on the verge of the horizon; while we, in our tiny craft, were gliding on through woody islands, rich in the first green drapery of summer. Perhaps the tranquil scene on which we looked impressed us the more from having been buffeted for some days previously on stormy lakes, to the navigation of which our little canoes were but ill adapted. This evening we set nets, and obtained a good supply of fish.

8th June.

Continued our course up Swan River against a tortuous stream with a strong current. The banks of the river, near Swan Lake, are of an alluvial soil, but so low that they must be occasionally overflowed. As we ascend they become higher, and the growth of timber indicates a soil of unsurpassed fertility; weather fine, but rather windy.

9th June.

Start at daybreak, and at 8 a. m. reach a building called the store, at which we find 40 carts, which the Hudson's Bay Company use in the fall to draw their supplies inland, but see neither men nor horses. Breakfast, and an observation taken, we continue our course, passing through a fine country. Gradually the river becomes more rapid, but we make, nevertheless, a fair journey, having came over 20 miles against a very strong current. To-day I travelled a long distance by land; the soil is good, but densely wooded, chiefly with poplar of a large size. In the low grounds, however, spruce and larch are abundant; and on the banks of the river maple is to be seen on the flats. The weather continues fine, but the mornings are rather cold.

10th June.

This morning we were awakened by a regular serenade from the birds: the woods here are positively alive with them. Continue our course, but are much delayed by the shallowness of the stream, and continuous rapids, full of large boulders. Having made a few miles, I leave the canoe, and march by land. The flats have given place to high rolling banks, and wide prairie openings appear among the forests. As we proceed the country becomes still more open; and to judge by the progress of the vegetation, and the black mould thrown up in countless hillocks by the moles, the soil must be very rich. Where land slips occur on the immediate banks of the river, they exhibit a face of yellow loam, or stiff clay, curiously stratified, and shewing the presence of minerals in the water which oozes from between the strata. To-day we saw some elk, but failed to get a shot; we also saw numerous bear tracks, but saw none of those interesting animals. The weather continues clear and fine.

11th June.

Dispatched an express at daybreak to Fort Pelly, in order, if possible, to get horses to take the baggage across to the Assiniboine by the time we should reach the carrying place; then send the men to pole or track up the rapids as best they might, while de Salaberry and I walk by land, in order the better to obtain courses and distances, and ascertain the nature of the soil, &c. We pass through a beautiful country, presenting about an equal extent of woodland and prairie. As we proceed, the openings become larger, and the wood less frequent. The valley seems to be about thirty or forty miles in width. To the west appear the Porcupine Hills, which separate it from the valley of Red Deer River, to the south, Thunder Mountain rises like a blue cloud in the distance, and to the east the smooth outline of the Duck Mountain is seen at intervals through the openings in the forest. The weather to-day was delightful, and the appearance of the country so pleasing, that we wandered too far, and, being unable to rejoin our party, had to sleep supperless and without covering.

12th June.

Rejoin the party, breakfast, and then leaving De Salaberry with the canoe, I take a man with me, and proceed by land. We continue to walk on for about 20 or 25 miles to the carrying place, in the hope of meeting the people who had been sent to Fort Pelly for horses. They do not come, and as we saw nothing to shoot, or at least that we could shoot, having but a riflle with us, we go to sleep fasting under the cover of some trees. Such a country as we have passed through to-day I have never before seen in a state of nature. The beautiful green of the rolling prairie, the trees rising in isolated groves, looking at a distance as if laid out by the hand of art, and the blue hills bounding the prospect, presented a picture pleasing in itself and highly interesting when considered in relation to the future. It required no great effort of the imagination in weary travellers to see civilization advancing in a region so admirably prepared by nature for its development, to picture herds of domestic cattle roaming over plains still deeply furrowed with the tracks of the buffalo, which with the hunters who pursued them had disappeared forever; or to plant cottages among groves which seemed but to want them, with the stir of existence, to give the whole the appearance of a highly cultivated country. The weather to-day has been as fine as we have had it for some days past.

13th June.

Astir at daybreak, load the rifle with shot, and go in search of game; come on some plover, which we knock over, roast, and have a delicious repast. Scarcely have we done when horsemen appear galloping down the opposite bank. They prove to be our messengers, bringing six horses from Fort Pelly, Mr. McDonald, the gentleman in charge, having kindly sent me his own horse to ride. This was very civil, seeing that he had never heard of us before, and knew nothing whatever about us. But the Hudson's Bay Company's officers are equally polite at all their stations. We mount and go in search of the canoe, and have the pleasure of galloping over the plains where we had such a weary walk yesterday. We fail to find the canoe, however, and camp on the prairie, having first hobbled the horses in the approved manner of the country. Weather to-day oppressively hot.

I conclude my remarks on this section by saying that the Indians report coal on the Rolling River—a tributary of Swan River—which has its source in the Duck Monntain—on Swan River itself, above the carrying place, and on Thunder Mountain. It has also, they say, been met with on Red Deer River. We found some specimens, but not *in situ*, in going up Swan River, and these are noticed in Mr. Billings' able remarks on the geology of the country.

I shall consider as the third grand division the vast prairie region extending from the broken ground just described westward to the sources of the Assiniboine, bounded on the south by the 49th parallel, and on the north by the Saskatchewan. Doubtless much of the country still farther to the west and northwestward might be comprehended in this as being of the same character, but I confine my remarks to what came under my observation, or of which I obtained reliable information.

Regarding this region in its general aspect, it is apparently level, or but slightly undulating, with an inclination to the eastward. It is, however, at a great elevation above the valley of Red River, even on the borders of the hilly tract, and gradually increases in altitude on proceeding to the westward, the sources of the Qu'Appelle taking their rise in a country probably 500 or 600 feet above the level of Red River. Through this high plain the streams run in valleys varying from one hundred and fifty to two or even three hundred feet below its general level. These valleys vary in width from a quarter of a mile to two or three miles, and have commonly a pretty uniform direction, but the streams wind through them in an exceedingly tortuous course. The plains, although at such an elevation above the streams, are much the same in appearance as those in the low alluvial valley of Red River, and present a soil apparently of as great fertility. Wood is not, however, so abundant, but on the immediate borders of the Assiniboine it is sufficiently plentiful to supply the wants of a new settlement.

The streams can be navigated, most of them, by light canoes, but they could only be rendered available for general traffic at an enormous outlay. Flat bottom boats may descend the Assiniboine at high water, but it would be an extremely difficult and tedious task to take craft of any kind up it, for, notwithstanding that it has a very tortuous course, sometimes winding about for three miles to make one in a straight direction, it is, nevertheless, very rapid, having a descent in some places of ten feet per mile. From the Rapid River downwards, in passing through the hilly region from the higher prairie plateau to the lower, it makes a descent of at least 300 feet in the distance of 50 or 60 miles. So that the difficulty of rendering it navigable to craft of considerable size may readily be conceived.

At the sources of the Qu'Appelle, a tributary of the Assiniboine, there is said to be a lake which during the spring floods discharges its waters two ways; that is, into the south branch of the Saskatchewan as well as into the Qu'Appelle, and some travellers who have passed that way imagine that they see in this a means by which a water communication might be opened through the valleys of the Assiniboine and Qu'Appelle to the Saskatchewan, and some others, without considering the configuration of the country, have adopted the same pleasing but delusive idea, the absurdity of which, in the present state of the country, will be apparent when it is considered that in the Qu'Appelle there is not water enough for the supply of a canal, that even if a supply could be introduced from the Saskatchewan there would still be the important item of at least 500 feet of lockage to provide for, and that locks would have rather an unstable foundation in the bottom of a soft alluvial valley like that of the Assiniboine, which must carry off the drainage of the high prairie plateau on either side, and which in many places is periodically overflowed between the higher banks which hem it in, affording no means of commanding or carrying off the surplus water.*

So difficult is it to navigate these streams, even in small canoes, that the Indians and traders prefer the use of carts, with which they travel over the plains with great facility.

But, if the rivers cannot easily be rendered navigable, the country is admirably adapted for railroads, and when settlement is introduced and trade has arisen, these will become an imperative necessity.

GENERAL REMARKS.

Taking the three sections of country just described, collectively, they embrace an area of 107,000 square miles or 68,056,000 acres, deducting 7,000 square miles for the area of the Lakes Manitouba, Winnipegoos, &c., and the barren lands bordering on Lake Winnipeg and the Saskatchewan, and there remains an area of 100,000 square miles or 64,000,000 of acres, chiefly of the prairie character, and in general fit for cultivation. In estimating the value of the western region, however, it must be borne in mind that this does not amount to a fourth part of the area, which, in point of soil and climate, is usually considered to be available for settlement. Far beyond the sources of the Assiniboine, on the slopes of the Rocky Mountains and northward to Peace River, there lies a vast region where the climate is on the average not inferior to that of Upper Canada. For, admitting that the summer isothermals have a northwest direction, as is incontrovertibly proved by Lorin Blodget and others, the climate of the valley of Peace River in 56° North latitude and 117° West longitude ought to be equal, if not superior, to that of Red River in the 50th parallel, and the observations of travellers seem to confirm this assumption. Sir Alexander McKenzie writing on the 10th of May, in reference to the valley of Peace River, says:—"The whole country displayed an exuberant verdure." Now, at Red River, on the same date last year, the trees were only just coming into leaf, and the green grass of the prairies beginning to shew itself above the withered herbage of the former year, so that Peace River would seem to have the advantage. But, without going so far north, there is a vast area where no dispute can be raised as to the climate, equally valuable with the region which I have endeavoured to describe.

In respect to the natural productions of the country in its present state, wood, as has been already stated, is in sufficient abundance, more especially in the eastern sections, to supply the wants of settlements for a long time to come. Oak and elm are to be had in small quantities, and a sufficiency of birch, larch and spruce might be obtained for building purposes. There is not a pine tree, however, so far as we could see, in the whole region through which we travelled, and the belief that there is arises from the circumstance of the natives calling spruce, cypress, &c., by the generic name of pine. The prevailing growth every where is poplar, and how that species of wood should be so prevalent, on soil so different from what it grows on in Canada, is due to the fires which so frequently sweep over the country. A prairie, or forest even, over which the fire has passed, is just prepared to receive the downy seeds of the poplar which in the month of June are constantly floating in the air. The Indians say, and I think there can be no doubt of the fact, that but for the fires the prairies would soon be overgrown with wood. Be this as it may, the rapidity of the growth of the poplar, once it has taken root in the rich soil of these plains, is truly astonishing.

*Since writing the above I have had the advantage of hearing Professor Hind's lecture on the subject to which it refers, but, even admitting that the whole volume of the South Branch of the Saskatchewan could be turned into the Qu'Appelle, it must not be supposed that locks could be digpensed with. It is possible, indeed, that in the valley of the Qu'Appelle itself, where the descent is represented as being very gentle, the current might not be too strong for steamers of great power. But, on the Assiniboine, from the Rapid River downwards, in making the descent from the higher prairie to the lower, where, as I have said, there must be a fall of 300 feet the accumulated mass of water would rush with the impetuosity of a mountain torrent. The plains of Red River would be converted into a sea, and the Settlement swept into Lake Winnipeg. Fortunately, this contingency is rendered highly improbable by the fact that to produce it a dam of 86 feet in height would be required across a river half a mile in width.

Stone of all kinds, fit for building purposes, is to be found on Lake Winnipeg; limestone appears on Red River, and is very abundant on the Manitouba and Winnepegoos Lakes; sandstone and limestone occur on the Assiniboine, about 150 miles west of Fort Garry, but farther to the west there did not appear to me to be fixed rock of any kind, that is, as far as we travelled.

In considering the means by which settlers could support themselves on being first introduced into the country, a very important article of consumption would be found in the fish, which are very abundant in the rivers and lakes; sturgeon, white-fish, pike, doré, and various other kinds are found in Lake Winnipeg. In Manitouba and Winnipegoos Lakes the white-fish literally swarm. On one occasion we caught with a very small net as many as sixty, equal in size and, I think, superior in flavour to the white-fish of Lake Huron. Ducks, geese, and aquatic fowl of all sorts frequent the marshes and lakes throughout the country. On passing along the reedy shores of Manitouba and Winnipegoos Lakes the ducks rose before us in a continuous cloud, for hundreds of miles. While descending the Assiniboine we had nothing to eat except what we shot, and in an hour or two at any time we could lay in a supply of ducks, geese and other fowl sufficient for two days. At certain seasons prairie fowl and snipe are numerous, and pigeons are sometimes very abundant. Moose deer, elk, and antelope are to be found on Swan River and some parts of the Assiniboine. Farther to the west the Buffalo roam in countless herds, and the pemican and dried meat made from their flesh are important articles of trade in the territory. A party of Red River hunters killed as many as four thousand of these animals on one excursion last summer.

Now, although the occupations of an agriculturist are not exactly compatible with those of a hunter, still the presence of so much that, in cases of necessity, might be made to contribute to the support of a settlement in its first stage should not be lost sight of.

GEOLOGICAL FORMATION OF THE COUNTRY.

No practical geologist having accompanied the party, I was careful to collect such specimens of fossils, shale, and coal as I conceived would best illustrate the geological formation of the region. These I submitted to Sir William Logan, and Mr. Billings, as already stated, and I would invite particular attention to the following most interesting communication from Mr. Billings.

Geological Survey of Canada,
Montreal, 21st Feb., 1859.

Dear Sir,—The fossils and rock specimens submitted by you for examination, are of great scientific importance, as they demonstrate the existence, in the country lying west of the Red River, of a formation not heretofore recognized within the British territories of the north-west. It has been long known, through the various papers published by Dr. Bigsby, Sir John Richardson, and others, that from the neighbourhood of the Lake of the Woods, a belt of silurian limestones and shales runs north-westerly to the vicinity of Great Bear Lake; but up to the present time we have had no data whatever upon which to found an opinion as to what might be the geological age of the vast region lying between this belt and the Rocky Mountains. Your collection furnishes us with almost indisputable evidence that a considerable portion of the territory belongs to the cretaceous period, or the great chalk formation so largely developed in the Old World. This one fact, which I believe to be now sufficiently established, is of the greatest value, as it affords a starting point, or foundation, upon which the materials collected by future explorers can be readily worked out.

The specimens were quite sufficient to enable me to determine the general question of their geological age, but as it was advisable to have also the opinion of scientific men who have made the cretaceous rocks their special study, I forwarded some of the fossils to Messrs. Meek and Hayden, who are now at Washington engaged upon the collections of the Government explorations of the United States. These gentlemen are the highest authorities in America on all points relating to the secondary and tertiary formations of the central portion of the continent. Their opinion is very cautiously given, but it is the more valuable on that account.

Dr. Dawson, Principal of McGill College of this City, examined microscopically several specimens of the fossil wood and lignite.

T. R. Jones, Esq., of the Geological Society of London, has decided that one of the small fossils from the limestone of Lake Winnipegoos (which I sent him in a letter) is a new species. All the others being silurian forms I have determined myself. It is not necessary in this communication to give teehnical descriptions of all the fossils in the collection, and, indeed, it would not be advisable to do so, for in most instances where species are defined without a careful examination and comparison of many good specimens, more or less error has been superinduced. I earnestly hope that another year's exploration may yield much new material, which will enable us to push further the work that has been so well commenced; I shall, therefore, for the present limit my report principally to the opinions of the several gentlemen who have been kind enough to furnish me with their assistance in the departments of the science in which they are so justly celebrated, with but a few general observations of my own. The following is Messrs. Meek and Hayden's letter:

SMITHSONIAN INSTITUTION,
Washington City, 5th Feb., 1859.

DEAR SIR,—The box of specimens sent by you interests us very much, as they seem to prove the existence of the cretaceous system in these far North-western territories. The *Ammonite* is undoubtedly identical with a species very abundant in the cretaceous beds of Nebraska. It is the form usually referred to *A placenta* of DeKay. The figures and descriptions of this species, published by DeKay and Morton, are not very satisfactory, as these gentlemen seem to have sometimes confounded it with another species since described by Prof. Tuomy under the name of *A lobatus*, which is sharply carinate on the dorsum. If you will look, however, at the figure first published by DeKay, (Annals N. Y. Lyceum Nat. Hist., Vol. II. p. 5,) you will see that he must have had before him, when he named his *A placenta*, the form with the flattened or grooved dorsum, which is common in New Jersey and Alabama. In Nebraska it occurs in formation No. 4, of the cretaceous series of that region.

The only question in regard to your specimen is, whether it may not have been carried by the Blackfeet Indians from some of the Upper Missouri localities. This tribe you know ranges from the head waters of the Missouri far north-west into the Hudson's Bay Territories; and in common with other Indians they are in the habit of carrying with them shells or any other bright object that may attract their attention or excite their curiosity. The matrix of the specimen you have sent is exactly like that in which most of our Nebraska fossils are enveloped, in formation No. 4, as you will see by the specimen of *Inoceramus sagensis* we send you. The shell itself is also precisely in the same state of preservation. Still it is quite probable that rocks belonging to the cretaceous system may occur in the region where the Indian says he obtained this specimen, which conclusion the other specimens from near Fort Garry seem to favor. If so, *Ammonites placenta* is one of the very species we would expect to find there, since it is known to have a great geographical range.

The fibrous material in the specimen No. 1, has exactly the structure of the external layer of the shells of *Inoceramus*, and is very similar to specimens in our possession from formation No. 3 of the Nebraska section, of which we send you a specimen with a few valves of *Ostrea congesta*, attached. We have always suspected these to be fragments of *Inoceramus*, flattened by pressure. The objection to this conclusion is the large size of some of the pieces, and the fact that no large *Inoceramus* is known to occur in this part of the series. The dark fragments in the specimen No. 1, appear to be fragments of fish bones.

No. 2 is unlike anything in our collection. It has the form and general appearance of a Dentalium, but on a closer examination it will be seen not to be hollow like the shells of that genus, but solid and composed of minute fibres radiating from a central or sub-central axis as in the genus *Balemnites* though it does not appear to have the semi-translucent horny appearance so common in that genus.

The small oyster in No. 3, seems to be different from *O. congesta*. We send you specimens of the latter, however, so that you can make a comparison.

The dark shale from the Assiniboine, 250 miles from Fort Garry, is undistinguishable from many specimens in our possession from No. 2 of the Nebraska section, along the Missouri above Big Sioux River, and from near the Black Hills. It also contains small scales of fishes, which we regard as identical with some we have from that formation. We send you specimens marked A. 1 and A. 2. These you will observe are almost exactly like that on your specimen, and lead us to think they belong to the same species and the same epoch.

The flinty specimens, we think, if organic, must differ from any thing in our collection.

Very truly yours,

MEEK & HAYDEN.

In explanation of such portions of the above letter as refer to the Nebraska section of the cretaceous rocks, I beg to state that Messrs. Meek and Hayden have carefully explored and mapped geologically a large tract of the north-west Territory of the United States, extending as far north as the boundary line or within one hundred miles of the Assiniboine. In Nebraska they find these rocks to exhibit the following series.

No. 5. 100 to 150 feet of gray and yellowish arenaceous clays, sometimes weathering to a pink colour. Great numbers of marine *mollusca*, with a few land plants, bones of *Mosasarus*, &c.

No. 4. 350 feet of bluish and dark plastic clay, containing numerous marine *mollusca.*

No. 3. 100 to 130 feet lead gray calcareous marl, weathering to light yellowish tint. Scales of fishes. *Ostrea congesta*, *Inoceramus*, *Problematicus*, &c.

No. 2. 90 feet dark gray laminated clay. Scales of fishes with a few small *Ammonites*, &c.

No. 1. 90 to 100 feet of yellowish sandstone and clay with water-worn fragments of lignite not known to belong to the cretaceous system.

In this section No. 1 is the lowest and the others lie above it in the order designated; No. 5 being the highest rock identified with the cretaceous formation; above No. 5 there is a deposit nearly 700 feet thick of tertiary rock with much lignite and numerous remains of huge land animals of extinct forms.

It is first worthy of notice that lignite occurs in No. 1, which is the bottom of the series, and also in the Tertiary which is the top of the whole group. Your specimens of lignite coal, therefore, not having been found in place, may be either of cretaceous or tertiary.

The Ammonite procured from the Indian belongs to No. 4, and that this formation does exist in the north-west I have some additional evidence in several fossils placed in my hands by Geo. Barnston, Esq., of the Hudson's Bay Company, after I had forwarded yours to Meek & Hayden. These were procured from a man who said he found them in the bed of the Saskatchewan. One is undoubtedly *Scaphites Nicolettii* and another *Nautilus DeKayi*, both characteristic of formation No. 4.

Taking the evidence of these fossils with the fact that all the cephalopoda except one, mentioned in Meek & Hayden's published works, occur in their Nos. 4 and 5, the opinion that these deposits do exist in the north-west may be advanced with a good deal of probability of its being sooner or later confirmed by positive proof. I judge a good deal from the condition of the fossils which I think had not been carried far before they came into the hands of yourself and Mr. Barnston.

The specimens of dark shale which, according to your memorandum, were "found on the Assiniboine, 250 miles from Fort Garry, where the banks or hills bordering the valley are composed of the material," are more satisfactory because they were taken from the bed.

I have carefully compared yours with those forwarded by Meek and Hayden, and feel quite satisfied that they are the same, both lithologically and palæontologically. I think it well established by these specimens that No. 2 of the Nebraska section does exist at the place on the Assiniboine where they were procured.

The specimens taken from the beds overlaid by sandstone 150 miles from Fort Garry being the small oyster, the fossil resembling dentalium, and also the fibrous substance, are all of doubtful species. More specimens should be procured. They appear to me to be cretaceous, but although we have in the museum of the survey a noble collection of English chalk fossils, I cannot find any that agree exactly with yours.

Judging from the aspect alone I have no hesitation in saying that they are either Jurassic or Cretaceous, and although the species cannot be determined, *yet they establish this important fact, that at a point on the Assiniboine,* 150 *miles west of Fort Garry, secondary rocks do occur.* That this fact is of great value any one may convince himself by connecting it with the existence of the belt of silurian rocks known to run from the Lake of the Woods by Lakes Winnipeg and Winnipegoos, north-westerly. This belt of silurian rocks consists of strata, which, upon the whole, dip towards the south-west, and must run under the locality on the Assiniboine. If, therefore, the true coal measures exist at all in that part of the country, they will be found between the western edge of the silurian belt and a line passing through a point 150 miles west of Fort Garry, and running north-west and south-east. The Assiniboine rocks are newer than the coal formation, and the silurian older, and thus by fixing the age of the former the coal question is narrowed down as it were to a tract of country 150 miles wide. I do not affirm that coal does exist there, but I think it a part of the country which should be submitted to a very careful geological investigation.

The specimens of tertiary coal collected by you are very different from that of the true carboniferous period. You are aware that in general it is not found in sufficient quantities to be of economic value. Beds, however, two feet in thickness are known to occur, and it is not beyond the limits of probability that some large deposit may be discovered which would furnish a great amount of fuel.

The following are the results of Dr. Dawson's examinations:

NOTE ON FOSSIL-WOOD AND LIGNITE FROM THE NORTH-WEST.

No. 1. *Coniferous wood mineralized by iron pyrites.*—It is not sufficiently well preserved to allow its minute structure to be seen.

No. 2. *Lignite.*—This specimen has the appearance of jet, but burns without flame, emitting a fetid odour. The structure, as seen in slices and in the ashes, consists of woody fibre without medulary rays, resembling coniferous wood, but very imperfectly preserved owing to the compression of the cell walls into a nearly homogenous mass. It much resembles in structure and appearance the lignite from Mackenzie River, examined by Bowerbank for Sir J. Richardson.*

No. 3. *Coniferous wood partially silicified.*—This shows layers of annual growth, and under the microscope wood cells with circular discs in one row and close to each other. The medulary rays are numerous and have each about fifteen rows of cells. It belongs to the genus *Peuce*, of Witham, and *Pinites* of Goeppert, and closely resembles *P. Ponderosus*, of the brown coal formation of Silesia.† This would not, however, prove the formation to be tertiary since wood of very similar character occurs in Jurassic deposits. Of the modern pines with which I have compared it, it most resembles the Balsam Fir. *Abies balsamea.*

J. W. DAWSON.

McGill College,
January, 19, 1859.

The coal of the north-west, as I have already stated, may be either tertiary or cretaceous. It occurs in numerous localities all over the great region lying between the belt of silurian rocks and the Rocky Mountains, and far south into the United States. Sir John Richardson says that tertiary coal formations occur on the flanks of the Rocky Mountains, the most southerly one being in the Raton Pass, in latitude 37° 15′ N., longitude, 104° 35′ W., and upwards of seven thousand fee tabove the level of the sea. Leaves of dicoty—ledenous trees obtained in these beds by Lieutenant Abert, in 1847, are figured in Colonel Emory's Report to Congress, (pp. 522–547.) Nuttal observed lignite beds associated with pink coloured pipe-clay on the Arkansas, near the 48th parallel. Sir Alexander McKenzie states that a narrow stripe of marshy, boggy, and uneven ground, producing coal and bitumen, runs along the eastern base of the Rocky Mountains, and he specifies latitude 52° N., longitude 112½° W., on the southern branch of the Saskatchewan, and latitude 56° N., longitude 116° W., (Edge coal Creek) in the Peace River, as places where coal beds are exposed. Mr. Drummond's specimens of coal with its associated rocks at Edmonton, (latitude 53° 45′ N., longitude 113° 20′ W.,) on the north branch of the Saskatchewan, and consequently between the places mentioned by Sir Alexander McKenzie. According to Mr. Drummond the coal was in beds varying in thickness from six inches to two feet, and interstratified with clay and sandstone. The samples he selected were precisely similar to the slaty and conchoidal varieties which are found at the mouth of the great Bear River, and the resemblance between the sandstone of the two localities is equally close. He also found a black tertiary pitch coal which breaks into small conchoidal and cubical fragments, which Mr. Small, a Clerk of the Hudson's Bay Company, who gave the first information of these beds, likened well to Spanish liquorice. At Edmonton the more slaty coal beds pass gradually into a thin, slaty, friable sandstone, which is much impregnated with carbonaceous matter, and contains fragments of fibrous lignite. Hand specimens cannot be distinguished from others gathered from the shale cliffs on the Arthabaska River. Highly bituminised shale, considerably indurated, exists in the vicinity of the coal at Edmonton, and clay iron-stones occur in the clay beds.

Coal beds have been observed on fire on the Smoking River and near Dunvegan on the Peace River." (Journal of a Boat, Voyage through Rupert's Land. By Sir J. Richardson, 1851; vol. 1, p. 197.)

The silurian specimens from Lake Winnipeg and Winnipegoos include several species which are new to science, but as before describing them it would be advisable to procure a greater number, in order to make the definitions as complete as possible, I shall not notice them further than to quote from Mr. Jones' letter what he says about the *Liperditia* from Lake Winnipegoos:

"Geological Survey, Somerset House,
"London, January 5, 1859.

"My Dear Sir,—The little *Leperditia* received in your letter, which came to hand on the third, is distinct from any species I have yet seen. In general form it resembles *Cytheropsis concinna* Ann, and mag. N. H. 3 ser., vol. 1, pl. 10, figs. 3 and 4; but it is a true *Leperditia*, and is five or six times larger. It is one of the narrowest and most cylindrical of the *Leperditia* that I know. *Cylindracea* would not be an inapt term if you like to use it.

"Yours very truly,
"J. R. JONES."

My time has been so much occupied with other work, that I have not been able to give all the specimens in the collection the attention they deserve. Should any more be procured and sub-

* Isbister. Journal of Geological Society. Vol. 11.
† Goeppert, Monographie des coniferie.

mitted to me, I shall endeavor to furnish a report that will include not only them but those of last year's collection, which have not yet been described. It would be well for the progress of geology if all explorers would take as much care in preserving the localities of their specimens as you have done, because it enables us to fix with certainty the geographical position of the outcrops of the formations. A few such points well established here and there throughout a large region such as that of the North-west, are in effect so many lengthened strides towards the determination of the structure of the whole area.

I have the honor to be,
Your very obedient servant
E. BILLINGS.

S. J. Dawson, Esq.,
Engineer in command of the
Red River Expedition,
Red River Settlement.

In the foregoing very valuable article it will be seen that Mr. Billings says:—"If the true coal measures exist at all in that part of the country, they will be found between the western edge of the silurian belt and a line passing through a point 150 miles west of Fort Garry, and running north-west and south-east." And recommends that this intermediate belt between the silurian and cretaceous formations should be submitted to a very careful geological investigation. Now, as I have before stated, the Indians report coal on the Duck Mountain, Thunder Mountain, and on Red Deer River, precisely in the tract which is recommended for investigation. On the Duck Mountain, especially, they say it exists in place and in large quantities. I think, therefore, that it would be advisable that a practical geologist should accompany the expedition next summer, it would add but little to the cost, while it would lead to the determination of a question of the highest importance. Sir William Logan, as I have stated in a previous communication, would willingly send with us a gentleman in every way qualified for the task, and has mentioned Mr. Richardson whose researches have already led to most important scientific results.

ADAPTABILITY OF THE REGION WEST OF RED RIVER FOR SETTLEMENT.

It is sometimes supposed by those who have not travelled on the prairies, that they present the appearance of a vast green lawn which the ploughshare could penetrate with ease, and in many places this is the case, but it is by no means generally so. The richest lands are often matted with roots, burrowed with holes, and sometimes hummocky and uneven. A constant war seems to be maintained between the fire and the forest, and where a prairie has been recently burned, two or three yoke of oxen might draw a plough with comparative ease, but where saplings have taken root a little preparatory work would be necessary. This is more particularly the case on the verge of the forests, although even there considerable spaces are sometimes met with without a tree or shrub of any kind. Once brought under cultivation, the prairie lands have the advantage of being free from stumps which so much embarrass the labours of the agricultuist in the new settlements of Canada. Another great advantage in settling in a prairie region, as compared with a country entirely wooded, is that cattle and horses may be at once introduced in any number, as hay and pasturage may be had to an unlimited extent. But for the little prairie wolves which infest the whole country, sheep might be introduced with profit, and, doubtless, as settlement advances the wolves will disappear.

I would here invite particular attention to the following remarks by A. J. Russell, Esq., of Ottawa. From these it will be seen how easy it is for a poor settler to establish himself in a country such as I have described, as compared to the difficulties which encompass him in the forest. Mr. Russell's long experience in these matters renders his opinions of the highest value.

His remarks are as follows:

"That part of our great western territory watered by the Red River and the Assiniboine, and lying between these rivers and Lake Winnipeg, as exhibited by the exploration of last spring, is of a character calculated to render it highly interesting to the public generally. Its value, however, as a field for colonization can be duly appreciated only by those who have a practical knowledge of the formation of new settlements.

"Though but a fraction of our western territory, it exceeds Great Britain in area, as it also apparently does in the fertility and general arableness of its soil. Possessing a climate equal on the average to that of the settled parts of Canada, it is evidently capable of sustaining in comfort millions of inhabitants. From its vast prairies in the south to the wooded part of it in the north-east, around and between its lakes, it presents a very desirable variety of prairie and woodlands; and to a very great extent that combination of them which offers to the settler the peculiar advantages of both, free from the disadvantages arising from the exclusive prevalence of either.

"Very important also is the great system of inland water communication of which the territory mentioned includes the central and most important part. Not only the four hundred miles of the navigable course of the Red River, falling into the south end of Lake Winnipeg, and almost connecting with the navigable waters of the Mississippi, and the four hundred miles, or probably double that extent, of the Saskatchewan falling into the north end of Lake Winnipeg (which exceeds Lake Erie in extent.) But, more particularly as regards this territory, the great interior system of water communication consisting of Lakes Manitouba and Winnipegoos, with their connecting waters, lying parallel to Lake Winnipeg and in continuous navigable connection with it, presenting by it and Red River an uninterrupted line of navigation from the northern extremity of this territory down into the heart of the State of Minnesota.

"Apart from the present advantage of these inland water systems as inlets for settlement, the physical geographer will at once recognise in them elements which, in accordance with all past experience, must render this region the most important in the northern interior part of the continent.

"The chief peculiar advantage which this territory presents as a field for settlement, lies in the combination it offers of prairie and woodlands. The full value of this advantage can be appreciated only by those who have had practical experience of the great and continued labor required to clear off and cultivate a new farm in a wooded country, and the obstruction it presents to the making of the roads necessary for the formation of new settlements.

"Much is said of the advantage of the superior supply of wood for fuel and fencing, afforded by wooded countries; but these are indefinitely over-estimated by many, in comparing the facilities for settlement offered by prairie lands and wooded countries respectively; such a comparison can be best approximated by reducing the matter to figures as far as possible.

"In some respects this will be easily done. Some of the difficulties presented by woodlands are very tangible, and the cost they occasion is well known, and by enumerating them on the one side of the account of comparison, we shall be better able to see how far, on the other hand, the superior facility of obtaining wood for fuel and fencing will go to balance the expense or lost labor they occasion.

"The first and most obvious cause of expense, in money, or labor is the necessity of clearing off the wood before the land can be even imperfectly cultivated, the average cost of which is three pounds five shillings an acre, but as the stumps still remain, an outlay of twenty-five shillings an acre may be set down as to be incurred afterwards in getting rid of them. Where the stumps are of pine or the land stony, the cost will be much greater.

"In general, pine stumps, if removed at all, will cost at least five shillings a piece, and some will cost twenty-five shillings.

"We have here as one item, at least four pounds ten shillings an acre of expense, to be incurred on account of the wood before the land can be brought thoroughly under the plough. This is the cost to those who can pay for the labor of skilled backwoodsmen, accustomed to the use of the axe, who can do twice as much of that kind of work as the emigrants from Europe, even though accustomed to other kinds of hard labor. To the farm labourer from Great Britain, whose time and industry, if applied to the cultivation of prairie land, would be even more valuable than that of the backwoodsman, the cost of clearing woodland in money's worth of his labor will be twice as much. If he be very young he may learn the use of the axe perfectly; if not, he will never learn to use it so as to be able to do as much work with it as the native backwoodsman.

"As by far the greater part of the emigrants who settle in the woods have to clear their farms by their own unskilled labor, admitting even that they become gradually more proficient, the cost to them, in their own labor, of clearing their farms and removing the stumps may, on a low estimate, be set down at five pounds ten shillings an acre. I do not here speak of the value which their labour in clearing would command. No one would give them such a price for it. I am speaking of the value of the labor unavoidably lost by them on account of the woods.

"Here we have then, to a family clearing a farm of a hundred acres in ten or fifteen years, a loss of five hundred and fifty pounds on account of the woods.

"The settler expends all this, and ten or fifteen years of the best of his life, in toilsome struggles to convert his farm into such proportions of open and wooded land as the settler on the partly wooded prairie lands finds his when first he goes to it. The latter can adopt a regular system of cultivation ten years sooner than the other. He can put as much land under the plough, and reap the fruit of it soon after commencing, as the former can do after ten or fifteen years of crushing toil in clearing land, which necessarily consumed much time which he would gladly have devoted to more extensive cultivation, and raising larger crops, had the woods not been an obstruction to his doing so. It is true he has had plenty of wood for fuel and fencing during the interval, so much so that he has had to burn up twenty times as much as he was able to use in any form. But it would be in the utmost degree absurd to suppose that he has benefitted thereby to the value of five hundred and fifty pounds beyond the settler on the prairie lands, for he, also, has had wood enough to serve his purposes, though he has not been affected with such abundance of it, and has had, perhaps, a greater distance to draw it.

"It must be borne in mind, however, that settlers, in a great part of the country under consideration, in the partly wooded regions, and near the streams where a luxuriant growth of wood extends, sometimes a mile, or even two miles in depth along them, would enjoy every advantage of a wooded country.

"To recur to the labor of clearing land, I have already explained that it is twice as great to the inexperienced, even though accustomed to other kinds of hard labor, as to the native backwoodsman. But how is it with the emigrant who has never been inured to labor, the unfortunate man of business, or professional man, the man who has lived hitherto in bodily ease by some sedentary occupation, and is perhaps advanced in life, but is forced by necessity to make a home in the backwoods? How does he get along with the heavy woods? When he goes up to attempt to fell a heavy tree, as thick as a hogshead, and as tall as a steeple, his heart must sink within him. He can make no more impression on it than a child, but yet it and others must be cleared away before he can raise the barest sustenance for himself and family. To him the clearing of a farm is painful and tedious toil, requiring the exercise of the utmost fortitude. What a relief it would be to him to be transported to a prairie settlement, near the banks of some wooded stream. How light the toil of raising the necessaries of life would be when relieved from the heavy additional labour of clearing the forest.

"Is it at all surprising then that so many European emigrants should pass through Canada to seek the prairie lands of the United States, where they can proceed at once with the regular cultivation of the land, in which many of them are well skilled, without undergoing the protracted toil of clearing land, to which they are unaccustomed? Even old and successful settlers in Canada have, to my personal knowledge, found it much to their advantage to do so.

"And would it not be desirable that our own prairie lands should be thrown open to them by the establishment of an efficient line of communication, and by the commencement of the organization and survey of lands for sale and settlement within the territory under consideration?

"But there is another heavy charge to enter in the account of comparison against settlement in wooded countries. That is the obstruction our dense forests present to the spread of settlement, and the expense that has to be incurred in making roads through them.

"We have abundance of vacant fertile lands, but how difficult it is to get at them, and who can doubt that they would be rapidly settled upon if it were not for the want of roads through the forest to let the settlers in. It takes an expenditure of more than a hundred pounds a mile to make a road through the woods as passable as the natural surface of the prairie by the innumerable routes it offers; and, when a hundred pounds a mile is spent in making a road through our forests, it gives access only to the land immediately on the sides of it. The moment the settlers strike from it to reach lands in the back concessions the obstacle is again encountered, and the expense of opening the roads commences anew. It is a moderate calculation to say that for every square mile of forest country settled, an expense in money or labor of £100 has to be incurred ultimately in making roads, or, what is worse, the settlers have to endure, in hardship and difficulty of communication, a much greater loss from the want of them.

"Here again the practical man only can duly appreciate the magnitude of the obstacle and the expense it entails; I speak from having been engaged in spending about forty thousand pounds in making roads through wooded countries.

"The facts I have mentioned may assist in showing more definitely the loss, or cost of the obstruction, which the forests in wooded countries entail upon the settler, in comparison with which the value of the difference of facility in obtaining wood for fuel and fencing, will be found to be insignificant where such difference does exist, which it evidently does not in a great part of the mixed prairie and woodlands of the territory under consideration. It is obvious, therefore, that this difference does not in any considerable degree balance the superior advantages presented by the prairie lands for immediate and extensive cultivation, with freedom from heavy and difficult labor in clearing the land.

"But the difficulty of access to this territory may be mentioned as an offset to the facilities which it offers to settlers. The briefest consideration, however, will make it apparent that the cost to settlers of getting into it, even by ordinary land travel, would form but a small part of the five hundred and fifty pounds to be sunk, as already mentioned, in the additional toil of clearing even a small farm in a wooded country.

"But it may be said that the territory is remote, and that the expense of opening a line of communication, partly by ordinary land roads, to connect the navigable waters available, and partly by improving the latter, through such an extent of uninhabited, and partly uninhabitable, country would be so great, compared with the cost of opening up our unoccupied territories much nearer home, as to be a serious objection on the score of economy against the settlement of the country.

"Now, far from this being true, the very reverse is the case, and the fact of its being so is one of the strongest points in favor of this territory. As to our unoccupied lands near home, their capacity is insignificant as compared with that of this region. Two-thirds of their extent at least are unfit for settlement, and a great part of the remaining third is far inferior in fertility.

From the obstruction which their wooded character, as before explained, presents to road making, it will take ten times the outlay in roads, either by Government or the settlers, to render the available lands they contain accessible for settlement, that would be required to open a communication with Red River.

And that place once touched the whole of our western prairie territories, from their open character, would be as accessible as would our forest lands nearer home if covered with a continuous net-work of colonization roads.

"The only objection of any importance whatever which can be adduced against this territory, as an advantageous field for immediate settlement, is its remoteness from any market for its produce. This is an objection which no practical man could overlook. It happens, however, that its value can be ascertained with tolerable accuracy by measurement upon the map. By that it will be seen that this territory is as near the navigable waters of the St. Lawrence (Lake Superior), as the less fertile, and worse wooded and watered territories of the United States, now proposed to be opened for settlement with sanguine prospects of success, are to Milwaukie and Chicago.

"Doubtless the same inducements will operate in favor of both, viz: the facilities for immediate and extensive cultivation, the absence of the profitless toil of the mere clearing of land, and the consequent opportunity of living an easier life, which is already in some degree observable in the small settlement of Red River.

"There is another objection to the immediate settlement of this territory, which seems invariably entertained, though groundless; that is, the assumption that it cannot be colonized successfully till the intermediate unoccupied regions north of Lakes Huron and Superior be filled up.

"With the great mass of experience before us, which we have on such matters, it is very singular that this objection should be considered of any importance.

"The unoccupied region north of Lakes Huron and Superior can as little affect the settlement of the Red River country as the thousand miles of the uninhabited shores of the St. Lawrence and Gulf affected the first settlement of Lower Canada. On a smaller scale we have had many similar instances since. The first settlement of the Eastern Townships; that of the Madawaska, on the River St. John; as well as the first settlements on the Ottawa, at Hull, &c.; and we might almost quote that of the interior Saguenay. All took place while there was no settlement for a great distance between them and the occupied parts of the Province. And, in the United States, we might cite the first settlement beyond the Alleghanies, with the fearful addition of Indian wars; and, far more conclusively, the overland emigration to Utah and Oregon, through regions more dangerous and inhospitable, and so much more remote, that our Red River settlement is merely the first station on what is now believed to be a more favorable route from the East to the West than that by which these emigrations took place."

A. J. RUSSELL.

REMARKS ON THE CLIMATE.

The climate of the Red River Settlement will compare not unfavorably with that of Kingston, Canada West. The spring generally opens somewhat earlier, and agricultural operations may be commenced sooner than at Kingston, but owing to the proximity of Lake Winnipeg, which is late of breaking up, the weather is always variable until the middle of May. The slightest breeze from the north-west, blowing over the frozen surface of that inland sea, has an immediate effect on the temperature. On the other hand, the fall is generally open, with mild, dry and pleasant weather. Last year the ice began to move on Red River on the 31st of March, and ploughing was commenced in the settlement on the 9th of April.

In the fall of 1857 the Red River froze over on the 9th of November, and it could be crossed on foot on the 16th, and this was said to be somewhat earlier than usual. The winter was mild throughout, except about the middle of February, when, as will be seen by the register, the thermometer indicated—37° Fah., but only on one occasion. The greatest average depth of snow on the prairies did not exceed one foot, while in the wooded region to the eastward near the Lake of the Woods, it might be about 1 foot 4 inches. The snow on the prairies evaporates even in cold weather with amazing rapidity, and with the first warm weather in the Spring it entirely disappears.

Cattle are seldom housed, but are commonly driven to the shelter of some wood, and left to forage for themselves, except in severe weather, when they are supplied with hay.

The winter which I spent in Red River Settlement may have been an unusually mild one, but that such winters are not unfrequent may be inferred from the following evidence given by Colonel Crofton, before a Committee of the House of Commons appointed to investigate the claims of the Hudson's Bay Company:—

Question 3197. (Mr. Roebuck).—Can you tell me when the spring or summer there (in the Red River Settlement) begins? The season opens about the first week in April, and closes about the middle of November; that is to say, the rivers, lakes and swamps freeze in the middle of November.

3198. That is about what occurs in Lower Canada?—I thought it was about that of Upper Canada; I may be wrong.

3199. Does the summer season close as early as the middle of November?—The summer season may be said to close in August, but the finest weather is what is called the fall, which extends from August to the middle of November.

3200. When does the permanent snow fall?– It commences in the latter part of November, and is not off the ground until the first week in April.

3201. Had you an opportunity of seeing any agriculture when you were there?—A great deal.

3202. What sort of crops did they grow?—Oats, barley, and wheat, chiefly, but all sorts of vegetables.

3203. Did the wheat ripen?—In 90 days from sowing.

3204. It ripened very perfectly? It was the finest wheat I ever saw.

3205. Was the soil fertile? Along the immediate banks of the rivers, and extending for, perhaps, the breadth of two miles, no finer loamy soil could be seen, with a limestone foundation.

3206. Is it geologically limestone?—All.

3207. And wherever limestone is, there is fertile land, is not there?—I think that is the consequence.

3208. Do you know how far the limestone extends; looking at that map? I have ascertained from servants of the Hudson's Bay Company that it extends, as a base of the whole prairie land, to the Rocky Mountains.

3209. So that, in fact, that part of the territory is fit for agriculture? Quite so.

3210. And would make a good colony?—It might maintain millions.

According to the Isothermal charts of Lorin Blodget, the lines of equal temperature for the summer should have a north-west direction from Red River. Now, admitting this theory to be correct, the climate of Red Deer River and Swan River, other circumstances being the same, should be equal to that of the Red River Settlement. But I am of opinion that it is superior, inasmuch as that these rich valleys, while they are at but a very slight elevation above the valley of Red River, are removed from the influence of the cold winds from Lake Winnipeg, which prejudicially affect the latter in the spring. As an instance of the change of climate which is produced by the difference of elevation in this region, I may mention that the vegetation, in the middle of June, was much further advanced in the valley of Swan River than at Fort Pelly, which is some distance further to the south, but at a greater altitude by some 400 feet.

While on this subject, I quote from Blodget's climatology some of his remarks on the climate of the north-west territories, which I am confident will be read with interest.

"By reference to the illustration of the distribution of heat we see that the cold at the north of the great lakes does not represent the same latitude further west, and that beyond them the thermal lines rise as high in latitude, in most cases, as at the west of Europe. Central Russia, the Baltic districts and the British Islands, are all reproduced in the general structure, though the exceptions here fall against the advantage, while there they favor it, through the immediate influence of the Gulf Stream.

"Climate is undisputably the decisive condition, and when

we find the isothermal of 60° for the summer rising on the interior American plains to the 61st parallel, or fully as high as its average position for Europe, it is impossible to doubt the existence of favorable climates over vast areas now unoccupied.

"This favorable comparison may be traced for the winter also, and in the averages for the year. The exceptional cold of the mountain plateaus, and of the coast below the 43rd parallel, masks the advantage more or less to those who approach these areas from the western part of the Central States, and from the coast of California; but though the distinct mountain ranges remain high at the north, the width of their base, or of the plateau from which they rise, is much less than at the 42nd parallel. The elevated tracts are of less extent, and the proportion of cultivable surface is far greater.

"It will be seen that the thermal lines for each season are thrown northward further on passing Lake Superior westward, in the charts of this work, than in those of the military report prepared by the author. At the time those were drawn the number of the observations beyond the limits of the United States were so small that the full expression was not given to the statistics then used, in the fear that some correction would ultimately be found to apply to them, reducing the extreme northward curvatures they indicated. But a further collection and comparison warrants the position now given to the thermal lines, placing them further northward than before, and extending them in a course due north-west from Lake Superior to the 58th parallel. For the extreme seasons, winter and summer, this accurate diagonal extension of the thermal lines across the areas of latitude and longitude is very striking. The buffalo winter in the upper Arthabasca, at least as safely as in the latitude of St. Paul's, Minnesota; *and the spring opens at nearly the same time along the immense line of plains from St. Paul's to McKenzie's River.*

"The quantity of rain is not less important than the measure of heat to all purposes of occupation; and for the plains east of the Rocky Mountains there may reasonably be some doubt as to the sufficiency; and doubts on the point whether the desert belt of lower latitudes is prolonged to the northern limit of the plains. If the lower deserts are due to the altitude and mass of the mountains simply, it would be natural to infer their existence along the whole line, where the Rocky Mountains run parallel, and retain their altitude; but the dry areas are evidently due to other causes primarily, *and they are not found above the 47th parallel in fact.* It is decisive of the general question of the sufficiency of rain, to find the entire surface of the upper plains either well grassed or well wooded; and recent information on these points almost warrants the assertion that there are no barren tracts of consequence after we pass the bad lands, and the coteaus of the Missouri. Many portions of these plains are known to be peculiarly rich in grasses; and probably the finest tracts lie along the eastern base of the mountains, in positions corresponding to the most desert-like of the plains at the south. The higher latitudes certainly differ widely from the plains which stretch from the Platte southward to the Llano Estacado of Texas, and none of the references made to them by residents or travellers indicate desert characteristics. Buffalo are far more abundant on the northern plains, and they remain through the winter at their extreme border, taking shelter in the belts of woodland on the upper Athabasca and Peace Rivers. Grassy savannas like these necessarily imply an adequate supply of rain; and there can be no doubt that the correspondence with the European plains in like geographical position,—those of eastern Germany and Russia,—is quite complete in this respect, If a difference exists, it is in favor of the American plains, which have a greater proportion of surface waters, both as lakes and rivers."

After remarking on the region west of the Rocky Mountains, he goes on to say—

"Next is the area of the plains east of the Rocky Mountains, not less remarkable than the first for the absence of attention heretofore given to its intrinsic value as a productive and cultivable region, within easy reach of emigration. This is a wedge shaped tract, ten degrees of longitude in width at its base along the 47th parallel, inclined north-westward to conform to the trend of the Rocky Mountains, and terminating not far from the 60th parallel in a narrow line, which still extends along the Mackenzie for three or four degrees of latitude, in a climate barely tolerable. Lord Selkirk began his efforts at colonization here* as early as 1805, and from personal knowledge he then claimed for this tract a capacity to support thirty millions of inhabitants. All the grains of the cool temperate latitudes are produced abundantly—Indian corn may be grown on both branches of the Saskatchewan, and the grass of the plains is singularly abundant and rich. Not only in the earliest exploration of these plains, but now, they are the great resort for buffalo herds, which with the domestic herds, and the horses of the Indians and the colonists remain on them and at their woodland borders throughout the year.

The simple fact of the presence of these vast herds of wild cattle on plains at so high a latitude, is ample proof of the climatological and productive capacity of the country. *Of these plains, and their woodland borders, the valuable surface measures fully five hundred thousand square miles.*"

To the above I may add, that the talented author in his isothermal chart, shewing the mean distribution of heat for the summer, places the line of 60° to the north of the Lake of the Woods, and that of 65° at Fort Garry. The same authority gives a summer of 95 days to Toronto, and of 90 to Cumberland House, in latitude 54° north, the extreme northern limit of the region to which my descriptions refer.

On proceeding to the south-eastward from Red River, the climate gradually becomes colder and more rainy. That a great precipitation of rain takes place at and near the high lands, which separate the waters flowing to Lake Winnipeg from those which run towards Lake Superior, is evinced by the magnitude of the rivers, as compared with the area which they drain. The climate, however, seems to be milder on the western slope of these high lands than on the eastern.

On the 12th and 13th of September we passed from Lac des Mille Lacs to Dog River, and found a marked difference in the progress of the season, notwithstanding the shortness of the distance. At Mille Lacs the tender foliage of the poplars near the water had just begun to change, but the woods in their general aspect were as green as in summer; while on Dog River, the rich tints of autumn appeared on the hill sides, and the rustle of falling leaves indicated how severe the weather had been on the eastern, as compared with the western declivity.

In closing my remarks on the climate, I would say that, as a general rule, the season during which agricultural operations can be carried on at Red River, is somewhat longer than in Canada, east of Kingston, while in winter the cold is more intense, although not uniformly so, than in any part of Canada west of Three Rivers. In regard to salubrity, there are no diseases, so far as I could learn, incidental to the country. Ague is unknown, and a population more healthy than that of the Red River Settlement cannot be met with anywhere.

THE RED RIVER SETTLEMENT.

So much has been written in respect to this settlement, that I need notice it but very briefly. In doing so, however, I shall endeavour to correct some misapprehension respecting the people of mixed origin, which might arise from the accounts disseminated regarding them.

The settlement commences about ten miles above Lake Winnipeg, and extends to the south for some fifty miles along the Red River, and to the westward for about seventy miles on the Assiniboine, there being, however, a long interval on the latter river, between the White Horse Plains and the Prairie Portage without any settlers. The population, by a census taken three years ago, was seven thousand, but including the settlers at the Prairie Portage, and the people who mostly live about Red River, without any fixed habitation, it may now amount to some ten thousand individuals.

The inhabitants of the lower section are mostly Indians, those

*At the Red River Settlement near the 50th parallel.

of the middle part chiefly retired traders and voyageurs, or descendants of the first Scotch settlers brought out by Lord Selkirk; while the upper part is almost exclusively peopled by a race of mixed origin, descended from the French Canadian voyageurs, and the native Indians.

That the settlement should have advanced but slowly is not to be wondered at, considering how far removed it is from the civilized world, but still there has been progress, and that of a most pleasing and satisfactory description, and I question if at this moment it would not compare favorably with any rural settlement of equal extent in Canada.

The people of mixed origin are generally leaving the customs of their Indian ancestry, and adopting the habits of civilised life. The Indians, in their part of the settlement, have built themselves houses, and cultivate the land. They send their children to school, and have embraced Christianity, to the observances of which they are singularly attentive. The people of unmixed European descent have excellent houses, and good farms, with horses, sheep, and herds of cattle.

In the settlement altogether, there are two libraries, nine churches, and eighteen schools. Of the Churches, five are Episcopalian, three Roman Catholic, and one Presbyterian. Of the schools, four are for the instruction of females, one of these at St. Cross being under the immediate patronage of His Lordship the Bishop of Rupert's Land, and there young ladies are instructed in French, Music and all the branches of education which it is usually thought necessary for them to acquire.

The Grey Nuns have a large establishment just opposite to the mouth of the Assiniboine, and another, a smaller one, at the White Horse Plains. These ladies devote themselves chiefly to the instruction of the children of mixed Canadian and Indian origin, and the effects of their zeal, piety, and unfailing industry are manifest in the social improvement of the race, for whose benefit they are content to lead a life of toil and privation.

There is still another school for the instruction of young ladies at the rapids, and I was informed that it, also, was an excellent establishment, under the direction of an accomplished lady from England.

At the collegiate school established by the Bishop of Rupert's Land, the higher branches of education are taught, and some of the young gentlemen of the settlement instructed there, have subsequently distinguished themselves in other lands, winning the highest degrees at Cambridge and elsewhere.

In the event of the communication being opened up to Red River, it will be a matter of no small importance to those who seek that remote region to know that establishments exist, where their children may have the advantage of an education scarcely inferior to what they might obtain in any country.

The demand for agricultural produce in the Red River Settlement being, of necessity, limited to the requirements of the fur trade, farming is not practised to that extent which it would otherwise be. And, if the settlers in a measure neglect this branch of industry, it is not fairly attributable to indolence of disposition so much as to absolute want of inducement.

In reference more especially to the population of mixed origin, or, as they are usually called, half-breeds, when it is considered that they are the descendants of hunters and voyageurs, that their fathers either lived by the chase or led a life of roving and adventure, the wonder will be that they have settled down so quietly, and evince such a tendency to exchange the wild freedom of Indian life for the dulness of a settled home and the quiet humanities of civilization.

As they live at present they generally grow enough for their own use, and they are possessed of cattle, sheep, and horses which demand some measure of attention; but they have also their hunting seasons, and after the crops are sown in spring, and when they are harvested in the fall, they form into bands and seek the pleasures and the dangers of the chase. Generally, they go southwestward to the plains of the Cheyenne or the Coteaus of the Missouri. In these long marches they are regularly organized, under the direction of officers duly elected to command. On seeing the buffalo, which usually go in large herds, they form into line and ride on at a hand gallop, until, at a word given by their leader, they dash forward among the bewildered animals, firing and loading from on horseback, with the most amazing rapidity. Sometimes the bulls give battle, and it not unfrequently happens that horse and rider are rolled on the ground. They have, however, more terrible enemies to encounter in the Sioux Indians, who hang on their path, with the view of possessing themselves of the scalp of any unfortunate straggler who happens to separate from the main party. They used formerly to have frequent encounters with these savages, but of late years the Sioux have learned to fear their sure aim and determined spirit, and confine themselves to such attacks as involve but little risk on their part, although they are annoying to the hunters, and compel them to be continually on their guard.

A successful hunt enables the settlers to live in comfort and abundance during the winter, and I doubt if they will relinquish the practice, or lose their fondness for the chase, until the buffalo have entirely disappeared. But this, I think, by no means evinces an indifference to the advantages of civilization. Formerly they lived entirely by the chase; now they follow it as affording an agreeable occupation, and an adjunct to their means of subsistence. With the produce of their farms they could not always obtain clothing, or the little articles of luxury, such as tea and tobacco, which are indispensable even to them; but with pemican and Buffalo meat they can, for these are always in demand at the Hudson's Bay Company's stores; so that hunting, after all, is a necessity of their situation as much as a passion. But even if it were the latter, that is that they could not resist the temptation; make the circumstances equal, and I question if they would be found to differ much in this respect from more civilized communities. And I doubt whether there is a community so sèdate, as to deny themselves the pleasure of shooting the wild cattle of the plains, if they were roaming by thousands within a few days' march of their dwellings, even if the pastime were attended with a little peril.

In physical appearance the half-breeds are far superior to either of the races to which they are allied. Among the *habitants* of Lower Canada they would look like a race of giants, and they are much more robust and muscular than the neighboring Indians. This is due, in a measure, no doubt, to their leading a life peculiarly favorable to the development of the human frame; to the nutritious food which they use, and the extreme salubrity of the climate in which they live. Physiologists might perhaps find other causes; but, be that as it may, they are a hardy, vigorous and active race.

In regard to the social condition of the settlement, crime is scarcely known, the only serious case which occurred, during my residence there, being that of a man who killed and appropriated to himself a cow, an offence, combined with the perpetration of which there were concomitant circumstances, which occasioned him to be imprisoned for six months.

The influence of the Missionaries, whether Protestant or Roman Catholic, is everywhere observable in the moral tone which pervades the community. During our residence in the settlement and on our exploratory excursions I employed many of the half-breeds, and was thrown of necessity so much among them that I had good opportunities of observing their character, and it is much to their credit and that of their instructors that I am able to say, that I never once heard an oath or an indelicate expression made use of among them. This is different from what may be too often observed among the lower orders of other communities. But with their good qualities they combine some others which are not so agreeable. In disposition they are proud, exceedingly sensitive, and ready to take offence. They will do anything to oblige, and fly to anticipate one's wants, but an order sternly given excites hostility at once. They are as volatile as children, and, if offended, would care little for marching off and depending on their guns for subsistance if they were even five hundred miles from their homes. Some of my assistants, who were young men, and did not trouble themselves much about the character and disposition of those who were under their command, had extreme difficulty in managing them; indeed, on one occasion, they were almost completely deserted, but for my own part I never had the slightest trouble or difficulty with them. Apart from the defects just mentioned, they are respectful to their superiors, and generally set about doing what is required of them with the greatest alacrity and enthusiasm. Active, however, as they are, they do not like continuous labour, and in travelling in the west, if I had the organization of a party, I would choose about an equal number of half-breeds and French Canadian voyageurs.

That the settlement is not retrograding but advancing, and that, far from falling back to the habits of their Indian ancestry, the people are rapidly improving in their social condition, may be gathered from the following remarks of the Bishop of Rupert's Land in his "Notes of the Flood." After remarking on the immediate distress which the flood had caused, his Lordship says:—

"Taking it, however, in another light, and asking how will the flood be felt? and the answer is very different. The distress will, we think, be small, compared with that on the former occasion. Much has been taken, but much more has been left. We arise from the flood in a very different condition from before. The settlement was then in its infancy, there were but few cattle; a single boat is said to have transported all in the Middle District in one forenoon. Now, each settler of a better stamp has a large stock of cattle: the one whose record of the first flood we had read at home had then but one cow; he has now, after all his losses in other ways, fifty or sixty head of cattle. Before, too, there was but little grain, and the pressure of want was felt even when the waters were rising. Their dependence throughout was on the scanty supply of fish, or what might be procured by the gun. Now, there is a large amount of grain in private hands, and, even with the deduction of the land which is this year rendered useless, a far larger number of acres under cultivation. In this light it is comparatively less severe: the whole of the cultivated land was then under water: nearly all the houses were carried off by it. It was, as many have called it, a cleaner sweep; but there were then few houses or farms below the Middle Church, or on the Assiniboine above the Upper Fort. The districts of the Rapids and Indian Settlement were still in the wildness of nature.

"Though there is, therefore, greater suffering and loss, there is greater elasticity and power to bear, and larger means to meet it."

"*All this may be attributed to the progress of civilisation, and the advancement of the people, in the interval.* Of this, a single visit to any of the encampments might have afforded sufficient proof. Instead of the bare, unfurnished tent, there was one supplied with every necessary, and generally with the stove to afford its warmth, with a large stock of domestic cattle feeding around.

"There was cooking, baking, and churning going on. Boat after boat was required to carry off the property; whereas before, nearly all might have been carried on their persons. The return, too, seems more hopeful. Before, a large number left, of those unattached to the soil, without any tie to their country. In the interval, a large population has sprung up, who are by birth bound to the land, and look to it as their home. Many are so connected that they feel it impossible to leave. Their family ties are here, the family branches are spread over the land, and root themselves in its very soil. Now to this we look as the strength and sinew of the country: a population contented and happy, I had almost said, proud of their land. The increased facilities of intercourse are also adding to this gradual improvement. To be brought more before the eye of others will inevitably have a beneficial tendency. The flood of 1852 will occupy a far larger space in the public mind than that of 1826. Instead of a few solitary settlers, unknown and almost forgotten by their fellow-men, they are now parts of a mighty system, linked more closely by sympathy and interest to other lands."

The importance of the Red River Settlement will not fail to strike those who have considered the subject of colonizing the region to the west thereof. Here is already a nucleus where the wants of settlers may be supplied in the first instance, and a population of ten thousand ready to welcome them and give them the advantage of their experience. At present, it is true, that the people raise but little more than suffices for their own wants, but let a demand arise and agriculture would soon be extended, and would continue to extend as that demand increased. To appreciate the full value of this little colony, it needs but to be considered how vastly the difficulty of introducing settlement would be enhanced if it were not there.

THE HUDSON'S BAY COMPANY'S ESTABLISHMENTS.

These cannot be passed over without some notice, in a description of the territory where they exercise so great an influence. In noticing them, however, I shall confine my remarks to what was open to observation, without reference to the political questions with which they may be associated, or the extent or nature of the trade which is carried on, as these are subjects beyond the scope of our instructions, and as, moreover, we could not with any degree of propriety have pried into the affairs of a Company, whose establishments were thrown perfectly open to us, and whose hospitality we so often experienced.

The first place of any consequence on the line of route is Fort William, at the mouth of the Kaministaquia, an establishment which derives peculiar interest from its having been at one period the great emporium of the North-west Company of Canada. At one time as many as three thousand people were sometimes assembled there, with rich loads of peltries from the interior, or merchandise from Montreal, but, since the amalgamation of the Companies, the trade of the interior has been diverted to Hudson's Bay, and Fort William has lost much of its importance. It is now the residence of a chief trader, and is still the centre of a considerable traffic in summer, Mr. McIntyre, the gentleman in charge, having established fisheries which afford employment to a number of people of mixed origin, who have settled a little higher up on the Kaministaquia. To Mr. McIntyre the various branches of the expedttion have been repeatedly indebted, not only for his kind welcome and hospitality, but for his readiness in supplying canoes and much of the material requisite for the journey across the country to Red River, and, latterly, he has done what lay in his power to aid the mail contractors in forwarding the mails to Fort Garry.

Fort Frances, the next establishment on the route, is situated at Rainy Falls, in the very heart of the Saulteux country, but it does not seem now to be a place of so great importance as it doubtless was when the trade of the interior passed that way. The officers at this post. Mr. Pether, Mr. Chatelains, and others, were always very kind and obliging.

The next post is at Rat Portage, but it is quite a small establishment as compared with the others.

Fort Alexander, at the mouth of the Winnipeg, the next in order, is the residence of a chief factor, but the trade to which it once owed its importance is now, I believe, diverted to Red River.

The Stone Fort, or Lower Fort Garry, on the Red River, is by far the finest establishment in the territory. A square area of some six acres in extent is enclosed with walls and bastions of stone. Within this enclosure are the Company's buildings, all of them most substantial edifices of stone. The stores are situated on either side, and in the centre stands the residence of the officers, a very imposing building, with verandahs running completely round it, and grounds in front laid off and planted with great taste. This establishment is in charge of Mr. Lilly, a young gentleman from Scotland, who, in addition to his duties as a trader, carries on extensive farming operations. Last spring he brought a considerable extent of new land under cultivation.

Upper Fort Garry, situated at the confluence of the Red River and Assiniboine, is a similar establishment, except that the space enclosed is not quite so large, and that there are a greater number of buildings. This is the residence of the chief factor of the district and Governor of Assiniboine. Here, also, the military are quartered. It is the principal commercial emporium of the settlement, and people of all shades, from the dusky Indian of the plains, to the fair complexioned native of the Hebrides, may be constantly seen thronging about the gates.

Besides the establishments above enumerated, we visited Pembina, Fort Ellice, Fort Pelly, and Swan River House, but reference has been made to all of these in previous reports.

On reflecting that since the amalgamation of the North-west Company of Canada with the Hudson's Bay Company, when they became one under the latter title, such establishments have gradually spread over half a continent, from Hudson's Bay to the Pacific, and from the boundary line to the Arctic seas, it is impossible not to admire the order and system which are everywhere observable in their management. It is a vast system of economy, carried out with the utmost sagacity and foresight in all its details; and a system, too, which seems to work equally well under circumstances widely different. In the back settlements of Canada, on the stormy shores of Labrador, among the warring tribes of the plains, or in the frozen regions of the north, it seems to be alike successful.

An organization so perfect can only be traced to the agency of superior management, and I am of opinion that the success of the united Companies is as much due to the high talent of the officers who have the direction of their affairs, as to other circumstances to which it is more frequently attributed; and there can be no doubt that the same judgment, care, and economy brought to bear on any pursuit, would meet with a very marked measure of success.

THE NATIVE INDIANS.—WHETHER THEY WOULD BE LIKELY TO OPPOSE COLONIZATION, &c.

In considering the project of colonizing the region to the west of Red River, the question naturally suggests itself as to whether settlement would be likely to be opposed by the native population, and whether if opposition arose, it would be of a formidable character.

Settlement would in the first instance spread over the territory which, with a soil well adapted for its development, was at the same time the most easy of access; and, in this respect, the valley of the Assiniboine, and the country bordering on the Manitouba and Winnipegoos Lakes, and their various feeders, would be the first to fill up; and throughout the greater part of this extensive region, even in its present state, a settler would be as safe as in the backwoods of Canada. The Assiniboine in one part of its course, that is, between the Rapid River and Brandon House, touches on the Sioux country, but upon the whole I think, with respect to this, the only part of the country under consideration where there is any danger of molestation from Indians, the accounts of their power and disposition to do harm are very much exaggerated. Taking the region to which my descriptions more particularly refer, the Indians are not more numerous than they are in some other parts of Canada, and with ordinary precaution no danger need be apprehended from them. Farther to the west, on the South Branch of the Saskatchewan, the Blackfoot Indians are said to be numerous and warlike, but, as their intercourse with the whites increases, their habits will no doubt change as those of other Indians have changed, and, judging from similar cases, it is but reasonable to suppose that ere settlement reaches them they will have ceased to be dangerous.

On the north branch of the Sakatchewan, near Fort's Pitt and Edmonton, the Indians are numerous, but from all I could learn I believe they are peaceably disposed. By this route the traders pass to and from the Columbia and the Pacific, and even emigrants have passed with their families without experiencing the slightest molestation.

That the Indians are diminishing in numbers in the region immediately to the westward of Red River does not admit of a doubt. But there is one pleasing circumstance connected with their partial disappearance, and that is that they have to a certain extent amalgamated with the whites. The population of origin thus mixed in the Red River Settlement, and in the country drained by the Assiniboine and Little Saskatchewan, cannot be less than six or seven thousand, which, taken together with the remaining natives of unmixed descent, is perhaps as great a number as ever occupied the territory.

To the eastward of Red River, at the Lake of the Woods and Rainy River, the Indians are said to be increasing of late years, and from all I could see, or learn, I believe this to be the case. And, as they occupy the country through which any line of communication between Lake Superior and Red River must pass, it becomes of the utmost importance to ascertain every particular regarding them, to cultivate a good understanding, and take such measures as would prevent the possibility of a collision with them. They are still in a state of primitive barbarism. Missionaries have been among them frequently, but have never succeeded in making much impression upon them, in the way of leading them to embrace Christianity. Their habits and customs are probably the same as they were at the time the first Canadian traders penetrated into the country. Every spring, on the opening of the navigation, they assemble at Fort Francis to celebrate the Dog feast and Scalp dance common to all the tribes in that direction. As many as 500, I have been told, sometimes assemble together on these occasions. Some of them come from Red Lake, within the United States boundary, others from Rat Portage, the Lake of the Woods, Rainy Lake and Lac des Bois Blancs.

Sir George Simpson estimates the entire population as follows:

Fort Frances,	1,500	
Fort Alexander,	300	
Rat Portage,	500	
White Dog,	100	
Lac du Bonnet,	50	
Lac des Bois Blancs,	200	
Shoal Lake (Lac Plat),	200	
		2,850

But this, I suppose, does not include those who come from the United States side. Nevertheless, when the extent of country over which this population is spread is considered, it cannot be considered very great even for an Indian population. But it is deserving of attention that the fact of their having abundance of food at certain seasons enables them to collect in numbers sufficiently great to be formidable if inclined to be troublesome. Sir John Richardson, who passed several times through their country, describes them as being "saucy, and independent of the Hudson's Bay Company, from the fact that they have abundance of sturgeon and great quantities of wild rice, so that they can feed themselves without having recourse to the supplies of ammunition or clothing with which the Hudson's Bay Company supply their Indians." This is the case, and I may add that their country is tolerably rich in furs for which, being near the frontier, they get a high price either from the Hudson's Bay Company, or from the American fur traders.

They are certainly of an independent, and I should say unmanageable disposition; and their natural ferocity is not lessened by their constant wars with the Sioux Indians, on the confines of the great prairies, to which every branch of the tribe occasionally sends its contingent. Though but few usually fall in these wars, they are always kept alive, and attended with all the circumstances of atrocity common to Indian feuds. The night attack on the slumbering enemy, the indiscriminate slaughter of men, women, and children, the scalping knife and the tomahawk, are as much characteristic of that warfare as they were of the early Indian wars of which we read. No later than last spring, eleven Saulteux Indians had encamped on an Island, on a lake near Crow Wing. They had been observed by a much larger party of Sioux; and in the morning eleven reeking scalps indicated how completely they had been surprised.

I mention these circumstances not with the view of conveying the impression that there is any present danger on the route by Rainy River, but simply to inform the Government of the character of the people with whom the country will be brought in contract in introducing settlement, and opening up the communication, and to point out the way in which I think the possibility of collision with them may be avoided. As already stated, the Indians who frequent Rainy River are in the habit of assembling in considerable numbers in summer. Now, admitting that their present pacific disposition should continue, and that a treaty were made with them by which they should agree to relinquish a certain portion of their lands, it is still to be considered that in introducing a mixed assemblage of colonists for settlement, or laborers for the construction of roads, there might be some among them who would not be slow to give offence to the Indians, who, in their turn, would be ready enough to retaliate. A collision once occurring, if it were of a serious nature, and no controlling power near at hand, the whole tribe would be at once in arms, and might, as it is to be apprehended they would, give a great deal of trouble. Fort Frances and the Rainy River are the only places where this could occur; and, in order to prevent the possibility of such a contingency, it would be necessary to have a military force at the former place of, say, as many troops as are now at Fort Garry; that is, something over a hundred men. From what I have observed of the character of these Indians, and after maturely considering the subject, I am convinced that *the moral effect which would be produced by the mere presence of such a force* would effectually prevent any act of violence or agression on their part, while, at the same time, it would be a protection to them against agression on the part of unruly settlers or laborers.

Fort Frances is in a position from which communication would be easy, either to the eastward or westward. Below it the navigation is unimpeded, by Rainy River and the Lake of the Woods, to within ninety miles of Fort Garry. Above it there is no interruption to the head of Rainy Lake, so that troops might, if necessary, be moved in either direction; but I do not think it would be necessary to move them, for their presence alone would be sufficient to ensure tranquility.

In my letter of the 21st August, 1858, from Fort William, I mentioned that I had had an interview with the Saulteux Chiefs at Fort Frances, the result of which was that they accorded their full permission to examine the country, but requested that some person might be sent to meet them on their assembling next spring, to explain the objects of the expedition, and whether it was intended to take up any of their lands for settlement, in which case they trusted nothing would be done until arrangements had been made with them. At this interview they were very friendly, and I gave them a letter which Pegwis, the chief of the Saulteux Indians at Red River, had prepared for me of his own accord, and entirely unsolicited on my part. The letter, written by a native schoolmaster, at the old chief's dictation, was as follows:—

(Copy.) THE CHIEF PUGISE,
(Pronounced Pegwis.)

My friends,—I hope you all that are to the east of this Colony will give the same respect as we have done to these gentlemen, giving them full permission so as to explore the country along the line of route. Knowing I being the oldest Chief, I have full confidence you will listen to my advice.

Your Chief, the
PUGISE ×
mark.

This letter I had read to them in their own language, and it evidently produced a favourable impression, although they do not look upon Pugise as so great a man, by any means, as their own principal chief.

In the event of taking up a portion of their lands for colonisation I have already had the honor to suggest that the payments should be made in the shape of yearly presents of such articles as would be most useful to them. This is the system adopted by the United States Government, and, when it is properly carried out, it is no doubt the best; but this is not always the case, as I had an opportunity of observing last Fall at the Grand Portage, where a branch of this same tribe receive payments for their lands. On arriving at that place on the 1st August I observed a number of Indians waiting for the agent with their payments. Two months later I again passed that way, and they were still there, with a considerable addition to their number, in daily expectation of the agent, as they had been when I first saw them. The Fall was setting in and it was time they should set out for their hunting grounds, but whether the agent came, eventually, or whether they had to return to the interior, disappointed of the pittance which they had made a long and laborious journey to receive, I am unable to say.

If I mention this circumstance it is that the evils arising from such a state of things may be kept in view in any arrangement which may hereafter be made with the Indians on the Canadian side. When they have to come a long distance, the time they lose from their other avocations is of more real importance to them than the pay which they receive, and when they are congregated with their wives and daughters, in large numbers, at some station where there are rival traders, as at the Grand Portage, and have to wait long, it will readily be believed that they are exposed to many demoralizing influences.

ROUTES BETWEEN LAKE SUPERIOR AND RAINY LAKE.

Pigeon River Route.

This route leaves Lake Superior at the Grand Portage Village, and, after passing through a high and hilly region, meets the canoe route from the Kaministaquia at Nequaquon Lake, or, as it is sometimes called, *Lac la Croix.* On reference to the accompanying plan, in profile, it will be seen that on leaving Lake Superior the country rises very rapidly, attaining a height at Mud Lake, just above Mountain Lake, of 1053 feet in a distance, by the windings of the stream, of about 40 miles. From thence westward it falls more gradually, Basswood Lake, which is on the opposite side of the water-shed, and distant from Mountain Lake about 80 miles, being still at an elevation of 661 feet above Lake Superior. In this high region the head-waters of four different rivers are crossed, namely, Pigeon River, Arrow River, Sageinaga River, and a branch of the Maligne. The Lakes are numerous, as they are everywhere in this district, and some of them large, but the streams which connect them are so small as to be barely navigable for moderate sized canoes; and in the event of a more perfect water communication becoming necessary, this being the highest land in the neighborhood, a summit in fact from which the waters run in every direction, there is no source of supply that could be made available. However, until some better sort of communication is opened, it is valuable as a route for very small and light canoes; as, although the extent of land carriage is great as compared with the Kaministaquia route, it is upon the whole shorter, and there are no rapids which could be at all embarrassing. But, though this much may be said in its favor, it is not so good as the Kaministaquia for large canoes, and a glance at the profile will shew that it can never be made practicable for larger craft than canoes. The Grand Portage itself is entirely within the United States territory, and from thence westward to Rainy Lake the canoe route forms the boundary line.

The Route by Dog Lake, Lac des Mille Lacs, and the River Seine.

This route has the advantage of a long extent of comparatively level country, and an abundant supply of water at the very summit of the water-shed. The country is lower, by 163 feet, at its highest elevation,—which is that of the pond at the west end of the Prairie Portage, 893 feet above Lake Superior,—than the Pigeon River route at Mud Lake, but the difference in altitude may fairly be reckoned from where the Savanne Portage strikes the Savanne River, which at that point is at an elevation of 835 feet above Lake Superior, shewing a difference of 218 feet in favour of this route.

To the east of the height of land Dog River and Lake present a reach of 35 miles, which might be made navigable for large vessels by simply throwing a dam across the outlet of

Dog Lake. To the west, and separated from this by a swamp from which the waters flow in either direction, a reach of 65 miles might be made navigable, in like manner, by throwing a dam across the River Seine, at the Little Falls, ten miles below Lac des Mille Lacs. The difference of level between the two reaches thus rendered available would then be about 100 feet, that on the western side being by so much the highest, and the distance between them but five miles, while the supply of water being abundant, and the ground low, a canal with locks might be constructed to connect the two, in which case there would be a navigable reach of one hundred miles across the summit of the water shed. A glance at the profile of the route will exhibit this much more clearly than language can explain it.

GENERAL REMARKS ON THE ROUTE FROM LAKE SUPERIOR TO RED RIVER.

In the first instance, before traffic has assumed such dimensions as to render canals and railroads necessary, the cheapest, and indeed the only way of opening the communication that can be adopted, is to place steamers or row boats on the navigable reaches, and make good land roads where the navigation is impracticable.

This being admitted, it remains for me to describe the extent of land road that would be necessary, and the navigable reaches that might be rendered available.

To commence at Lake Superior, a land road would be required from Thunder Bay to Dog Lake, as the navigation of the Kaministauqia is utterly impracticable, except for canoes, and could only be rendered otherwise at an enormous outlay. Dog Lake is distant from Lake Superior 22½ miles, and at a higher elevation by 718 feet, a difference of level which renders a canal out of the question, notwithstanding that the supply of water in the Kaministaquia would be ample. The only way of reaching it, therefore, is by land, and the surveys have progressed so far as to shew that a good line may be obtained in a distance of 28 miles.

Next follows the reach through Dog Lake and Dog River, which, allowing for bends, is equal to 35 miles; but, to render this available for large vessels, a dam would be necessary across the outlet of Dog Lake, which would have the effect of throwing back the water to Cold Water Lake, at the eastern end of the Prairie Portage.

From this reach to the Savanne River there would be a land carriage of five miles, through an easy country.

From thence to the Little Falls, on the River Seine—about ten miles below Lac des Mille Lacs—the distance is about 65 miles, which might be rendered navigable, in one unbroken reach, by means of a dam at the Little Falls, just referred to.

From the Little Falls to Rainy Lake the River Seine presents a succession of short navigable reaches, alternating with falls and rapids, the total distance in a direct line being 67 miles, and the fall about 350 feet, that is without including the descent at the Little Falls. Having its source in large lakes the Seine is not subject to great floods, and does not seem at all to overflow its banks; immediately below Lac des Mille Lacs it has an average width of about one hundred feet, and gradually increases in volume until it reaches Rainy Lake. By means of lock and dam it might be rendered navigable, but the amount of lockage would involve too great an expenditure for the present state of the country, or any traffic that could be looked for, for a long time to come. I would, therefore, propose improving the navigation by means of dams of the simplest construction thrown across the river where they would produce the best effect, as marked on the accompanying plan. In this way a broken navigation of 59 miles, between the Little Falls and the twelve portages, which latter occur close to Rainy Lake, might be converted into five navigable reaches by means of six dams, while none of the Portages would be of a greater length than from 50 to 200 yards; at the twelve portages a land road of about seven miles in length would be required. If the Seine were improved in this way the distance from the Little Falls to Rainy Lake, by land and water, respectively, would be nearly as follows:

	Land carriage, yards.	Navigable miles.
Portage, past dam at Little Falls ...	200	
1st. Navigable reach to Long Rapid.		13
Carrying place at Long Rapid.	200	
2nd. Navigable from Long Rapid to Island Falls		8
Carrying place at Island Falls.	100	
3rd. Navigable to two Island Falls...		12
Carrying place at do do..	50	
4th. Navigable to High Falls.......		1
Carrying place at High Falls..	200	
5th. Navigable water to the twelve portages		25
In all...	750 yards.	59 miles.

The total distance from the Little Falls to the twelve portages would thus be 59 miles of navigable water in five different reaches, with an aggregate length of land carriage between them of only 750 yards, added to which about seven miles of land road would be required to pass the twelve portages at Rainy Lake. All the dams would have a foundation on solid rock, and wood of the best quality for their construction is everywhere abundant.

For this part of the route, boats such as the Hudson's Bay Company use could be most advantageously employed, as they are easily drawn over a short portage, and might be rendered still more so, in this case, by means of a simple slide or inclined plane made of logs over which they could be quickly hauled.

If, however, it should be considered that a land road would afford a more rapid means of communication, in the event of mail service having to be performed, it should, also, be constructed. For expedition, merely, as regards rapidity of travelling, the land road would be the best, while the river would afford a more easy means of carrying heavy articles.

From the twelve portages, through Rainy Lake, and a part of Rainy River to Fort Frances, the distance is fifty miles; and in this reach there is no obstruction to the navigation.

At Fort Frances there is a fall of 22 feet, which might be overcome by wooden locks. The land carriage, however, being only 300 yards, can involve no great difficulty for the present. Thence to the western extremity of Lac Plat there is no obstruction to the navigation except at two little rapids in Rainy River which a steamer of moderate power could stem with ease: the entire navigable distance in this reach being 158 miles.

From Lac Plat to Fort Garry, a route available for a land road can be obtained in a distance of 91½ miles; and the country being level and favourable, this road might be constructed at a moderate outlay.

By opening the communication in this way the total distance from Lake Superior to Red River Settlement, by land and water would be as follows:

	Land carriage, miles.	Navigable miles.
From Thunder Bay to Dog Lake..	28	
Through Dog Lake and River to the Prairie Portage		35
Land road past Prairie and Savanne Portages to Savanne River.....	5	
Through Savanne River, Lac des Mille Lacs and the River Seine to the Little Falls		65
Broken navigation on River Seine.		59½
Land carriage past the twelve portages on River Seine.......	7	
From the Seine to the western extremity of Lac Plat navigable with only one break at Fort Francis....................		208
Thence to Fort Garry by land....	91½	
Total...	131½	367½

In all one hundred and thirty-one miles and a half of land road, and three hundred and sixty-seven miles and a half of navigable water.

The foregoing does not represent the distance in a direct line, but the extent to be travelled making allowance for the tortuosities of the route.

In regard to the means of transport which could be most economically and advantageously used.

Waggons or carts would be required on the road between Thunder Bay and Dog Lake.

On Dog Lake and River boats, such as already referred to as being used by the Hudson's Bay Company, or even a steamer might be employed.

At the Prairie Portage, carts or waggons would be necessary.

On the Savanne River, Lac des Mille Lacs, and the River Seine as far as the Little Falls, after the dam was constructed at the last mentioned place, there would be an unbroken reach of 65 miles, and on this section it would be advantageous to have a small steamer.

On the 59½ miles of broken navigation, on the River Seine, between the Little Falls and the Twelve Portages, boats, as already explained, should be used, while at the land road past the twelve portages, carts or waggons, as on the other sections of road, would be necessary.

From the River Seine to Fort Frances a steamer would have a clear run of fifty miles.

From Fort Francis to Lac Plat, steamers would have an uninterrupted run of 158 miles.

From the latter place to Fort Garry no provision would have to be made, as the means of transport are to be had in abundance at the Red River Settlement.

If the communication were opened in the manner above set forth, the journey from Lake Superior to Red River might be performed in about three days, that is, allowing that steamers could be maintained at an average speed of ten miles an hour on the navigable reaches, that the land roads might be passed over at the rate of five miles an hour, and that the interrupted navigation of the Seine could be accomplished at the rate of four miles an hour.

This estimate, and it cannot be considered a high one, would give the following result as to distance and time:

FOR LAND ROADS.

From Thunder Bay to Dog Lake............	28	miles.
Across prairie and Savanne Portages........	5	
Past twelve portages on Seine	7	
From Lac Plat to Fort Garry..............	91½	
		131½
Broken navigation on Seine................		59½

NAVIGABLE WATER.

Through Dog Lake and River	35	
Lac des Mille Lacs and Savanne River......	65	
From twelve portages R. Seine to Lac Plat ..	208	308
Total distance...................	499	miles.

Which might be travelled thus:

	Hours.	Minutes.
131½ miles land road at 5 miles per hour......	26	18
59½, or say 60 miles broken navigation at 4 miles per hour	15	00
308 miles navigable water at 10 miles per hour	30	48
Total........................	72	6

that is, in three days as nearly as may be. But if mail service had to be performed with great celerity, there can be no doubt that with a proper equipment of horses and carriages, a speed of seven miles an hour might be kept up on the land roads; while on the navigable reaches, with boats of sufficient power, an average rate of twelve miles an hour, or even more, might be maintained.

According to a rough estimate which I have made, the total cost of opening the communication in the manner which I propose, would not exceed fifty thousand pounds, but until the surveys now in progress are completed, and the necessary measurements taken, it is impossible for me submit a very precise estimate in detail.

It has been said, indeed strongly urged, by interested parties, that the route through the State of Minnesota by St. Paul and Pembina, would afford the best means of communication with the Red River Settlement. But I do not see how this opinion can reasonably be ascertained. Under any circumstances there would be a greater extent of land carriage by the Minnesota route, I should say twice or perhaps three times as great as by the Canadian route, so that the time occupied in travelling it would be greater, unless a railroad were constructed, and it will be borne in mind that the extreme western limit of the United States railway system does not as yet approach within seven hundred miles of the Red River Settlement.

The communication, therefore, which it is proposed to open, would afford advantages superior to any line which can be adopted in Minnesota except a line of railroad, and it is reasonable to believe that ere the United States Government can construct such a work, through an unsettled or but thinly peopled country, the trade of the Western territories will have become so much developed as to warrant the construction of railroads between the navigable reaches on the Canadian route.

When the circumstances of the country would admit of the outlay, a continuous railroad—195 miles in length—might be made between Lake Superior and Rainy Lake, and another, of 91½ miles, between Lac Plat and Fort Garry. If this were done, and two locks constructed at Fort Frances, the Red River Settlement would be within less than two days journey of Lake Superior, there being 195 miles of railroad at the eastern end of the route, and 91½ at the western, with an intermediate reach of 208 miles of navigable water.

This would bring Fort Garry within five days' journey of Toronto.

As to the immediate results which might be reasonably anticipated from opening the communication, in the manner which I have proposed, as a preliminary step towards works of a more extended nature, I may enumerate the following:

1st. In the first place the trade of the Red River Settlement would be at once transferred to this part of Canada.

2nd. It is reasonable to suppose that when the Hudson's Bay Company discovered that they could carry on their trade more economically by this route than by Hudson's Bay, they also would adopt it, thus bringing through the country the greater portion of a trade amounting to nearly half a million sterling yearly.

3rd. It would be the first step towards a route through Canada and British Columbia. Once at Red River, there is navigable water with but little interruption to the base of the Rocky Mountains; and through these it appears that Capt. Palliser has recently discovered easy passes, within British territory. From thence westward to Frazer's River the distance is, comparatively, not great. It is, therefore, reasonable to believe that if the route were opened to Red River it would soon be continued all the way to Frazer's River and the Pacific, and as it is the shortest that can be adopted, it would no doubt become the highway of an emigration to the gold regions, the extent of which no one can forsee.

4th. Another, and by far the most important consideration is, that by opening this route a vast extent of fertile land would be thrown open to colonisation, and this is of peculiar interest to Canada at present. It is a well known fact that an emigration is constantly going on from Lower Canada to the prairies of the Western States. Now, the *Rivière Rouge* and the *Nord Ouest*, from the time that the Canadian voyageurs occupied the country, have been familiar words in Lower Canada, and if the route were once opened there can be no doubt but that Canadian emigrants would prefer a land with which they are so much connected by old associations, where a kindred people are ready to receive them, and where they would have the inestimable advantage of living under British laws, to a country where they would not understand the language, and where most of those of them who do emigrate become mere hewers of wood and drawers of water, to a people who have sharpness enough to turn their simplicity to account.

THE PROGRESS OF THE SURVEY AT PRESENT BEING CARRIED ON BETWEEN LAKE SUPERIOR AND RAINY LAKE.

I have just received despatches from my assistants Mr. L. A. Russell and Mr. J. F. Gaudet. It appears that up to the time they wrote the surveys had been prosecuted with great energy. Mr. Gaudet had completed a cursory exploration of Dog Lake, and made an instrumental survey of Dog River, the Muskaig Lake and the Savanne River, as far as Lac des Mille Lacs. At the time he wrote (23rd Dec.), he was about to commence the survey of the River Seine, as explained in his letter, which I annex, with a plan of his surveys.

Mr. L. A. Russell's operations are detailed in his letter, which I also annex with the plan to which he refers. It seems he has been successful in finding a good line for a road between Thunder Bay and Dog Lake, near an Indian path a little to the west of Current River. This is important, as the region to which his explorations have been confined is the roughest on the whole line of route.

Mr. Gaudet has his head quarters on the Savanne River, and when he wrote was about to set out on the survey of the River Seine.

Indian Mission,
Fort William, January 3rd, 1859.

Sir,—I send a sketch shewing the Current River line, and, also, the Indian winter route which starts from Thunder Bay, about half way between the mouth of the Kaministaquia and Current River, then following the valley of the second river north of Fort William, comes out on Dog Lake, about a mile and a half west of the Current River line.

A better road site can be found along the side of this little river than anywhere about the line, as the latter crosses all the hills, ravines and swamps, among which the said river takes its rise, as well as those at the sources of another little river or brook between the first and Current River.

On the Indian track the country is less hilly and not so swampy. There is a mile or so of bad swamp at the outset, but it can be avoided by starting a short distance to the north, thus coming on a ridge of high land which runs down close to the shore of the lake. The next bad place is where the track first crosses the river, which here cuts its way through a range of granite hills of but inconsiderable height, and through them I think a good pass could be found as they are no worse than those on the Current River line.

There would be three or four little bridges, but they would cross mere brooks, the banks of which are low and good. In the immediate vicinity of Dog Lake there would be some rough ground, but that has to be passed through whatever be the route.

With the exception of the places that I have mentioned the country is pretty level, and the swamps few and short. The soil on the level ground and in the bottom of the swamps is a clayey sand. Under this sand is granite, and indeed, here, as in many other places,, there is but little soil over the rock.

The woods are white birch, poplar, spruce, and pitch pine, on the high grounds; and spruce, larch, and cedar in the lower places.

On the sketch the red dotted line winding about the black one is where a road would have to be taken were it necessary to make it on the Current River line. Between the 2nd and 3rd mile posts is a bad swamp that would have to be passed straight through, as it runs to the river on the east side, and on the other side extends even further, terminating in rocky and broken ground ; it would cost more to take a road round, on either side, than it would to make it across. Near the 6th mile post is another swamp, to avoid which there is a sharp turn to the right ; and the hill at the sixth post is so steep that we have to wind round its base to the left again. Between the 7th and 9th the dotted line crosses the same brook four times, but it is so small that it merely requires culverts. At the 11th post a swamp occurs, through which we pass straight, as in the one between the 2nd and 3rd, and for the similar reason, that it would cost more to get a good road round it than to make it on the line. The three brooks between the 12th and 13th mile posts are small, the width of the largest being not more than 12 feet. From the 13th to the 18th posts, the ground is good, but about half way to the 19th commence the ranges of hills which lie around Lake Pijké and Hawk Lake. The best way from this is along a valley which comes out about three quarters of a mile west of the discharge of Hawk Lake.

On the hills to the north of this valley are occasional large white pines ; there are also some on the 5th mile of the line, on the shore of Current River, about 4 miles from its mouth, and on the shores of Hawk Lake.

The country to the north-east of the line is very rough. At the end next Thunder Bay, Current River, having a general course of north, winds about among steep rocky hills, which sometimes rise straight up from its edge ; from the top of one of these, about 6 miles from its mouth, we could see its course for a long distance through an exceedingly rough country. It is full of rapids and falls pouring through clefts of up-heaved granite and slate. Opposite the second mile of the line it passes through slate ; but higher up, through granite. About the 14th and 15th miles the ground to the N. E. of the line is more level, till near Lake Pijké, when it gets uneven again.

The banks of that lake are high, and wooded with white birch, poplar, and spruce, but the soil is of the same light sand as elsewhere on the line, and the hills are rocky ; its discharge into Hawk Lake is a small and rapid stream, which falls, perhaps, nearly a hundred feet between the two lakes.

The shores of Hawk Lake are much higher than those of Lake Pijké, rising in steep rocky hills from the water's edge ; the north shore of the eastern end of the lake is a wall of cliffs, about 150 feet in height, which then rises higher in wooded hills behind. Hawk Lake discharges itself into Dog Lake about a mile to the west of the line by a rapid and shallow brook, which runs from its western extremity. On it are two little lakes, and Indian Portages out of one into the other, and then into Dog Lake.

I am now going to run a line from the Mission to the thirteenth mile on the Current River line ; that finished, I will locate the road from Thunder Bay to Dog Lake. I run the line first, as it will give me a much better knowledge of the country.

As I understand my instructions, I am to locate the road on the best ground I can find. I will, therefore, take it along the valley of the second little river north of the Kaministaquia, as shown in the sketch.

I would have run the line from the Mission to the 8th or 9th mile posts, were it not that I know the country through which it would pass to be bad ; it would run through the swamp, on the two little rivers behind the Fort, and the ground about the 9th mile at the line is very rough.

I take with me five men, the number you allowed me when I should be running lines. They are Jos. Whiteway, J. Smith, and three of the Mission Indians. I have, as yet, got only one of the Indians at 3s. 9d. per day. I may, perhaps, have to do with three men, as the people of the mission are asking exorbitant wages.

We have had a mild winter hitherto and a good deal of snow, it is now about two feet in depth. The greatest cold that has occurred was on the 8th December, at sunrise, when the thermometer indicated 25° below zero; and on the 17th December, at sunrise, on Hawk Lake, when it shewed 27° below zero. The temperature at the mission was at the same time 19° below zero. On the 15th November the people first crossed the Kaministaquia on the ice.

Thunder Bay is not yet frozen across, but the small bays are nearly all frozen.

Mr. McIntyre expects a Red River Mail this month, and will forward it at once. I will write by that mail.

A mail has arrived but did not bring us any letters.

Wishing you a happy new year,

I remain,

Yours truly,

(Signed,) L. A. RUSSELL.

To S. J. Dawson, Esq., C. E.
in command of the Red River Expedition,

Savanne Depot,
23rd December, 1858.

Sir,—Having now all the surveys in this section completed, as you desired me, I send you plans of my operations.

They are executed in a very rough manner, but it would be losing too much valuable time to do them otherwise.

After scaling through to Jourdain's Rapids, I ran the line connecting the two points. You will perceive, on reference to the plan that the two first miles on the line are swampy, but not bad. I dug down at several places, and found that the average depth of black mould over clay is from 6 to 18 inches. From the 2nd mile post to the 7th anything more beautiful cannot be asked for; it is a cypress ridge, soil sand. The remaining $1\frac{3}{4}$ mile is a gradual descent to the river. The timber changes to poplar and birch, and we meet with a few rocks but not many; however, these can be avoided by making a short detour.

Dog River, from Jourdain's Rapid to Dog Lake, is 37 miles in length.

On upper Dog River, from where Muskaig River branches off, there is but one portage, 17 chains long and about 15 feet fall. The river has more the appearance of a lake than a river, varying from two to five chains in breadth.

On Muskaig River you will see by the sketch that there are nine portages and five rapids from its confluence with Dog River to Muskaig Lake. This river is exceedingly small and rough.

Muskaig Lake is an extensive sheet of water, a great deal larger than I expected to find it from the size of the river. Where the portages occur the average breadth of the river is from 15 to 20 feet, through cliffs of granite rock. From the head of the 5th portage, with the exception of the rapids that occur between it and the Muskaig Lake, it has a general width of about two chains, winding through a low swampy country timbered with tamarac.

All the islands in Muskaig Lake, with the exception of the two at the outlet, are sketched in. It had been my intention to survey the whole, but the survey of the shore occupied a longer time than I had made calculation for.

Savanne River, from the depot to Mille Lacs, is $18\frac{1}{2}$ miles in length.

While engaged in making my plans I sent all the men, with the rest of the provisions, to Mille Lac, making small *cachettes* at intervals, so that I may usé them as I go along. I shall leave in the morning and join my party at the mouth of the Savanne River.

It is my intention to scale the north-west coast of Mille Lacs straight through to your *cachette*, and when I reach it I shall continue down Rivière La Seine, and leave the south-east portion of Mille Lacs till I return.

I think you will agree with me that it is better for me to complete as much of the lower section with what provisions I have now, so that when I return it will be closer for me to portage provisions to Mille Lacs than to the lower end of Rivière la Seine.

The two Indians have up to date taken about 500 rabbits. Fish we have had on every occasion, and have salted half a barrel.

We make with the rabbits a sort of pemican, by boiling the meat with a little lard, which, when frozen and chopped up, is very portable, and makes excellent soup, which is a great saving of pork.

I believe I have written every thing that may interest you; I shall therefore close my letter, and trusting that the work, so far as it has gone, and the arrangements I have made, may meet with your satisfaction.

I remain,

Your humble servant,

(Signed,) J. F. GAUDET.

S. J. Dawson, Esquire,
&c. &c.

THERMOMETRICAL REGISTER.

Up to the 25th of March, 1858, the following register was kept near Fort Garry—lat. 49 ° 55′ N.; long. 97 ° 21′ W.,—and from that time up to the 6th July, at Dr. Bunn's Cottage—lat. 50 ° 1′ N.; long. 97 ° 20′ W.,—the approximate altitude of both places above sea level, being 700 feet, or thereabout.

In the fourth column, the minimum temperature of the previous night is set down at the hour of observation in the morning. Owing to the other occupations of the party, a few days are omitted, and from the same cause no observation was made in the middle of the day during the month of November and a part of December.

The Register at Fort William was kept by Chief Trader McIntyre, the officer in charge of the Hon. H. B. Company's establishment at that place.

The observations were all made with Fahrenheit's thermometer.

RED RIVER SETTLEMENT—THERMOMETRICAL REGISTER,

SEPTEMBER, 1857.

Day of the Month.	Hour of the Day.	Temp. at the hour of observation.	Minimum Temperature.	Direction of the Wind.	Remarks.
16	7 a m....	52.00	49.00	N W	Clear, a light breeze.
	1 p m....	60.00	....	N W	Do do
	7 p m....	35.00	....	W	A fresh breeze.
17	7 a m....	36.00	33.00	S W	Clear, a high wind.
	1 p m....	45.00	....	S W	Do do
	7 p m....	41.00	....	S W	Clear, a fresh breeze.
18	....	.....	....	....	No observation.
19	....	.....	....	....	Do
20	....	.....	....	....	Do
21	7 a m....	45.00	36.00	S S W	Slightly clouded, a gentle wind.
	12 Noon ..	69.00	....	S S W	
	6 p m....	55.50	....	S S W	
22	7 a m....	41.00	36.00	S	Do do
	12 Noon ..	77.00	....	S	A high wind.
	6 p m....	77.50	...	S	
23	7 a m....	62.50	56.25	N W by N	Cloudy, a brisk wind.
	12 Noon ..	75.00	....	N W by N	
	6 p m....	71.00	....	N W by N	
24	7 a m....	44.00	38.00	S	Clear, a very high wind.
	12 Noon ..	70.00	....	S	
	6 p m....	68.00	....	S	Cloudy, a brisk wind.
25	7 a m....	45.00	37.50	S	Do do
	12 Noon ..	70.00	....	S	Clear, a high wind.
	6 p m....	61.00	....	S	Clear, a pleasant wind.
26	7 a m....	47.00	39.00	S	Clear, a high wind.
	12 Noon ..	65.00	....	S	Clear, a very high wind.
	6 p m....	59.00	....	S	Do do
27	7 a m....	42.00	34.00		
	12 Noon ..	61.00	....	S S W	A fresh breeze.
	6 p m....	52.00	....	....	A high wind.
28	7 a m....	41.00	33.00	N W	A fresh breeze.
	12 Noon ..	51.50	....	N W	Do
	6 p m,...	46.00	....	N W	A pleasant wind.
29	7 a m....	40.00	32.50	S	Clear, a pleasant wind.
	12 Noon ..	49.00	....	S	A high wind.
	6 p m....	42 00	....	S	Do
30	7 a m....	39.00	33.00	S	Do
	12 Noon ..	51.00	....	....	A fresh breeze.
	6 p m....	47.00	....	....	A pleasant wind.

OCTOBER, 1857.

Day of the Month.	Hour of the Day.	Temp. at the hour of observation.	Minimum Temperature.	Direction of the Wind.	Remarks.
1	7 a m....	67.00	61.00	....	Calm.
	1 p m....	70.00	....	S W	A light breeze, clear.
	7 p m....	63.00	....	S W	A light breeze, clear aurora.
2	....	....	....	N	Nearly clear, a pleasant wind.
3	....	....	....	S	A very high wind.
	....	....	....	....	Smoky in middle of day.
	....	....	....	....	Aurora, lunar corona.
4	7 a m....	63.00	57.00	S	Overcast, rained a little.
	1 p m....	65.00	....	....	
	7 p m....	59.00	....	....	Aurora, bright.

RED RIVER SETTLEMENT—THERMOMETRICAL REGISTER,

OCTOBER, 1857—(*Continued.*)

Day of the Month.	Hour of the Day.	Temp. at the hour of observation.	Minimum Temperature.	Direction of the Wind.	Remarks.
29	8 a m....	29.00	27.00	W N W	A light breeze.
	1 p m....	34.00	....	W N W	Do
	7 p m....	31.00	....	NW by N	A fresh breeze.
39	8 a m....	30.50	31.00	N W by N	A pleasant wind.
	1 p m....	33.00	....	....	
	7 p m....	31.50	....	....	Clear, calm.
31	8 a m....	34.00	27.25	S by E	A high wind.
	1 p m....	44.00	....	....	
	7 p m....	36.50	....	S by E	A light breeze, snow and rain.

NOVEMBER, 1857.

Day of the Month.	Hour of the Day.	Temp. at the hour of observation.	Minimum Temperature.	Direction of the Wind.	Remarks.
1	8 a m....	32.50	32.00	N W	A high wind, snowing.
	6.30 p m.	34.75	....	N W	A high wind, two inches of snow fell.
2	8 a m....	32.50	31.00	N W by W	A fresh breeze.
	6.30 p m.	36.00	....	....	Calm, clear; lunar corona at 2 p. m.
3	7.30 a m.	34.00	31.00	N W	A pleasant wind.
	6 p m....	33.50	....	N W	Overcast, calm nearly.
4	7.30 a m.	35.50	30.50	N W	Overcast, hazy, a fresh beeze.
	8 p m....	35.50	....	....	Overcast, calm.
5	8 a m....	34.50	28.00	....	Foggy, calm.
	7.30 p m.	34.00	....	N W	Overcast, a fresh beeze, snowing.
6	8 a m....	30.50	28.00	....	Overcast, calm.
	7 p m....	30.00	....	....	Cloudy, calm, appearance of snow.
7	10 a m....	30.50	26.10	....	Cloudy, calm; about 4 inches of snow in all.
	7.15 p m.	29.50	....	....	
8	7 p m....	24.00	....	N N W	Overcast, a light breeze; geese flying south.
9	10 a m....	20.50	18.75	N W	Overcast, a fresh breeze, snowing slightly.
	7 p m....	19.00	....	N W	Overcast, snowing a little.
10	8 a m....	19.50	17.00	W	Overcast, snowing slightly.
	7.30 p m.	19.00	....	W by N	A fresh breeze.
11	7.15 a m.	16.00	15.00	....	Calm.
	7.15 p m.	14.00	....	W by N	A high wind. snowing.
12	7.30 a m.	19.50	14.00	W by N	A fresh breeze.
	7.15 p m.	10.00	....	...	Clear, calm; floating ice on river.
13	7.45 a m.	13.00	9.00	....	Overcast, calm; ice taken on river.
	7 p m....	12.50	....	....	Partially clouded, calm.
14	8 a m....	18.00	12.50	W	Overcast, a pleasant wind.
	7.30 p m.	24.09	....	W by N	Overcast, a pleasant wind.
15	8.30 a m.	19.50	17.00	....	Overcast, calm.
	8.30 p m.	29.50	....	S	Overcast, a pleasant wind.
16	8 a m....	30.00	20.00	S S W	Clear; good crossing on the ice.
	7.30 p m.	30.00	....	S W by W	Overcast, a pleasant wind.
17	8 a m....	20.00	19.00	N W by W	A high wind.
	8 p m....	13.50	....	N W	Beautiful aurora—Southern corona.
18	8 a m....	10.00	9.50	W by N	Partially clouded, a high wind.
	6.30 p m.	3.00	....	N W by W	Clear, a high wind; faint aurora.
19	7.30 a m.	5.50	1.50	N W	Clear, a high wind.
	6.30 p m.	6.00	....	N W	Clear, a light breeze; bright aurora.
20	7.30 a m.	4.00	3.50	....	Overcast, calm.
	9.30 p m.	23.00	....	N W by W	Overcast, a high wind.

5	7 a m....	55.00	46.90	W	A brisk wind.
	1 p m....	69.00	....	....	Fires on the plains.
	7 p m....	....	....	....	
6	7 a m....	52.00	45.00	S	Large fires on the plains.
	1 p m....	55.00	....	S	A brisk wind.
	7 p m....	48.00	....	....	
7	7 a m....	50.50	42.00	S	Very smoky, a brisk wind.
	1 p m....	61.00	....	....	
	7 p m....	54.00	....	....	Fires on plains continue.
8	7 a m....	67.00	41.00	S W by S	A stiff breeze—rained a little at 10.
	1 p m....	70.00	....	S W by S	Very light breeze.
	7 p m....	65.00	....	S S W	A fresh breeze.
9	7 a m....	63.00	62.25	S	A high wind—rained about two hours.
	1 p m....	63.50	....	S S W	A light wind.
	7 p m....	56.50	....	....	Calm.
10	7 a m....	46.00	42.00	....	Do
	1 p m....	55.00	....	S W	A light wind.
	7 p m....	52.00	....	....	Calm.
11	7 a m....	45.00	43.00		
	1 p m....	46.00	...		
	7 p m....	49.00	....		
12	7 a m....	41.50	35.00		
	1 p m....	51.00	....		
	7 p m....	45.50	....		
13	7 a m....	47.50	42.00		
	1 p m....	....	....		
	7 p m....	61.00	....		
14	7 a m....	38.50	34.00		
	1 p m....	50.00	....		
	7 p m....	46.00	....		
15	7 a m....	33.00	32.00		
	1 p m....	45.00	....		
	7 p m....	43.00	....		
16	7 a m....	40.00	32.50		
	1 p m....	44.00	....		
	7 p m....	38.00	....		
17	7 a m....	31.00	24.50		
	1 p m....	34.00	....		
	7 p m....	31.00	....		
18	7 a m....	28.00	28.00		
	1 p m....	36.00	....	....	Calm with a hazy atmosphere.
	7 p m....	22.00	....		
19	7 a m....	20.00	17.00		
	1 p m....	30.50	....		
	7 p m....	28.00	....		
20	7 a m....	28.00	22.50		
	1 p m....	50.50	...		
	7 p m....	42.00	....		
21	7 a m....	28.00	26.00		
	1 p m....	55.00	....		
	7 p m....	45.00	....		
22	7 a m....	39.00	36.00		
	1 p m....	49.00	....		
	7 p m....	44.00	....		
23	7 a m....	35.00	34.50		
	1 p m....	52.00	....		
	7 p m....	51.00	....		
24	7 a m....	40.00	32.00		
	1 p m....	41.00	....		
	7 p m....	40.00	....		
25	7 a m....	36.50	29.00		
	1 p m....	47.00	....		
	7 p m....	46.00	....		
26	7 a m....	39.00	36.00	S S E	A high wind—raining
	1 p m....	44.00	....	S	A high wind—raining slightly.
27	8 a m....	34.00	34.00	N	A high wind—rained all night.
	1 p m....	49.00	....	N N W	A high wind—snowing a little.
	6 p m....	31.00	....	N N W	A high wind—still snows.
28	8 a m....	27.00	27.00	N W	A fresh breeze.
	1 p m....	31.00	....	N W	Do
	6 p m....	29.00	....	N N W	A high wind.

21	8 a m....	3.00	2.50	N W by W	Partially clouded, a gale.
	8 p m....	-10.00	....	....	
22	8 a m....	-12.00	-15.00	....	Clear, calm.
	7 p m....	-1.00	....	....	Clear, calm.
23	8 a m....	8.00	-1.00	W	Cloudy, a fresh breeze.
	7.45 p m.	4.00	....	N W	Cloudy, a fresh breeze.
24	8 a m....	00.00	-2.00	....	Clear, calm.
	7.30 p m.	4.00	....	....	Clear, calm.
25	7.30 a m.	10.00	3.00	....	Clear, calm.
	7.30 p m.	20.00	....	N W	Clear, a gentle wind.
26	8 a m....	31.00	....	....	Cloudy, calm.
	8.30 p m.	34.00	....	N W	Overcast, a fresh breeze.
27	8 a m....	34.00	20.00	....	Cloudy, calm, a heavy fog.
	9 p m....	25.00	....	W	Cloudy, very light wind.
28	8 a m....	21.00	21.00	....	Cloudy, calm, snowing.
	7.30 p m.	19.00	....	....	Cloudy, calm, snowing.
29	10 a m....	25.00	20.00	W	Cloudy, a light breeze.
	10 p m....	18.00	....	N W	Cloudy, a high wind, snowing.
30	8 a m....	18.00	17.00	W	Cloudy, a fresh breeze.
	11 p m....	9.00	....	....	Hazy, calm.

DECEMBER, 1857.

Day of the Month.	Hour of the Day.	Temp. at the hour of observation.	Minimum Temperature.	Direction of the Wind.	Remarks.
1	8 a m....	16.00	12.00	....	Cloudy, calm.
	10 p m....	14.00	....	....	Cloudy, calm; snowing a little.
2	8 a m....	10.00	8.00	....	Cloudy, calm.
	9 p m....	7.50	....	....	Cloudy, calm; snowing slightly.
3	8.30 a m.	11.00	7.00	....	Cloudy, calm.
	9 p m....	5.00	....	....	Do do
4	8.30 a m.	-8.00	-12.00	....	Cloudy, calm; white frost.
	7 p m....	-1.50	....	....	Cloudy, calm; bright aurora.
5	8.30 a m.	22.00	-2.00	....	Cloudy; very light wind.
	7 p m....	25.00	....	S	Do do
6	10 p m....	4.00	....	....	Clear, calm.
7	8.30 a m.	17.00	4.00		
8	8.30 a m.	10.00	0.00	....	Cloudy; a fine breeze.
	7 p m....	16.00	....	....	Cloudy; a pleasant wind.
9	8.30 a m.	18.00	16.00	....	Cloudy, calm.
	7 p m....	14.00	....	....	Do do
10	8.30 a m.	14.00	12.00	....	Do do
	7 p m....	16.00	....	....	Do do Aurora.
11	8.30 a m.	0.00	-6.00	....	Do do
	7 p m....	12.00	....	....	Clear, calm.
12	8.30 a m.	18.50	10.00	S by E	Cloudy; a light breeze.
	9 p m....	20.00	....	....	Clear, calm.
13	8.30 a m.	16.00	15.50	....	Do do Mirage this morning.
	9 p m....	16.00	....	....	Do do
14	8.30 a m.	7.50	7.00	....	Do do
	9 p m....	16.00	....	....	Cloudy, calm.
15	8.30 a m.	13.00	13.00	....	Do do Fog.
	8.30 p m.	25.00	....	....	Do do Rains and freezes.
16	8.30 a m.	20.00	18.00	....	Do do
	9 p m....	9.00	....	....	Clear, calm. A brilliant aurora at 9 p. m.
17	8.30 a m.	3.00	3.00	W	Cloudy; a pleasant wind.
	9 p m....	7.00	....	....	Cloudy, calm.
18	8.30 a m.	0.00	-3.00	....	Clear, calm.
	9 p m....	7.00	....	....	Do do
19	8.30 a m.	11.00	7.00	....	Do do
	9 p m....	18.00	....	....	Do do
20	8.30 a m.	20.00	16.00	....	Do do
	9 p m....	14.00	....	...	Cloudy, calm.

RED RIVER SETTLEMENT—THERMOMETRICAL REGISTER,

DECEMBER, 1857—(*Continued.*)

Day of the Month.	Hour of the Day.	Temp. at the hour of observation.	Minimum Temperature	Direction of the Wind.	Remarks.
21	8.30 a m.	14.00	10.00	N W	Cloudy; a high wind.
	9 p m....	15.00	....	....	Do do
22	8.30 a m.	28 00	6.00	....	Cloudy, calm.
	9 p m....	3.40	....	S W	Cloudy; a pleasant wind.
23	8.30 a m.	....	11.00	W	Cloudy; a high wind.
	9 p m....	15.00	....	....	Clear, calm.
24	8.30 a m.	3.00	-4.00	....	Cloudy, calm.
	9 p m....	0.00	....	....	Do do
25	8.30 a m.	1.00	-6.00	....	Do do
	9 p m....	-8.00	....	....	Clear, calm; aurora.
26	8.30 a m.	-7.00	-11.00	....	Clear, calm.
	9 p m....	14.00	....	....	Cloudy; snow drifting.
27	8.30 a m.	17.00	12.00	....	Clear, nearly calm.
	9 p m....	9 00	....	....	Do do
28	8.30 a m.	2.00	0.00	....	Cloudy, calm.
	8 p m....	-5.00	....	....	Cloudy, calm; snowing.
29	8.30 a m.	-2.00	-8.00	....	Clear, calm.
	8 p m....	12.00	....	....	Do do
30	8.30 a m.	11.00	10.00	....	Do do
	8 p m....	2.00	....	....	Do do
31	8.30 a m.	3.00	-5.00	....	Do do
	8 p m....	5.00	....	....	Do do

JANUARY, 1858.

Day of the Month.	Hour of the Day.	Temp. at the hour of observation.	Minimum Temperature.	Direction of the Wind.	Remarks.
1	8.30 a m..	12.00	-1.00	N E	Cloudy, a high wind.
	8 p m	-7.00	....	N E	Do do
2	8.30 a m..	10.00	-21.00	N E	Do do
	8 p m.....	19.00	....	N E	Do do
3	8.30 a m..	23.00	19.00	W	Clear, blowing a gale.
	8 p m.....	35.00	....	W	Do do
4	8.30 a m..	-2.00	0.00	....	Clear, calm.
	8 p m.....	-12.00	....	....	Do
5	8.30 a m..	-14.00	-16.00	....	
	8 p m.....	1.00	....	....	
6	8.30 a m..	-4.00	-5.00	....	Cloudy, calm.
	8 p m.....	-15.00	....	....	Clear, calm.
7	8.30 a m..	-7.00	-18.00	....	
	8 p m.....	7.00	....	....	
8	8.30 a m..	7.00	-5.00	....	
	2 p m.....	13.00	....	....	
	8 p m.....	6.00	....	....	Clear, calm, bright Aurora.
9	8 a m.....	-16.50	-18.50	....	Clear, calm.
	2 p m.....	12.50	....	....	Cloudy.
	8 p m.....	12.00	....	S	Do a strong wind.
10	8 a m.....	10.00	9.00	....	Clear, calm.
	2 p m.....	26.00	....	....	Do
	8 p m.....	6.00	....	....	Do
11	8 a m.....	-7.00	-9.00	....	Do
	2 p m.....	7.50	....	....	Do
	8 p m.....	2.00	...	N E	Cloudy, a high wind.

RED RIVER SETTLEMENT—THERMOMETRICAL REGISTER,

FEBRUARY, 1858.—(*Continued.*)

Day of the Month.	Hour of the Day.	Temp. at the hour of observation.	Minimum Temperature.	Direction of the Wind.	Remarks.
2	8 a m....	-15.00	-15.50	W	Clear, a high wind.
	2 p m....	-6.00	....	....	
	8 p m....	-10.00	....	....	
3	8 a m....	-17.00	-18.50	W	Clear all day, a fresh breeze.
	2 p m....	15.00	....	....	
	8 p m....	-13.00	....	....	
4	8 a m....	-18.00	-21.00	....	Clear, a light wind.
	2 p m....	4.50	....	....	Cloudy, a light wind.
	8 p m....	-1.00	....	....	Cloudy, snowing slightly.
5	8 a m....	3.50	-2.00	W	Do do
	2 p m....	14.00	....	W	Clear, a gentle breeze.
	8 p m....	14.50	....	W	Cloudy, a gentle breeze.
6	8 a m....	14.00	10.00	W	Clear, a light wind.
	2 p m....	15.00	....	....	
	8 p m....	9.00	....	....	Cloudy, calm.
7	8 a m....	-10.00	-13.05	W	Clear, a fresh breeze.
	2.¾ p m..	-6.00	....	....	Do do
	10 p m....	-16.00	....	....	Do do
8	9 a m....	-17.00	-22.00	....	Clear, calm.
	2 p m....	-4.00	....	....	Do
	8 p m....	-9.00	....	....	Do
9	8 a m....	-18.50	-18.50	W	A high wind, snowing.
	2 p m....	-21.00	....	W	A high wind, drifting.
	8 p m....	-26.00	....	W	Do do
10	8 a m....	-21.00	-29.00	W	Clear, a light wind.
	2 p m....	-11.00	....	W	Do do
	8 p m....	-20.00	....	....	Clear, calm.
11	8 a m....	-26.00	-27.50	....	Do
	2 p m....	-6.00	....	....	Do
	8 p m....	-16.00	....	....	
12	8 a m....	-24.00	-27.00	....	Do
	2 p m....	0.00	....	N W	Overcast, a light wind.
	8 p m....	-2.75	....	N W	Overcast, a light wind, snowing.
13	8 a m....	1.50	-1.00	N W	Cloudy, a light wind, snowing.
	2 p m....	-2.00	....	N W	Cloudy, a high wind.
	8 p m....	-10.00	....	....	
14	8 a m....	-22.00	-22.00	N W	A high wind.
	2 p m....	-15.50	....	N W	Do
	8 p m....	-24.00	....	....	Clear and calm.
15	8 a m....	-34.00	-37.00	....	Do
	2 p m....	-18.00	....	....	Do
	8 p m....	-23.00	....	....	Cloudy, calm.
16	8 a m....	-10.00	-25.05	....	Do
	2 p m....	-6.50	....	S E	Cloudy, a light wind.
	8 p m....	-7.00	....	S E	Do do
17	8 a m....	-7.00	-10.00	N	Cloudy, a light wind.
	2 p m....	4.00	....	....	Cloudy.
	8 p m....	-3.00	....	....	Cloudy, calm.
18	8 a m....	0.00	-9.00	N W	Cloudy, a light wind.
	2 p m....	9.00	....	....	Snowing a little.
	8 p m....	2.00	....	N	A light wind, snowing a little.
19	8 a m....	-1.00	-8.50	....	Cloudy, calm.
	2.20 p m.	13.00	....	....	Do
	8 p m....	10.00	....	....	Do
20	8 a m....	4.00	4.00	W	Cloudy, a gale of wind.
	2 p m....	3.50	....	W	Do do
	8 p m....	-4.00	....	W	Do do
21	8 a m....	-15.00	-18.50	....	Clear, calm.
	2 p m....	-3.00	....	....	Do
	8 p m....	-10.50	....	....	Do

Day of the Month.	Hour of the Day.	Temp. at the hour of observation.	Minimum Temperature.	Direction of the Wind.	Remarks.
12	8 a m	6.00	2.00	W S W	Cloudy, a pleasant wind.
	2 p m.....	1.00	....	W S W	Cloudy, a fresh breeze.
	8 p m.....	-1.00	....	W	Clear, a light breeze.
13	8 a m	-24.00	-28.50	....	Clear, calm,
	2 p m.....	-13.00	....	....	Do
	8 p m	-18.00	....	....	Do
14	8 a m.....	-14.00	-18.00	....	Do
	2 p m.....	-1.00	....	....	Do
	6½ p m....	-3.00	....	N	Cloudy, a light breeze.
15	8 a m.....	-11.00	-11.00	....	Clear, calm.
	2 p m.....	-6.00	....	....	Do
	8 p m.....	-15.00	....	....	Do
16	8 a m.....	0.00	-17.00	....	Do
	2 p m.....	2.00	....	....	Do
	8 p m.....	-5.00	....	....	Do
17	8 a m	-10.00	-17.00	....	Do
	2 p m.....	1.00	....	....	Do
	9 p m.....	-4.00	....	W	Clear, a light breeze.
18	8 a m.....	5.00	-8.00	....	Clear, calm.
	2 p m.....	15.00	....	....	Do
	8 p m.....	9.00	....	....	Do
19	8 a m.....	2.00	0.00	....	Do
	2 p m.....	19.00	....	....	Do
	8 p m.....	11.00	....	....	Do
20	8 a m.....	5.00	2.00	....	Do
	2 p m.....	19.00	....	....	Do
	8 p m.....	6.00	....	....	
21	8 a m.....	1.00	-2.00	E	Do
	2 p m.....	18.00	....	....	Do
	8 p m.....	12.00	....	....	Do
22	8 a m.....	25.00	12.00	....	Do
	2 p m.....	37.00	...	S	A light breeze, rained a little in the afternoon.
	8 p m.....	35.00	....	....	
23	8 a m.....	32.00	31.00	....	Calm.
	2 p m.....	32.00	....	....	Do
	8 p m.....	31.00	....	....	Do
24	8 a m.....	19.00	19.00	....	Cloudy, calm.
	2 p m.....	16.00	....	....	Do
	8 p m.....	10.00	....	....	Do
25	8 a m.....	10.00	10.00	....	Overcast, calm.
	2 p m.....	17.50	....	....	Do
	8 p m.....	9.00	....	....	Do
26	8 a m.....	6.00	4.00	....	Overcast, snowing a little.
	2 p m.....	10.00	....	....	
27	8 a m.....	13.00	13.00	....	
28	8 a m.....	0.00	-1.00	....	Do do
	2 p m.....	-1.00	....	....	
	7 p m.....	-3.00	....	....	
29	10 a m....	1.00	-8.00	....	
	2 p m.....	-1.00	....	....	
	8 p m.....	-3.00	....	....	Clear, a calm all day.
	8 a m.....	4.00	-8.50	....	
	2 p m.....	10.00	....	....	
	8 p m.....	6.00	....	....	Do do
	8 a m.....	9.00	7.00	....	
	2½ p m....	15.00	....	....	
	8 p m.....	9.00	....	....	Light snow.

FERUARY, 1858.

Day of the Month.	Hour of the Day.	Temp. at the hour of observation.	Minimum Temperature.	Direction of the Wind.	Remarks.
	8 a m....	4.00	4.00	....	Clear, calm, snowed a little during the night.
	2 p m....	-1.00	....	....	
	m....	-6.00	....	W	Clear, a high wind.
22	8 a m....	-10.00	-19.00	....	Calm, Cloudy.
	2 p m....	10.50	....	....	Do
	8 p m....	5.50	....	....	Do
23	8 a m....	8.00	4.00	S	Cloudy, a light wind.
	2 p m....	24.00	....	....	Cloudy, snowing a little.
	8 p m....	18.00	....	W	Cloudy, a light wind.
24	8 a m....	17.00	14.00	N W	Clear, a light wind.
	2 p m....	26.00	....	....	Clear, calm. Snow melting in the sun.
25	9 a m....	25.00	15.50	....	Do
	2 p m....	37.50	....	....	Cloudy, calm.
	8 p m....	30.00	....	....	Do
26	8 a m....	34.00	28.00	....	Do
	2 p m....	39.00	....	W	Cloudy, a light wind.
	8 p m....	27.00	....	W	Blowing a gale.
27	8 a m....	12.00	11.00	W	Cloudy, a high wind.
	2 p m....	8.00	....	W	Clear, a high wind.
	8 p m....	2.00	....	....	Cloudy, calm.
28	9 a m....	-9.00	-11.00	N	Clear. a high wind.
	2 p m....	-2.00	....	N W	Cloudy, a light wind.
	8 p m....	-7.00	....	....	Cloudy, calm.

MARCH, 1858.

Day of the Month.	Hour of the Day.	Temp. at the hour of observation.	Minimum Temperature.	Direction of the Wind.	Remarks.
1	8 a m	-10.00	-19.50	....	Clear, calm.
	2 p m	5.00	...	S	Clear, a light wind.
	8 p m	5.00	....	....	Snowing a little.
2	8 a m	-8.00	-14.00	W	Clear, a light wind.
	2 p m	1.00	...	....	Clear, a light wind.
	8 p m	-8.00	....	....	Clear, a light wind, Aurora.
3	8 a m	2.00	-8.50	....	Cloudy, calm.
	2 p m	14.00	....	....	Do.
	8 p m	-4.00	....	....	Do.
4	8 a m	2.00	-4.00	....	Do.
	2 p m	21.00	....	....	Hazy, calm.
	8 p m	21.50	....	S	Cloudy, a high wind.
5	8 a m	23.00	19.00	....	Cloudy, calm, snowed in the night.
	2 p m	32.00	....	....	Cloudy, calm.
	10 p m....	20.05	....	....	Do.
6	8 a m	15.00	5.00	....	Do.
	2 p m	31.00	....	....	Snowed for an hour, very slightly.
	9 p m	23.00	....	....	Cloudy and calm.
7	8 a m	21.00	18.00	....	Do.
	2 p m	30.00	....	....	Do.
	9 p m	24.00	....	....	
8	8 a m	30.00	21.00	....	Do.
	2 p m	37.00	....	....	Do.
	8 p m	29.00	....	....	Do.
9	8 a m	28.00	26.00	S	Cloudy, a light wind.
	2 p m	35.50	....	....	Clear, a light wind.
	8 p m	31.00	....	S	Cloudy, a light wind.
10	8 a m	30.00	27.00	N W	Snowing, a high wind.
	2 p m	30.00	....	W N W	Do. do.
	8 p m	10.00	....	W	Clear, a light wind.
11	8½ a m....	22.00	....	....	Cloudy and calm.
	2 p m	38.00	....	....	Clear and calm.
	8 p m	19.00	....	....	Do.
12	8 a m	22.00	11.00	....	Do.
	2 p m	40.00	....	....	Cloudy and calm.
	8 p m	38.00	....	S	Clear, a light wind, bright Aurora.
13	8 a m	34.50	30.00	....	Clear and calm.
	2 p m	41.00	....	....	Do.
	8 p m	34.00	....	....	Do.
14	8 a m	35.50	25.00	....	Do.
	2 p m	51.00	....	....	Cloudy, hazy.
	8 p m	40.00	....	....	Thunder and lightning, rain.

RED RIVER SETTLEMENT—THERMOMETRICAL REGISTER,

MARCH, 1858.—*Continued.*

Day of the Month.	Hour of the Day.	Temp. at the hour of observation.	Minimum Temperature.	Direction of the Wind.	Remarks.
15	8 am	41 00	40.00	E	Cloudy, a high wind.
	2 pm	45.00	....	....	Cloudy, calm.
	8 pm	39.00	....	....	Heavy fog, a drizzling rain.
16	8 am	34.00	32.00	N	Snowing a little, a light wind.
	2 pm ..	35.00	....	....	Snowing a little.
	8 pm	33.00	....	....	Cloudy, calm.
17	8 am	27.00	26.00	N E	Cloudy, a light wind.
	2 pm	31.00	....	....	Cloudy, calm, plenty of crows.
	8 pm	26.00	....	....	Cloudy, calm.
18	8.35 a m ..	26.00	24.00	....	Do.
	2 pm	32.00	....	....	Do.
	8 pm	29.00	....	....	Do.
19	8 am ..	27.00	24.00	....	Hazy, calm.
	2 pm	37.00	....	....	Geese and Ducks appear, also Hawks, &c.
	8 pm	31.00	....	....	Raining a little at 9, p. m.
20	8 am	31.00	30.00	N W	Cloudy, a high wind.
21	....	....	....	....	No observations.
22	....	....	....	N W	Stormy, a high wind; River rising.
23	8 pm	34.00	....	....	Fine clear day, calm.
24	8 am	35.0	...	....	Clear, calm.
	2 pm	40.00	....	....	
	8 pm	31.05	....	....	
25	8 am	34.00	22.50	....	Ice breaking on Red River.

APRIL, 1858.

Day of the Month.	Hour of the Day.	Temp. at the hour of observation.	Minimum Temperature.	Direction of the Wind.	Remarks.
8	7 am....	34.00	34.00	N E	Clouded, a light breeze.
	2 pm....	41.00	....	N E	Cloudy, very brisk wind, river falling.
	8 pm....	38.00	....	N E	Cloudy, gentle wind.
9	7 am....	32.00	31.00	N	Brisk wind.
	2 pm....	42.06	....	N	Do
	8 pm....	33.05	....	....	Calm.
10	7.10 am.	37.00	33.00	S	A gentle wind.
	2 pm....	43.50	....	S	A brisk wind.
	8 pm....	39.50	....	S	A gentle wind.
11	8 am ...	36.00	32.00	N	Do
	2 pm....	40.50	....	N	A brisk wind.
	8 pm....	33.00	....	N	A high wind.
12	7.10 am.	33.00	31.00	N	A brisk wind.
	2 pm....	41.00	....	N	A gentle wind.
	8.10 pm.	35.50	....	....	Clear, calm, faint aurora.
13	7.30 am.	40.00	28.00	S E	Clear, gentle wind, frogs begin to croak.
	2 pm....	50.50	....	S E	Clear, brisk wind, farmers ploughing.
	8 pm....	43.00	....	S E	Clear, gentle wind, faint aurora.
14	8 am....	40.00	36.00	S E	Cloudy, very high wind.
	2 pm....	38.00	....	S W	High wind.
	8 pm....	25.00	....	N	Clear, gentle wind.
15	7 am....	30.00	28.00	N W	A high wind.
	2 pm....	26.00	....	N W	Do
	8 pm....	24.00	....	N W	A gentle wind.
16	7.10 am.	22.00	19.00	W	A high wind.
	2 pm....	32.00	....	W	H very high wind.
	8 pm....	28.00	....	....	A gentle wind.

RED RIVER SETTLEMENT—THERMOMETRICAL REGISTER,

MAY, 1858—(*Continued.*)

Day of the Month.	Hour of the Day.	Temp. at the hour of observation.	Minimum Temperature.	Direction of the Wind.	Remarks.
7	7.10 am.	43.00	30.00	....	Calm.
	2 pm....	58.00	....	....	Do
	8 pm....	43.00	....	N	High wind.
8	7 am....	41.00	30.00	N W	Clear, gentle wind.
9	7 am....	37.00	21.00	....	Calm, clear.
	2 pm....	48.00	....	N	Very brisk wind.
	8 pm....	41.00	....	N	A gentle wind.
10	7 am....	46.00	29.00	W	A brisk wind.
11	7 am....	49.00	29 50	S	Very high wind.
	2.30 pm.	52.00	....	W	Do do
12	7 am....	31.00	24.00	N W	Do do
	7 pm....	42.00	....	N	High wind.
13	7 am....	31.00	29.50	N	high wind, snowing a little.
	2 pm....	31.50	....	W	Very high wind, snow storm.
	8 pm....	32.00	....	W	High wind, snow ceased at 5, p. m.
14	7 am....	31.50	26.50	N	A gentle wind.
	2 pm....	32.50	....	W	A moderate wind.
	8 pm....	30.00	....	W	A high wind.
15	7 am....	31.50	26.50	S	Clear, gentle wind.
	2 pm....	41.00	..	S	A very high wind.
	8 pm....	33.00	....	S	Do do
16	7 am....	42.00	28.00	S	Cloudy, light wind.
	2 pm....	55.00	....	S	Light wind.
17	8 pm....	40.00	....	S	Gentle wind.
18	7 am....	31.50	29.00	W	Overcast, gentle wind; a little rain.
	2 pm....	39.50	....	W	Gentle wind.
	8 pm....	35.00	....	....	Calm.
19	7 am....	35.00	26.50	W	High wind.
	2 pm....	....	....	S W	Overcast; raining; high wind.
	8 pm....	42.00	....	....	Clear, calm.
20	7 am....	42.00	34.00	....	Calm.
	2 pm....	63.00	....	S W	Gentle wind.
	8 pm....	57 00	....	S W	Clear, a gentle wind.
21	7½ am ..	53.00	37.00	S	Calm, brisk wind.
	2 pm....	68.00	....	S	Hazy, high wind.
	8 pm....	62.50	....	S	High wind.
22	7 am....	56.00	56.00	S	Gentle wind.
	2 pm....	....	....	....	Do
	8 pm....	54.00	....	N	A brisk wind.
23	7 am....	48.00	42.00	N	Gentle wind.
	2 pm....	62.00	....	N	Hardly perceptible wind.
	8 pm....	55.00	....	....	Calm, clouded.
24	8 am....	62.00	46.00	S	Gentle wind.
	2 pm....	70·00	....	S	Do
	8 pm....	62.00	....	S	Clear, gentle wind.
25	7¼ am....	61.00	48.00	....	Clear, calm
	2 pm....	72.00	....	S	Clear, hardly perceptible wind.
	8 pm...	62.00	....	S	Clear, gentle wind.
26	8 am....	65.00	46.00	S	High wind.
	2 pm....	71·00	....	....	Very high wind.
	8 pm....	51.00	....	N W	Light wind; rain.
27	7 am....	52.00	43.00	S E	Gentle wind; rain.
	2 pm....	70 00	....	....	Gentle wind.
	8 pm....	58.00	....	S W	Do
28	7 am....	55.00	45.00	S W	Thunder and lightning; hail.
29	8 am....	57.00	44.00	W	Gentle wind.
	2 pm....	62.00	....	W	High wind.
	8 pm....	55.00	....	W	Gentle wind.
30	7 am....	52.00	44 00	W	High wind.
	2 pm....	53.00	....	N	Gentle wind.
	8 pm....	46.00	...	N	Overcast; gentle wind; a little rain.

Day of the Month.	Hour of the Day.	Temp. at the hour of observation.	Minimum Temperature.	Direction of the Wind.	Remarks.
17	7 a m....	33.00	24.00	....	Clear and calm.
	2 p m....	46.50	....	S	Some ice from Assiniboine floating down.
	8 p m....	39.00	....	....	Calm.
18	7 a m....	40.00	33.00	....	Do
	2.30 p m.	42.00	....	....	Do
	8 p m....	38.00	....	....	Do
19	7 a m....	40.00	34.00	....	Do
	2 p m....	55.75	....	....	Clear, calm.
	8 p m....	48.00	....	W	Clear, a gentle breeze.
20	7 a m....	45.00	34.00	W	A gentle wind.
	2 p m....	58.00	. ..	W	A brisk wind.
	8 p m....	42.00	....	....	Calm, faint aurora,
21	7 a m....	37.50	32.00	....	A brisk wind.
	2 p m....	31.00	....	W	A gentle wind.
	8 p m....	48.25	....	W	A high wind.
22	7 a m. ..	20.00	19.00	N W	Do
	2 p m....	35.50	....	N W	Hazy, a gentle wind.
	8 p m....	25.00	....	....	Clear, high wind, lunar corona.
23	7 a m....	20.50	17.00	W N W	Clear, a gentle wind.
	2 p m....	43.00	....	W N W	Clear, a brisk wind.
	8 p m....	23.00	....	....	Clear, a gentle wind.
24	7 a m....	24.00	15.00	E	Do do
	2 p m.. .	37.00	....	E	A gentle wind.
	8 p m....	33.00	....	....	Light hazy appearance, calm.
25	7 a m....	35.00	29.00	S	A gentle wind.
	2 p m....	49.00	....	S W	A brisk wind.
	8 p m. ..	38.00	....	N	A gentle wind.
26	7 a m....	35.00	32.00	N W	Do
	2 p m....	50.50	....	W	Do
	8 p m....	44.00	....	S E	Do
27	8 a m....	45.00	40.00	S	A very high wind.
	2.10 p m.	59.00	....	S	A gale.
	8 p m....	54.00	....	S S W	A light wind.
28	7 a m....	50.50	40.00	S	Hazy, a gentle wind.
	2 p m....	65.00	....	S W	A gentle wind.
	8 p m....	56.00	....	....	Calm.
29	7.5 a m..	38.00	38.00	W N W	A high wind.
	2 p m....	47.00	....	N	A storm.
	8 p m....	41.00	. ..	W N W	A high wind.
30	7 a m....	40.00	27.50	W	A gentle wind
	2 p m....	45.00	....	N	Clear, a strong wind.
	8 p m....	33.00	....	N W	Clear, a gentle wind.

MAY, 1858.

Day of the Month.	Hour of the Day.	Temp. at the hour of observation.	Minimum Temperature.	Direction of the Wind.	Remarks.
1	7 a m....	33.00	28.50	....	Calm, clear.
	2 p m....	48.75	....	W	Clear, hardly perceptible wind.
	8 p m....	42.00	....	....	Calm.
2	7 a m....	49.00	33.00	S	Clear, gentle wind.
	2 p m....	58.50	....	S	Clear, very brisk wind.
	8 p m....	50.00	....	S	Clear, gentle wind.
3	7 a m....	48.50	38.50	S	A gentle wind.
	2 p m....	61.00	....	S	Do
	8 p m....	53.00	. ..	S	Hazy, gentle wind.
4	7 a m....	50.00	38.00	S	Hazy, very light wind.
	2.20 p m.	73.00	....	....	Very smoky, calm.
	8 p m....	51.00	....	....	A gentle wind.
5	7 a m....	50.50	38.75	N	A brisk wind, smoky.
	2 p m....	64.00	....	N	A high wind.
	8 p m....	51.00	....	N	Cloudy, a stiff breeze.
6	7.5 a m..	45.00	39.00	N E	High wind.
	2 p m....	45.00	....	W	A moderate wind.
	8 p m....	40.50	....	N W	A pleasant wind.

Day of the Month.	Hour of the Day.	Temp. at the hour of observation.	Minimum Temperature.	Direction of the Wind.	Remarks.
31	7 a m....	55.00	40.00	S	Clear, a gentle wind.
	2 p m....	72.00	....	S	A gentle wind.
	8 p m....	62.00	...	S	Do

JUNE, 1858.

Day of the Month.	Hour of the Day.	Temp. at the hour of observation.	Minimum Temperature.	Direction of the Wind.	Remarks.
1	9 a m ...	61.00	48.00	S	Clear; gentle wind.
2	7 a m....	45 00	36.00	N W	A gentle wind, clouded.
3	7 a m....	56.00	44.00	W	Do do
	2 p m....	62.00	....	N	Do do partially clouded.
	8 p m....	53.00	....	N	A high wind.
4	7 a m....	52.00	40.00	N	A light wind.
	8 p m....	63.00	....	S W	A gentle wind; rain.
5	7 a m....	61.00	50.00	W S W	A gentle wind.
	2 p m....	67.00	....	W	Do do
	8 p m....	52.00	....	N W	Clear; hardly perceptible wind.
6	7 a m....	61.00	43.00	S	A gentle wind.
	2 p m....	63.00	....	S W	A brisk wind.
	8 p m....	47.00	....	N W	A high wind; sky clouded.
7	7 a m....	47.00	32.00	...	Calm.
	2 p m....	58.00	....	....	Do
8	7 a m....	50.00	37.00	....	Calm, clear.
	2 p m....	66.00	....	S W	A gentle wind.
	8 p m....	54.00	....	....	Calm; partially clouded.
9	7 a m....	48.00	44.00	N	Clear; a gentle wind.
	2 p m....	55.00	....	N W	A high wind; clouded.
	8 p m....	47.00	....	N W	A gentle wind.
10	7½ a m...	51.00	32.50	....	Clear, calm.
	2 p m....	61.00	....	W	Clear; a brisk wind.
	8 p m....	53.00	....	....	Clear, calm.
11	7 a m....	56.00	39.00	....	Do do
	2 p m....	71.00	....	....	Do do
	8 p m....	....	....	....	Do do
12	7 a m....	58.00	56.00	W	A gentle wind.
	2 p m....	71.00	....	W	Do do a few clouds.
	8 p m....	55.00	....	....	Clear, calm.
13	7 a m....	58.00	44.00	W	Clear; a gentle wind.
	2 p m....	79.00	....	W	A high wind.
	8 p m....	56.00	....	N	A very high wind.
14	7 a m....	51.00	43.50	....	Calm.
	2 p m....	65.00	....	S W	A gentle wind.
	8 p m....	54.00	....	W	A gentle breeze.
15	7 a m....	52.00	40.00	W	A gentle wind.
	2 p m....	65.00	....	W	Clear; a brisk wind.
	8 p m....	62.00	....	.. .	Clear, calm.
16	7 a m	54.00	40.00	S W	A gentle wind; a little rain.
	2 p m....	70.00	....	N	A high wind.
	8 p m....	58 00	....	N	A high wind.
17	7 a m....	56.00	54.50	N	Do do
	2 p m....	60.00	....	N	A very high wind.
	7 p m....	56.00	....	N	A high wind.
18	7 a m....	53.00	51.50	....	Do do
	2 p m....	63.00	....	N	A heavy mist; moderate wind.
	8 p m....	59.00	....	....	Drizzling rain; calm.
19	7 a m....	57.00	52.00	N W	A brisk wind.
	2 p m....	64.00	....	W	Do do
	8 p m....	57.00	....	N	A gentle wind.
20	7 a m....	61.00	49.00	....	Clear, calm.
	2 p m....	77.50	....	N	Clear; a high wind.
	8 p m....	75 00	....	S	A gentle wind.
21	7 a m....	69.00	63.50	....	Calm; rain, also thunder.
22	7 a m...	72.00	59.00	S	A gentle wind.
	2 p m....	80.00	....	S E	Hardly percep. wind; thunder & rain in afternoon.

RED RIVER SETTLEMENT—THERMOMETRICAL REGISTER,

JUNE, 1858—(*Continued.*)

Day of the Month.	Hour of the Day.	Temp. at the hour of observation	Minimum Temperature.	Direction of the Wind.	Remarks.
23	2 p m....	76.00	....	....	Calm.
	8 p m....	65.00	....	....	Do
24	7 a m....	67.00	50.00	....	Clear, calm.
	2 p m....	86.00	....	....	Do do
	8 p m....	70.00	....	....	Do do
25	7 a m....	67.00	60.00	....	Calm.
	2 p m....	77.00	....	S	Hardly perceptible wind.
	8 p m....	69.00	....	N W	A gentle wind.
26	8 a m....	70.00	60.00	S W	Do do
	2 p m....	77.00	....	S	Clear; a gentle wind.
	8 p m....	64.50	....	....	Clear, calm. Barley in the ear.
27	7 a m....	62.00	51.00	....	Calm. Ripe strawberries.
	2 p m....	62.00	....	N	A gentle wind.
28	2 p m....	69.00	....	N	Light wind.
	8 p m....	72.00	....	S E	A gentle wind.
29	7 a m....	64.00	56.00	....	Overcast, raining; gentle wind.
	2 p m....	88.00	....	....	Calm.
	8 p m....	74.00	....	....	Calm, clear.
30	8 p m....	72.00	....	....	Do do.

JULY, 1858.

Day of the Month.	Hour of the Day.	Temp. at the hour of observation.	Minimum Temperature.	Direction of the Wind.	Remarks.
2	8 a m....	64.00	51.00	....	Clear, calm.
	2 p m ...	79.00	....	S	Clear, a gentle wind.
	8 p m....	65.00	....	....	Clear, calm.
3	7 a m....	56.00	51.00	N	A brisk wind.
	2 p m....	62.00	....	N	Clear, a brisk wind.
	8 p m....	55.00	....	....	Clear, calm.
4	7 a m....	53.00	47.00	S	A great storm; thunder and rain.
	2 p m....	74.00	....	S	Do
	8 p m ...	76.00	....	S	Do
5	7 a m....	76.00	57.00	S	A brisk wind.
	2 p m....	87.50	....	S	A storm.
6	7 a m....	69.00	64.00	....	Calm.

FORT WILLIAM—THERMOMETRICAL REGISTER,

AUGUST, 1857.

Day of the Month.	Hour of the Day.	Temp. at the hour of observation.	Direction of the Wind.	Remarks.
3	1 p m....	72.00	W, gentle.	Close, appearance of thunder.
	8 p m....	65.00	W do	Do
4	8 a m....	65.00	N W	Cloudy.
	12 noon ...	70 00	N W	
5	8 a m....	68.00	N W	
	7 p m....	65.00	E	Appearance of rain.

FORT WILLIAM—THERMOMETRICAL REGISTER,

AUGUST, 1857—(*Continued.*)

Day of the Month	Hour of the Day.	Temp. at the hour of observation.	Direction of the Wind.	Remarks.
30	7 a m....	65.00	N W	Cloudy.
	1 p m....	67.00	N W	Do
31	7 a m....	58.00	N E	Clear.
	1½ p m....	61.00	....	
	7 p m....	59.00	S E	Do

SEPTEMBER, 1857.

Day of the Month.	Hour of the Day.	Temp. at the hour of observation.	Direction of the Wind.	Remarks.
1	7 a m....	58 00	E, light.	Foggy, with light clouds.
	7 p m....	69.00	S W	Thunder and rain.
2	7 a m....	65.00	S W	Do do
	3 p m....	66 00	S W	Do do
	6 p m....	60.00	N E	Heavy clouds.
3	7 a m....	63.00	N	D do
	12 noon ...	71.00	N	
	6 p m....	70.00	W	Appearance of rain.
4	7 a m....	65.00	N W	Rain.
	7 p m....	62.00	N	Rain ceased, still cloudy.
5	7 a m....	61.00	N W	Clear, much rain last night.
	12 noon ...	63.00	N W	Clear.
	6 p m....	60.00	N W	Fine weather.
6	7 a m....	57.00	S E	Do
	1 p m....	60 00	S E	Overcast.
	7 p m....	58.00	S E	Do
7	7 a m....	56.00	S E	Heavy clouds.
	1 p m....	63.00	S E	Appearance of rain.
	6 p m....	61.00	S E	Do do
8	8 a m....	61.00	S E	Do do
	1 p m....	66.00	E	Do do
	6 p m....	60.00	E	Heavy fog.
9	7 a m ...	55.00	E	Do light rain.
	1 p m....	57.00	E	Do drizzling rain,
	7 p m....	58.00	E	Do do
10	8 a m....	65.00	W	Rain and thunder during the night.
	2 p m....	63.00	W	Rain all day.
	6 p m....	61.00	N	Still raining.
11	7 a m....	51.00	N	Much rain during the night.
	2 p m....	46.00	N E, a gale.	Continues to rain.
	6 p m....	47.00	NE very brisk	Poured rain all day.
12	7 a m....	52.00	W, light.	Rainy sky.
	1 p m....	53.00	Do	
	6 p m....	55.00	Do	Appearance of more rain.
13	8 a m....	53.00	W	
	6 p m....	54.00	S E	
14	8 a m....	55.00	W	Much rain last night.
	12 noon ...	60.00	W	
	6 p m....	55 00	W	
15	8 a m....	53.00	W	
	1 p m....	54.00	W	
	6 p m....	52.00	S E	Overcast, appearance of rain.
16	8 a m....	50.00	E	Rain.
	6 p m....	54.00	E	

6	8 a m....	66.00	S, gentle.	Cloudy, appearance of thunder shower.
	1 p m....	67.00	S, brisk.	Do do
	6 p m....	66.00	Calm.	Appearance of shower.
7	8 a m....	65.00	E, gentle.	Clear, appearance of fine weather.
	2 p m....	71.00	Do	Clear, fine weather.
	7 p m....	68.00	W, fresh.	Do do
8	7 a m....	63.00	N do	Clear.
	2 p m....	66.00	N do	Clear, fine weather.
	7 p m....	64.00	S E	Cloudy, appearance of rain.
9	8 a m....	62.00	N E	Dry, cloudy.
	1 p m....	64.00	N E	Thunder and rain.
	7 p m....	63.00	E	Thunder clouds.
10	7 a m....	63.00	S E	
	1 p m....	62.00	S E	Sultry, heavy rain.
	7 p m....	62.00	E	Do do
11	7 a m....	60.00	E	Dry, foggy.
	1 p m....	64.00	E	
	7 p m....	61.00	E	Thunder clouds, appearance of rain.
12	7 a m....	60.00	E	Thunder clouds.
	1 p m....	63.00	E	Do
	7 p m....	61.00	N E	Heavy rain and thunder.
13	7 a m....	64.00	S E	Heavy clouds, appearance of rain.
	1 p m....	71.00	W, brisk.	
	7 p m....	67.00	Do	Clear, fine weather.
14	7 a m....	67.00	N W	Clear, fine.
	1 p m....	64.00	W	Do
	6 p m....	62.00	N W	Do
15	7.30 a m.	62.00	N W	Do
	1 p m....	64.00	N W	Do
	7 p m....	61.00	S E	Do
16	8 a m....	57.00	W	Do
	1 p m....	61.00	W	Sultry, thunder clouds.
	7 p m....	60.00	E	Appearance of rain.
17	8 a m....	65.00	S E	Do do
	1 p m....	65.00	S E	Light clouds.
18	6 a m....	59.00	E	Overcast, raining.
	3 p m....	62.00	E	
	6 p m....	60.00	E	
19	7 a m....	60.00	N W	Clear.
	12 noon ...	63.00	S E	Thunder clouds.
	6 p m....	62.00	S E	
20	7 am....	59.00	S E	Clear.
	12 noon ..	64.00	W	Shower of rain.
	6 p m....	61.00	W	Foggy, rain finished.
21	7 a m....	56.00	N E	Do
	2 p m....	57.00	N E	Do
	7 p m....	57.00	N E	Do
22	7 a m....	59.00	N	Heavy clouds.
	12 noon ..	62.00	....	Heavy clouds.
	6 p m....	59.00	W	Clear.
23	8 a m....	63.00	S W	Heavy clouds, showers of rain.
	1 p m....	64.00	S W	Raining.
	7 p m....	61.00	S E	Do
24	8 a m....	65.00	S E	Heavy clouds, showers of rain.
	1 p m....	61.00	S E	Do do
	6 p m....	67.00	S E	Do do
25	7 a m....	61.00	S E	Do do
	1 p m....	72.00	S E	Do do
	7 p m....	69.00	S E	Do do
26	7 a m....	61.00	S E	Do do
	1 p m....	68.00	S E	Do do
	6.30 p m.	65.00	S E	Thunder, with heavy rain.
27	7 a m....	61.00	N	Thunder, with rain.
	1 p m....	65.00	N	Thunder storm, with rain.
	6 p m....	62.00	N	
28	8 a m....	58.00	N W	Heavy clouds, showers.
	1 p m....	61.00	N W	Do do
	7 p m....	60.00	N W	Do do
29	8 a m....	62.00	N W	Clear, appearance of fine weather.
	1.30 p m.	65.00	N W	

17	7 a m....	50.00	N	Very cloudy.
	1 p m....	51.00	N	Do
	6 p m....	49.00	N	
18	7 a m....	47.00	N	Frost during the night,
	12 noon...	48.00	N	Clear.
	6 p m....	44.00	S E	Do
19	7 a m....	40.00	S E	Ice formed during the night.
	2 p m....	49.00	S E	
	7 p m....	46.00	E	
20	8 a m....	46.00	W	Cloudy, slight frost last night.
	1 p m....	50.00	S E	
	6 p m....	47.00	S E	
21	7 a m....	44.00	S E	Rain.
	1 p m....	51.00	W	Raining.
	6 p m....	46.00	N	Still raining.
22	7 a m....	38.00	N W	Severe frost last night.
	6 p m....	45.00	N W	Cloudy.
23	7 a m....	46.00	W	Clear.
	2 p m....	54.00	W	Do
	6 p m....	54.00	S E	Do
24	7 a m....	49.00	S E	Do
	6 p m....	56.00	N W	Cloudy, appearance of rain.
25	8 a m....	54.00	W	Thunder storm and rain last night.
	1 p m....	56.00	W	Clear.
	6 p m....	54.00	W	Do
26	7 a m....	51.00	E	Thunder storm, with hail and rain.
	1 p m....	58.00	E	
	6 p m....	57.00	N W	Cloudy.
27	8 a m....	57.00	W	
	4 p m....	56.00	W	
	6 p m....	55.00	W	
28	8 a m....	48.00	W	Heavy clouds.
	2 p m....	53.00	W	
	6 p m....	50.00	N W	Do
29	7 a m....	47.00	W	Overcast.
	1 p m....	52.00	W	Appearance of rain.
	6 p m....	48.00	N W	Thunder and rain.
30	7 a m....	45.00	N W	Do do
	2 p m....	50.00	S E	Cloudy.
	6 p m....	47.00	S E	Do

OCTOBER, 1857.

Day of the Month.	Hour of the Day.	Temp. at the hour of observation.	Direction of the Wind.	Remarks.
1	7 a m....	40.00	N E	Cloudy.
	1 p m....	44.00	N E	Do
	6 p m....	43.00	N E	Do
2	7 a m....	43.00	E	Heavy clouds.
	3 p m....	44.00	S E	
3	7 a m....	38.00	S E	
	1 p m....	48.00	S E, high.	
	6 p m....	46.00	S E	
4	8 a m....	43.00	S E	
	12 noon ..	48.00	S E	
	6 p m....	46.00	S E	
5	7 a m....	42.00	W	
	2 p m....	56.00	W	
	6 p m....	54.00	W	
6	8 a m....	50.00	E	Heavy clouds.
7	7 a m....	45.00	E	
	1 p m....	60.00	E	
	6 p m....	56.00	S E	
8	1 p m....	58.00	S E	
	5 p m....	55.00	S E	

FORT WILLIAM—THERMOMETRICAL REGISTER.

OCTOBER, 1857—(*Continued.*)

Day of the Month.	Hour of the Day.	Temp. at the hour of observation.	Direction of the Wind.	Remarks.
9	7 a m....	53.00	E	
	2 p m....	52.00	E	
	6 p m....	50.00	E	Appearance of rain.
10	7 a m....	50.00	E	
	3 pm....	57.00	S E	
11	8 a m....	52.00	S E	Heavy clouds.
	1 p m....	54.00	S E	
	6 p m....	51.00	N E	
12	7 a m....	47.00	W	Heavy clouds, slight rain.
	1 p m....	56.00	W	
	6 p m....	50.00	W	
13	7 a m....	47.00	W	Clear, fine weather.
	12 noon ..	56.00	W	
	6 p m....	56.00	W	
14	10 a m....	50.00	W	Do do
	5 p m....	49.00	N W	Do do
15	7 a m....	44.00	N W	
	1 p m....	49.00	N W	
	6 p m....	45.00	E	Cloudy.
16	7 a m....	44.00	W	
	1 p m....	45.00	W	
	6 p m....	43.00	W	
17	8 a m....	43.00	N W	Rain and snow last night.
	1 p m....	44.00	N W	
	5 p m....	41.00	N W	
18	10 a m....	39.00	N W	Slight showers of snow.
	1 p m....	41.00	S E	
	6 p m....	42.00	S E	
19	8 a m....	34.00	N	Snowing.
	1 p m....	35.00	N	Ground covered with snow.
	5 p m....	31.00	N	
20	7 a m....	31.00	W	Clear.
	5 pm. ..	35.00	S	Heavy clouds.
21	8 p m....	33 00	S	Clear.
	2 p m....	37.00	S	Do
	6 p m....	35.00	S	Do
22	7 a m....	32.00	S	
	5 p m....	35.00	E	
23	8 a m....	35.00	S E	
	2 p m....	38.00	E	
24	7 a m....	39.00	S W	
	2 p m....	39.00	S W	
25	8 a m....	37.00	N	
	2 p m....	38.00	N	
	5 p m....	35.00	N	Cloudy.
26	7 a m....	34.00	N E	
	6 p m....	35.00	N E	Slight shower of snow.
27	7 a m....	37.00	S E	
	5 p m....	38.00	S E	Appearance of snow.
28	7 a m....	38.00	N E	
	3 p m....	45.00	N E	
29	7 a m....	35.00	N E	
	5 p m....	32.00	N	
30	7 a m....	32.00	N	
	5 p m....	35.00	N W	
31	7 a m....	31.00	N	
	2 p m....	36.00	N	
	5 p m....	34.00	N E	

FORT WILLIAM—THERMOMETRICAL REGISTER.

NOVEMBER, 1857—(*Cantinued.*)

Day of the Month.	Hour of the Day.	Temp. at the hour of observation.	Direction of the Wind.	Remarks.
13	8 a m....	24.00	N W	
	2 p m....	23.00	N W	
14	8 a m....	19.00	W	
	4 p m....	24.00	W	
15	8 a m....	25.00	W	
	3 p m....	26.00	S W	
16	9 a m....	26.00	S W	
17	8 a m....	24.00	S W	
	4 p m....	27.00	W	
18	12 noon ...	21.00	W	
	4 p m....	17.00	N W	
19	9 a m....	15.00	N W	
20	10 a m....	16.00	N W	
	3 p m....	17.00	N W	
21	10 a m....	16.00	N W	
	4 p m....	-10.00	W	Light snow.
22	10 a m....	-0.00	W	
	4 p m....	-2.00	S W	Cloudy.
23	8 a m....	-2.00	S W	
	12 noon ...	7.00	W	
24	12 noon ...	2.00	Calm.	Clear and cold.
25	8 a m....	-2.00	W	
	4 p m....	5.00	W	
26	9 a m....	5.00	W	
	2 p m....	20.00	S	
27	10 a m....	23.00	S	Rain.
28	8 a m....	24.00	S	
	3 p m....	34.00	E	
29	9 a m....	34.00	E	Cloudy; rained all night.
30	9 a m....	34.00	S	
	4 p m....	34.00	S	Heavy clouds in the south.

DECEMBER, 1857.

Day of the Month.	Hour of the Day.	Temp. at the hour of observation.	Direction of the wind.	Remarks.
1	3 p m....	27.00	W	
3	10 a m....	25.00	....	Cloudy, slight snow.
	2 p m....	25.00	N W	
3	8 a m ...	19.00	W	Clear.
	2 p m....	22.00	....	Do
4	9 a m....	20.00	N	
5	10 a m....	16.(0	N W	Snowing.
	2 p m....	18.00	N W	
6	2 p m....	24.00	S E	
7	2 p m....	26.00	S E, a storm	Snowing hard.
8	10 a m....	24.00	N W	Cloudy.
9	10 a m....	24.00	N W	Snowing.
	3 p m....	23.00	N W	
10	10 a m....	21.00	N W	
	2 p m....	22.00	N W	
11	9 a m....	22.00	N W	
12	10 a m....	23.00	S W	Clear, dry, mild.
	3 p m....	25.00	S W	

NOVEMBER. 1857.

Day of the Month.	Hour of the Day.	Temp. at the hour of observation.	Direction of the Wind.	Remarks.
1	8 p m....	35.00	W	
2	8 a m....	34.00	N W	
	5 p m....	34.00	N W	
3	12 noon ...	35.00	W	
	5 p m....	32.00	W	
4	8 a m....	35.00	W	
	3 p m....	38.00	E	Blowing a storm.
5	8 a m....	40.00	E, high.	Raining very heavily.
	3 p m....	41.00	W	
6	8 a m....	38.00	N W	Appearance of snow.
	1 p m....	39.00	N W	
7	1 p m....	35.00	N W	
	5 p m....	36.00	N E	Snow and rain together.
8	4 p m....	33.80	N W	Do do
9	7 a m....	30 00	S W	Nine inches of snow on the ground.
	5 p m....	30.00	W	Rain.
10	10 a m....	30.00	W	
11	7 a m....	26.00	S W	
	1 p m....	26.00	S W	
	5 p m....	24.00	S W	
12	8 a m....	25.00	N W	Snowing.
13	8 a m....	20.00	S W	Light clouds.
	3 p m....	24.00	S W	Dry.
14	10 a m....	21.00	S W	Dry, cloudy.
	3 p m....	24.00	N	
15	9 a m. ..	25.00	N W	
	3 p m....	29.00	N W	
16	10 a m....	27.00	W	
	1 p m....	28.00	W	
17	1 p m....	29.00	W	
18	9 a m....	19.00	N W	
	2 p m....	21.00	N W	
19	8 a m....	15.00	N, light.	Clear.
	12 noon....	14.00	....	
20	1 p m....	21.00	S W	
21	9 a m....	20.00	N	
	1 p m....	22.00	....	
22	8 a m....	16.00	W	Cloudy.
	12 noon...	15.00	....	Do
23	12 noon...	24.00	....	
24	12 noon...	14.00	N W	
25	9 a m....	12.00	N	
26	11 a m....	10.00	....	
	4 p m....	8.00	S	
28	9 a m....	17.00	W	
29	9 a m....	19.00	E	Snowing.
30	9 a m....	16.00	W	Do
	4 p m....	17.00	W	
31	11 a m....	17.00	W	

LETTER FROM THE BISHOP OF ST. BONIFACE,

RED RIVER SETTLEMENT.

The following communication has just been received from the Right Reverend the Bishop of St. Boniface, in the Colony of Assiniboia. His lordship, it will be observed, gives a very interesting account of the educational and religious establishments in his diocese, and of the gradual advancement of the half breed Indians in civilization and refinement. The absence of great crimes among them, and the facility with which little differences are arranged, speak volumes in favor of those who, regardless of personal considerations, devote their lives to their moral training and instruction. Bishop Taché has been in the country for fourteen years, and has travelled far and wide among the native tribes, and in the exercise of the duties of his high office must have had the best opportunities of observing their habits and character. His remarks are therefore of the greatest value, and will, I am confident, be read with deep interest:

RED RIVER, Feb. 7th, 1859.

DEAR SIR,—I have just received your pleasing letter of 29th December last. Not only the excellent report which you are preparing of the soil, climate, and other matters relating to the physical and geographical character of the vast country explored by you will, I am sure, be hailed with pleasure in Canada, where a strong desire prevails of obtaining information on the subject; but anything which is calculated to throw new light on, and add to, the knowledge which is now so much in request, will find ready acceptance there. I have therefore much satisfaction in contributing the information which you request of me. I regret only that the shortness of the time will not allow me to make that information as complete as might be necessary to give an exact idea of what is passing here, even to persons who have some slight knowledge of the country. The efforts made by the Catholic Clergy to extend the empire of Jesus Christ over my immense diocese, have two distinct and very different fields, which may be thus defined:

1. The ministration afforded to the Catholic population of Red River and the neighbourhood, in which, excepting some slight difference, our functions resemble those of the priests in new parishes in Lower Canada.

2. The care of the missions in the interior of the country, which are altogether heathen missions, entitled to particular attention, in order to their being perfectly understood, and yielding to the evangelical husbandman a rich harvest of merits and a glory which the world does not sufficiently regard, simply because the Kingdom of Jesus Christ is not of this world. To this division I shall attach myself in the details which it is my intention to supply.

§ 1. Aids afforded to the Catholics of Red River, and reflexions concerning them.

The Catholic population of Red River is distributed into what we conventionally term parishes. These are four in number: 1. St. Boniface—2. St. Norbert (on the banks of the Red River)—3. St. Francis Xavier—4. St. Charles (on the banks of the Assiniboine.) The boundaries of these parishes, as shown in the map accompanying the Report of Professor Hind, are tolerably exact.

St. Boniface.—Here was the first resting-place of the earliest missionaries who visited the Hudson's Bay Territory after the Conquest of Canada by the English. At the request, and with the aid of Lord Selkirk, Messrs. J. N. Provencher and T. J. N. Dumoulin landed at Fort Douglas on the 16th July, 1818, and commenced their labor of self-devotion and self-sacrifice, bestowing their first cares on a few old Canadian *voyageurs* and their families, half breeds, who were still unconverted. In 1820, was laid the foundation of the first religious edifice. This poor chapel, which was of wood, and was nevertheless destined afterwards to serve as a Cathedral, was consecrated to religious worship under the patronage of St. Boniface. The parish to which it belongs has now a population of 1400 souls. It is here that I peform the duty of a *Curé*, being assisted in it by the Rev. P. Lefloit. Two other missionaries, the Rev. Messrs. Mestre and Moulier, who arrived last summer, are passing the winter with us, awaiting the opening of the spring to be enabled to proceed to one of the posts, to which they are summoned by their pious zeal. The Church of St. Boniface, my Cathedral, is certainly not externally a monument of elegance, as you may have observed, but it is nevertheless indisputably the finest building in the country. It is 100 feet in length, 45 in breadth, and 40 in height. Its two towers, surmounted by belfries covered with tin, display the emblem of our redemption on their airy summits, at a height of 100 feet from the ground. A fine and well matched peal of three bells (weighing upwards of 1600 lbs.) adds to the joyous zest and the pomp attendant on our religious fêtes. The internal decorations of this Church are a subject of surprise to strangers traversing the wilderness, who are far from expecting to find here so much of art and good taste. In rear of the Cathedral, and overshadowed by its walls is the dwelling of the Bishop. This is a large house, in which the prelate is happy in being able to share with his clergy, the brothers of his schools, and some orphans, the small portion of his revenue which is not employed in other good works. The Parish of St. Boniface has the advantage of possessing excellent schools. Until May last, the boys' school of the brothers of the Christian Doctrine was kept in the bishop's house; but since that time, the brothers and their pupils have taken possession of a large house built for them, a few hundred feet from the cathedral. Here I cannot refrain from expressing the feeling of gratitude which warms my heart at the sight of that house. In 1852, a terrible fire destroyed a considerable portion of the City of Montreal. Scarcely had the fiery and destructive blast passed away when the soft and refreshing breath of charity stirred the still smoking ashes, and wafted to the Red River a lesson too advantageous to its friends to be lightly forgotten. The faithful at Montreal, at the instance of their chief pastor, in order to appease by their good works the anger of Heaven, shewn by so terrible an infliction, desirous moreover of fulfilling the conditions imposed by the Sovereign Pontiff to entitle them to the indulgences of the Jubilee, made a collection for the benefit of Red River, which produced the sum of £364. This was to be expended in the foundation of a school. To this liberal act of charity we are indebted for the means of erecting the beautiful house in which 58 boys are now receiving their education. On the other side of the cathedral, at an equal distance from it is the "very spacious Convent" mentioned by Mr. Hind in his report. I am sorry that the learned professor, in speaking of this Convent, found nothing to speak of more interesting than the cultivation bestowed on the garden in front of it. The cabbages and other vegetables produced in it are no doubt very fine, nevertheless the interior of the house, and the extensive benefits dispensed by its inhabitants, are matters still more worthy of admiration and comment. This Convent belongs to the Sisters of Charity, known in Canada as the Grey Nuns. These heroic followers of Christ, animated by a degree of courage and devotion which Catholicism alone can inspire, journeyed to this remote spot to dispense among the youth of their own sex, and among the population in general, the blessings which they diffuse wherever they go. The amount of good effected in this establishment, the mild and salutary influence exercised by the sisters in the bosoms of families, the thousands of good deeds which the eye of the world sees not, because the charity which prompts them seeks the veil of humility, can never be fully told. In this Convent young ladies are boarded, and receive, I do not say a passable education, but an education suitable for the middling classes of the most civilized countries. They are now twenty in number. This school, in which the intellect is trained to a knowledge of the beautiful, the useful, and the agreeable, while the education of the heart is not neglected, is, I have not the least doubt, by far the best in the whole colony of Red River. Even those persons who are afraid to acknowledge it openly, are fully aware of it. Besides the boarders, the sisters maintain and educate (in a separate apartment, unconnected with that of the boarders) fifteen poor orphan girls. This admirable work is achieved by sacrifices of all kinds, self imposed by the adopted mothers of these poor children; and what is here beheld will forcibly illustrate the triumphant working of charity even without extraneous aid. In addition to the education thus afforded to thirty-five inmates of their house, a day-school is kept for all the little girls of the parish who are desirous of instruction. Moreover, besides the labor thus bestowed on education, the Sisters of Charity are unwearied in the services and attentions which they minister to suffering followers of Jesus Christ. The following short note may give an idea of the good done by them in this behalf:

	Sick persons attended.	Visited at their own abodes.
From Oct., 1856, to Oct., 1857.........	175	210
do 1857, do 1858.........	157	130

	Wounds dressed.	In hospital.
From Oct., 1856, to Oct., 1857.........	53	21
do 1857, do 1858.........	58	21

A knowledge of these facts should accompany Professor Hind's remarks, at page 366 of his report, in order to enable you to form an idea of the salubrity of the climate, in accordance with the rule laid down by that gentleman. To these numerous, important, and nearly gratuitous services rendered to the population of this parish by the sisters at St. Boniface, we must add the strong influence which they exercise over the community. To their lessons of industry, economy, &c., we must attribute the very perceptible change which has taken place in the population since the arrival of these generous nuns. The internal decorations of the church and other objects of art and taste, are the work of the inmates of this vast convent, around which certain persons can see nothing but well cultivated vegetables.

2. St. Norbert.—This new Parish has a population of 700 souls. The Rev. P. Lestane is the Pastor of this little flock. The chapel, built of wood, is an edifice of 90 feet in length, by 33 in breadth. At one end are the apartments of the Priest, the teacher, (who, in this case also, is happily a brother of the Christian Schools. This good brother has 31 children to teach. The girls, 29 in number, are confided to the care of the Sisters of Charity, who have their dwelling near the Church; a happy symbol reminding us that the only true and solid education is that which is based on religion.

3. St. Francis Xavier.—This Parish dates from 1824. The population consists of nearly 1200 souls, permanent residents, not including several hundreds of hunters who pass the year on the prairies, but at certain times visit the colony. Some no doubt take up their abode in other parishes, but most of them stop short at St. Francis Xavier. This parish is known by the name of the Prairie of the White Horse. The old wooden chapel, now too small for the population, threatens to fall. We have accordingly begun to collect materials for a new church. The senior of our missionaries, Mr. Thibeault, my Grand Vicar, is the *Curé* of this parish. He has exercised his sacred office in this diocese for twenty-six years, speaks the Cree language very well, and understands that of the Sauteux. This knowledge is indispensable in his situation, as a great number of those who require the exercise of his sacred functions know no other language. The Parish of St. Francis Xavier has only one school, which is kept by the Sisters of Charity. Here 13 boys only and 26 girls receive instruction.

4. St. Charles.—This Parish contains only 210 persons. There is neither church nor chapel. Divine service is performed in a house which serves as a temporary chapel. The congregation have neither school nor resident Pastor. When it is practicable, a priest proceeds thither from this place to officiate. In other circumstances, the worshippers are obliged to go to St. Boniface or St. Francis Xavier.

LAKE MANITOBA.

To these details I may be permitted to add a few words concerning another establishment, commenced last summer at the extremity of Lake Manitoba, for the convenience of 30 or 40 families who are settled there. This small community have built a chapel. Mr. Gascon resides at the post, and has so much the greater merit in his good deeds, that although connected with the colony, he derives no advantage from the connection.

I have thus, my dear Sir, given you a full account of the religious comforts afforded to the Catholic population of Red River. Far then from being neglected in this respect, they possess a full measure of this essential blessing. Considering the sparse character of the settlements, the schools would need to be increased in number in certain districts, but the absence of any law relative to education, and of zeal in the people themselves, renders it utterly impossible to do more. I venture to assert what all reasonable and impartial persons must, in view of what is done, acknowledge, that the result far transcends the means which we can command. The truth is that, but for the unselfish zeal of some who devote themselves without fee or earthly reward to the arduous and meritorious task, it would be absolutely impossible to keep up the schools. So far, scarcely one child in ten has paid for his schooling, although the charge does not exceed ten shillings per annum, and I am certain that if we insisted on the payment of even that trifling sum, many of the scholars would leave the schools, such is the carelessness and indifference of the parents in that respect, notwithstanding our oft repeated entreaties and the sacrifices which are made in that behalf. This indifference concerning the education of their children and neglect of the many advantages afforded them is a standing reproach which may be justly cast on our population. I do not say that this is their only fault. Alas! the history of mankind, study it where we may, gives us much food for regret on the score both of omission and commission. I am desirous here to correct an error sorely injurious to us, into which nearly all the writers on this country have fallen. A fixed idea seems to have been entertained by them generally, at least a stereotyped expression of it appears in all their works, intended to establish an humiliating and unjust comparison between the half breeds of French Canadian parentage, and those of a different origin. I have neither the wish nor the time to controvert all the absurd things that have been said on this subject. One word only I shall say relative to this idea, as it is propounded in one of the best books which I am acquainted with on this country. In his Arctic Searching Expedition, Sir John Richardson remarks at page 273-4: "In character the half breeds vary according to their paternity; the descendants of the Orkney "laborers" being "generally steady, provident agriculturists of the Protestant "faith; while the children of the Roman Catholic Canadian "Voyagers have much of the levity and thoughtlessness of their "fathers, combined with that inability to resist temptation, which "is common to the two races from whence they are sprung." It is not necessary either to have been the inventor of gunpowder or to have enriched the domain of science by any important discovery, in order to detect, at a glance, the tendency of these remarks, and the spirit which dictated them. For my own part,

I have been in the country nearly fourteen years, I have come in contact with half breeds of every extraction, and I have as yet failed to discover the grounds for any such distinction. I repeat once more, that my affection for our Canadian half breeds does not make me look upon them as perfect or free from defects, but it is also true that I have not found any greater degree of the same perfection among the others.

Every nation has its weak side, but counterbalanced by qualities which others do not possess, and in judging of a people it is not wise to select one particular point in its character, which may be defective, it is necessary, on the contrary, to estimate the good or bad inclinations of that people, as a whole, making every allowance for the circumstance in which it is placed, the centre in which it lives.

Looking at the matter in this light, I have arrived at the firm conviction, in spite of multiplied assertions of the contrary, that our Canadian half breeds are no worse than the others. I belong myself (and I thank heaven for it) to this French Canadian race, the subject of these remarks, and if, as the learned and respected author of the above mentioned work affirms, we had shared "the inability to resist temptation," I should have given a proof of my origin by not resisting a temptation which I have often experienced. This temptation consists in the publication of certain memoirs relating to the history of this country, which would indeed be the saddest page in French Canadian history, but, at the same time, a proof amounting to a demonstration, that even in this matter my fellow-countrymen have played but a secondary part, and that our race has never been sufficiently favored to enjoy a monopoly of crime any more than any other monopoly. Besides the complete tranquillity and freedom in which we spend our days; our ignorance of the precautions everywhere else considered indispensable to ensure security of life and property, and the facility with which any little differences which may arise, are appeased; the utter absence of great crimes, though we have no police to preserve order, (it is well known to every one that the tranquillity we enjoy is in no way owing to the presence of the troops;) the horror excited by the simple recital of what occurs in other places; the polite, affable and hospitable manners of our people, notwithstanding the little instruction they have received; does not all this prove, in spite of all that has been said and written against them, that the French Canadian half breed race are a gentle, moral and honest people? Were I called upon now to enumerate their defects, the list would no doubt be a long one; no one is better aware of it than I am, for no person is more anxious to remedy the evil; but I have no desire whatever to follow the example of some who lavish their praises on those of the same origin and belief with themselves, and have nothing but contempt for others. There is one thing certain, here as elsewhere, some of those who have been the greatest calumniators of their fellow-men have, in fact, been guilty of shameful crimes and monstrosities, which those whom they do not fear to overwhelm with contempt and insult, were incapable of committing.

As to myself, there is one thought which occupies my mind in the midst of all the movement and agitation at present existing with reference to the Red River country. As you say, "the country is open, the tide of emigration is about to pour into it." I am an exile, though a voluntary one, from my country, cut off from every thing that I held dear in the world before coming here, exposed equally with all those who surround me to the inconveniences resulting from our isolated position, and it is, no doubt, very gratifying to me to see the distance, which separates me from my native land, as it were shortened, and to be brought nearer to the friends whom my heart continues to love so much, to see my adopted country on the eve of enjoying the advantages intended to be procured for it, and yet in view of all this, I experience something like a feeling of pain, for it appears to me that while making a great gain, we shall perhaps also lose much. At all events I fancy that we are about to be deprived of much of the tranquillity which we now enjoy; this unbroken quiet may appear very insipid to those who are habituated to the bustle of business or the noisy stir of great cities, but it possesses an especial charm for the mind of one who has lived for many years in solitude and away from all agitation. Undoubtedly, and I cannot lose sight of the fact, there prevails throughout the country a general feeling of uneasiness about our present political condition, but notwithstanding this, my views are too narrow and contracted to enable me to anticipate any great beneficial results. I am well aware that the system proposed for us is, in the abstract, very far superior to our present system, which is in fact an anomaly, and yet I very much fear that it will be a long time before we shall reap any benefit from it. Whatever may be the result, the movement is an actual fact, and we must cease to be what we have hitherto been, an exceptional people. Red River will no longer be the oasis in the desert, it is about to lose this distinctive and poetical characteristic, and become as it were the mere suburb of some great city. A large number of strangers have already reached us during the last summer, and this immigration cannot fail to increase. I am a French Canadian in heart as well as by origin, and I should undoubtedly prefer to see our lands occupied by some of our *brave* and *respectable inhabitants* from Lower Canada. In the present condition of the Red River Settlement, those who have large families are not the persons who should come; we are more in want of arms than of mouths. A company of soldiers and the exploring parties who visit the settlement suffice, as it is, to create a famine. The price of many of the necessaries of life has doubled since last year, and although the harvest was pretty good, and hunting and fishing abundantly successful, nevertheless, there is an extreme scarcity of every thing. Young *married* men, farmers or tradesmen, would have the best chance of success here. Although I have said that I should prefer to see our lands occupied by settlers from Lower Canada, it is not that I wish to advise my fellow countrymen to leave their homes; far from it. On the contrary, it seems to me that our beautiful country is worthy of retaining and rich enough to support, all those who are born upon her soil. For my own part, had not motives of a higher order determined my will, had not a voice stronger than the voice of nature and of blood sounded in my ears, never could I have consented to break the link that bound my heart to the land of my birth. The bread of exile is so bitter, the strange land so barren, even in the height of its fertility, that the word "good bye" has always seemed to me the most painful in the vocabulary of the human heart; I am, therefore, very far from wishing to encourage the Canadians to emigrate, but if, owing to special and exceptional reasons, they are compelled to remove from their native land, if they are determined to take up the pilgrims' staff, rather than see them take the direction of the United States, I prefer to have them come here. Here their faith, at all events, will not be exposed, and if they be deprived of some material advantages, these will be compensated for by others of a higher order. Their children will here find masters and mistresses to enlighten their minds, and at the same time to train their hearts to virtue. Daily experience enables me to promise them zealous priests, who will be true fathers to them, and who, here as they did at home, will preach to them in their own mother tongue, the goodness of their God and the love which they owe him. Another thing which I can promise them is the affectionate interest which is, and ever will be, entertained for them, by the pastor whose flock they will come to join. The Bishop of St. Boniface, a Canadian-like themselves, their brother, therefore, and their friend, will be really rejoiced to have it in his power to assist them, and willingly engages to devote to the welfare of the new comers, as well as to that of the rest of his people, everything at his disposal. Besides, no one has a better right to occupy the valley of the Red River, or even the valley of the Saskatchewan, than the Canadians of French origin. Our fathers were the hardy champions of civilization who first penetrated these regions, influenced by motives far superior to the low interests of commerce; these brave, and skilful discoverers came at the call and in the company of the missionaries, to raise the standard of the cross in the vast prairies of the West. Evil days have since come upon the beautiful portion of the American continent which they reclaimed from barbarism, but after a century of struggles and constancy, our

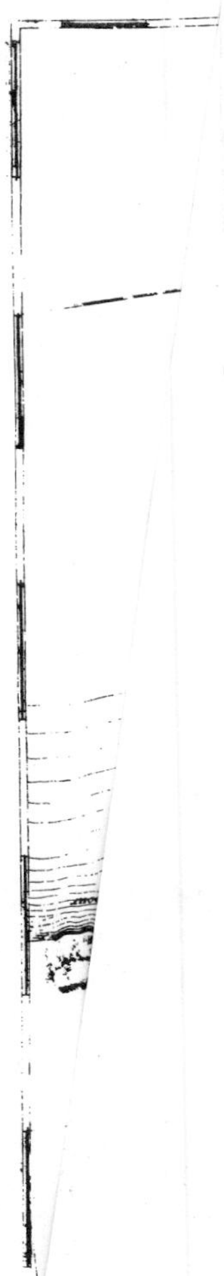

nationality shines out before an astonished world and nothing can be more natural than that our brethren should once more take possession of the land discovered by their ancestors, and consecrated by them as destined to witness at a future time the regeneration of the unfortunate tribes whom they found inhabiting it. Besides, the chain which connects Canada, and especially French Canada, with the Red River, has never been broken. Ever since the occupation of our country by the power which protects us so well to-day, the intrepid and skilful *voyageurs* have still continued to be recruited from amongst our fellow countrymen. They occupy an inferior position, under the circumstances, but their services have been acknowledged as indispensable. At a later period, French Canadians have here acquired claims, of which they cannot without injustice be deprived. There is one name deserving of the most prominent and honorable mention from every historian of the colony of Assiniboia, it is the name of a French Canadian, Mgr. J. N. Provencher, first apostle and first Bishop of the diocese, who succeeded in gaining the love and respect of all, the father of the poor whom he suffered so much in relieving, and the friend of the rich, who admired his virtues. In many instances the paternal influence of this worthy prelate prevented the commission of crimes which would have caused the ruin of this colony, while the wisdom and firmness of his counsels prevented the adoption of measures calculated to bring about the same misfortune. Without wishing to detract in any way from the merits of those to whose good offices it stands indebted, I think it is but just to assert that Mgr. Provencher has been one of the main props of this colony, and that but for him it would have been destroyed on more than one occasion.

Would that the limits of this letter but allowed me to give free vent to my gratitude, as chief pastor of the Catholic population of this colony, for the benefits with which our generous country has overwhelmed us, but I am not writing a book, and I must restrain within my heart the thoughts that struggle for utterance.

*** The remainder of this letter was not in time to appear with this Report.

RETURN

To an Address from the Legislative Assembly to His Excellency the Governor General, dated the 20th instant, praying His Excellency to cause to be laid before the House, "Copies of all Reports and communications of the Assiniboine and "Saskatchewan Exploring Expedition, under the charge of Professor H. Y. Hind, "during the year 1858."

By Command,

C. ALLEYN,
Secretary.

Secretary's Office,
Toronto, 29th April, 1859.

COPY OF INSTRUCTIONS.

No. I.

Secretary's Office,
Toronto, 14th April, 1858.

Sir,—During the last week I communicated to you, verbally, instructions in reference to the proposed Expedition to the neighborhood of the Red River during the present year.

2. It has been decided, as you are aware, with a view to keep down as much as possible the expenditure this year, to dispense with the services of Mr. Gladman as its general manager.

3. The exploration party this year will consist of two divisions, one to be placed under your direction and control, and the other under the direction of Mr. Dawson.

4. His Excellency in Council has been pleased to place under your charge the Topographical and Geological portion of the Exploration, respecting which full instructions will be given in another letter, while Mr. Dawson will continue to perform the same duties as last year, viz.: those of Surveyor, &c.

5. The estimate of the probable expenditure of the Expedition, submitted by you on the 6th instant, was laid before His Excellency in Council, and has been approved of by them, and I have accordingly now to direct you to be guided as much as possible by that estimate in engaging your assistants, hiring your men, as well as in the other necessary expenditures of the Expedition.

6. It is hardly necessary to say that His Excellency relies upon your exercising a due economy in all matters connected with the Expedition.

7. As soon as you have completed your contemplated party, you will furnish me with a schedule, giving the names of all the persons composing it, and stating their rates of pay, and the dates from which their pay is to commence. Such a schedule will be necessary to supply the Auditor with the means of auditing your accounts.

8. Having organized your party, you will lose no time in repairing with them to Red River, taking with you the supplies (referred to in the estimate) required for Mr. Dawson.

9. On your way to the Red River, you will take possession of the canoes, provisions, and other articles belonging to the Government, either at Collingwood or Sault Ste. Marie. These, with the men intended for Mr. Dawson, you will deliver over to that gentleman when you meet him, either at Red River or on his way back.

10. You are to consider all the articles and materials of any description belonging to the Canadian Government, connected with the late Expedition, as available for the purposes of the present Expedition, and you and Mr. Dawson may therefore divide them between you in whatever way you may think most advantageous. Such articles, if any, as may not be required by either of you, should be left in the custody of some trustworthy person to await the orders of the Government.

11. As soon as you shall have put Mr. Dawson in possession of the men and canoes intended for him, each of you will be held separately responsible for the expenses of his own party. You will therefore be careful to keep an accurate account of your expenditure.

12. The Auditor General of Public Accounts will give you any information you may require as to the most convenient mode of making out and furnishing your accounts, &c.

13. On your return from Montreal I shall be prepared to give you your instructions with reference to the localities in which your explorations are to be conducted, and as to the objects to which your attention is to be more especially directed.

I have the honor to be, Sir,
Your obedient servant,
T. J. J. LORANGER,
Secretary.

To H. Y. Hind, Esq.
Toronto.

No. II.

Secretary's Office,
Toronto, 27th April, 1858.

Sir,—I have the honor to communicate to you the instructions promised in the last paragraph of my letter to you of the 14th instant, for your guidance in connection with the branch of the expedition to the west of Red River, which has been committed to your charge.

2. The instructions contained in that letter will suffice for your guidance up to the time of your arrival at the Red River settlement, and the present instructions therefore have reference merely to your operations after having left that settlement.

3. The region of country to which your explorations are to be then directed is that lying to the west of Lake Winipeg and Red River, and embraced (or nearly so) between the rivers Saskatchewan and Assiniboine, as far west as "South Branch House," on the former river, which latter place will be the most westerly point of your exploration.

4. It will be your endeavor to procure all the information in your power respecting the Geology, Natural History, Topography and Meteorology of the region above indicated.

5. As to the general character of the Geological portion of your labors, it is unnecessary to add anything to the instructions communicated to you last year, and which, so far as this point is concerned, will serve for your guidance for the present season.

6. There are, however, two matters to which I am to request you to direct your particular attention, namely, the Salt region in the neighborhood of Lake Manitobah adverted to in your report for last year, and the deposit of Tertiary Coal or Lignite, reported to exist in the valley of Mouse River.

7. It is most important that you should ascertain, by actual examination, as far as possible, the existence, extent and character of these deposits.

8. In ascending or descending the different rivers you may have occasion to explore, it is advisable that you should note with care, their breadth, depth, rate of current, and the probable quantity of water discharged by them at different points, and at different seasons of the year; their facilities for navigation by boats or steamers, and whether they overflow their banks to any extent at any season of the year.

9. The general aspect of the whole region should be carefully described. The character of the timber and soil observed, and the general fitness of the latter for agricultural purposes ascertained as far as may be from observation and inquiry.

10. It is desirable that your Meteorological observations should be made with the maximum and minimum thermometer, and with the wet and dry bulb. The temperature of the rivers, lakes and springs should also be recorded, and the rain-fall observed.

Any reliable information you can obtain as to the quantity of snow precipitated during the winter, would also be of interest.

11. Your topographical explorations should be made with reference to the construction of a map (as complete as possible) of the region explored, on a scale of two miles to one inch—and your operations should be conducted in view of a possible extension, at some future time, of the exploration, so as to embrace the entire valley of Lake Winipeg and its feeders.

12. With a view to illustrate the Natural History of the country, you will avail yourself of such opportunities as may present themselves to collect any objects that may be useful for that purpose.

13. Any Geological or Natural History specimens which you may have collected during your explorations, may be left by you at Red River, on your return, with the other property of the Government belonging to the expedition, to await the orders of the Government, with the other articles referred to in the tenth paragraph of my letter of the 14th inst.

14. I am to add that His Excellency, having every confidence in your judgment and discretion, does not wish to trammel you with more detailed instructions, and that you are left at liberty to make any other exploration, in addition to those particularly named therein, should you, upon information obtained in the locality, deem it desirable for the general purposes of the expedition.

15. It is hardly necessary to state that you will be held responsible for the conduct, diligence, and fidelity of the party under your charge.

16. With a view to distinguish your branch of the expedition for the present year, it will be convenient to designate it as the "Assiniboine and Saskatchewan Exploring Expedition;" by this title therefore you will describe it in your Reports.

I have the honor to be, sir,
Your obedient servant,
(Signed,) T. J. J. LORANGER,
Secretary.

Henry Y. Hind, Esq.,
Toronto.

REPORTS OF PROGRESS.

No. I.—LAKE SUPERIOR TO RED RIVER.

Red River Settlement,
3rd June, 1858.

Sir,—I have the honor to inform you of my arrival at the Red River Settlements yesterday afternoon, after a canoe voyage of twenty-three days from the west end of the Grand Portage, Lake Superior.

It affords me much pleasure to be able to state that no accident or difficulty of any description occurred during the voyage, and upon a careful review of our supplies, instruments, and personal baggage, the fracture of one small thermometer represents the only injury sustained.

The arrival of this expedition at Red River in advance of Sir George Simpson, has excited some surprise in the Settlements. The well known rapidity with which that distinguished traveller has for many years been accustomed to accomplish the voyage between Lake Superior and Red River, *viâ* the Kaministiquia, may render desirable a more detailed description of the old Northwest Company's route we followed, than would otherwise appear to be necessary.

We have all enjoyed excellent health, and were providentially assisted by very favourable weather, which, though at times stormy and cold, did not retard our progress for many hours at a time. On our arrival at Moose Lake, May 12th, (*vide* accompanying map) a glistening sheet of solid ice overspread its surface, and seemed to threaten a long delay; but by noon on the following day, under the influence of a hot sun and a gentle breeze, lanes of water opened, through which we succeeded in passing the canoes, and on the evening of the same day a high wind accompanied by rain completely broke up the ice in the higher Lakes, and opened the communication.

On the Winnipeg we encountered violent thunder storms, with hail and heavy rain, succeeded twenty-four hours afterwards by a boisterous snow storm; but happily the direction of the wind was generally in our favour, and aided our progress.

We remained one day at Fort Frances, with a view to repair the canoes, rest the men, and celebrate her Majesty's Birthday.

The Iroquois from Caughnawaga worked admirably. They were easily controlled, and fully maintained the excellent name they have acquired for hard-working, patient voyageurs.

When we started from the east end of the Grand Portage the baggage of the expedition weighed considerably over six thousand pounds, and the labour of carrying it, in addition to the canoes, over the Portages, was necessarily great, and occasioned severe sores on the shoulders of some of the men, which were submitted to with characteristic good nature. The storage of Mr. Dawson's supplies in Fort Frances seemed to be a great encouragement, and when relieved of this duty our progress was remarkably rapid.

We camped off the mouth of Red River seven days after leaving Fort Frances, and might easily have reached the Settlements on the first day of June, but in view of our rapid voyage from Rainy Lake I did not think it necessary to press the guide we therefore waited for a few hours at Fort Alexander, and enjoyed the very generous hospitality of Mr. Sinclair, the gentleman in charge.

The exact time the expedition spent in canoes between Lake Superior and Red River, after deducting the delays at the Forts before mentioned, was twenty-one days and six hours, as opposed to twenty-seven days and six hours by the Kaministiquia Route last year. The average daily progress was twenty-eight and a half miles against twenty-five miles in 1857.

The Grand Portage made to overcome the falls of Pigeon River, one hundred and twenty feet high, has been often cited as the chief obstruction to the Pigeon River Route. Its length is eight miles fifteen chains. The road is dry, and in comparison with some of the portages on the Kaministiquia route, in good condition. It is passable for an ox team, which is employed by the people in charge of the American trading post in forwarding their supplies.

I endeavoured to procure the ox cart and team to transfer the heavy baggage from the east to the west end of the Portage, but although the cart was available the team was not; one ox having died during the winter, and the other was in such a miserable condition that he could scarcely draw the cart itself.

The passage of the Grand Portage consequently occupied five days instead of two, and in making a comparison between the two canoe routes to Lake Winnipeg these facts must be borne in mind. In 1857 the Red River expedition landed at Fort William on the 31st of July, and reached the Settlements on the 4th of September, having been thirty-four days on the road, or forty from Toronto. This expedition reached Grand Portage on the 5th of May, and arrived at the Stone Fort 2nd June, a period of twenty-eight days, or thirty-four from Toronto. The Grand Portage lying within the Territory of the United States loses all interest as the terminus of a Canadian route. But that part of the water communication which forms the boundary line, and the country between Arrow Lake and Fort William, seems to acquire importance in proportion to the extension of our knowledge respecting its capabilities and resources.

The waters on the rivers and lakes on the east side of the height of land, the Lake Superior water-shed, were high, while those on the west side, or the tributaries to Lake Winnipeg, unprecedentedly low. In many of the Lakes recent water-marks, four and five feet above the present level, were frequently observed. This remarkable lowness of the water is attributed by the half-breeds and Indians to the very small quantity of snow which fell on the western slope during the last winter.

It is important to bear in mind that the voyage of this expedition to Red River was made under the great disadvantages inseparable from unusually low water, and whatever superiority the route appears to possess over that of the Kaministiquia by Fort William, will be much more apparent in ordinary seasons, when the lake and river levels are from two to five feet above their present altitude. The following brief sketch of the route is not intended to anticipate any results which may be furnished by Mr. Dawson's proposed exploration during the ensuing summer and autumn; but as it was made under peculiar circumstances, and in the spring of the year, it will serve to supply a blank which would be felt if a hot summer with little rain-fall should reduce the water levels much below their ordinary height in the autumnal months; an event which will not be deemed improbable when their present condition is known.

From Lake Superior to the Lake of the Woods our course lay on the boundary line between British America and the United States, as laid down upon the authorized lithographed map furnished by the Crown Land Department.

The accompanying chart is based upon that survey, and the observations of Mr. Dickinson assisted by Mr. Fleming are marked in red ink.

Chart No. 2 shews the Penawa River, down which the expedition voyaged, with a view to ascertain if it possessed any advantages over the old route by the Winnipeg. As the information obtained is not likely to possess any practical value in its bearing on the subject of a boat communication, I would wish it to be regarded merely as one of the small additions to our geographical knowledge of this country which opportunities occasionally enable us to make, and which it is desirable to secure, so long as they do not interfere with the general objects of the expedition.

The part of the Pigeon River Route to which this notice refers, commences at Arrow Lake, a fine expanse of water in connection with White Fish Lake, lying in a north-easterly direction, and within 30 miles of the Kaministiquia.

From Arrow Lake, a short portage brings us into Rose Lake on the course of the old North West Company's route, following the boundary line.

The portages between Rose Lake and the Height of Land are short and low, while the Height of Land portage is not 500 yards long, and does not rise above fifty feet. The passage from the St. Lawrence water-shed to that of Lake Winnipeg is short, easy and dry, incomparably superior to the Prairie portage, and the Great Savanne on the Kaministiquia Route. An inspection of the map will shew that in consequence of the very low state of the water this year, numerous small rapids were formed in the rivers connecting Gun Flint Lake with Lake Seiganagah. In ordinary seasons these rapids are passed without difficulty, but this year they involved the portage of a portion of the baggage and the letting of the canoes down them by rope.

From Lake Seiganagah* an Indian route passes into Little Seiganagah Lake, which connects with Sturgeon Lake on the route passed last year. The Little Seiganagah is a favourite wintering place of numerous families of Indians; it abounds with fish, and near its shores the winter road to Fort William runs.

Between Knife Lake and Birch Lake there are two routes, one coinciding with the boundary line, the other passing in a north-westerly direction by the dotted line shewn on the map, which we followed, making however two portages instead of one, but escaping some rapids.

From Nequaquon Lake one route passes into the Nameukan River, and another, turning south, follows the boundary line through Loon's Narrows and then north into Nameukan Lake. Our guide preferred going by Loon's Narrows, fearing that the always dangerous Nameukan Rapids would be almost impassable for heavily laden canoes, on account of the low stage of the water.

* Seiganagah or "Full of Islands."

In Loon's narrows we found a shallow river with a strong current and many boulders, and in making the north-westerly turn, instead of the broad channel shewn on the map, a very tortuous, sluggish and shallow stream, led us into the south arm of Sand Point Lake.

The banks of Loon's narrows showed that in ordinary seasons plenty of water is found in the river to admit of loaded canoes or boats without difficulty, and the delineation of this part of the route on the accompanying map, must be regarded as representing the narrow valley occupied by the river during periods of high water.

Sand Point Lake is connected with the Nameukan Lake by a broad channel, and it is at this point that the route through Loon's Narrows coincides with the more northern route and follows the boundary line through Rainy Lake to Fort Frances.

My own impression of the Pigeon River Route as compared with the one pursued last year is very favourable, but as you will be placed in possession of all particulars by the exploration of Mr. Dawson, I refrain from further notice of this valuable line of communication.

On my arrival at the Middle Settlement, where Mr. Dawson and his party reside, I found Mr. Russell in charge of the house and effects, Mr. Dawson with the other members of his party having started some days previously for the Saskatchewan, whence they are not expected to return until the end of June; I have therefore placed Mr. Russell in possession of the canoes and men intended for Mr. Dawson, and am now engaged in organizing a party to proceed immediately up the Assiniboine.

I beg to enclose Mr. Dickinson's remarks on the route, and maps marked,

No. 1. Pigeon River Route.
" 2. The Penawa.
" 3. A chart of the whole route, shewing the camping places, with corresponding dates.

I have the honour to be, Sir,
Your obedient servant,
(Signed,) HENRY Y. HIND.

The Hon. T. J. J. Loranger, M. P. P.
Provincial Secretary.

No. II.—Mr. DICKINSON'S REPORT ON THE PIGEON RIVER ROUTE.

Red River Settlement,
8th June, 1858.

Sir,—I beg leave to submit the following description of the Pigeon River route, compiled from the notes and observations taken by myself and Mr. Fleming, according to your instructions.

The accompanying map is a copy of part of the map made by David Thompson for the Boundary Commissioners; the notes in red ink being those taken by us.

The heights and distances were only estimated approximately, it being thought not necessary to make use of instruments for the purpose, as a complete exploration of the route is to be made hereafter.

Our observations more particularly commenced at Arrow Lake, as the head of this lake is the terminus of the proposed road to Point des Meurons, near Fort William, and in the case of its being made, (and it is most desirable that it should be if possible,) the route between Grand Portage Bay and Arrow Lake would not be made use of. However, a short description of it may not be thought unnecessary.

Grand Portage Bay, where formerly was the chief depot of the North West Company, affords a sufficiently safe harbor for small vessels, being very shallow, however, for some distance out from the shore.

At the head of the bay commences the Grand Portage, which is eight miles thirteen chains in length; without any difficulty and with very little expense it might be made suitable for waggons, but at present it is only a rough foot-path. As it and Grand Portage Bay are altogether within the United States territory, it is perhaps needless to propose any improvements that might be made in them. This portage is unavoidable, as Pigeon River for sixteen miles from its mouth is quite unnavigable, from the numerous falls and rapids in it.

From the end of this portage there is one and a half miles of still water to Partridge Portage, which is four hundred and forty-five yards in length. The path is on the American side of the boundary line, as it is also at many other places along this route. In these cases paths should be sought for on British territory, and which could be obtained, as well as we could observe, without much difficulty.

Above Partridge Portage the river is deep and wide, with a moderate current for three and a half miles; but from this for one mile to the semi-décharge the river is shallow and the current very strong; so much so, that canoes have to be poled up.

At this semi-décharge the path is on the British side, and is short but rough. When the water is high, no semi-décharge is required; but at the time we passed, the water here and in all the rivers and lakes was peculiarly low, the high-water mark appearing to be four feet above the present level.

The distance to the next semi-décharge is two miles, in which length there are no obstructions.

The second semi-décharge is about thirty chains long; in going down stream the portage need not be made; the path is on the American side.

Between this and Fowl Portage, a distance of three and a half miles, the river is quite navigable.

Fowl Portage is two thousand yards long, and is pretty level except at the west end, where it is very precipitous. The boundary line runs along the path, as it does also at some other portages, but the paths could be all easily made on British territory, due precautions having been taken that the boundary line be not obliterated.

We here enter on Fowl Lake, which is four and three quarter miles long; in the middle there is a narrow strait about ten chains wide and thirty chains long, part of it being rather shallow; the other parts of the lake are one mile wide on an average.

At the end of it is Moose Portage, seven hundred and twenty-one yards long; the path, which is the boundary line, could be easily improved or removed to one side.

Moose Lake is four and a half miles long, with an average width of half a mile; it is very deep, and is never frozen over till late in the season, and the ice is not broken up till long after that in the other lakes.

Great Cherry Portage is the next; it is eight hundred and forty-four yards long, leading to a small lake quarter of a mile long, at the end of which is Mud Portage, two hundred and sixty-five yards long; and between it and the lesser Cherry Portage there is another small lake fifteen chains long.

On these three portages the boundary line, as it appears from the map, runs on the paths, although the lakes are connected by creeks. The paths are tolerably good, but better could be easily made and solely on British territory.

We then come to the beautiful Mountain Lake, which is seven and three quarter miles long and three quarters of a mile broad, deep, and navigable for boats of any size.

Watap portage, five hundred and thirty-nine yards long, lies between it and Watap Lake; the path is the boundary line.

Watap Lake is a narrow strip of water five and three quarter miles long and about twelve chains wide, sufficiently deep throughout the entire length for any kind of craft.

The Great New Portage is two thousand five hundred and seventy-nine yards long; it is rather uneven, and is crossed by some small creeks; the boundary line is on it, but judging from the nature of the ground, a good path could be made on British land.

We now arrive at Rose Lake, which is separated from Arrow Lake by a narrow neck of land, across which a portage must be made.

Arrow Lake is sixteen and a half miles long, and has an average width of one mile; but as we did not visit it, its character cannot be described.

Rose Lake is three miles long, and averages three quarters of a mile across; it is deep, and well sheltered on all sides.

At the end there is a portage which is not shewn on the original map; it is only twenty yards long, and on the American side.

Mud Lake is two and a half miles long and a quarter mile wide, and from three to four feet deep, with a soft muddy bottom; the water having the peculiar property of retarding the canoe, similar to that of the Viscous Lake on the Kaministiquia route.

Between it and the next lake there is another portage which is not named or described on the original map; it is three hundred and eighty yards in length, and is the boundary line, the present path being tolerably good and level.

South Lake is the last on the east side of the Height of Land; it is two and three quarter miles across to the Height of Land Portage; the lake is about three quarters of a mile wide, and not more than four feet deep along the canoe route, the bottom consisting of very soft mud.

The Height of Land Portage is four hundred and sixty-eight yards long, and is one of the best on the route; a good road might be made without the slightest difficulty, there being plenty of pine and other good materials for the purpose close by.

We enter a lake now which is the head of the Winnipeg water-shed; having no name it may be considered part of Gun Flint Lake, with which it is connected by a strait two and a half miles long and varying from three to ten chains in width. The traverse across this lake is one and three quarter miles long. Near the middle of the strait there is a semi-décharge not noticed on the original map; it is but twenty yards long with about four feet fall; when the water is high the rapid could be run by canoes even when loaded.

Gun Flint Lake from the end of the narrow strait to Little Rock Portage is seven miles long, and has an average width of one mile; it is a fine open sheet of water of considerable depth. Before arriving at Little Rock Portage there is a rapid of two feet fall, down which the canoes were lowered by ropes; the rapid is caused by boulders of various sizes in the bed of the stream, but which might doubtless be removed.

Little Rock Portage is only thirty-three yards long; it is, as its name implies, over a rock, which is very steep on the west side.

From this to Mill Fall Portage is a mile; the river is about six chains wide; at the end there is a rapid with a fall of three feet, the channel being filled up very much with boulders, so much so, that the canoes were let down with great difficulty.

Mill Fall Portage of one hundred and ten yards in length, is over a very rugged rock on the American side.

The next portage is a quarter of a mile further on; it is five hundred and nine yards long, over an island; the path is very good and level except at the ends, where it is rather steep and the landings are bad, but could be easily improved, as indeed the landings at all the portages might be and without any considerable cost, as the materials for doing so can be obtained without difficulty.

This river or chain of lakelets is twelve miles long from Gun Flint Lake to Lake Seiganagah; for four miles below the last mentioned portage it is full of large boulders, which make the navigation of it difficult; there are in this length six rapids, varying from five feet to one foot fall, at four of which the canoes had to be carefully let down by ropes.

From thence to the semi-décharge of one hundred yards in length and five feet fall, which is one mile from the end, the navigation is good. At the mouth of this river there is a portage which is not shewn on the original map, neither are any of the rapids between this and the last portage.

This portage is thirty yards long, over a rocky point on the American side.

We now enter Lake Seiganagah, the route through which follows the boundary line or nearly so, and is nine miles in length. The greatest length of this lake is twelve miles, and the greatest width six miles. It is full of islands, from which it derives its name, affording good shelter to canoes, at the same time not impeding the navigation for large boats.

After passing through a short channel twelve chains wide we enter Swamp Lake, which is two and a quarter miles long and averaging thirty chains wide; in it there is a small portage twenty yards long; the channel being only about three feet wide and very shallow. The water in the western portion is higher by about one foot than that in the other; the waters of Lake Seiganagah must therefore find an exit elsewhere.

Swamp Portage is four hundred and twenty-three yards long, on which is the boundary line; the path is very good, except at the east end, where it is swampy, the landing there being exceedingly bad; however, here as elsewhere, there is no reason why a good one might not be made very easily.

Cypress Lake, the next we enter on, is a long narrow lake five and a quarter miles long by a quarter of a mile wide, and of sufficient depth. There is a portage at the end forty-seven yards in length; with little labour and expense this portage might be done away with, or at least made a semi-décharge; the present path, however, is very good, and is on the British side of the boundary line.

Knife Lake, the next on the route, is of a very irregular shape; the course follows the boundary line for eight miles, when it then diverges to the north. When the water is high the course may continue along the boundary line the whole way as it is shorter than the other; but when the water is low the narrow channel is full of rapids, and becomes unfit for the navigation of large canoes, and then the northern course on British territory, as shown by the red-dotted line on the map, which we took, is much preferable.

The first portage, three quarters of a mile from the boundary line, is rather bad; it is seven hundred yards long, and is very rough and hilly; but a better one might no doubt be made.

After passing through a lakelet three quarters of a mile long by a quarter of a mile wide, we come to the next portage, which is a short one, sixty yards long.

The upper portion of Birch Lake is then entered, and the course is continued in a south-westerly direction for four and a quarter miles till it reaches the boundary line, along which it afterwards goes.

Half a mile further on the Carp Portage is reached; it is two hundred and fifteen yards in length; the path is very good indeed, but forms the boundary line, as it appears from the map, although there is a channel close by which would have been the more natural one.

The course through the other portion of Birch Lake is four miles long, along which the water is deep enough for any kind of boats.

At the entrance to Basswood Lake there is a portage one hundred and ninety-six yards long on the British side of the boundary line; the path requires but very little labour to make it quite good.

Basswood Lake, perfectly navigable for small steamers, is a large lake of most irregular form, and containing many islands. The usual course through it lies along the boundary line, and is seventeen and a half miles long. At the end there are two rapids of considerable fall, a quarter of a mile asunder, which are avoided by portages; the first one, of one hundred and ninety yards in length on the American side, is pretty good; as far as could be seen there appears no reason why the portage might not be made on the opposite side. The next portage, the Fir, is three hundred and fifty yards long; the path is the boundary line; it is over very rocky ground, and rises considerably in the middle. After passing this portage there are in the first mile two rapids not shewn on the original map; one of three feet fall, the other of two feet, but which are easily run.

The channel is not more than about ten chains wide, and continues of this width for seven and a half miles.

Two miles below the last rapid is a portage one hundred and sixty-six yards long over a high rocky point on the American side; there was no apparent objection why a portage path might not be constructed on the other side in British territory.

Three miles further down the channel Crooked Lake, which fully deserves its name, is fairly entered on. The course follows the boundary line through it, and by its windings is fourteen miles in length, though the absolute length of the lake is but ten. The navigation is somewhat intricate, at the same time quite sufficiently good for even boats of large dimensions.

Curtain Fall Portage is one hundred and eighty-three yards long; the path is rather bad, being carried over a hill, and is very rough indeed; it is on the American side. Just below it there is a rapid of about three feet fall, which is run by canoes without difficulty. Iron Lake, the next in succession, is a small lake full of Islands; the usual canoe route through it, which is also the boundary line, being four and a half miles long; the water was of good depth the entire way; at the east end, where it becomes like a river, there is a strong current, but which canoes or boats when ascending can easily overcome.

Bottle Portage, which is four hundred and forty-eight yards long, might be easily made one of the best on the route, the ground is very level except at the ends, which are rather steep, but by a little management the path could be made with the proper inclination.

There is a long stretch twenty-two miles in length now of navigable water through Nequawquon Lake, the course following generally the boundary line, except about the centre of the lake, where it keeps to the south of the large island.

At the south-west end there is a portage two hundred and seventeen yards long on the American side; very little is required to make a good path on British territory.

After passing through a small lake four miles long by the course, a narrow channel called Loon's Narrows, is entered; one and three quarter miles from the commencement there is a portage of two hundred and sixty-three yards on the American side; it appeared as if a shorter one and on as level ground could be made on the British side.

Half a mile from this, there is another portage sixty-seven yards long, also on the American territory, which might be transferred, as the other to the north of the boundary line. Below these portages the current is very strong, and at the bend the river is very shallow, and the bed covered with small boulders, which however could be easily removed and the channel deepened.

As the water was very low at the time we passed along, it was confined to a channel from two to four chains in width for a distance of 6 miles meandering through a valley which in times of high water is covered as shown on the map.

Sand Point Lake may be said to commence here; it is nine and a half miles long, and down the centre is the course and boundary line; it is free from any kind of obstruction to good navigation; it is connected with Nameukan Lake, by a strait sixteen chains wide.

From this through Nameukan Lake there are two courses to Rainy Lake, the one following the boundary line by the Kettle Falls Portage of 127 yards in length, the other by the eastern channel, by which the Expedition went last year and which we took this; on it there are two very short portages, and the course is much shorter than the former.

I have refrained from offering the necessary suggestions for the improvement of this route, as I understand Mr. Dawson is to make a complete exploration and survey of it this year, and who will therefore be better able to form an opinion as to its capabilities and required improvements.

However, from even the cursory examination I was enabled to make, it appears greatly superior to the Kaministiquia route.

It is sixty-three miles shorter than the other. There are fewer portages, all much shorter with the exception of the Grand Portage, and none of them are nearly so bad as the Savanne, Prairie or Great Dog Portages. There are very much fewer rapids, and which are all more easily run. Excepting Pigeon River, it consists of a chain of lakes the whole way connected by short channels, in few of which only the current is at all strong.

I think that with a comparatively small outlay, the route could be made navigable for large row boats, and that on many parts of it, small tug-steamers could be advantageously employed.

I remain, yours truly,

(Signed,) JAMES A. DICKINSON.

Henry Y. Hind, Esq.,

&c. &c. &c.

No. III.—FORT GARRY TO FORT ELLICE, *viâ* THE LITTLE SOURIS.

Fort Ellice, Rupert's Land.
9th July, 1859.

Sir,—In the letter I had the honor to address to you from Red River on the 3rd June last, I stated that after making the necessary preparation, I should immediately commence the Exploration of the valley of the Assiniboine River. The distrust, and even dread with which the Sioux Indians are regarded by the Red River hunters, made it necessary to secure the services of a strong party for the Exploration of the Little Souris or Mouse River, where Tertiary coal was reported to exist. In consequence, however, of the failure of last year's autumn buffalo hunt, and the ravages of the grasshoppers at Prairie Portage, and elsewhere in the Settlements, most of the able-bodied men fitted for the exigences of a journey into the Indian Country, had left the settlements a few days before my arrival, either for the Buffalo Plains or for St. Paul; and it was with some difficulty that I could procure eight men and the necessary provisions for a three months journey, but by the 14th of June, the Expedition was *en route* for the interior.

After arriving at St. James' Church, on the Assiniboine, I proceeded with Mr. Dickinson to ascertain the position of the Big Ridge bounding the valley of the Assiniboine, and followed its windings for a distance of seventy or eighty miles, until it is cut by Portage River near Lake Manitobah opposite Prairie Portage. Mr. Fleming proceeded with the carts and canoes by the hunters' road to Prairie Portage, making on his way a section of the Assiniboine River, and ascertaining by numerous trials its rate of current, volume of water, &c.

The Assiniboine valley south of the Big Ridge, on the north side of the River, comprising an area exceeding half a million acres, was described in my Report of last year, as possessing a soil of remarkable excellence; the results of a more particular examination during the present season fully bear out the favorable opinion previously formed.

After reaching Prairie Portage we proceeded on the north bank of the Assiniboine as far as the mouth of the Little Souris River. During this part of our journey we occasionally stopped for half a day to make the necessary Astronomical observations, to measure the valley of the River, and make sections of its banks.

The impressions with which I returned to Toronto last year, regarding the extent of forest on the banks of this River, confirmed as they appeared to be by all descriptive accounts I received from residents at Red River, led me to suppose that the Assiniboine flowed for about eighty miles from its mouth through a vast level Prairie timbered only at the points or bends of its course. I was much astonished to find, that this is true only as regards the *north* bank of the River, the south bank being occupied by forest, which commences some thirty miles from Fort Garry, and covers the country westward for a distance exceeding sixty miles, with a depth varying from three to twenty-five miles. We frequently saw this vast forest from hills on the north side of the River covering a tract of country which could not be less than twelve or fifteen miles in breadth, and with a good telescope, the Prairie between it and an extension of Pembina Mountain or Ridge, was traced. I have ascertained that the forest contains some fine timber, and is well known to Indians who hunt there during the winter, but the trails of the buffalo hunters avoid it and keep to the open Prairies; hence its existence even is unknown to many of the residents at Red River, and the buffalo hunters, always shunning it, have but little knowledge of its timber resources.

It is my intention on returning to the settlements to penetrate through this forest in two or three directions, with a view to ascertain its character, as far as time will allow.

It is needless to dwell upon the great importance of so abundant and unexpected supply of serviceable timber, within one or two days' journey of a very extensive and fertile, arable country, and on the banks of a navigable river, within a day's march of Fort Garry.

The country on the north side of the Assiniboine between Prairie Portage, and the mouth of the Little Souris for a distance of several miles back from the river, is poor and scantily timbered. The prairies on the Little Souris are also light, and the deep valley of that river contains but little timber. At Snake Creek, numerous specimens of drift Lignite were found, and after a few hours' exploration, favorable indications led me to have a section of the river's bank exposed, by making a cutting at right angles to it, with a view to shew the stratification. Here, no less than four distinct beaches of a former lake were brought to light, each beach bearing numerous rounded and polished boulders and pebbles of drift Lignite, varying from two to fifteen inches in diameter, but no trace of the Lignite in place was seen on the Little Souris north of the 49th parallel. The beaches just referred to, were several times noticed, further up the river; they are accompanied by a bed of ferruginous sand, above which, several extensive deposits of Bog Iron Ore, and Shell Marl were found.

Having determined, if time will permit, to return to the settlements *viâ* the Assiniboine in canoe, I forbear for the present from referring to the geology of its rock exposures, further than to state, that what I have already seen leads me to think it will repay an attentive and careful exploration.

Having reached the 49th parallel, the expedition proceeded up the banks of Red Deer's Head River for about fifteen miles, and then crossed over a treeless prairie, sixty miles broad, towards Fort Ellice.

The hill sides in the valley of the Little Souris River, were scored with tracks of buffalo, and every where we saw the bois de vache of last year, but it was not until arriving at the Two Creeks, in the Assiniboine valley, that we killed a buffalo bull. The buffalo this year, are far south, and the hunters have suffered much distress on that account. Yesterday we saw three bulls at a considerable distance from us, they are considered to be the pioneers of numerous herds which are anxiously looked for by the people of the Fort, who are almost altogether destitute of provisions.

Every where we find grasshoppers. On the Assiniboine, the brood of this spring is yet unable to fly, but when traversing the treeless prairie between Red Deer's Head River and the Assiniboine, innumerable hosts of grasshoppers were flying northward in the direction of the wind. At times they would cast a shadow over the prairie, and for several hours one day, the sky from the horizon, to an altitude of thirty degrees, acquired an indescribably brilliant ash-white tint, and seemed faintly luminous as the semi-transparent wings of countless millions of grasshoppers towards the north and north east, reflected the light of the sun.

On Monday, July 12th, I propose to start for the Saskatchewan by the Qu'Appelle or Calling River, returning to the settlements by the end of August.

The weather on the whole has been very favorable, but in the early part of our journey, thunderstorms, for many days in succession, caused three or four hours delay during their continuance. We have had seventeen thunderstorms in twenty-three days; nearly all were of a violent character, with hail, heavy rain and boisterous winds.

We did not see any Indians before our arrival at Fort Ellice. On the Red Deer's Head river, an attempt was made in the night to stampede the horses, which was fortunately frustrated by the distant neighing of a horse reaching our ears, and giving us time to take precautionary measures, but the tracks of hostile Indians, close to our camp, were found in the morning.

This letter is written in the expectation that some hunters may soon be returning, *via* Fort Ellice to Red River for supplies, who will be instructed by Mr. McKay, the gentleman in charge at Fort Ellice, to place it in the post-office at Fort Garry.

I have the honor to be, Sir,
Your obedient servant,
(Signed,) HENRY Y. HIND.

The Hon. T. J. J. Loranger, M.P.P.,
Provincial Secretary,
Toronto, C. W.

No. IV.—FORT ELLICE TO THE SOUTH BRANCH OF THE SASKATCHEWAN, THENCE TO FORT A LA CORNE AND RED RIVER.

Red River, 10th September, 1858.

Sir,—On the 18th of July, or nine days after the date of the Report which I had the honor to address to you from Fort Ellice, we arrived at the Qu'Appelle Mission, recently established on one of the Lakes which distinguish that part of the Qu'Appelle or Calling River valley.

From the 19th of June to the 18th of July, it was found necessary or advantageous to preserve the party composing this expedition united, but having arrived in the Cree country, to the north of the prairies generally occupied by bands of Sioux and Assiniboine Indians, I found it desirable to form three divisions, with a view to traverse and examine the country hereafter described.

The Mission of the Qu'Appelle Lakes is situated about halfway between Fort Ellice and the South Branch of the Saskatchewan. From this point Mr. Dickinson, with two men, proceeded in a small canoe down the Qu'Appelle River, to its junction with the Assiniboine, thence on horseback to Fort Pelly, where he met Mr. Hime with four men, who after having examined Long Lake, some 50 miles west of the Qu'Appelle Mission, travelled across the country to Fort Pelly, with Mr. Dickinson's carts and supplies.

The third division of the party, comprising myself, Mr. Fleming, and two men, sailed or tracked up the Qu'Appelle Lakes and River to the Grand Forks, a distance of 50 miles, where three men, with our supplies, met us at the appointed time; we then followed the valley of the Qu'Appelle River to its source, and passed on through a continuation of the same valley, to the South Branch of the Saskatchewan by the "River that turns," flowing westerly.

We struck the South Branch at the Elbow, and launched our three fathom canoe on that magnificent river, down which Mr. Fleming and I drifted for 250 miles, until we came to the junction of the North and South Branches of the Saskatchewan.

The supplies, with four men and a Cree guide were sent across the country to Fort à la Corne, opposite the Nepoween

Mission, about 18 miles below the Forks. Two days were occupied in examining part of the Coal Falls on the North Branch, above the Forks; after which we joined the carts on the ninth of August at Fort à la Corne. Here I made another division, sending Mr. Fleming with two men in a canoe to Cumberland—thence to proceed down the Saskatchewan, and by the west coast of Lake Winnipeg to Red River. Taking the carts and four men, I followed the course of Long Creek against the current, running parallel to the South Branch, for a distance of 50 miles,—then turning in a south-easterly direction, travelled across the country to the Touchwood Hills and thence to Fort Ellice, where, after an absence of forty-three days, I met Mr. Dickinson and his party, within three miles of our appointed rendez-vous.

After Mr. Dickinson's arrival at Fort Pelly, he proceeded with Mr. Hime to examine the flanks of the Dauphin Mountain, from Swan River to Rapid River or the Little Saskatchewan, a tract of country comprehending the greater portion of the north eastern water-shed of the Assiniboine. After our union at Fort Ellice we proceeded to Red River *viâ* the White Mud River, which flows into Lake Manitobah, and arrived at the Settlements on the 4th of September, nearly three months from the date of our departure.

Mr. Fleming has not yet returned, and I am now preparing to go in a canoe with a supply of provisions to meet him, in case the southern wind should prevent him from advancing.

The importance of ascertaining the true character of the Qu'Appelle Valley became more evident as we proceeded westward and met with Indians and a few half-breeds, whose accounts and descriptions seemed to agree in the general statement that a great valley, a mile or a mile and a half broad, and from one hundred to three hundred feet deep, did exist, running in a course nearly due east and west, between the South Branch of the Saskatchewan and the Assiniboine.

The Qu'Appelle River rises within sixteen miles of the Saskatchewan, as shewn on the accompanying map. Its course is first northerly for several miles, through a narrow gully which widens into a deep valley before it reaches the Qu'Appelle valley proper. About four miles west of the Qu'Appelle, and running in a direction nearly parallel to it, a river called by the Crees of the Sandy Hills "The River that turns," flows into the same great valley, and pursues for twelve miles a westerly course when it falls into the South Branch at the Elbow; this is evidently the Heart River of Thompson's Map. By the united action of these Rivers, and other agents to be described in full in my general Report, a great valley stretching from the Saskatchewan to the Assiniboine has been excavated. This valley has a greatest breadth of about one and a half, and a least breadth of about half a mile at the Sandy Hills; its greatest depth below the Prairie is between three hundred and four hundred feet, its least depth one hundred and thirty feet. Between the Qu'Appelle River and the "River that turns," there is a space of about four miles occupied by ponds in the valley, which unite into a shallow lake in the spring and send their waters at the same time to the Assiniboine and the Saskatchewan. With a view to determine the height of the Qu'Appelle, where it enters the great valley, above the South Branch, we levelled from one River to the other, and found a difference in sixteen miles of eighty-six feet. The Qu'Appelle is here about ten feet broad and one and a half deep. The "River that turns," nearly of the same dimensions, and the South Branch of the Saskatchewan about half a mile broad with a channel ten feet deep. These altitudes and distances are given in round numbers, but they will be accurately expressed in accordance with repeated measurements in my general Report. In order that the waters of the Saskatchewan might flow down the Qu'Appelle Valley into the Assiniboine, a rise of eighty-six feet in twelve miles would have to be overcome, and I am persuaded from indubitable evidence that this has not occurred during modern times. During very wet seasons, in the early spring months, the whole valley of the Qu'Appelle from within fourteen miles of the South Branch of the Saskatchewan, is converted into a narrow, shallow Lake, all the way to the Assiniboine, a distance exceeding two hundred and fifty miles, with a current of perhaps one mile per hour; and from the "River that turns," to the South Branch, a distance of twelve miles, an impetuous torrent occupies the valley, leaving along its course many indications of its violence and force. In the spring of 1852, ever remarkable in this country for its extreme humidity, a canoe might have passed from the Saskatchewan to the Assiniboine by rising eighty feet in twelve miles; then descending about two hundred feet, in a distance of perhaps two hundred and fifty miles to the Assiniboine. The Qu'Appelle Lakes east of the mission are briefly described in the accompanying Report by Mr. Dickinson. The Lakes west of the mission are four in number; the depth of three of them is about fifty feet, the last or Salt Lake near the height of land is very shallow, and does not contain in the summer months drinkable water.

From the first Fork, vide accompanying map, another great valley similar in all respects to that of the Qu'Appelle River, stretches in a north-westerly direction, and for forty or fifty miles is occupied by water, forming a long, narrow lake, varying from three-quarters of a mile to two miles in breadth; this is called by the Crees, the Long Lake, also the Last Mountain Lake; it is connected with the Saskatchewan by a broad excavated channel, similar to that occupied by the "River that turns." Long Lake abounds in fish, but there is very little timber to be found on its steep cliff-like banks.

The South Branch of the Saskatchewan is a noble river, varying in width from half a mile to three hundred yards, for a distance of 100 miles from the Elbow; it then gradually contracts its channel and changes its character from a river full of sand-bars and mud-flats, pursuing a comparatively straight course, to a rapid and uniform torrent of water, sweeping down the narrow but deep valley it has excavated, from one bank to the other in magnificent curves, until it joins the North Branch.

The country on the south side of the South Branch as far as the Moose Woods is a light prairie; there is very little timber to be seen, and all of small dimensions; the same may be said of the Qu'Appelle valley; open prairie on either side, or prairies covered with clumps of aspen. In the numerous gullies which give variety to the steep banks of both the Qu'Appelle and Saskatchewan valleys small timber is invariably found. The main Saskatchewan is a river of very imposing magnitude. Like the South Branch it occupies a narrow, deep valley, varying in width from one and a half to three miles, extending a few miles below the Nepoween Mission. It flows in grand curves from side to side, and its general level is about 300 feet below the country through which it has excavated its channel, after which it enters the low region.

We have made many sections of the South Branch, Main Saskatchewan and Qu'Appelle, &c., and numerous trigonometrical measurements of their valleys, and noticed continually the rate of currents, volume of water, character of banks, &c. &c., all of which will be embodied in the general Report. In the large expanse of country over which our explorations have extended, the area of land of the first quality, namely, of black vegetable mould reposing on gravel or clay is far more extensive and important than we anticipated. It is distributed as follows:—

1. On the South Branch of the Saskatchewan ; from the Moose Woods to the Nepoween Mission, and according to the description of half-breeds familiar with the country, a soil of equal excellence extends to the valley of Swan River. The immediate banks of the Saskatchewan are of a poor, sandy or gravelly soil, but on the prairie plateau three miles from the river, the rich soil commences, and in the part over which I passed, has a breadth of sixty miles. 2. The Touchwood Hill range, having an area exceeding 1,000,000 acres; for beauty of scenery, richness of soil, and adaptation for settlement, this is by far the most attractive area west of the Assiniboine. 3. The soil is of first quality in the valley of Swan River, and over the whole of the east watershed of the Assiniboine, with the exception of the country near its banks. 4. The valley of White Mud River is generally fertile and inviting. Until the maps which will accompany the general Report are prepared, it is impossible to give an approximate calculation of the area of available arable land, but I may here say, that the ratio which land of excellent quality bears to land of indifferent or worthless quality in the regions just referred to is largely in favour of the former.

The Riding Mountain as described in Mr. Dickinson's report is timbered with large aspen. On the level country drained by the Saskatchewan, from the Moose Woods to the Nepoween Mission the timber is small, but on the Touchwood Hill range there are some fine aspen forests.

I have succeeded in finding numerous rock exposures on the Qu'Appelle and South Branch of the Saskatchewan, which will enable me to produce a geological map of a large portion of the country briefly described.

I start immediately to meet Mr. Fleming, and then propose to visit the east flank of Dauphin or Riding Mountain, and the Salt Springs on Dauphin River and Lake.

Mr. Dickinson will examine the country south of the Assiniboine with a view to ascertain the extent and character of the forest to which allusion was made in my report from Fort Ellice.

I have the honor to be, sir,
Your obedient servant,
(Signed,) H. Y. HIND.

Hon. T. J. J. Loranger, M.P.P.
Provincial Secretary,
Toronto.

No. V.—MR. DICKINSON'S REPORT ON THE QU'APPELLE VALLEY EAST OF THE MISSION.—FORT PELLY TO THE RAPID RIVER.

Red River,
6th September, 1858.

Sir,—The following Report contains a short description of those parts of the country which I have examined according to your letter of instructions, dated Fort Ellice, July 12th, 1858, together with a brief notice of some of my operations from July 20th, the day we parted at the Church of England Mission, Qu'Appelle Lake, till we met at Fort Ellice on August 23rd. After our separation at the head of the river issuing from the lake at the Mission, I took a section of the bed of the river and ascertained the rate of the current, and then proceeded down it to the next lake, which is the second of those called the Fishing Lakes, as fish are much more abundant in these than in those lakes further down the Qu'Appelle Valley.

The character of this portion of the river which connects these two lakes together, being exactly similar to that of all other parts of it, one general description will suffice, together with special descriptions of a few places where there are differences.

The river varies in width from one to one and a half chains, and in depth from two to five feet; the average rate of current taken from several trials being one and a quarter mile per hour. The river is most wonderfully tortuous throughout its entire length; for ever being deflected from one side of the valley to the other, so that it is much more than double the length of the valley; several, indeed most of the bends, are so very sharp that it was with much difficulty the small canoe, only two and a half fathoms long, could be steered safely round them and prevented from running in on the banks, the current at some of them being two miles per hour.

The second of the "Fishing Lakes," the one which I first came to, is about three and a half miles long and three quarters of a mile broad; it is more than seven fathoms deep everywhere I tried it, even within a few yards of the shore. The river flowing from this to the next lake is but half a mile long. The name of the lake in Cree is, "Pa-ki-tah-wi-win," in English the "Fishing Lake," called so *par excellence* from the great quantities of fish it contains at some periods of the year. It is about six miles long and three quarters of a mile wide, which is about the average width of the valley. I tried the depth of it in several places along the course I took, which was down the middle of it, and found it to vary from five to eleven fathoms.

Having made a section of the river and ascertained the rate of current, I proceeded down it to the next lake called the "Crooked Lake," or in Cree "Ka-wa-wa-ka-mac," where I arrived in the forenoon of the 23rd. The general character of this portion of the river is the same as I have given before, but at some places here and there it varies from it. In two places, each about a quarter of a mile long, the river is full of sand and gravel bars, the depth of water over them being only about nine inches. In another place the current exceeds three miles an hour, to ascend which would indeed be a tedious and difficult task. Half way between these two lakes I took measurements for calculating trigonometrically the width and depth of the valley. The results of these and other measurements and observations will be shewn on the maps. In round numbers I may say, however, that the valley appears to be from two hundred and fifty to three hundred and fifty feet deep, and from half a mile to one in width. The average height of the immediate banks of the river over the present level of water was about six feet, the high water mark being eight feet over the same level. The greater portion of the valley is therefore always liable to be flooded, which I believe is the case every spring.

The middle of the valley between the bends of the river is mostly covered with willows, with here and there a few young sugar maple. The south slope of the valley is thickly covered throughout with small aspens, the balsam poplar growing well also in some places, while the north slope is quite bare of trees, which I found to be caused by the fires which almost every year sweep along this side of the valley, for I saw in several places the remains of burnt trees, and in the hollows and deep recesses of the slope the young oak shoots springing up from the half burnt roots.

On this side for the whole way there is a track along which the Indians travel constantly during the year, which accounts for the numerous fires.

"Crooked Lake," the most beautiful of the Qu'Appelle Lakes which I have seen, is upwards of eight miles in length, and from half a mile to one mile in width. There are several long points running out from the shore, on which grow oak, elm, ash, and poplar; none of them very large however, but which would be useful for various purposes. There was no place where I sounded less than four fathoms deep. The water in this lake, as well as in the others, was at this time rendered very disagreeable by the great quantity of confervæ, covering nearly the whole surface and to some depth, now decaying and rotting under the hot sun.

At the commencement of the next portion of the river flowing out of this lake there is a very rapid current, or rather a series of small rapids, for two miles and a half, and the river is if possible more winding than ever, and is at some places only forty feet wide. The rest of it, cross-sections of which I took at different points, as far as the next lake, resemble in its character the general description of the river. In the evening of July 24th I reached the lake called "Round Lake," the Indian name of which is "Ka-wah-wi-ya-ka-mac;" it is the last of the chain of lakes in descending the river.

It is four and a half miles in length, and is about one mile broad in the widest part. Owing to a long point of land running out from the south side of the valley, about one mile and a half from the head of the lake, part of it looks nearly round, from which it derives its name. It is, in all places where I sounded it, more than four fathoms deep, except at the mouth of the river and one hundred yards from it where it was only two feet. The south slope of the valley is here as densely covered as before with young poplar, and with patches of young oak, elm and ash, and the north slope is burnt as usual by the devastating fires. Two miles down the river from the lake, the bed is thickly strewed with boulders for about one hundred yards, where the current is very strong, making the navigation, even for a small canoe, rather intricate. The Indians call this place the "Stony Barrier," or, as it is in the Cree language: A-si-ne-pi-che-pee-ya-kan.

Between this point of the Qu'Appelle River and its confluence with the Assiniboine, there were two places, one on each side of the valley, where the slopes were exposed; on examining them, I found shale in position, but very much decomposed. These places will be marked on the map hereafter. After a long search I found but one fossil shell, which I enclose to you, together with specimens of the rock. At many places I ascended the sides of the valley to see the country on both sides, and found it to be generally level prairie, of light sandy loam, with scattered clumps of willows and small poplars. Several small creeks, the principal of which are the Big and Little-cut-arms and the Scissors creek, flowing in from both sides, gradually increase the depth of the river, but not its width, six feet being now the average depth. The river, twisting and turning about in every direction, is continually cutting out new channels, forming sometimes a most intricate maze. As it approaches the Assiniboine, the Qu'Appelle valley gets wider, and the slopes flatter, on which grow more and better timber, on the south side particularly; it consists of elm, ash, aspen, balsam, poplar and maple, all mingled together, with an underwood of willows, dogwood, hazel and roses. I arrived at the mouth of the river (a section of which I took,) at 6, a. m., July 27th. Having left one man in charge of the baggage at the landing place, I hastened to Fort Ellice with the other, and sent him back with a cart which Mr. McKay kindly lent me to fetch it. The next day I was delayed several hours trying to procure a guide who knew the track on the west side of the river from this to Fort Pelly, and in consequence was not able to start till late in the afternoon. Mr. McKay kindly sent men to assist me in crossing the Qu'Appelle River, which was accomplished without any loss, and with but one accident,—my horse receiving rather a bad cut when getting up the bank of the river, which was very soft and covered with broken trees. We camped for the night on the north side of the valley; this side is composed of fine loose sand intermixed with small boulders. From this to the Wolverine Creek, a distance of about 15 miles, the land is light sandy clay, in many places pure sand, covered principally with a low growing creeper, bearing berries like the juniper; the grass is very short and scanty, and the aspens, which are the only trees, are very small. Further on, the country improves very much as to its soil and vegetation, but it abounds with marshes, swamps and ponds of various sizes, round which grow willow and young aspens, and this is for about 60 miles.

From thence to Fort Pelly the country is densely covered with aspens from 5 to 15 feet high, and willows of different kinds; there are open spaces to be seen now and then, where the wonderful luxuriance of the vegetation is beyond description. Lakes and ponds are very numerous throughout, encircled with large aspens and balsam poplars.

There are several rivers and creeks flowing into the Assiniboine, into which many of these marshes and swamps might be easily drained. White Mud River, which is the largest of them, is 70 feet wide, 4 feet deep, and very rapid, so rapid that it was with much difficulty we forded it.

I arrived at Fort Pelly on August 1st, where I found Mr. Hime and the others of my party. Next day I took observations for latitude and variation of compass, and in the afternoon, accompanied by Mr. Macdonald, who was in temporary charge of the Fort, inspected the farm which the company have here. The crops had been beautiful at the beginning of the season, but have been all, excepting the potato, completely devoured by the grasshoppers. The next day I rode to Swan River, by the valley of Snake Creek, with Mr. Macdonald and Mr. Hime. This beautiful valley contains all the requirements necessary for a settlement. The timber is very plentiful and of a good size; there is no pine, however, but the balsam spruce, which the people here mistook for it, is abundant, and averages 2 feet in diameter at 5 feet from the ground. There is some tamarack also, tall and straight, from 1 ft. 6 in. to 2 feet in diameter. The balsam and aspen poplar grow to a large size, and are everywhere to be had. The land, for the most part, is good sandy loam, and is traversed by numerous creeks.

Snake Creek is about 13 feet wide, and 1 ft. 6 in. deep, it yields plenty of fish, as also do one or two small creeks running into it. Swan River is from 90 to 100 feet wide and 14 feet deep, its current is very rapid, being about three miles an hour; it is very winding where the Snake Creek joins it, and I believe is so all along. The valley, which is from 80 to 100 feet below the general level of the country, is most rich and fertile, but almost altogether filled up with trees, such as poplar, balsam spruce and willows. The next day, August 4th, we left Fort

Pelly, and proceeded along the base of the Duck Mountain, a part of the chain of mountains called the Dauphin; properly speaking it is a high ridge between the Assiniboine River and Lake Manitoba. The ground rises gradually from the river towards the summit of the so-called mountain, which appeared about three miles distant, and is thickly covered with poplar, so thick that the forest is nearly impenetrable.

The land for a few miles is rather light, but it then becomes much better, and for the whole way to the Little Saskatchewan or Oak river, (the eastern limit according to your letter of instructions to this line of exploration,) the land may be said to be good sandy loam.

In a short report as this must necessarily be, I cannot give descriptions of the different portions into which this side of the valley of the Assiniboine may be divided, but taking it as a whole, I may say, that in fertility of soil, timber, and water power, it surpasses all other parts of the country I have seen. I made several attempts to reach the summit of the Mountain, particularly that part called the Riding Mountain, but was baffled each time by the extraordinary thickness of the wood of young poplars, among which there were lying the half-burnt remains of older trees concealed by the long grass, vetches, convolvuli and innumerable other plants.

I cannot pass by, however, the valley of the Little Saskatchewan without making a special note of it. We reached it on 11th August, and the next day I was able, fortunately, to take observations for latitude, &c., for early in the afternoon the sky became cloudy and a thunder storm came on; next morning accompanied by Mr. Hime, who has been giving me great assistance in making the survey, I rode on horseback up the valley, we could only go, however, fifteen miles, as the trees and underwood became then so marvellously dense as to make it quite impassable for horses.

The valley is about eighty feet below the general level of the country, the bottom of it is from half a mile to one mile wide, through which the river winds its way flowing rapidly and uniformly; it is about forty feet wide, and at this time was five feet deep. There is no appearance of the valley ever being flooded, the willows which grow along its banks being green and luxuriant down to the ground.

There are large open flats occurring frequently on both sides of
the river, w... here the richness of the grass and beauty of the
various flowers p... ...rove the great fertility of the soil, places marked
out by nature to be ... cultivated and inhabited by man; there is
abundance of good sized ... poplar and balsam spruce, sufficiently
large for building and farming ... purposes.

I followed the ... ...lley down to its junction with
the valley of the ... course of the va... ...or the greater part of the
way it is rich and fertile ... Assiniboine, and ... ...d adjoining. Within a
few miles of the Assiniboine ... as is also the lan... ...anges considerably,
the soil is much lighter and the country ch... ...' smaller, and at
the junction of the vallies ... trees fewer and ... ...deed, being
sandy and gravelly clay ... the country is very poor in... ...lders of
various sizes. ... abounding with granite bou...

I returned then by the same way to the track called "The Lower Road" from Red River to Fort Ellice, to where it crosses the Little Saskatchewan, and where I had left the greater number of my party.

From thence I proceeded by this track to Fort Ellice, stopping one day at Shoal Lake in order to make a survey of it; as this track joins the White Mud Road about eighteen miles from the Little Saskatchewan, which we travelled back together from Fort Ellice to Red River, I need not give you any description of the country through which it passes.

Yours truly,

(Signed,) JAMES A. DICKINSON.

H. Y. Hind, Esq.,
&c. &c. &c.

No. VI.—RED RIVER SETTLEMENTS TO THE SALT REGION ON WINNIPEGO-SIS LAKE, THENCE TO THE SUMMIT OF THE RIDING MOUNTAIN —THENCE TO THE SETTLEMENT.

Red River Settlement,
8th November, 1858.

Sir,—I have the honour to report the result of an exploration of the Salt Region on Winnepego-sis Lake, and of the country traversed since the 18th September (the day of my departure from Red River), to October 31st. Accompanied by Mr. Fleming, I skirted the West Coast of Lake Winnipeg, in a Red River freighter's boat, with a crew of seven men, as far as the mouth of the Little Saskatchewan River. Our progress through the southern half of Lake Winnipeg was delayed by contrary winds, which, however, afforded me time and opportunity to collect numerous specimens in illustration of the rocks exposed on the islands and coast, and to accumulate materials for a geological map of the country.

Numerous rock exposures, showing sandstones, limestones, and shale of Silurian age, are met with some sixty miles north of the mouth of Red River. On some of the islands, the exposures are, geologically, of great interest; but, with the exception of sandstone fit for building purposes or the manufacture of grindstones, and of yellow ochre of fine quality, in a silicious limestone rock, no economic materials of particular interest or value were seen.

The west coast of Lake Winnipeg, after passing Grindstone Point, is very deeply indented with bays, whose extremities cannot always be seen from the traverse between the points at their outlets. Frequent soundings showed sixty feet to be the greatest depths in the part of the lake we visited—twelve to twenty-four feet being the general depth within two miles of the shore. In no point seen do the rocky escarpments exceed sixty feet in altitude; but when they are found having that elevation, they present a succession of wild, picturesque, and rugged scenes.

The lowest rock, often at the water's edge, is a sandstone, very friable, and easily disintegrated by waves and atmospheric agents. Above this a limestone, beautifully stratified, and of a hard and compact character, occasionally projects for many feet, the beach below being strewed with large masses, which have fallen off from time to time. In the shaly portion, numerous nodules of iron pyrites occur, assimilating the forms of shells, spheroids, discs, &c. Both the limestone and sandstone are nearly destitute of fossils, but the shale contains certain forms in great abundance, in a very fragile condition. The rocks on the
...t coast of Lake Winnipeg, and on many of the islands, are
wes... ...ous, while the east side is wholly Laurentian. The Lau-
fossilife... ...' fossiliferous rocks often approach one another; but I
rentian and ... enough to find on the east side, the fossilifer-
was not fortuna... the Laurentian.
ous rocks reposing o... Region lay up the Little Saskatchewan;

Our course to the Salt ... from Lake Manitobah into Lake
a fine, broad river, leading

Winnipeg, and forming the chief outlet by which the drainage water of a very large tract of country finds its way to the sea. The Little Saskatchewan flows for sixteen to eighteen miles through a flat country, between clay banks which never exceed thirty feet in altitude. The river is rapid, and in some parts shallow, its channel being often obstructed by boulders, although it nowhere opposes an obstacle to the passage of craft drawing less than two and a half feet water. This river issues from St. Martin's Lake, a sheet of water about thirty miles long and sixteen broad. The rocks in St. Martin's Lake possess some remarkable geological relations. Near the narrows, at its eastern extremity, are two gneissoid islands, and close to them one of metamorphosed sandstone, with the tilted strata of sandstone inclined at an angle but a few degrees from the vertical. West of these gneissoid islands, and about half a mile distant from them, Sugar Island discloses cliffs of metamorphosed sandstone, inclined at an angle of 45 degrees, and dipping N. 70 W. This sandstone contains some very obscure fossil remains, in which the stems of encrinites were thought to have been recognized.

The occurrence of metamorphosed Silurian strata, even on a small scale, is of very great interest. The gneissoid rocks were traversed by quartz and felspathic veins; but although a careful search was made for the precious metal, none was found.

Sugar Island is named from the ash-leaved maple, which grows there, and furnishes a supply of sugar to the Indians who inhabit this part of the country. About six miles west of Sugar Island, horizontal and undisturbed limestone, highly fossiliferous, is seen exposed in cliffs about sixteen feet high on Thunder Island, so named in remembrance of a thunder storm of great violence, accompanied by hail and rain, which detained us on the afternoon of September 28th. St. Martin's Lake is very shallow, and in many parts thickly set with weeds. By the action of ice, long semi-circular accumulations of boulders have been driven up in shallow places, forming reefs, which soon become islands, or, connecting with the main land, cut off large portions of the lake, and give rise to the formation of marshes and swamps in their rear. The effect of this is gradually to diminish the size of the lake on one side, and probably to increase it, though not to the same extent, in another direction. These constant changes were observed on a larger scale, some weeks later, in Winnipego-sis and Dauphin Lakes, and will be fully discussed in my general Report. Their relation to the past history and probable future of an extensive portion of the country included within the salt region, is very instructive and curious. St. Martin's Lake receives the waters of Partridge Crop River, which flows for the most part through a flat limestone country, not ten feet above the present level of the lake, and often not five feet above the river; many parts, indeed, being even now nothing more than extensive wide spread marshes, through which the river meanders.

At the upper end of Partridge Crop River, the Mission of Fairford is established, where I was very hospitably entertained by the Rev. Mr. Stagg. The present prospects of this Mission are at first sight encouraging; but, when the number of years during which Missionary labour has been directed to the Indians frequenting Partridge Crop River and the neighbouring country is considered, perhaps no more hopeful results among adults have been obtained, than can be discerned at other stations of bygone reputation and worn-out resources.

We entered Lake Manitobah on the 29th September, and fortunately found some fine rock exposures on the east coast, which will enable me to carry on the succession of rocks in their order of occurrence. A few days sailing and pulling brought us to the mouth of Water Hen River, which we ascended, and entered Water Hen Lake, then passing on to Winnipego-sis Lake, we arrived at the Salt Springs, about six miles N. W. of Moss River, on the 5th October. We spent two days at this place, occupying the time in making a plan of the works and springs, and examining the surrounding country. It may be sufficient here to state in relation to the manufacture of salt, that the method employed is of the rudest and most primitive description, nevertheless the salt obtained is abundant in quantity and excellent in quality. Wells to the depth of five feet are sunk near the spot where a little bubbling brine spring is found. I saw several of these springs at some distance from the wells, which, to the number of twenty-six, had already been opened. The brine is carried in buckets to the evaporating pans, which are of iron, about five feet long, two feet broad, and sixteen inches deep, placed on rough stones so arranged as to form the sides of a rude furnace below the kettles. The salt is removed by wooden shovels from the pans as fast as it accumulates, and is stored for transmission to Red River without further purification. From each pan about two bushels of salt on an average can be procured daily during the long days of summer. Wood for fuel is close at hand, and of brine an unlimited quantity could doubtless be procured by boring. When a well does not yield brine freely enough, another is dug near to it; none of them however are more than five or six feet deep, and no attempt at boring or deep-sinking has been made; the supply of brine being sufficiently abundant for all present purposes. No rock exposures are found at or near the springs. The soil in which the wells are dug is a stiff yellow clay, very retentive, and holding drift boulders of limestone, with a few of the non-fossiliferous rocks. From the general aspect of the country there can be little doubt that boring would bring an abundance of brine to the surface. Large areas of so called salt ground, that is of ground absolutely barren and often covered with efflorescent salts, are plentifully distributed over the country bordering Winnipego-sis Lake; and the existence of various brine springs is well known to Indians and half-breeds from Swan River to beyond the Assiniboine, a distance exceeding two hundred and fifty miles in an air line. At several places salt has been and is now manufactured, or is known to occur as a thick crust on the ground, north and south of the salt springs just described. These are, the Salt Springs of Swan River, and of Duck River at the foot of Duck Mountain; the springs at Salt point, Winnipego-sis Lake; at Crane River, Manitobah Lake, and at the Scratching River, South of the Assiniboine. It will be shewn in my General Report that the salt-bearing rocks probably extend from near the Saskatchewan to beyond the 49th parallel in a general north and south direction, and it is extremely probable that with boring, brine could be found in workable quantities over a very extensive area of country in the direction indicated above.

Leaving the Salt Springs we ascended Moss River, and after some delay, owing to the shallowness of the water and the occurrence of rapids involving portages, we reached Dauphin Lake. The elevation of this extensive sheet of water above the sea level is about seven hundred feet. Its length may reach twenty-one miles but its breadth does not exceed twelve It receives several tributaries which rise in the Duck or in the Riding Mountain, none of them capable of receiving a freighter's boat for more than seven miles from the Lake. To the west of Dauphin Lake lies the imposing range of the Riding

Mountain, the nearest point of its summit being about seventeen miles distant from the shore of the lake.

North-east of Dauphin Lake is the Duck Mountain, a high range of table-land similar in its external aspect to the Riding Mountain. From the imposing appearance which the Riding Mountain presents from Dauphin Lake, and the singular relation it bears to the level marshy plain from which it rises, I thought it would be highly advisable, if possible, to reach the summit. Several difficulties were urged by the Indians we met against the ascent, chiefly on account of the swampy and boggy character of the level country at its foot. They stated that no difficulty would be found in passing through the valley between the Riding Mountain and Duck Mountain, by an Indian pitching track. It appeared, however, important that an ascent should be made in as direct a line as possible from Dauphin Lake, to the nearest and highest point; and with this object I set out with Mr. Fleming, four men, and an Indian, on the 8th October. The statement of the Indians respecting the existence of formidable swamps and bogs was quite true, and it was with some difficulty we got through them. On the evening of the first day we encamped at the foot of the mountain, having accomplished a distance of twelve and a half miles. In the afternoon of the second day we reached the summit. The latter part of the ascent was very steep, through a forest containing very fine white spruce, aspen, poplar and birch. The Riding Mountain at its eastern exposure forms the abrupt termination of a series of elevated table lands which rise one above another from the south and west by distinct steps, commencing within ten miles of the Assiniboine; its breadth is consequently about forty miles; its altitude above Lake Dauphin fully exceeds 1,000 feet, which makes it nearly 1,700 feet above the sea. The whole of its rise above Dauphin Lake is embraced within five and a half miles, but its greatest rise is included within a mile and a half. The eastern escarpment of the Riding Mountain bears the aspect of an ancient sea coast, once abrupt, afterwards by atmospheric influences rounded, abraded, and sloped. The last rise is very steep, showing a cliff bank of drift clay with boulders, about two hundred and fifty feet high, terminating in a sharp well defined margin at its summit, from which the country slopes very gently westward.

Only one rock exposure was met with during the ascent; this occurred at an elevation of about 400 to 600 feet above Dauphin Lake, and I was at once enabled to identify the formation with its extension on the Little Souris, the Assiniboine below Fort Ellice, and the Qu'Appelle or Calling River. It belongs to the Cretaceous group, and, by its presence on the Riding Mountain, settles the question of the occurrence of coal of Carboniferous age between this range and the South Branch of the Saskatchewan.

The result obtained by the ascent of the Riding Mountain has been of great interest in a geological point of view, since it has unlocked, in a great measure, the geology of this region of country. Such bold eminences as the Riding and Duck Mountains, uprearing their eastern flanks to an altitude exceeding 1000 feet above the surrounding country, naturally gave rise to many conjectures as to their origin and composition. They are probably nothing more than the remains of vast Cretaceous and Tertiary table lands, stretching from the Saskatchewan Valley to the Laurentide Mountains, which have escaped denudation; and the uniform dip of the strata, wherever seen, appears to show that no disturbance has taken place since the Devonian period.

The forest on the summit of the Riding Mountain is very fine, vindicating the soil and climate of Rupert's Land from the sweeping detractions which have been urged against them. I beg to subjoin the circumference, five feet from the ground, of a few trees within 50 yards of our camp on the Riding Mountain :—Aspen, 4 ft. 6 in., 4 ft. 6 in., 4 ft. 1 in., 3 ft. 9 in., 5 ft.; White Spruce, 7 ft. 3 in., 5 ft. 6 in., 6 ft. 6 in., 6 ft.; Birch, 3 ft. 6 in., 3 ft.; Poplar, 4 ft. 9 in., 4 ft. 6 in. These trees represent, as far as observations permitted, the general character of the forest on the summit plateau of the Riding Mountain.

During the night of our encampment a snow storm came on, and in the morning six inches of snow warned us to hasten to lower and more genial regions. We accomplished the return to the boat on Dauphin Lake on the afternoon of the fourth day; but I regret to say that the constant wading through ice-cold water for many hours together, in crossing the swamps, disabled two of the men, who suffered much pain in the head and limbs, until partially relieved by bleeding, vomiting, and warm applications.

The character of the region between Manitobah Lake and the Riding Mountain remained to be ascertained, in order to complete a general outline of a topographical sketch of the country. With some difficulty I prevailed upon an Indian to guide me from Dauphin Lake, in as straight a line as possible, to the H. B. Co's. post on Lake Manitobah, a distance of 70 miles from our camp. I then placed the boat in charge of Mr. Fleming, instructing him to meet me at the Manitobah post as soon as possible. With a half-breed, and an Indian as guide, I proceeded across the country, fortunately without knowing its character beforehand, or I should scarcely have ventured on such a fatiguing journey at so late a season of the year. For thirty miles we had to wade through marshes and bogs, separated by low ridges; in fact the distance named may be said to be made up of marsh, bog, ridge, marsh, bog, ridge, in most wearisome succession. We had horses to carry our provisions and bedding, but the bogs were so bad that, in order to get the horses through them, we were compelled to carry the load ourselves. A thin crust of ice, ¼ inch thick, was formed over their surface the night after our start, which added in no slight degree to the fatigue of the journey. Upon our arrival at the post I was very hospitably received by Mr. McKenzie, the gentleman in charge.

The greater part of the country lying between Manitobah Lake and Dauphin Lake, between Dauphin Lake and the Riding Mountain, and between the southern part of Winnipego-sis Lake and the Duck Mountain, may be considered as having recently emerged from the former extension of the Lakes first named. This emergence has resulted probably from the lowering of the waters of the lakes by drainage, and not by a rising of the land. The Little Saskatchewan is not the only outlet from Manitobah Lake into Lake Winnipeg; and before these outlets were eroded to their present depth, the waters in Lakes Dauphin and Manitobah were evidently about 10 or 15 feet above their present level. This is shown by the lowest beach round Lake Dauphin, which, on the west side, is well preserved, about 7 miles distant from the present shores. Between Dauphin Lake and Lake Manitobah the ancient coast of the latter, for a long period of time, is about 20 miles due west from the H. B. Co.'s post, and it follows the shores of the lake until lost in the general rise of the prairie near White Mud River. I find the impression prevailing among Indians and half-breeds familiar with the general outline of this region of country, that the lakes are fast lowering their level, and although they agree in the popular error of supposing here, as elsewhere, that there is a rise and fall every seven years, yet the fall is considered to be greater than the rise. If the drainage of many thousand square miles of swamp and marsh in this part of the country should ever become a question of national interest, I know of no enterprise of the kind which could be executed with so little cost of time or labor, and promise at the same time such wide spread beneficial results.

Commencing about 15 or 20 miles south of my track, as shewn on the map which accompanies this report, the country is represented to be dry, and to contain large areas of land fit for agricultural purposes. This statement, received from persons familiar with its general character, is partly confirmed by the observations we were able to make when on White Mud River, in September. Our course will be seen on the map which accompanied the last report I had the honor to address to you.

From the 17th to the 28th October, while awaiting Mr. Fleming's arrival, I was employed in examining the country in the neighborhood of the Manitobah post, and as far as Manitobah Island, from which the lake takes its name. I spent four days on this island, which has acquired celebrity from the superstitious belief of the Indians, that it is the abode of a kind of "Manitou" or fairies. Limestone is here exposed in cliffs 15 feet high, on the north side; it contains but few fossils, is extremely hard, and produces, when struck with a hammer, a distinct ring, so that when the waves beat on the shore, and strike on the shingle at the base of the cliff, a loud musical sound, not unlike the ringing of a large number of distant church bells, is produced. Limestone, of a very compact and fine grained description, occurs in massive layers a few feet from the ground; many small pieces, well adapted for lithographic purposes, can be procured, but I fear, in an economic point of view, the value of the rock as a source of lithographic stone, in large slabs, is inconsiderable, on account of the occurrence of the forms of shells which have been replaced by crystalline carbonate of lime, of a softer description than the matrix.

From Manitobah post we proceeded by the east coast of Lake Manitobah to Oak Point, where we exchanged our boat for horses and carts, and started for Red River, via Shoal Lake, where we arrived on the 31st October.

On the 18th December, Mr. Dickinson set out to explore the country between the Assiniboine and the 49th parallel, in accordance with instructions, of which a copy is herewith transmitted. I beg to refer you to Mr. Dickinson's Report for an account of the results of his exploration. The examination of the country east of Red River was undertaken with a view to place you in possession of a summer reconnaissance of that important district; Mr. Dawson's exploration having been made during the winter months, when the swamps and bogs were frozen.

The map which accompanies this report is based upon Thompson's map, with such alterations as the time at our disposal enables us to make. It is only intended to illustrate, for the present, the general features of the country, as well as to show our several tracks and the area traversed. The dotted red line indicates the general direction of the tracks followed; but the traverses made from time to time are not represented; these, with the soundings—(upwards of 350 by the lead)—are necessarily reserved for the General Report, and its accompanying maps and charts.

Mr. Hime occupied the period of his stay on Red River in executing a number of photographs of scenery, churches, buildings, Indians, &c., which will form an interesting collection.

I am glad to be able to state, that during this last exploration, as in the summer expedition to the South Branch of the Saskatchewan, no accident or untoward event of any description has occurred to interfere with our progress or lessen its results.

In inspecting the accompanying map, I beg to refer you to the one which accompanies the report dated September 10th, from which the connection between the two explorations will be apparent.

I have the honor to be, Sir,
Your most obedient Servant,
(Signed,) HENRY Y. HIND.

Hon. T. J. J. Loranger, M.P.P.,
Provincial Secretary,
&c. &c. &c.

Red River Settlement,
16th September, 1858.

Dear Sir,—One of the alleged drawbacks to the settlement of the valley of Red River and the Assiniboine, is the scarcity of timber fit for building purposes. You will remember that during our journey up the Assiniboine, in June last, we frequently saw an extensive forest, stretching for many miles in a southerly direction, on the right or south bank of the river. It is very desirable that the nature and extent of the forest should be determined, and the character of the timber composing it ascertained. As soon, therefore, as you can complete your preparations, I would wish you to determine the limits or boundaries of the forest referred to, and by making frequent traverses or intersections, ascertain the general character of its timber.

As far as is consistent with the safety of your party, you will also examine the country between the Assiniboine River and the 49th parallel, west of Red River; and, if time permits, the country east of Red River, and between German Creek and the 49th parallel.

I am, dear Sir,
Yours truly,
(Signed,) H. Y. HIND.

Jas. A. Dickinson, Esq.,
&c. &c. &c.

No. VII.—THE COUNTRY EAST AND WEST OF THE RED RIVER, NORTH OF THE 49th PARALLEL.

Red River Settlement,
2nd November, 1858.

Dear Sir,—In accordance with your letter of instructions, dated September 16th, I proceeded with my party, on the 18th, to examine those various portions of the country therein specified.

As the country east of Red River—extending to the Lake of the Woods—is quite unknown except for a few miles back from the river, to any but to those Indians who have there their hunting grounds, I was anxious to procure one of them as a guide. Having succeeded in doing so after some little delay, I was obliged to examine this part of the country first, as the Indian guide was about to leave the settlement in a few days for his winter quarters, and if I had not secured his services immediately, would have failed in doing so afterwards.

Considering that one of the objects of this exploration should be that of seeing where a *summer road* could be most easily made from Red River to the Lake of the Woods, that being now a subject of great interest among the settlers, who were about sending a party out for that special purpose, I thought it advisable first to go long the straight picket line made by Mr. Dawson last winter, in which direction, I understand, he reports that a road can be made for some miles, in order that I might be able to institute a comparison between this and any other portion of the adjacent country through which the Indian might guide me.

The first day I was able only to go about fourteen miles—two-thirds of this distance at least being through marsh and wet prairie.

The general course was along the picket-line, from which I was obliged to diverge frequently—sometimes a mile or more, but always keeping it in view—in order to avoid, when possible, the wide marshes through which it passes. The next day I continued in the same direction, and having reached a point opposite the 22nd mile-post, on the picket-line, I could go no further, being stopped by a swamp or quagmire, impassable for horses, or even men, extending in front for many miles, and on both sides as far as the eye could reach. Though taking advantage of all the dry places within reach, ten miles of the course I took lay through marsh and wet land, and five miles at least through swamp. There are a few small clumps of young aspens along the line, and low willows in some of the marshes; but far away towards the north may be seen some clumps of larger trees.

The land is, for the most part, a rich loam, with a sub-soil of sandy clay; but the difficulty, or rather the impossibility of draining the numerous swamps and marshes, and the want of timber, render this tract of country unfit for settlement; and for the same reasons, the difficulty of constructing a suitable road through it would be very considerable, and the expenses enormous.

Judging, then, that I had seen enough of this part of the country for my purposes, I retraced my steps to the settlement; from which I set out again, under the guidance of the Indian, who promised to conduct me by the only dry path towards the Lake of the Woods, as far as the boundary of his hunting grounds.

On the morning of the 23rd, I proceeded along the south side of "la Rivière Seine," or German Creek, which flows into the Red River a little below its junction with the Assiniboine. There are farm-houses and a good road along it for a distance of five miles, when the Indian's track then begins, which keeps close to the valley of the Creek for eight miles, between it and the marsh, which is shown on the map.

This dry space varies from half a mile to a quarter mile wide, crossed by two small sluggish creeks, which if widened and deepened would effectually drain the marsh. There is plenty of good timber along the valley, consisting of poplars, elm and black ash, with small oaks. Leaving the German Creek here on our left, we went along a low ridge about one foot above the level of the marsh, and varying in width from fifty to one hundred yards; it runs in a south easterly direction for about three miles, and then widens out on the left as far as I could see, and on the right to half a mile. At this point we were about three miles from German Creek which we lose sight of now for some time. Continuing in the same direction for three miles more through beautiful, rich grass, with clumps of aspens on the left and high willows on the right, we came to a creek called Oak Creek, which is about two chains wide, but so still and sluggish that it rather resembles a long lake. Our course then lay along it nearly due east for two and a half miles, when the creek then turns to the south. This would be an admirable place for a settlement, the land being as rich as any in the whole country, and there being a large supply of oak, averaging 1 ft. 6 in. in diameter, and poplars suitable for fencing.

On the south side of Oak Creek the open prairie stretches away to the horizon, the greater part of that which was within view being dry, there being only a few patches of wet land. Leaving Oak Creek we went through a country of this character for about nine miles in a south easterly direction, our track winding, however, a little to avoid the wet places, a few of which we had to cross; none of them, however, being more than seven or eight chains wide and easy of crossing. There are numerous clumps of small aspens and willows in every direction. We then proceeded nearly due east for about seven miles, German Creek being from one and a half to two miles on the north, a beautiful and rich prairie lying between us and it, and on the south one mile distant runs a well wooded ridge, parallel with our course; then turning to the south east we wound round numerous large clumps of aspen from five to thirty feet high, and willows for seven miles, when we came to a rising ground so densely covered with young aspen and fallen timber that it was impossible for carts to go further; we therefore left them here and made packs of a few things for the horses to carry. Here the land becomes of a lighter description, being of a light sandy and clay loam. The timber has been all burnt; the ground was so thickly strewed with the fallen logs that it was with much difficulty the horses could travel. Two miles further on we came to the banks of German Creek; its valley here is from fifteen to twenty chains wide, and about forty feet deep; it is full of excellent timber, elm, oak, poplar and black ash, all large enough for building purposes. The creek, which is here very rapid, is thirty feet wide and about one foot six inches deep. We follow its course now for twenty-seven miles, never being more than half a mile away from it. The country through which we passed is for the most part covered with trees of various kinds growing in large clumps, balsam poplar, aspen, tamarack, balsam spruce, cedar and oak. The whole country has been burnt some years ago; the remains of the timber everywhere to be found indicate that there was once a vast forest of large trees.

The Indian guide now said he had come to the boundary of his own country and could not bring me further, and though I tried to induce him by every means, he remained firm to his resolution. He was unwilling for some time even to give me a description of the country beyond; but finally I procured from him the following account:

At half a day's journey on snow shoes, or a distance of fifteen miles from where we were, there is a mountain or ridge thickly covered with trees stretching towards the Lake of the Woods. A part of this intervening space is a swamp in which grow tamarack, cedar and spruce; the remainder is dry ground covered with small aspens and willows. Passing along the "mountain" you come to a marsh which extends to the "Lake of the Woods;" but through it there flows a river up which large canoes could come within the hearing of a gun-shot, or about two miles from the mountain. The entire length of the way I had come was seventy miles; fifty miles, at least, of this distance being fit for settlement, and throughout the whole of it a road could be made without the slightest difficulty and at little cost. If time and means had permitted, I would have pushed through to the lake, but under the circumstances I considered it better not to attempt it.

From the description given by the Indians of the country, and which I think may be relied on as correct, I am of the opinion that a road can be easily made through it.

I returned by the same track as I came by for some distance, when I crossed German Creek, at a place about thirty five miles from its mouth, and then continued along the north side of it.

At this crossing place there are two or three houses, the commencement of a settlement which is likely to be quickly extended.

On the 1st of October I set out again to examine the country between the Assiniboine and the 49th parallel; and more particularly the forest which was said to extend for so many miles to the south from the river at Prairie Portage.

Proceeding along the road to St. Paul, I turned off from it where it crosses "La Rivière Salé," (or Stinking River,) and went by the hunters' track on the south side of the river, along which it goes for thirty miles, cutting across the large bends of the valley, which is very winding, and through which the river meanders in a remarkable manner.

The country lying between it and the Assiniboine is very marshy, and is covered with willows and clumps of small aspen. In the valley and along both sides grow oak and elm and some fine ash, many trees two feet in diameter—they extend the whole way up the river. On the south side there is a prairie apparently as level and boundless as the ocean; the grass on it is most beautiful and luxuriant, indicating the richness of the soil.

The valley is about 20 chains wide and 40 feet deep; there are many salt springs in it, which make the water in the river quite brackish, from which it derives its name. The river higher up opens out into small lakes, and rises from a marsh which is very extensive. The track here joins the hunters' track from the White Horse Plain, it turns to the south, in which direction it goes for about 12 miles, whence turning nearly due south for 15 miles, it crosses "La Rivière des Isles de Bois," a river 15 feet wide and two deep, flowing into the Scratching River. This portion of the country is all a level prairie, the greater part of it being wet and marshy, except near this river where it is quite dry for five miles; the land is a rich sandy loam, yielding most luxuriant grass. On both sides of the river there is a skirting of trees, chiefly oak, averaging 1 ft. 6 in. in diameter.

The buffalo-hunters when they have crossed this little river begin to keep a sharp look-out for the Sioux, and to take their usual precautions.

The track continuing in the same direction crosses a prairie twenty miles wide.

This prairie is of light sandy soil, with clumps of aspen and willows growing here and there; it is intersected by many small valleys, in all of which, with one exception, the creeks that formed them, are now dried up. The valley of "La Rivière Tabac" is 7 chains wide and 20 feet deep, there was but very little water at this time in the creek, but in spring time there is a rapid flow.

The prairie on the south and west is bounded by what is generally called the "Pembina Mountain," which is rather a series of steps rising up from the prairie below to one above. There are three steps from 20 to 25 feet high, together with a gradual ascent for two miles; the whole of it is thickly strewn with boulders of granite. This "Mountain," which consists of clay, gravel and sand, runs in a south-easterly direction, from a little above Prairie Portage to Pembina. Where we crossed it there is no timber, but on both sides it is well covered, particularly on the south, where the trees seemed large and good. Here the forest is said to begin which reaches to the Assiniboine, but with the exception of some oak on the mountain, there is no good timber, nothing but young aspen from 20 to 30 feet high, growing very close together, forming a dense thicket.

On reaching the summit of the "Mountain," the track turns to the west across a prairie called "the round prairie," which is perfectly level and open for six miles; on the north and south it is bounded by woods of poplars. On its western limit, within a few hundred yards of the track, there is a conical hill about two hundred feet high, called the "Calf's Tent"; rather a remarkable looking object, rising as it does so abruptly from out the level plain and alone.

We then crossed an undulating prairie, ten miles wide, covered with willows and clumps of aspen, from 20 to 40 feet high; the soil is a rich sandy loam. This part of the country is quite destitute of water; there are no creeks, and the ponds which are said to be generally full of water were now quite dry; from twelve o'clock one day till two o'clock the next, we could find none.

Here commences the hilly district; its highest hills which can be seen so well from the banks of the Assiniboine, are called the "Blue Hills." The general direction of its eastern boundary is dearly S. W. by N. E. The track now turns towards the north-west. The country it traverses for 13 miles may be described generally to be an undulating or rolling prairie, studded with numerous conical and dome-shaped hills from 50 to 150 feet high, some covered with willows and aspens, and some quite bare. They are all composed of sand and gravel mixed with clay, and having on their flanks many granite boulders.

Running parallel with our track for some miles is a valley, ten chains wide and twenty feet deep, called "La Grande Coulée," in which there is no water; and we crossed many smaller ones, also dry, connecting with it.

Here I left the track, and went in a northerly direction to the thick poplar woods, the "Le Grand Bois" of the French half-breeds, which seemed 6 or 7 miles away, but on arriving there I found it to consist only of large clumps of aspens and poplars, which at a distance looked like a dense and continuous wood, as it is commonly supposed to be by the buffalo hunters. The trees, though high, only average about nine inches in diameter.

I made several traverses hereabouts, and found that at distances from one to three miles back from the open prairie, the wood becomes densely thick, quite impenetrable in many places.

The trees are all small, none greater than one foot in diameter; they are of the poplar species, with here and there a young oak or a sugar maple.

On my return to the Hunters' track we passed by a pretty lake about three miles long and half a mile broad, surrounded by a close mass of poplars and willows. We came upon the track at a point about four miles to the west of where we had left it, and followed its windings through the hills, still going to the north-west. There are here many isolated hills, as well as chains of hills running in every direction. The low ground is generally marshy, through which gently flow several small creeks, all emptying themselves into a stream on our left, which we cross seven miles further on.

This stream is six feet wide and two feet deep, it flows in a valley fifty feet deep and about twelve chains wide. The ground here is much covered with granite boulders and fragments of shale.

Observing this broken shale throughout the whole of the hilly district to be lying about in every direction on the surface and often turned up by the badgers, I searched on the hill-sides and along the valleys for solid rock but could find none. I suppose it to be, therefore, from its similarity in appearance, drift from the rocks on the Little Souris and other places towards the north where it was found to exist. The country now becomes more hilly than before, and is completely covered with low willows; oaks, and poplars, single and in clumps, grow plentifully on all sides. There are several small lakes, on some of which were large flocks of white swans. The main woods on the right are here from five to six miles distant. This whole region was once upon a time an extensive forest of oak, for everywhere the remains of them are to be found. On the left there are large clumps of balsam poplar, forming for several miles almost a continuous forest. We crossed another of those valleys here so numerous, called "Le Grand Coulé de la Grosse Butte," deriving its name from a large conical hill about two hundred feet high. The valley varies in width from twenty to thirty chains, and is about eighty feet deep, but appearing much deeper in many places by reason of the hills adjoining it.

The sides are very precipitous and the bottom is quite level and all covered with beautiful grass; there is no creek flowing through it, or even the appearance of any recent one. Two miles up in it toward the north there is a small lake and another valley branching off from it, which we crossed four miles further on; in it there is a small creek six feet wide and one foot six inches deep. The track turning to the north soon comes close to "Le Grand Coulé de la Grosse Butte," and continues along it for nine miles. The scenery is now very wild and beautiful; the valley, the bottom of which is eighty feet below the general level of the country, cuts through ranges of hills, many of them one hundred and fifty feet high, and winds round the bases of others, some bare and rugged and some covered with poplars. There are many lakes of various sizes, which add considerably to the picturesque beauty of this peculiar region, the favourite haunt of the moose and red deer. Travelling on for five miles more we reach the top of a hill, when suddenly bursts on our view a vast undulating prairie stretching away to the Assiniboine and Little Souris. The track, which had been very faint for some time, here became quite invisible. It was thought advisable therefore to return to where another one had been seen branching off, some six or seven miles back. Having regained it we followed it for eighteen miles, still among the "Blue Hills," crossing the low ridges and windings through the valleys between the high hills, several of them three hundred feet high, and around us were many pretty lakes; we then came upon the open prairie.

From this across to the Assiniboine is thirteen miles. The prairie is thickly spread over with low willows, and is swampy in many places; there are but a few clumps of young aspen to relieve its bleak and dreary aspect.

The valley of the Assiniboine where we crossed it, forty miles above Prairie Portage, is about one mile and a quarter wide; its sides are much broken and indented.

The poplars and oaks, of which it is full, are all young, none exceeding fifteen feet in height, and there are no trees of any kind along either side for many miles. The river is at this point ten chains wide and three feet deep, and has a hard, gravelly bottom, so that we forded it very easily. On the north side of the river are the Sand Hills, through which we passed last June. The forest, whose southern limits I have ascertained, extends twenty miles above Prairie Portage, along the river, where it then dies away. I remained at Prairie Portage three days, making explorations of the forest, and obtaining information concerning it from some people who were well acquainted with it. I found that the good timber grows merely along the river, in width from half a mile to three miles: beyond that the wood is exactly what it is on the south side. Here and there among the young poplars are solitary oaks at long intervals, many of them two feet in diameter, the remnants doubtless of a fine forest. About eight miles back from the river there is a large clump of balsam spruce, but which are all small. The following is a list of the different trees and their dimensions, which form the band of good timber along the river: Oak, 2 ft. in diameter; aspens, 2 ft.; balsam poplars, 2 ft 9 in; elm, 1 ft. 3 in.; basswood, 2 ft. 6 in.; ash (very few), 1 ft. There is an abundant supply of oaks, straight and tall, 1 ft. 6 in. in diameter; and of balsam poplars, 2 ft. On the Pembina Mountain there is some good timber, including tamarack, not found elsewhere, but which only averages, I am told, 9 in. in diameter.

Yours very truly,

(Signed,) JAMES A. DICKINSON.

Professor H. Y. Hind,

&c. &c. &c.

ON THE

QU'APPELLE, OR CALLING RIVER,

AND THE

DIVERSION OF THE WATERS OF THE SOUTH BRANCH OF THE SASKATCHEWAN DOWN ITS VALLEY,

WITH A VIEW TO THE CONSTRUCTION OF A STEAM-BOAT COMMUNICATION FROM FORT GARRY, RED RIVER, TO THE FOOT OF THE ROCKY MOUNTAINS.

CONTENTS.

ON THE QU'APPELLE, OR CALLING RIVER; AND THE DIVERSION OF THE WATERS OF THE SOUTH BRANCH OF THE SASKATCHEWAN DOWN ITS VALLEY, WITH A VIEW TO A DIRECT STEAMBOAT COMMUNICATION FROM FORT GARRY, RED RIVER, TO NEAR THE FOOT OF THE ROCKY MOUNTAINS.

TORONTO, Feb. 3rd, 1859.

SIR,—I venture to submit the following notice of the Qu'Appelle River, in anticipation of a more detailed description, which will be furnished in my General Report.

I have the honor to be,
Your obedient servant,
HENRY Y. HIND.

The Hon. C. Alleyn, M.P.P.,
Provincial Secretary,
&c. &c. &c.

---

1. The discoveries of gold in British Columbia have invested with great interest the facilities for communication which exist between the Atlantic and Pacific seaboard, north of the 49th parallel. Gold on the Pacific slope of the Rocky Mountains, within British Territory, will probably induce a large emigration to that quarter, and speedily create great commercial activity.

2. The valley of the Mississippi being separated from the northern Pacific Territories of the United States by an extensive region presenting extraordinary difficulties in the initiation of a commercial or even emigrant route, until the construction of a railway removes the obstacles, public attention in the north-western States of the Union and in Canada has been directed to the valley of the Saskatchewan, and the feasibility of employing it as a link in a great chain of communication between the Mississippi and St. Lawrence on the one hand and the western slope of the Rocky Mountains on the other.

3. For the sake of the valley of the Saskatchewan alone, great efforts have been made and are making in Canada to establish a communication between it and Lake Superior, which, for commercial purposes, is in effect the same at the St. Lawrence or Atlantic. These efforts will probably receive a great impetus, now that the discoveries of gold in British Columbia are confirmed; while the area over which the precious metal is known to be distributed, leads to the inference that its occurrence in quantity sufficient to create a powerful Pacific Colony, with great rapidity, is no longer a matter of doubt.

4. During the past summer, when returning from the South Branch, I met several parties of American emigrants, who were proceeding to Frazer's River *viâ* Carlton House and the North Branch of the Saskatchewan. One party was well furnished and equipped by an influential Company at St. Paul, whose objects and proceedings have been published in pamphlet form. Some of the emigrants are wintering at Red River Settlement, purposing early in the spring to follow in the track of the party I met. Others are now organizing in the north-western States, to journey to the "Mines" by the same route. It is apparent that a strong effort will be made to establish a North-Western Emigrant Land Route to the Pacific, by the people of the north-western States of the Union. The Missouri route is too difficult and hazardous at present, and that by the North Branch of the Saskatchewan is the one adopted.

5. In the prospectus of the Canadian North West Transportation Company the line of steam communication proposed is through Lake Winnipeg and the North Branch of the Saskatchewan. Lake Winnipeg is now proposed to be gained from Lake Superior, by taking advantage of the navigable reaches of water on Dog Lake, Milles Lacs, Rainy Lake, Rainy River, and the Lake of the Woods, with intervening roads and portages. As the country through which this route passes is an inhospitable region, with few areas fit for cultivation, as far as the west side of the Lake of the Woods, the proposed communication will probably not be open for rapid transit, without enormous outlay is incurred, for a period of several years. Up to the date of my departure from Red River last year, no communication had been effected, in *summer time*, between the settlements and the Lake of the Woods, except in canoe, although every effort was made to pass through the formidable bogs and swamps which intervene. This important link on the proposed line of route is still a *terra incognita* for a short distance.

6. The projectors of the navigation of Red River below Breakenridge, in the State of Minnesota, look also to the North Branch as offering the most favourable means of reaching the foot of the Rocky Mountains. They are constructing a steamer on Red River, and propose to connect, by a line of stages, with Crow Wing and St. Paul. Crow Wing is within one hundred and twenty miles of Lake Superior City, and a travelled summer road already exists between them. As no impediment is known to exist in the navigation of Red River for steamers of shallow draft, the close of this summer will witness, no doubt, the navigation of the Red River of the North by steam; and its connection with Lake Superior on the one hand, and the Mississippi on the other, by travelled roads. This connection can be maintained during the winter months, from Crow Wing or St. Paul. In these projects, so rapidly approaching completion, the North Branch of the Saskatchewan is the route to be followed to British Columbia. In a word, public attention seems to be almost exclusively directed to Lake Winnipeg and the North Branch.

7. One of the results of this Exploring Expedition to the South Branch of the Saskatchewan last year has been to ascertain the practicability of constructing, at a very small cost, when compared with a railroad, a communication for steamers of considerable size to near the foot of the Rocky Mountains, by an undescribed route, which starts from Fort Garry or any navigable part of Red River, proceeds up the Assiniboine to the mouth of the Qu'Appelle or Calling River, then up the Qu'Appelle valley to the South Branch of the Saskatchewan, then up the South Branch to Bow River. Bow River is an affluent of the South Branch issuing from the Bow River Pass, one of the best in the Rocky Mountain range. In order to convert this route into a steamboat communication without any serious in

terruption,* the diversion of the waters of the South Branch down the Qu'Appelle valley is involved.

8. In September last I communicated to you under date, Red River, Sept. 10th, the results of an exploration of the Qu'Appelle River valley. I have now the honor to submit the plans of that exploration on a scale of two inches to one mile. My instructions authorised me to make a survey on a scale of two miles to one inch, but in consequence of the great importance of this valley, and of the subject to which it refers, I have preferred to send you plans for the information of the Government on a much larger scale, without however intending them to stand in place of those which will accompany the general report.

9. I now proceed to show the relation of the Qu'Appelle valley to the South Branch of the Saskatchewan and Assiniboine Rivers, and to trace the results of sending the waters of the South Branch down the Qu'Appelle valley into the Assiniboine, thence into Red River, past Fort Garry, and finally into Lake Winnipeg.

10. The valley of the Qu'Appelle River joins the Assiniboine five miles above Fort Ellice, and by the windings of the river valley about 360 miles from Fort Garry. It is 270 miles long, and appears to be a former continuation of the South Branch, in a direction nearly due east, to the low regions now occupied by Lakes Manitobah and Winnipeg. Its western extremity issues from the South Branch at the Elbow, or the point where that river from a due easterly course, suddenly takes and preserves for 250 miles a northerly course, until it joins with the North Branch.

11. The narrowest breadth of the bottom of the Qu'Appelle valley is half a mile; its greatest breadth about one mile and a half. Its shallowest part is about 120 feet below the level of the prairie, and its greatest depth is between 350 and 400 feet. It cuts a gently sloping plain, extending from the South Branch to the Assiniboine. The surface of this plain is slightly undulating, and at its western extremity sand hill ranges and sand dunes in process of formation, occupy extensive areas.

12. The highest part of the bottom of the Qu'Appelle valley is only 85 feet above the South Branch at its summer level, and from 75 to 78 feet above it during the spring elevation of its waters. This occurs at a point distant 11½ miles from the junction, where a lake is found, which discharges itself both into the Saskatchewan and Assiniboine. Before connecting with the Assiniboine, it falls 284 feet in 256 miles, or 1 ft. 1 in. per mile. The difference of level between the South Branch, at one end of the Qu'Appelle valley and the Assiniboine at the other, is only 200 feet.

13. In its long, deep, and narrow course there are eight lakes, having an aggregate length of seventy miles. Most of these lakes abound in white fish of great size and the finest quality. They are connected with Long Lake, as shown in sheet No. 10, occupying another valley running north-westerly, a counterpart of that of the Qu'Appelle, inosculating with it at the Grand Forks, and with the South Branch some thirty miles north of the Elbow. Long Lake is forty miles long, similar, as far as I saw of it, to Buffalo Pound Hill Lake (shown on Sheet No. 11), in the Qu'Appelle valley. It occupies a deep, narrow, excavated valley, not exceeding a mile and a half to two miles broad, and from 300 to 400 feet deep.

14. Numerous measurements of the depths of the Fishing Lakes, showed them to hold from forty to sixty-six feet of water. These depths were maintained with great regularity. Timber ceases in the valley about 168 miles from the Assiniboine. It appears again at the Moose Jaws Forks, 194 miles from the Assiniboine, and occurs again in small quantities at the Sandy Hills, near the height of land. Moose Jaws Forks is well wooded for a considerable distance: it comes from the Grand Coteau de Missouri, whose blue outlines are distinctly visible from this point of the Qu'Appelle valley.

15. Without considering here the question whether the South Branch did ever pass down the valley now occupied by the insignificant Qu'Appelle, I propose to glance at the kind of work which would be required to send its waters through this magnificent channel, into that of the Assiniboine, and thence past Fort Garry into Lake Winnipeg. And I may here remark, that almost every spring, the whole of the Qu'Appelle valley is flooded from the Height of Land to the Assiniboine. We frequently found water-marks eight feet above the level of the river in August last; *so that there does occur, for a few weeks or days each year, when the snow melts, a continuous water communication from Fort Garry to near the South Branch,* similar to what would be produced if the Saskatchewan were diverted down the valley of the Qu'Appelle. In 1852, it was converted into a lake from the Sandy Hills to the Assiniboine.

16. A dam, 85 feet high, and 600 to 800 yards long (a few miles lower down the length of the dam would be much less) across the deep narrow valley in which the South Branch flows, below where the Qu'Appelle valley joins it, would send its waters down the Qu'Appelle valley, thence down the Assiniboine past Fort Garry, and thus establish a splendid and probably uninterrupted navigation, for steamers of large size, for a distance exceeding six hundred miles. Beyond the point I reached, the South Branch was reported to me, by the half-breeds who have visited it, to contain no impediment as far as the mouth of Bow River, a distance westward of 300 miles. By the Crees of the Sandy Hills, who hunt on the Qu'Appelle and the South Branch, I was also assured that no rapids or impediments of any description, beyond changing mud and sand-bars, exist between the Elbow and Bow River. The magnitude of the South Branch at the Elbow, and the character of the country through which it flows, lead to the inference that at the mouth of Bow River it is still a large and navigable stream.

17. Whether it would be a matter of economy to construct a dam, forty, fifty, or sixty feet high, across the South Branch, and make a cutting through the Height of Land in the Qu'Appelle valley, corresponding to the altitude of the dam, is an engineering question I am not competent to discuss. It may be here remarked that the hill sides and the valley of the Qu'Appelle, for a distance of six miles from the South Branch, are covered with large boulders, and would furnish an abundant supply of that kind of material. Large and water-worn trees of many species were observed on the sand-bars and mud-flats of the great river, evidently brought by the stream from some distance above.

18. It will be asked whether injurious consequences to the settlements on Red River and the Assiniboine might not ensue from the passage of so large a body of water, during spring freshets, down the valleys of those rivers. The answer to this question is rendered remarkably simple, by the peculiarity of

* A short break might be required about 150 miles from Red River, to overcome a very rapid descent of the Assiniboine, but this break would occur a short distance from a splendid agricultural country, and would involve a portage road over a sandy tract of between twenty and thirty miles. It is, however, probable that steamers of high power would be able to ascend the current at the spot referred to.

the valley of the Assiniboine just before it merges into the open low prairie country, two miles above Prairie Portage. Here the river glides in an excavated trench about 16 feet below the prairie, but in times of very high floods it sends water across the prairie, down the broad, shallow valley of Rat Rivulet, into Lake Manitobah. Rat Rivulet rises in the Bad Woods, west of Prairie Portage, within two or three miles of the Assiniboine; and the ridge which divides it from the river is an imperceptible rise in the prairie which the eye can scarcely detect. A shallow cut through the gentle rise separating the Assiniboine from Rat Rivulet would permit all flood waters to flow into Lake Manitobah, and protect the settlements on Red River from any danger of being flooded.*

19. The country drained by the South Branch above the Elbow, is very little known. The descriptive accounts I received from half-breeds who have traded with and resided among the Blackfeet Tribe of Indians occupying this region, were very encouraging as regards the Bow River; especially in respect of climate, and the timber which covers the eastern slope of the Rocky Mountains. They represent it as far more attractive and delightful, in every way, than the region drained by the North Branch and its tributaries, which, being cut by the 54th parallel of latitude, is three degrees further north, and thus suffers from many of the disadvantages of climate belonging to its geographical position.

20. As an instance of the difference in climate between the North and South Branch, I may mention that, in August last, we found the Mesaskatomina berry ripe, luscious, and in the greatest profusion on the Qu'Appelle and South Branch, growing on trees 16 to 20 feet high, whereas on the North Branch, ten days afterwards, they were found scarcely ripe, on small stunted bushes from five to seven feet in altitude. I had an opportunity of conversing with men who had resided for years among the Blackfeet, and who had wandered backwards and forwards from Bow River to the Columbia, through Bow River pass; from their descriptions I infer that, in point of soil and climate, the eastern slope of the Rocky Mountains, unwatered by Bow River and Red Deer River, is well adapted for a grazing country.

21. The advantages to be derived from the suggested diversion of the waters of the South Branch down the valley of the Qu'Appelle, are numerous and highly important.

(1.) The distance between Fort Garry and the foot of the Rocky Mountains, would be shortened by at least 400 miles.

(2.) The route would be a steamboat navigation, probably with one short break on the Assiniboine, from Breakenridge, on Red River, or any point on Lake Winnipeg, to the foot of the Rocky Mountains.

(3.) Batteaux might drift from Bow River to Fort Garry without discharging cargo, or even touching land.

(4.) The season of navigation would be eight to ten weeks longer than by the North Branch The ice does not often leave the head of Lake Winnipeg before the 10th of June. The South Branch might be reached from Fort Garry, through the Qu'Appelle valley, by the 10th day of May, often by the 1st of May.

(5.) The proposed route passes through the most promising and fertile part of Rupert's Land, namely, the valley of the Assiniboine. The whole western flank of the Riding Mountain would then become available for settlement, as well as the fertile area south of the Qu'Appelle, as far as the Mission, 119 miles from its mouth. The Touchwood Hill Range, on account of its proximity to Long Lake, would acquire the importance which its wonderfully rich and fertile soil promises for it.

(6.) The best pass through the Rocky Mountains would be approached by the most direct route, and be, in fact, a continuation of that route.

(7.) The dangerous and circuitous navigation of Lake Winnipeg avoided, the Grand Rapids surmounted, and the yet apparently unknown difficulties of the Coal Falls, just above the forks of the Saskatchewan, overcome. The "Coal Falls" are situated on the North Branch, they consist of a series of rapids for 18 miles, and are much obstructed by boulders, many of which are exposed during low summer levels. In the South Branch, for a distance of 250 miles, I saw no rapid which might not be ascended with ease by any river steamer, and at the Elbow it is a finer stream than the North Branch is at the Grand Forks.

(8.) The route from Lake Huron, via Lake Superior and Lake of the Woods, would lie in a line nearly straight to the Rocky Mountains.

22. These observations apply exclusively to a steamboat route, which is necessarily limited to the summer months. But in the initiation of any permanent postal route across the continent, north of the 49th parallel, the means of establishing a winter communication must not be omitted. If possible, the summer and winter route should coincide, and pass through areas of country fitted to invite settlement, and become centres of civilization in this vast unpeopled wilderness.

23. The line of route by the Assiniboine, Qu'Appelle, and South Branch, is admirably fitted for a postal communication, which could be carried on during summer and winter, by horses and dogs, at a minimum speed of one hundred miles a day. This might be easily accomplished by the establishment of post stations in localities where they would become centres of population in the midst of fertile areas. Such areas

* In Mr. Dawson's Report, dated Toronto, 22nd February, 1859, the following foot-note is inserted:—"Since writing the above, I have had the advantage of hearing Professor Hind's lecture on the subject to which it refers; but, even admitting that the whole volume of the South Branch of the Saskatchewan could be turned into the Qu'Appelle, it must not be supposed that locks could be dispensed with. It is possible, indeed, that in the valley of the Qu'Appelle itself, where the descent is represented as being very gentle, the current might not be too strong for steamers of great power. But on the Assiniboine, from the Rapid River downwards, in making the descent from the higher prairie to the lower, where, as I have said, there must be a fall of 300 feet, the accumulated mass of water would rush with the impetuosity of a mountain torrent. The plains of Red River would be converted into a sea, and the Settlement swept into Lake Winnipeg." It is necessary to mention, in relation to this paragraph, that I exhibited at the lecture referred to, a map on a scale of two miles to one inch of the country between Lake Manitobah and the Assiniboine, showing the valley of Rat Rivulet, and the means to be adopted to prevent the injurious consequences which might be supposed to arise from the passage of so large a body of water as that of the South Branch of the Saskatchewan in conjunction with the Assiniboine, past the Settlements at Red River during spring freshets. Mr. Dawson appears to have forgotten this map and the explanations which accompanied its production, otherwise he would not have imagined "*the plains of Red River converted into a sea, and the Settlements swept into Lake Winnipeg.*" This forgetfulness is the more remarkable, since it appears that Mr. Dawson was familiar with this old watercourse of the Assiniboine long before the above description of the Qu'Appelle valley was published, for I find, on page 6 of his report, under date "4th July, 1858," the following paragraph:—"In speaking of navigable lines that might be made available, I should mention that at the Grand Portage there is said to be an *old water-course, by which the Assiniboine, in all probability, has at some period discharged its waters into the Manitobah Lake.*" "The accumulated mass of water, rushing with the impetuosity of a mountain torrent," is a poetical description of a phenomenon which is approached every spring when the snow melts; but it leaves no traces of its occurrence beyond water-marks on the sides of the deep broad valley in which the Assiniboine flows, and on the trees which cover a large portion of the flats. (See paragraph No. 15, of this communication.)

PRELIMINARY REPORT.

TORONTO, March 28th, 1859.

SIR,—I have the honour to address to you a Preliminary Report on the results of the Assiniboine and Saskatchewan Exploring Expedition to accompany the topographical maps of the region explored. These maps are constructed upon a scale of two miles to one inch in compliance with your instructions dated 27th April, 1858.

I have sent to you from time to time, during the past summer and autumn, reports on the progress of the Expedition. These reports were as follows:

No. 1. Dated Grand Portage, Lake Superior, May 5th.

No. 2. Dated Red River Settlement June 3rd. Including a Report on the Pigeon River Route, by Mr. Dickinson, C. E., with the following maps.

1. Map of the Pigeon River Route.
2. A general Map of the whole Route.
3. A track survey of the Pennawa River.

No. 3. Dated Fort Ellice; July 9th 1858.

No. 4. Dated Red River Settlement, September 10th, 1858, including a report on the track Survey made by Mr. Dickinson, with one map showing the extent of country traversed by the Expedition.

No. 5. Dated Red River Settlement, November 8th, 1858, including a report by Mr. Dickinson on a track Survey south of the Assiniboine &c., with a map showing the extent of country traversed by the Expedition.

On February 3rd, 1859, I had the honour to submit to you a communication " On the Qu'Appelle or Calling River and the diversion of the waters of the South Branch of the Saskatchewan down its valley, with a view to the construction of a steamboat communication from Fort Garry, Red River, to near the foot of the Rocky Mountains;" with, 1st, a map of the Qu'Appelle River valley from the South Branch of the Saskatchewan to the Assiniboine River, on a scale of two inches to one mile.

2nd. A map of the country between Prairie Portage on the Assiniboine and Lake Manitobah.

I now beg leave to describe the general features of the whole country explored, as delineated upon the large map which accompanies this outline of the results attained during the past year.

AREA TRAVERSED.

The country traversed by the Expedition is embraced between the 49th and 54th parallels of latitude and the 96th and 107th degrees of longitude. The lines of Exploration crossed an area of about 80,000 square miles, or nearly equal to that of Great Britain. The form of this area is similar to that of a parallelogram, being bounded on the south by the 49th parallel, and a line drawn from the point where the Little Souris River cuts it, to the Elbow of the South Branch of the Saskatchewan.

On the east it is bounded by the west coast of Lake Winnipeg, on the north by the main Saskatchewan and on the west by the South Branch of that River. The longest diameter of this parallelogram from Pembina to the Grand Forks is about 450 miles, and its transverse diameter slightly exceeds 330 miles.

SURFACE FEATURES.

The whole country, from the South Branch of the Saskatchewan to the valley of the Assiniboine, slopes in an easterly direction, with a general inclination of about one foot in a mile. This slope is continued throughout the valley of the Assiniboine to Red River, after a rather abrupt descent near where the Assiniboine makes its easterly bend.

North-east of the Assiniboine the country rises almost imperceptibly for a distance of 15 to 35 miles, as far as the base of a series of hill-ranges lying parallel to the general direction of the river valley before it makes its easterly bend; it then rises by successive steps and sloping plateaux to a summit altitude of about one thousand feet above Lake Winnipeg, or sixteen hundred feet above the sea.

These hill-ranges are known by the names of the Riding Mountain and the Duck Mountain. On their eastern and south-eastern flanks they show an abrupt and broken escarpment, and within the space of 5 to 15 miles the country sinks from 1600 to 680 feet above the sea, or within 80 feet of the level of Lake Winnipeg.

At the foot of these hill-ranges, and east of them, lie the great Lakes Winnipego-sis and Manitobah, which are separated from Lake Winnipeg by a low, marshy, and nearly level tract, having an elevation rarely exceeding eighty feet above it.

A line drawn through the largest expanse of Lake Winnipeg, another through Lakes Manitobah and Winnipego-sis, a third through the upper part of the Assiniboine Valley, and a fourth through that of the South Branch of the Saskatchewan, from the Elbow to the Grand Forks, would be nearly parallel to one another, maintaining a direction nearly due north and south, the deviation being in favor of N. W. by N. and S. E. by S. It may be further observed that the Main Saskatchewan, from the Grand Forks to Cedar Lake, and the southern portion of the Assiniboine, flow through valleys also nearly parallel to one another, and at right angles to those before enumerated.

This uniform distribution of lake and river valleys is determined by the direction of the hill and ridge ranges which characterize the country. The South Branch of the Saskatchewan, below Red Deer's River, is separated from the Missouri by the Grand Coteau du Missouri. A continuation or spur of the Grand Coteau comes on the Qu'Appelle River at the height of land about 18 miles from the Elbow of the South Branch. Here it is called the " Eyebrow Hill Range," by the Crees. It appears to terminate suddenly in the form of an isolated hill about 400 feet above the plain, called " The Lumpy Hill of the Woods," a few miles beyond the point where the South Branch takes its easterly turn to join the North Branch at the Grand Forks.

The South Branch flows for fully two hundred miles below the Elbow at the foot of this continuation of the Eyebrow Hill Range, in a northerly direction, and its deep excavated valley appears to lie at an average distance of twelve miles from it. This range is cut by several narrow, deep valleys; and from the small lakes or ponds which occupy their summits, water, during spring freshets, flows to the Saskatchewan and Assiniboine.

The valley of the Qu'Appelle River is a singular and important instance of this interlockage. A general description of this valley is given in my communication dated February 3rd, 1859. Within 50 miles south-west of the Grand Forks, and a short distance south of the Lumpy Hill of the Woods, there is another deep valley in the dividing ridge, from whose summit lakelets, water flows in the spring to the South Branch, a distance of 10 or 12 miles, and also to the Main Saskatchewan, which it reaches below Pine Lake, a distance exceeding 160 miles. One other interlockage between the South Branch and the valley of the Assiniboine will be noticed in the description of the valley of the Qu'Appelle River.

Besides the imposing Riding and Duck Mountains, the Touchwood Hills may be enumerated as very important and striking in a region whose marked characteristic is that of a gently sloping plain. These hills lie between the head waters of the Assiniboine and the South Branch; the elevation of the highest peak, the Heart Hill, probably does not exceed 700 feet above the general level of the Great Plain. The course of this range is from north-east to south-west, and it forms the most prominent of several ranges which lie parallel to one another. West of the Touchwood Hills the continuation of the range is known by the name of the Last Mountains, and at its base is found one extremity of the Last Mountain Lake, which occupies a valley forty miles long, and is narrow and deep, like that of the Qu'Appelle River.

South of the Assiniboine the Turtle Mountain is a prominent and important feature. It is cut by the 49th parallel. The Blue Hills of the Souris serve to destroy the general sameness of the prairie level on the river after which they are named, while the Blue Hills south of the Assiniboine, and east of the little Souris River, offer perhaps the wildest and most picturesque scenery in the area here referred to. The Porcupine Hill, Thunder Mountain, and Pasquia Hill were not included within the area explored. They are eminences which lie between the Grand Rapids of the Sackatchewan and the head waters of the Assiniboine, all of them probably forming at a former epoch a continuation of a vast table land, now broken into detached mountain ranges by denudation.

LAKES AND RIVERS.

Prominent among the physical features of this region are the vast expanses of water which occupy the larger portion of its eastern area. Lake Winnipeg is 300 miles long, and in several parts more than 50 miles broad. Lakes Manitobah and Winnipego-sis together are nearly of the same length, and the broadest part of the first named is not less than 35 miles across. Nearly the whole country between Lake Winnipeg and its western rivals is occupied by smaller lakes, so that between the valley of the Assiniboine and the eastern shore of Lake Winnipeg fully one third is permanently under water. These lakes, both large and small, are shallow, and in the same water area show much uniformity in depth and coast line. Several hundred soundings in Lakes Winnipeg and Manitobah showed a greatest depth of 64 feet, which is exceeded by that of the Qu'Appelle Lakes in the valley of the Qu'Appelle or Calling River. Some of the smaller lakes are of dimensions which entitle them to notice. Such are St. Martin's Lake with an area exceeding 300 square miles; Water-hen Lake; Ebb and Flow Lake, and Dauphin Lake, both covering an area of more than 150 square miles.

West of the Assiniboine we have the Qu'Appelle Lakes, situated in the Qu'Appelle valley, eight in number, and with an aggregate length of 70 miles. Besides these, the last Mountain Lake before mentioned is 40 miles long, and varies from ¾ of a mile to 2 miles in width. The Qu'Appelle Lakes are very deep, 11 fathoms or 66 feet having been recorded.

North-east of the Touchwood Hills there are numerous large lakes, having areas varying from 120 to 130 square miles. Some of these are strongly impregnated with saline ingredients, and are the haunts of innumerable hosts of geese and other aquatic birds. On the south-east flank of the same range and throughout the plain stretching towards the Assiniboine, lakes and ponds are everywhere distributed.

The western flank of the Riding Mountain is dotted with small lakes, ponds and marshes; the same remark applies to a large area south of the Assiniboine and east of the Little Souris.

Lake Winnipeg receives the waters of numerous rivers, which, in the aggregate, drain an area of about 400,000 square miles. The Saskatchewan (the river that runs swift) is its most important tributary. The South Branch, 18 miles below the Elbow and 584 miles from its mouth is 600 yards broad. The rate of the current is here 2¾ miles per hour; the greatest depth is 10 feet in the main channel; the mean depth across being 4.6 feet. There are channels on both sides of the river, one being 6 and the other 10 feet deep. After passing the Moose Woods about 90 miles from the Elbow the river channel is much contracted, its current is uniform and swift, varying from 2¾ to 3¼ miles per hour; mud and sandbars disappear, and it flows between high banks of drift clay, with a boundless, treeless, arid prairie or plain on either hand. At the Moose Woods, where the river is very broad and sandbars numerous, the paddles of canoes have touched the bottom from one side to the other with the ordinary stroke of the voyageurs; this occurred during a season of low water. At the time of our visit in August last, Indians were crossing on horseback from the right to the left bank above the Elbow, the depth not exceeding four feet. Before joining the North Branch the current becomes very strong, often from 3½ to 4 miles an hour. The river winds between high precipitous banks, forests of oak, elm, ash, aspen and birch cover the low points, the opposite hill banks being clothed chiefly with birch and aspens. Groves of spruce show themselves on approaching the North Branch, but the soil on the prairie plateau maintains the most luxuriant growth of vetches, roses and berry-bearing bushes of different kinds wherever the aspen forests have been burnt and open areas formed. From the Elbow to the Grand Forks the distance is 250 miles, and in general, throughout the last 50 miles of its course the South Branch flows through a thinly wooded country, but possessing a soil of great depth and fertility.

The main Saskatchewan opposite Fort à la Corne is 320 yards broad, 20 feet deep in the channel, and flows at the rate of 3 miles an hour. The mean depth across the river here is 14 feet, but it is in the memory of those living at the fort, when the river was crossed on horseback during a very dry season.

About 158 miles below Fort à la Corne, near Tearing River,

the main Saskatchewan is 330 yards broad, 22 feet deep in the channel, has a mean sectional depth of 20 feet, and flows at the rate of 2 miles an hour. 291 miles below the Grand Forks the main Saskatchewan enters Cedar Lake, 30 miles long. Issuing from this large body of water it expands into a small lake, but soon again contracting its channel, the Cross Lake Rapids come into view; these rapids have a fall of 5¾ feet. Hudson's Bay Company's boats of 4 or 5 tons are tracked up them with half cargo, but loaded boats descending, run the rapids. The length of the portage involved in ascending the river is 230 yards. The Saskatchewan now enters Cross Lake, and after issuing from this elongated expanse of water begins a rapid course to Lake Winnipeg, with a current often 3 and sometimes 3½ miles an hour. The head of the Grand Rapids is about 4 miles from the mouth of the river. The length of the portage is 1 mile 7 chains. The rapids below the portage are about 1½ mile long, so that the total length of the Grand Rapids exceeds 2½ miles. The fall from the west to the east end of the portage, as ascertained by levelling, is 28½ feet. The fall below the portage is estimated to be 15 feet, consequently the total fall is about 43 feet. The Grand Rapids are run by Hudson's Bay Company's loaded boats; in ascending from the foot of the rapids to the east end of the portage boats are tracked or towed up with half cargo; they are then run back again, and again tracked up with the other half of their freight. From east to west end of the portage boats are tracked up on the south side of the river, with a load of fifteen pieces, (1350 lbs.,) the remainder of the freight is carried over the portage. The distance from the Grand Forks to the mouth of the Saskatchewan is 342 miles; the distance from the Elbow of the South Branch to the mouth is 603 miles.

The Saskatchewan receives several affluents on its south side which are important only on account of the fertile tracts of country they drain.

Long Creek rises within ten miles of the South Branch, and following the same northerly direction, empties itself into the Saskatchewan near Fort à la Corne, after a course of about 40 miles.

Carrot or Root River rises near the head waters of Long Creek, and flowing in an easterly direction to the north of the Birch Hills, empties itself, after a course of 170 miles near the Pas.

About 110 miles in an air line south from the Grand Rapids, and 136 miles by the Canoe route along the coast, Lake Winnipeg receives the Little Saskatchewan or Dauphin River, through which Lakes Manitobah and Winnipego-sis discharge themselves. During ordinary summer levels, the Dauphin River offers no impediment to small steamers of light draft; it thus forms a valuable and direct communication between the vast water areas which it links together. It flows through a flat and swampy country offering very few inducements or indeed opportunities for settlement. The Mission of Fairford is situated on that part of this River which lies between St. Martin's Lake and Lake Manitobah, having been removed to its present position from the lower part of Dauphin River in consequence of the occurrence of destructive floods, the surface of the country not being above eight feet over the summer level of the River. Dauphin Lake is connected with Lake Winnipego-sis by Moss River, navigable in high water by Red River freighters' boats. The tributaries received by Dauphin Lake scarcely require notice here, although they may become useful as affording means for transporting the valuable spruce of the Riding and Duck Mountain to Lake Manitobah; the most important of these tributaries is the Valley River, which separates the Duck from the Riding Mountain.

Lake Winnipego-sis receives the Red Deer River and Swan River, which open communication to an important tract of country east and north east of the head waters of the Assiniboine. The south-western extremity of Lake Manitobah is distinguished by the extent and richness of the prairies which at a higher lake level it has assisted in forming. The White Mud River which meanders through them may be classed among the most valuable of the lesser tributaries of the Great Lakes of the Winnipeg basin.

At its southern extremity, Lake Winnipeg receives the Red River of the North, which, together with its important affluent, the Assiniboine, unwaters an area of extraordinary fertility and extent, already partially described in my report on the Red River Expedition in 1857.

The Assiniboine joins Red River in Lat. 49° 54′. At the confluence of these rivers Fort Garry is situated. It rises in lat. 51° 40′ and pursues a south easterly course for a distance of about 260 miles parallel to the basins of the Great Lakes on the east of the Riding and Duck Mountains. Within 18 miles south of the 50th parallel it takes a sudden bend to the east, which direction is preserved until it falls into Red River, a distance of about 240 miles from the great bend at Lane's Post. 22 miles from Fort Garry the Assiniboine is 120 feet broad, (June 28th, 1858,) with a mean sectional depth of 6 feet. Its greatest depth here is 7½ feet and the rate of its current is 1½ miles an hour. Near Prairie Portage, 67 miles from Fort Garry, the speed of the current is 2 miles an hour and its fall as ascertained by levelling is 1.18 feet in a mile. At its junction with the Little Souris, an affluent which it receives 140 miles from its mouth, the breadth of the River is 230 feet, its greatest depth 12 feet and its mean sectional depth 8.6, the speed of its current being 1¼ miles an hour. It thus appears that this River is considerably larger 140 miles from its outlet, than 22 miles from the same place. Even at Fort Ellice 280 miles from its junction with Red River, the Assiniboine is 135 feet wide, 11.9 feet deep in the channel, with a mean sectional depth of 8 feet, and a current flowing at the rate of 1¾ miles an hour; in other words this River, 280 miles from its mouth carries a larger body of water than at a point 22 miles from it.

The following table shews the quantity of water which the Assiniboine carries at three different points, distant respectively in round numbers, 22 miles, 140 miles, and 280 miles from its outlet by the windings of the river valley, but not by the windings of the river itself, which will be at least double the length of the river valley.

Volume of Water in the Assiniboine.

	Cubic Feet per hour.	Distance from outlet at Fort Garry.
Lane's Post	5,702,400.	22 miles.
Mouth of Little Souris	12,899,040.	140 "
Opposite Fort Ellice	9,979,200.	280 "

It thus appears that the volume of water in the Assiniboine is nearly twice as large at Fort Ellice as 258 miles lower down the river, if the foregoing table affords sufficient data on which to rest an opinion. It is very probable that the character of the season would modify these results in different years. The measurements were not made simultaneously, and the rainfall in the neighbourhood of the Touchwood Hills and in the region about Fort Pelly was represented to be more in the extreme, than is usual during the summer months. But judging from the appearance of the river bank, and the statements of Indians and half-breeds, familiar with the summer level at the localities where the sections were made, there is no reason to suppose that its waters were in excess of their ordinary summer level. It is therefore very probable that evaporation during a long and tortuous course through an open valley, is adequate to diminish the volume of water in the Assiniboine very much in excess of the supply which it receives from tributaries or springs during its course to Red River.

East of Prairie Portage the Assiniboine flows through a flat, open, prairie country not sixteen feet below its general level where it is cut by the stream. The whole country rising in steps above or west of the Portage, the Assiniboine has excavated a deep broad valley in which it meanders with a rapid current.

At the mouth of the Little Souris, or Mouse River, this valley is 880 yards across and eighty-three feet below the general level of the prairie. At Fort Ellice its valley is one mile and thirty chains broad, and two hundred and forty feet below the prairie.

The Assiniboine receives numerous and important affluents. On its eastern water-shed are the Two Creeks, Pine Creek, Shell River, Birdstail River, and Rapid River or the Little Saskatchewan. The distances of the rivers from Fort Pelly, which may be considered as lying at the head of the bateau navigation of the Assiniboine, will be noticed hereafter when the country they unwater is described. From its western water-shed it receives the White Sand River from the Touchwood Hills; the Qu'Appelle or Calling River, inosculating with the South Branch of the Saskatchewan; Beaver Creek, a small rivulet on which Fort Ellice is situated; and the Little Souris or Mouse River, from the Grand Coteau de Missouri. The Crees of the Sandy Hills on the South Branch, state that Elbow Bone Creek, an affluent of the Qu'Appelle River, inosculates by a deep valley with the Mouse River, or an arm of it, and is connected continuously with the Assiniboine, winding round the northern flank of the Grand Coteau de Missouri.

The Qu'Appelle or Calling River falls into the Assiniboine about five miles below Fort Ellice. At its mouth this stream is 88 feet broad, 12 feet deep in the main channel, and shows a mean sectional depth of eight feet; its current is at the rate of 1½ miles an hour. The valley in which it flows inosculates with the South Branch of the Saskatchewan at the Elbow. It is 270 miles long, and 70 miles from the Assiniboine about one mile broad (78 chains), and 310 feet below the prairie, which stretches north and south from its abrupt edges as far as the eye can reach. At the Qu'Appelle Mission, 119 miles from the Assiniboine, the valley is one mile and a quarter broad and 250 feet deep. The river here is 48 feet wide, 6 feet deep in the channel, with a mean sectional depth of 3 feet 6 inches, and a current of one mile an hour. The lakes at this point have a depth of fifty-seven feet, so that the total excavation below the prairie on either hand is 307 feet.

Near the first or Qu'Appelle Forks the valley is one mile and one third broad, and 220 feet deep. At the east end of Sand Hill Lake, 239 miles from the Assiniboine and 31 miles from the South Branch, the valley is one mile and five chains broad, with a depth of 140 feet below the prairie. Eight miles from the west end of Sand Hill Lake, or 15 miles from the Saskatchewan, the valley is one mile and 70 chains broad and 150 feet deep. At the height of land where it has been invaded by Sand Dunes from the west and south west, it is still nearly one mile broad (73 chains), and 110 feet deep, estimated from the well defined edge of the valley, where a low escarpment of rock still uncovered by the advancing sand of the Dunes, serves to mark its limit and the power of the forces which excavated it. The level of the prairie dotted with Sand Hills and Dunes is some thirty feet above the edge of the rock noticed above.

The Little Souris, or Mouse River, joins the Assiniboine 140 miles from Fort Garry, by the windings of the river valley, and 116 by the buffalo hunters' trail. At its mouth the Little Souris is 121 feet broad, 3 feet 6 inches deep in the channel, with a mean sectional depth of 2 feet 4 inches, and a current of half a mile an hour. Its valley, at the Back-fat Creek, 25 miles from the Assiniboine, is one mile and a half broad (8016 feet), and 225 feet deep, with a level prairie on either hand. Near Snake Hill, 61 miles from the outlet, the valley is only 110 yards broad, and 66 feet deep, with open prairie on both sides. The river here is 100 feet broad, and 4 feet deep in the channel. At this spot several beaches of a former lake were exposed in making a cutting in the bank, with a view to ascertain the nature and extent of the deposits of Tertiary coal or Lignite which the occurrence of numerous water-worn masses of that material, in the bed of the river and on its banks appeared to indicate. In its passage through the Blue Hills of the Souris, the river has excavated a ravine or valley between four and five hundred feet deep, making a sudden turn from a due easterly course to one almost northerly, and avoiding what appears to be an ancient channel but slightly elevated above its present level. This old channel pursues a straight course to Pembina River, with which, on the authority of half-breeds familiar with the country, it is said to be connected. The length of the Little Souris, within British teritory, is 106 miles. A short distance south of the boundary line it receives the Red Deer's Head River, a small stream about 18 feet broad, within a few hundred yards of its junction with the Souris.

WOODED AND PRAIRIE LAND.

The western and south-western slopes of the Riding and Duck Mountains support heavy forests of white spruce, birch, aspen, and poplar. The trees are of a large size, and often exceed 1½ and 2 feet in diameter, with an available length of 30 to 50 feet. On the summit plateau of the Riding Mountain the white spruce is the largest tree; here it attains dimensions, and is found in quantity sufficient to give to this region a great economic value. The wooded area over which timber consisting of the four kinds of trees enumerated, is found on the Riding and Duck Mountains, has a length of 120 miles, with a breadth exceeding 30 miles. The affluents of the Assiniboine will serve during spring freshets, to bear these valuable forest productions to areas which will probably first attract settlement, and where they will be most required.

In the valley of the Assiniboine is an extensive and valuable forest of oak, elm, ash, maple, poplar, and aspen, with an average breadth of 4 miles; its length is about 30 miles. The flats and hill sides of the deep eroded valley through which this river flows above Prairie Portage, sustain a fine forest, in which aspen, oak, birch, elm, and maple appear to prevail in numbers corresponding with the order in which they are enumerated; but this forest does not extend beyond the excavated valley of the river or its tributaries. All the affluents of the Assiniboine flow through deep ravines, which they have cut in the great plain they drain; these narrow, deep valleys are well clothed with timber, consisting chiefly of aspen and balsam poplar, but often varied with bottoms of oak, elm, ash, and the ash-leaved maple. On the west side of the main river, the valleys of the tributaries, such as the Little Souris and the Qu'Appelle River, are timbered continuously for a distance of 30 to 70 miles from their outlets, and at intervals further up stream. On the Qu'Appelle River good timber is found as far as the Mission; but in progressing westward it is seen gradually to diminish in size, and finally to disappear altogether.

The Touchwood Hill Range, together with small parallel ranges, such as the Pheasant Mountain and the File Hill, averaging 20 miles in length by 10 in breadth, are in great part covered with aspen forests, but the trees are generally small. At the Moose Woods, on the South Branch of the Saskatchewan, forests of aspen begin to appear; they continue, with occasional admixtures of birch and oak, more rarely of oak and elm, as far as the Grand Forks; here the spruce becomes common, and, with aspens, occupies the excavated valley of the main Saskatchewan for many miles. The hill-bank, with the plateau on the south side of the river, for a distance of three or four miles south, sustain the banksian pine, which disappears as the soil changes from a light sand to a rich and deep vegetable mould, supporting detached groves of aspen and clumps of willows.

On the Little Souris, especially in the neighbourhood of the Blue Hills, the country is fertile and beautiful, but the areas adapted for settlement lose much of the value which would otherwise belong to them from the absence of wood. West of the Souris is a boundless, treeless prairie; so that in crossing from Red Deer's Head River to Fort Ellice it was found necessary to carry wood for fuel for a distance of 60 miles. This prairie extends to the South Branch and beyond it. At Sand Hill Lake, on the Qu'Appelle, timber is so scarce in the river valley and gullies leading to it, that we were compelled to use the bois de vache for fuel. The South Branch, from the Elbow to the Moose Woods, flows through a treeless region, as far as relates to the prairie on either side; but in the ravines leading to the river detached groves of small timber occur. The boundary of the prairie country, properly so called, may be roughly shewn by a line drawn from the great bend of the Little Souris, or Mouse River, to the Qu'Appelle Mission, and from the Mission to the Moose Woods, on the South Branch. South and west of this imaginary line, the country, as a whole, must be ranked as a level or slightly undulating, treeless plain, with a light and sometimes drifting soil, occasionally blown up into dunes, and not, in its present condition, fitted for the permanent habitation of civilized man; the narrow vallies of the streams which unwater it, such as Plum Creek, Moose Jaws Creek, as well as some low valleys of comparatively limited area being excepted. There can be no doubt that, if the annual fires which devastate these prairies were to cease, trees would rapidly cover them in most places. Everywhere young aspen and willows show themselves in groves where "fire" has not "run" for two or three seasons. A few years of repose would convert vast wastes, now treeless and barren, into beautiful and fertile areas. East and north of this dry prairie region there is a large expanse of cultivable land, which I now proceed to describe more in detail.

AREAS FIT FOR SETTLEMENT.

Valley of the Assiniboine.

Issuing from the Duck Mountain are numerous streams which meander through a beautiful and fertile country. This area may be said to commence at the Two Creeks, ten miles from Fort Pelley, thence on to Pine Creek fifteen miles further. The vegetation is everywhere luxuriant and beautiful, from the great abundance of rosebushes, vetches and gaudy wild flowers of many species. After passing Pine Creek the trail to Shell River pursues a circuitous route through a country of equal richness and fertility. Shell River is 42 miles from Pine Creek, and in its valley small oak appear, with balsam poplar and aspen, covering a thick undergrowth of raspberry, currant, roses and dogwood. Between Shell River and Birdstail River, a distance of 39 miles, the country is level and often marshy, with numerous ponds and small lakes, but where the soil is dry the herbage is very luxuriant, and groves of aspen thirty feet high vary the monotony of the plain.

Between the trail and the Assiniboine the soil is light, and almost invariably as the river is approached it partakes of a sandy and gravelly nature, with boulders strewn over its surface.

The flanks of the Riding Mountain are covered with a dense growth of aspen and poplar, and cut by numerous small rivulets. From Birdstail River to the Little Saskatchewan, or Rapid River, a distance of thirty-three miles, the same kind of soil, timber, and vegetation prevail. About one hundred miles from its mouth the Rapid River issues from the densely wooded flanks of the Riding Mountain through a narrow excavated valley filled with balsam poplar, and an undergrowth of cherry and dogwood, with roses, convolvuli, vetches, and various creepers. The slopes are covered with poplar eighteen inches in diameter. Descending the river, groves of poplar and spruce show themselves, with thick forests of aspen and balsam poplar covering the plateau on either hand. The river is here forty feet wide, with a very rapid current. Before it makes its easterly bend the ash-leaved maple shows itself in groves, and on both sides is an open undulating country, attractive and fertile, with detached clumps of young trees springing up in all directions. The region unwatered by the Rapid River continues beautiful and rich until within twenty-five miles of the Assiniboine, so that it may with propriety be stated, that for a distance of seventy-five miles this river meanders through a country admirably adapted for settlement. Ponds and lakes are numerous, wild fowl in great numbers breed on their borders, and the waters of the Rapid River abound in fish. Canoes and bateaux may descend it from the point where the Exploration terminated to its mouth, a distance of one hundred miles. It will probably become important as a means of conveying to the settlements on the Assiniboine and Red River supplies of lumber from its valley and the Riding Mountain.

From the Rapid River to White Mud River the distance is thirty-three miles, and the country continues to preserve the same general character with respect to fertility and fitness for settlement which has now been traced out for a space of 164 miles. White Mud River flows into Lake Manitobah at its south-western extremity. This river unwaters an extensive area of the richest prairie land, similar in all respects to the White Horse Plains on the Assiniboine, or the rich wastes on Red River. White Mud River is connected with Prairie Portage by an excellent dry road, the crossing place being about eighteen miles from the Portage. The river banks are well timbered with oak, elm, ash, maple, aspen, and balsam poplar. It possesses valuable fisheries, and communicates by an uninterrupted canoe navigation with Lake Manitobah for a length of thirty miles. The soil on its banks and far on either side is of the finest quality. At the mouth of the river a fishing establishment has been maintained by the people of the Portage for several years.

The valley of La Rivière Salè has a general direction parallel to that of the Assiniboine, and about sixteen miles south of it. The country between the two rivers is wet and marshy, with large areas covered with willow thickets and clumps of small aspen. South of the valley of the first named river, the prairie is magnificent and not surpassed by any area of equal extent on Red River.

The area of the region well adapted for settlement on the east and north of the Assiniboine, and in the valley of La Rivière Salè may be assumed fully equal to 3,500,000 acres. In the valleys of Mouse River, the Qu'Appelle River and White Sand River, the area of land likely to invite settlement does not exceed one million acres. The Lakes in the valley of the Qu'Appelle River are important, they abound in fish, among which white fish are numerous, large in size and of excellent quality; the grey and red suckers, pike and pickerel are also abundant.

Valley of the Saskatchewan.

1. The country between the Lumpy Hill of the Woods and Fort à la Corne, or the Nepoween Mission, including the valley of Long Creek and the region west of it, bounded by the South Branch and the Main Saskatchewan. This area may contain about 600,000 acres of land of the first quality.

2. The valley of Carrot River and the country included between it and the Main Saskatchewan, bounded on the south by the Birch Hill range. There is a narrow stripe on the great river, about five miles broad, where the soil is light and of an indifferent quality. The area of available arable land probably does not exceed 3,000,000 acres.

3. The country about the Moose Woods on the South Branch.

4. The Touchwood Hill range.

5. The Pheasant Hill and the File Hill.

The aggregate area of these fertile districts may be stated to extend over 500,000 acres.

If we assume that the prairies of Red River and the Assiniboine east of Prairie Portage, contain an available area of 1,500,000 acres of fertile soil, the total quantity of arable land included between Red River and the Moose Woods on the South Branch of the Saskatchewan will be as follows:

	Acres.
Red River and the Assiniboine Prairies east of Prairie Portage	1,500,000
Eastern water-shed of the Assiniboine and La Rivière Salè	3,500,000
Long Creek and the Forks of the Saskatchewan	600,000
Between Carrot River and the Main Saskatchewan	3,000,000
The Touchwood Hill range, the Moose Woods, &c., &c	500,000
Carried forward	9,100,000
Brought forward	9,100,000
Mouse River, Qu'Appelle River, White Sand River	1,000,000
The region about the head-waters of the Assiniboine, including the valley of Swan River	1,000,000
Total area of arable land of first quality	11,100,000

or eleven million, one hundred thousand acres.

Of land fit for grazing purposes, the area is much more considerable, and may with propriety be assumed as fully equal in extent to the above estimate of the area of arable land.

East of the Riding and Duck Mountains.

In a former report I have shown that the country east of the Riding and Duck Mountains when taken as a whole will furnish a very insignificant field for settlement and civilization. Where the soil is dry, the limestone rock approaches in general so near to the surface, as to be exposed whenever small trees are blown down or the soil is penetrated to the depth of 6 or 8 inches. With respect to the greater portion of the area I visited on the shores of Lake Winnipeg, Lake Manitobah, the Little Saskatchewan, Moss River, Dauphin Lake, and St. Martin's Lake, together with the region between Lakes Winnipeg and Manitobah, always excepting the southern shore of the latter lake, I am of opinion that it is not generally fitted for settlement. In my report from Red River, dated November 8th, I have described more at length the natural features of a large portion of this region from practical information obtained during a journey on foot exceeding one hundred miles in length, from the summit of the Riding Mountain to Manitobah House, on Lake Manitobah.

GEOLOGICAL FEATURES.

During an exploration extended over half a year, and embracing a very wide area of country, numerous rock specimens and specimens of organic remains, have been collected. Most of these were brought to the Red River settlements at too late a period to admit of their being taken to St. Paul before the spring of 1859. By far the larger portion of the collection I have made is still at Red River. I shall, therefore, confine myself, at present, to a very general outline of the geological features of the country.

The most striking peculiarity in the arrangement of the different formations, from Red River to the South Branch, and from the 49th parallel to the main Saskatchewan, is their undisturbed and horizontal condition. With two or three exceptions to be noticed hereafter, no appearance of local disturbance was observed throughout the whole region traversed. The rocks dip, generally, with a very gentle inclination from the north-east to the south-west. Sometimes it is not only impossible to detect any dip by the eye, but the level fails to show the smallest deviation from perfect horizontality. The result of very careful levelling on the Little Souris failed in one instance to shew any dip. The same observation applies to some exposures on Lake Winnipeg and Lake Manitobah. Lake Winnipeg is excavated in Silurian formations; Lake Manitobah and Winnipego-sis partly in Silurian, and partly in rocks of Devonian age. Fossils were collected in numerous localities on the east coast of Lake Winnipeg, and on the islands of Lake Manitobah and Winnipego-sis. From the Saskatchewan at the Grand Rapids, to Red River, exposures of

Silurian rocks are everywhere numerous on the west shores of the Great Lake. About forty miles due south of Snake Island, in Winnipego-sis Lake, there are exposures of a light ash-colored shale, exactly similar, in its lithological aspect, to those on the Little Souris, and a small tributary of the Rapid River. They occur at an altitude of 400 to 600 feet above Dauphin Lake. The country between these exposures and Lake Manitobah, as well as in a direction south-east to Red River, is nearly horizontal, and all rock exposures seen were in an undisturbed condition. The ash-colored shale is undoubtedly of Cretaceous age, and is a continuation of the horizontal beds on the Little Souris, holding *Inoceramus* in great abundance, and of large size.

On the Little Souris the Cretaceous rocks are exposed for a distance of 50 miles. They are loaded with nodules and concretions, holding abundance of carbonate and oxide of iron. The Blue Hills south of the Assiniboine are covered with the debris of this rock. It appears 10 feet below the level of the Prairie, at the mouth of the first of the Two creeks below Fort Ellice, affluents of the Assiniboine. It is also seen on a small tributary of the Rapid River, and in several places on the Qu'Appelle, east of the Mission, and on the east flank of the Riding Mountain. In a former report I have mentioned that brine yielding springs occur from Swan River to La Rivière Salè, a distance of 230 miles. Whether the salt-bearing rocks belong to recognized members of Devonian age, is a question yet undetermined ; but, as the whole of the fossils which I have collected will be submitted, when they arrive, to Mr. Billings, the palæontologist of the Canadian Geological Survey, their precise position will then be determined. It is sufficient, at present, to state that salt springs occur on the east flank of Dauphin Lake, within 10 miles of the outcrop of the Cretaceous rocks on the flanks of the Riding Mountain, which leads to the inference that the Carboniferous group is totally wanting in the region where it might be supposed to exist, between Lakes Manitobah and Winnipego-sis, and the range of high land forming the eastern water-shed of the Assiniboine.

On the Qu'Appelle, sixteen miles from the South Branch of the Saskatchewan, a greenish colored arenaceous rock occurs, destitute of fossils, but intersected with veins of selenite, and holding a large number of concretionary masses. Many of these concretions have fallen into the bed of the river, or are exposed in its banks, where the Qu'Appelle comes from the Eyebrow Hills and enters the Great Valley. Many concretions in the rock referred to were three, four, and five feet in diameter, very hard, and, when broken with a sledge hammer, portions often "peeled" off like the coats of an onion. The selenite generally occurs in fragmentary portions about 6 inches long, but the veins are easily traced for many feet, most commonly in a vertical direction. In an admirable paper on the Cretaceous strata of the United States, by the distinguished Palæontologist of the New York State Geological Survey, James Hall, Esq., reference is made to the report of Mr. Nicollet, on the Cretaceous formations of the upper Missouri. In section C of Mr. Nicollet's subdivision of the rocks of that region, the formation is described as "a ferruginous sand, of a yellowish colour, containing masses resembling septaria and seams of selenite." On the South Branch, a few miles north-west of the "River that turns," there is an extensive exposure of a yellowish ferruginous sand, holding septaria and concretions, with seams of selenite. The rock is Cretaceous, and I think it probable that it is identical with formation C of Mr. Nicollet. If so, it constitutes one of the uppermost members of the Cretaceous system ; and the coal in situ, noticed in a former report as occurring about 80 miles south-west of the Qu'Appelle or Calling Mission, will most probably be of Tertiary age. I think, however, that the fossils collected on the Saskatchewan, and throughout the entire region explored, will be amply sufficient to establish the true position of the rock formations over the greater part of the country visited. It is sufficient for present purposes, to mention that the addition which has been made to our geological knowledge of this country may be thus briefly stated :

1. The eastern flanks of the Riding and Duck Mountains as far as the Pasquia Hill, form the present eastern limits of the Cretaceous rocks of this region.

2. The Cretaceous rocks occupy the whole of the country from the Riding and Duck Mountains and Pasquia Hill, to the South Branch of the Saskatchewan.

3. The Cretaceous rocks are seen in situ, undisturbed and nearly horizontal, at an altitude not exceeding 400 to 600* feet above rock of Devonian age, recognized in situ 30 miles to the east.

4. Brine Springs, similar in all respects to the Brine Springs issuing from Devonian rocks in situ, occur within ten miles east and north east of the outcrop of the Cretaceous rocks on the east flank of the Riding Mountain.

5. The Riding Mountain in its former extension probably covered the area now occupied by the great lakes, from which it has been removed by denudation.

6. The Cretaceous rocks probably repose on the brine-bearing rocks of Devonian age on the flanks and east of the Riding mountain, and as far north as the Pasquia Hill.

7. It is not probable that any outcrop of the Carboniferous rocks will be found to exist in the eastern part of the valley of the Saskatchewan. The Lignite or Coal of the Souris appears to be of Tertiary age.

With reference to the Lignite on the Little Souris, it may be here stated, that a very careful search was made for it in position, but without success. A cutting into the bank just above where a fine exposure of Cretaceous rocks occurs, holding *Inoceramus* from four to nine inches in length, showed no less than five distinct beaches, in each of which numerous water worn masses of Lignite from three inches to one foot in diameter were discovered. In several places the accumulation of lignite boulders was very extensive, and might become of economic value. But in no instance was the Lignite observed in place on the Souris. The boulders were generally found in a highly ferruginous sand; when burned they emitted a strong sulphurous odour, showing the presence of iron pyrites. The "grain" of the wood could be perceived with the greatest ease when large masses were broken open, and not unfrequently particles and strings of amber were found in the interior. The specimens I have brought to Toronto have cracked on becoming dry in many directions; they will, however, serve to illustrate the character of the singular accumulation of boulder lignite in the valley of the Little Souris.

Until I have had an opportunity of submitting my collection of fossils, illustrating the rock formations of the country to Sir William Logan and Mr. Billings, I refrain from giving expression

* The section exposed on the flanks of the Riding Mountain was on the side of a gully 200 feet deep. The exposure was traced from top to bottom. The bottom of the gully is about 400 feet above Dauphin Lake, and 420 feet above the last exposure of limestone seen on Moss River.

to any further views respecting the geological features of the region explored. I think I am in possession of sufficient materials upon which a tolerably accurate geological map of the country from the Great Lakes to the South Branch of the Saskatchewan can be constructed. But as this is a work involving much cautious enquiry, and the co-operaton of gentlemen thoroughly acquainted with the fossils of the secondary rocks, some months must elapse before a geological map can be prepared.

CLIMATE.

In a communication, dated 2nd February, "On the Qu'Appelle or Calling River Valley," I introduced some remarks on the climate or rather seasons of the South Branch, in comparison with the North Branch at the Forks and Fort à la Corne. The impression conveyed by the progress of vegetation in these far separated parts of the country led to the opinion that the period of flowering and of ripening fruit on the South Branch at the Elbow was two or three weeks in advance of similar periods on the North Branch. The vegetable productions in the gardens attached to Fort à la Corne, with a brief notice of the periods of planting and gathering will show that the climatic adaptation of the North Branch near the Grand Forks is not of a character unfavorable to agricultural operations. As this subject is one of great importance I have ventured to introduce some extracts from the journal of the Fort, which are both interesting and valuable.

On the 7th August, in the garden attached to Fort à la Corne (about 18 miles below the Grand Forks) potatoes were in flower, and the tubers of early varieties of the size of hens' eggs. Cabbages were well formed. Beet roots and carrots quite ready for the kitchen. Indian corn in silk, from seed which was grown in the garden last year. Peas ready for gathering.

No disease has yet been noticed in the potatoes; and the grasshoppers, that scourge of the country south of the Touchwood Hills, have not made their appearance at Fort à la Corne.

In the garden attached to the Nepoween Mission, under the charge of the Rev. Henry Budd (a zealous missionary of native origin), all the vegetables gave promise of fair and remunerative crops. The potatoes were superb; turnips, both Swedes and white, remarkably fine; Indian corn, from seed grown on the spot last year, in silk; wheat rather too rank in the stalk—it measured 5 ft. 3 in. in length to the ear, which was well formed but green, and it seemed doubtful whether it would ripen. Mr. Budd speaks very favourably of the soil, climate, and extent of land available for agricultural purposes. Both the mission and the fort are situated within the excavated valley of the Saskatchewan, and are not, in my opinion, so favorably placed for farming purposes as they might be in the Valley of Long Creek. The river, however, is the great highway, and, during the season, affords an abundant supply of sturgeon.

Extracts from the Journal at Fort à la Corne, Saskatchewan River. Lat. 53·27; *long.* 104·30 *W.*

1851.

Oct. 25. Ice made its appearance in the river.

1852.

April 8. Ice solid for the season of the year.
" 12. Ice started.
" 13. Ice drifting and lodging on the banks.
Ap'l. 21. Ice drifting and disappearing along the banks.
" 22. Garden operations commenced.
May 14. First sturgeon caught
" 24. Planted potatoes.
Oct. 11. Finished taking up potatoes.
" 25. Fishing season ended.
" 26. Snow.
Nov. 3. Ice floating in the river.

1854.

Apr. 14. River broke up. On the 15th nearly clear of ice.
" 28. Garden operations commenced.
May 1. First sturgeon caught.
" 8. Preparing potato fields.
" 13. Potato planting.
Oct. 2. Gathered turnips.
" 3. Taking up carrots.
" 10. Commenced taking up potatoes at the mission (190 kegs), turnips, carrots, cabbages—large and good.
" 11. Cabbages taken up.

1855.

May 24. Turnips sown.
Sep. 12. Hard frost over night.
" 27. Took up potatoes—poor crop, much destroyed by grubs.
" 29. Hard frost. A little ice seen at the gates.
Oct. 1. Women digging potatoes.
" 2. Do. do. do.
" 3. Taking up turnips.
" 22. Ice on the edges of river.

1856.

April 2. Hard frost last night.
" 4. Water making its appearance on the edges of the river.
" 7. Froze hard last night.
" 9. Ice made a start.
" 17. Ice drifting.
" 23. Fall of snow during the night.
" 23. Nets set. One sturgeon caught.
" 25. Hard frost.
May 2. Garden operations commenced.
" 10. Storm of snow.
" 12. Planted potatoes.
" 14. Sowed Swedes.
Sep. 16. Slight frost last night.
Oct. 2. Commenced taking up potatoes.
" 22. Hard frost during night.
" 23. Severe frost during night.
" 26. Snow in night.
Nov. 11. River full of ice.

1857.

April 9. Water appearing on the edges of the river. Snow shoes required everywhere.
" 16. Ice started to-day.
" 24. Snowed without intermission the whole day.
May 3. Ice drifting all last night.
" 5. River full of ice.
" 12. Planted potatoes and onions.
" 20. Planting potatoes. Three sturgeons caught.
June 2. Hard frost last night.
" 30. Starvation is staring the people in the face. Have caught no sturgeon for some time back.*

* A common record in the journals at the different posts in Rupert's Land. The cause must be referred to the habits of the people, their occupation, &c., and not to the capabilities of the country.—H. Y. H.

1858.
Apr. 21. Ice drifting. Large quantities of ice on the banks.
May. 1. Clearing up of north garden.
" 7. Preparing potato ground. First sturgeon caught.
" 12. Planted potatoes.
" 17. Slight fall of snow.
" 18. Wind from N. and cold. Think we are going to have a second winter.

In the General Report of the Expedition which is already well advanced, I shall have an opportunity of describing not only the topographical and geological features of the country in detail, but also the habits and customs of the Indian tribes with whom we came in contact; the condition and prospects of the Missionary Stations; the Forts and Posts of the Honble. Hudson's Bay Company: the character and influence of the Fur trade; the history and progress of the devastating host of grasshoppers which we traced for more than 600 miles in the prairie region, &c. &c. &c.

I have much pleasure in having this opportunity of expressing my warmest thanks to Sir George Simpson, not only for the letters of introduction with which he favoured me to the officers of the Honble. Hudson's Bay Company's service in Rupert's Land, but also for his personal efforts when at Fort Garry, to facilitate the progress of the expedition by every means in his power. The assistance rendered by Sir George Simpson was of the greatest use to me, and the kind and courteous manner in which it was granted increases my indebtedness to him.

From the officers of the Hon. Hudson's Bay Company's service in charge of the different posts, I received without any exception, kind attention and valuable assistance. To Mr. McTavish, Chief Factor, in charge of Fort Garry, Mr. Lily of the Stone Fort, Mr. Sinclair, Chief Factor, then in charge of Fort Alexander, Mr. McKenzie of Manitobah House, Mr. McKenzie of Pembina; the gentlemen in temporary charge at the Touchwood Hills, Fort Ellice, Fort Pelly, Fort à la Corne, and Cumberland House,—I beg to express my grateful thanks. I shall elsewhere have an opportunity of recording many friendly acts, which would be out of place in a preliminary report.

The aggregate distance travelled by the Expedition in the region marked out for exploration, was as follows:

On horseback	2392	miles.
In small canoes	1263	"
In freighters' boat	685	"
On foot..	111	"
Aggregate distance	4451	miles.

In journeying to Red River, where the exploration commenced, the route followed was by the Great Western Railway to Detroit; thence by steamboat to the Grand Portage, Lake Superior. From Grand Portage the voyage to Red River was made in north canoes, a distance of 636 miles. Returning, we travelled in dog carioles from Fort Garry to Crow Wing, a distance of 410 miles, by the winter road; thence by stage to La Crosse, on the Mississippi; and from La Crosse to Toronto by rail.

I have the honor to be,
Your obedient servant,
HENRY Y. HIND.

The Hon. Charles Alleyn, M.P.P.,
Provincial Secretary.

TABLE SHOWING THE DIMENSIONS OF VALLEYS AND RIVERS.

Name and Place.	Width.	Depth.		Rate of Current.	Remarks.
		Greatest.	Mean.	mil. p. hr.	
Red River, Middle Settlement	480 feet.	18 feet.	12 feet.	1¼	
Assiniboine River, Lane's Post	120 "	7.6 "	6 "	1½	
" " Prairie Portage	....	....	....	2	Falls 1.18 feet per mile.*
" Valley, Junction with Little Souris	2550 feet.	....	83 feet.	....	The depth below the prairie applies to the North Bank of the river only, the slope on South being very gradual.
" River	230 "	12 "	8.6 "	1¼	
" Valley, at Fort Ellice	1 m. 30 ch.	....	240 "	....	
" River	136 feet.	11.9 feet.	8 "	1¾	
Little Souris River, at its mouth	121 "	3.6 "	2.4 "	½	
" " Valley, at Back-fat, or Mussel Creek	8016 "	....	255 "		
" " " near Snake Hill	330 "	....	66 "	....	This depth applies to west side only, east side rises very gradually to about same level. Lignite 45′ above river
" " River, at Snake Hill	100 "	4 feet.			
Qu'Appelle Valley, seventy miles from Assiniboine	78 chains.	....	310 feet.		
" " at Mission	1 m. 20 ch.	....	250 "		
" " near Grand Forks	1 m. 30 ch.	....	220 "		
" " each end of Sandhill Lake	1 m. 5 ch.	....	140 "		
" " eight miles from west of Sandhill Lake	1 m. 70 ch.	....	150 "		
" " between Height of Land and Saskatchewan	73 chains.	....	110 "		
Qu'Appelle River, at its mouth	88 feet.	12 feet.	8 "	1½	
" " two miles from mouth	66 "	7.6 "	6.6 "	1½	
" " at Mission	48 "	6 "	3.6 "	1	Falls 56′ (7 inches) per mile.*
Saskatchewan, South Branch, (28 miles from Qu'Appelle Valley)	1848 "	10 "	4.6 "	2½	Channels on both sides 6 to 10 feet deep.
" Main River, Fort a la Corne	967 "	20 feet.	14 feet.	3	
" Near Tearing River	980 "	22 "	20 "	2	Falls 16′ (2 inches) per mile.*

* Fall determined by the Level.

GENERAL REPORT

AND

NARRATIVE OF THE EXPEDITION.

GENERAL REPORT.

CHAPTER I.

FORT GARRY TO THE MOUTH OF THE LITTLE SOURIS RIVER.—THE MOUTH OF THE LITTLE SOURIS TO THE BOUNDARY LINE.

The Start—Supplies—Prairie Ridges—The Big Ridge—Pigeon Traps—Stony Mountain—Birds—Saline Efflorescence—Character of the Big Ridge—The Assiniboine—Grasshoppers—Ojibway encampment—Archdeacon Cochrane—Prairie Portage—Cliff Swallow—Thunder Storms—Ojibways—The Bad Woods—Assiniboine Forest—River-Rabbits—Sandy Hills of the Assiniboine—Latitude—Dimensions of Valley—Variation of Compass—Sand Dunes—Aspect of Country—Hail storm—Balsam Spruce—Pine Creek—The Little Souris—Grasshoppers—Fish—Sioux—Cretaceous Rocks—Blue Hills—Pembina River—Backfat Lakes—Vast Prairie—Prairie Fires—Horizontal Rocks—Inoceramus—Guelder Rose—Lignite—Ancient Lake beaches—Sand Dunes—Oak Lake—Souris Sand Hills—Nighthawk—Bog Iron Ore—Floods in 1852*—Grasshoppers, Infinite multitude of—Appearance of the Sky, of Prairie—Little Souris Valley—Tracks—Turtle Mountain—Sioux—Character of Prairie—Souris Lakes—Boulders—Mandan Village—Character of the Souris south of the* 49*th parallel.*

On the morning of the 9th June, 1858, the Half-breeds engaged for the Expedition into the Prairie country west of Red River, assembled at our temporary quarters in the Settlement, and began at once to load five Red River carts and a waggon of American manufacture, with two canoes, camp equipage, instruments, and provisions for a three months journey. At noon the start was made, and the train proceeded to Fort Garry,* a distance of eight miles, to take in a supply of flour and pemican. We camped about half a mile from the Fort and took an inventory of our baggage, and made such regulations and arrangements as are considered necessary at the commencement of a long journey through a country partly inhabited by hostile tribes of Indians, and not always affording a supply of food even to skilled hunters.

The whole party consisted of thirteen individuals besides myself, namely: Mr. Dickinson, Surveyor; Mr. Fleming, Assistant Surveyor; Mr. Hime, Photographer and Assistant Surveyor; six Cree half-breeds, a native of Red River of Scotch descent, one Blackfoot half-breed, one Ojibway half-breed, and one French Canadian. Our provisions consisted of one thousand pounds of flour, four hundred pounds of pemican, one thousand rations of Crimean vegetables, a sheep, three hams, and tea for three months, with a few luxuries, such as pickles, chocolate, a gallon of port wine, and one gallon of brandy. Each cart was loaded with about 450 lbs. weight, and the waggon with double that amount. The canoes of birch bark, 18 feet long, weighed 150 lbs. each. At the White Horse Plains, 22 miles from Fort Garry, we purchased an ox to serve as a *dernier resort* in case we should not meet with buffalo; and at Prairie Portage, the last settlement on the Assiniboine, I engaged the services of an old hunter of Cree origin, who had been from his youth familiar with Indian habits and stratagems. This addition increased the party and material, before we left the last settlement, to fifteen men, fifteen horses, six Red River carts, one waggon, and one ox.

Leaving our camp early on the morning of the tenth, we ascertained by levelling the altitude of an ancient lake ridge, near to St. James' Church, to be eleven feet above the prairie at Fort Garry, and about two miles from it. These ridges are common in the prairies of Red River, and do not necessarily point to an ancient lake margin. It is probable that most of them were formed under water. They may be traced for many miles, but are sometimes lost in the general rise of the prairie.

The ancient boundaries of Lake Winnipeg, when its waters were about 90 feet above their present altitude and occupied the whole of the country now covered by Lakes Manitobah, Winnipego-sis, and Winnipeg, with the intervening low land, is well defined in one direction by the Big Ridge, which on one side or another of Red River is easily traced for more than three hundred miles; it is shown on the map. On arriving at St. James' Church, we separated into two divisions, Mr. Fleming and Mr. Hime, with the carts and waggon, proceeding to Lane's Post on the Assiniboine, 22 miles from Fort Garry, while Mr. Dickinson and myself, with two half-breeds, struck in a north-westerly direction across the prairie to Stony Mountain, and thence to the Big Ridge, having arranged to meet at Prairie Portage.

In a wheat field opposite St. James' Church were several pigeon traps, constructed of nets 20 feet long by 15 broad, stretched upon a frame; one side was propped up by a pole 8 feet long, so that when the birds passed under the net to pick up the grain strewed beneath, a man or boy concealed by the fence withdrew the prop by a string attached to it, and the falling net sometimes succeeded in entrapping a score or more of pigeons at one fall. Near the net some dead trees are

* "The mean of five observations at Upper Fort Garry, at the mouth of the Assiniboine, for latitude, three meridian by altitude of the Sun and two by Polaris, gave for the latiude 49° 53′ 24″. Mr. Calhoun, who was attached to Major Long's expedition in 1823, made it 49° 53′ 35″, but according to a record in the possession of one of the officers of the Fort, Lefroy placed it in latitude 49° 58′." Owen's Geological Survey of Wisconsin, Iowa, and Minnesota, p. 180.

placed for the pigeons to perch on, and sometimes stuffed birds are used as decoys to attract passing flocks.

In pursuing our course to Stony Mountain we endeavoured to follow the ridge before alluded to, but after tracing it for several miles it became imperceptibly blended with the level prairie. Several ridges were crossed after we lost the first, but in all cases they died away after having preserved their rounded form for two or three miles. Stony Mountain is a Limestone island of Silurian age, (?) having escaped the denuding forces which excavated Red River valley. It is about four miles in circumference, its highest point is 66 feet above the prairie level. Horizontal layers of Limestone, holding very few and obscure fossils, project on its western cliff-like sides. Its eastern side is gently sloping, and some ten feet from the summit the remains of an ancient lake beach is well preserved. Viewed from a distance Stony Mountain requires little effort of the imagination to recall the time when the shallow waters of a former extension of Lake Winnipeg washed the beach on its flank, or threw up as they gradually receded, ridge after ridge over its level floor, where now are to be found wide and beautiful prairies, covered with a rich profusion of long grass.

Leaving the Stony Mountain our course lay westerly through a wet prairie to the Big Ridge. Gray cranes, ducks, and plover were numerous on the marshy tracts, and in every little bluff* of aspen or willow the beautiful rice birds were seen or heard. Where we camped on the edge of a lake near the the foot of the Big Ridge, bittern, grackle, and several varieties of duck flew to and fro in alarm at our invasion of their retreats. On the flank of the Big Ridge the Cinnamon or solitary thrush was noticed; but most common of all was the Tyrant flycatcher, who endeavoured to hold undisputed sway over the bluff he had selected as his home. Near and west of Stony Mountain many small barren areas occur covered with a saline efflorescence. They may be traced to the Assiniboine and beyond that river in a direction nearly due south to la Rivèire Salé, and the 49th parallel. These saline deposits are important, as they in all probability serve, as will be shown hereafter, to denote the presence of salt bearing rocks beneath them, similar to those from which the salt springs of Swan River, Manitobah Lake, and La Rivière Salé issue.

Early on the morning of the 17th, we ascended the Big Ridge. Its elevation above the Prairie is about 60 feet; on its south side it slopes gently to the Prairie level, on its north side is a Plateau well wooded with aspens. The view from its summit extends far and wide over the Assiniboine Prairies. On the south flank, and skirting its base, are groves of aspen and balsam poplar, with scattered oak trees and willow bushes. The pasturage in the open glades is of the first quality. The ridge is quite level and about 80 to 100 feet broad, destitute of trees, slightly arched and composed of gravel forming an excellent road. Here and there it is cut by rivulets draining the marshes in the Plateau on its northern side. As it approaches Prairie Portage its apparent elevation diminishes, until at the Portage River it is no longer discernible. We traced it for a distance of 70 miles. It will be mentioned further on that this ridge, or one formed at the same period, is again seen west of Manitobah Lake, near the Hudson Bay Company's Post, Manitobah House. It continues to preserve there the same characters of horizontality, uniform outline, gravelly formation and admirable suitability for the purposes of a road which have been noticed in connection with its extension north of the Assiboine and east of Red River. For many miles, ties for a Railway might be laid upon it without a pebble being removed, and the only breaks in its continuity occur where streams from the Plateau and higher grounds in the rear have forced a passage through it. It follows, however, the south and western contour of Lakes Winnipeg and Manitobah, and passes through a country not likely to be first selected by a large body of settlers. It is important, in so far that it forms the boundary of land of the first quality which occupies the low Prairie Valleys of the Assiniboine and Red River. Soundings in Lake Manitobah showed such a uniform depth of eighteen feet for a distance exceeding 60 miles along its south eastern coast that, if its bed were exposed, it is probable that it would, in process of time, also become a rich and extensive prairie country, with its present beach, distinctly visible as its old boundary. Indeed, the aspect of this drained country for several miles beyond the Big Ridge, both on the Assiniboine and Red River, is similar to the undrained marshes, ridges and bogs which exist on the west coast of Lake Manitobah, and points to a very gradual but constant draining of this region.

* The Half-breeds call little groves of aspens or willows in the prairies "bluffs."

We reached Prairie Portage in the evening, where we joined the main party. The Assiniboine at Lane's Post (June 16th) is 120 feet broad; its turbid water flows at the rate of one mile and a half per hour. A few miles west of Lane's Post, the saline efflorescence before noticed as occurring in patches on the prairies and forming small barren areas, is no more to be seen; it consists of chloride of sodium and sulphate of magnesia, with a little chloride of calcium.

Grasshoppers were first observed at Lane's Post this year, they were the brood from the eggs deposited by a swarm which alighted on the White Horse Plains in September last. At Prairie Portage we found an Ojibway encampment in which were some of the refractory personages who had hitherto resisted the humane and unceasing efforts of Archdeacon Cochrane to christianize them. Among the various methods tried by the Archdeacon to induce them to settle and farm, the first preliminary to the progress of christianity among wild Indians, that of presenting the most docile with an ox and plough and teaching them to use it, was the least successful. At the first good opportunity, or during a time of scarcity, the ox and plough would be sold to the highest bidder for very much less than it cost. A prômise to add another ox at the end of a year if the first gift was faithfully preserved was of no avail,—the charms of the buffalo plains were too tempting or the seduction of gambling too powerful to be withstood, notwithstanding the most solemn heathen promises. The school, however, gives better hope, and no doubt the rising generation, both Indian and half-breed at Prairie Portage, will form a thriving, industrious and christian community.

Prairie Portage is very delightfully situated 65 miles west of Fort Garry, on the banks of the Assiniboine. The prairie here is of the richest description, towards the north and east, boundless to the eye. The river bank is fringed with fine oak, elm, ash, and ash leaved maple; on the south side is a forest from three to six miles deep; the river abounds in sturgeon and gold eyes, and, within 18 miles, there is a splendid fishing station on the coast of Lake Manitobah, where the Portage people take vast numbers of white fish every fall. The old water course of the Assiniboine, near the Portage, now a long narrow lake, fringed with tall reeds, teems with wild fowl and grackle, among which we frequently noticed and procured specimens of the yellow-headed blackbird.

Prairie Portage will become an important settlement, not only on account of the vast extent of fertile country which surrounds it, but because it lies in the track of the buffalo hunters proceeding to the Grand Coteau and the South Branch by way of the Souris River. It is also near to the fertile country unwatered by White Mud River, and the road to the south western flanks of the Riding Mountain passes by the Portage. The current of the river is very uniform here, careful levelling showed that it fell $1\frac{18}{100}$ inches a mile; its speed is two miles an hour. The cliff swallow (*hirundo fulva*) had built its nests in great numbers on the banks of the river, which are about 16 feet above the level of the water; I counted no less than thirteen groups of their nests within a distance of five miles, when drifting down in a canoe. The cliff swallow was afterwards seen in great numbers on the Little Souris, the South Branch of the Saskatchewan, and the Qu'Appelle River.

The first of a series of thunder storms which lasted for some weeks visited us this afternoon (17th). The warm rain fell in torrents and thoroughly wet all who were exposed. Pigeons were flying in vast numbers across the Assiniboine, and the black tern was numerous in the prairies near the settlement. In descending the river for a few miles to inspect its banks, we had occasion to pass by a fish weir where a number of Ojibways from the camp near the Portage were watching with spears in their hands for sturgeon. They took no notice of us as we passed, being too busily engaged, but on our return to the encampment we found them waiting with fish to barter for tobacco and tea. We made them a few trifling presents, and, by way of recompense, sustained during the night the loss of a fine cheese, which after curiously eyeing during supper, they had modestly asked for a morsel to taste. They found it excellent, no doubt, and quietly in the dead of night, opened the basket in which it had been placed and abstracted it. In future, when Indians were around, all eatables and articles they might covet were properly secured, and the cheese proved to be our only loss during the exploration.

Leaving Prairie Portage on the morning of the 19th, we took the trail leading to the Bad Woods, a name given to a woody district about 30 miles long, by the buffalo hunters in 1852, who, in consequence of the floods of that year could not pass to their crossing place at the Grand Rapids of the Assiniboine by the Plain or Prairie Road. There were four hundred carts in the band and the hunters were compelled to cut a road through the forest of small aspens which form the Bad Woods, to enable them to reach the high Buffalo Prairies. This labour occupied them several days, and will be long remembered in the settlements in consequence of the misery entailed on the children and women.

The trail lay for three miles through a continuation of the low prairies of the Assiniboine, until a sudden ascent of 20 to 25 feet introduced us to a different kind of country, the plateau beyond the Big Ridge, which here crosses the river, and forms the lowest or first step of the Pembina Mountain. The physical features of this boundary to a great table land will be noticed at length in the sequel. The soil continues poor and sandy for several miles, supporting clumps of aspen with a few oak in low places. The view across the Assiniboine reveals in the distance the Blue Hills, and between them and the river is a vast forest, which a subsequent exploration in the autumn showed to consist for two to three miles nearest to the Assiniboine, of oak, elm, ash and aspens, beyond this limit the forest is almost entirely composed of aspens of small growth.

Grasshoppers were observed in great numbers, and the first humming bird was seen here. The banks of the river showed recent water marks twelve feet above its present level, willow and other trees overhanging the stream being barked by the action of ice during spring freshets at that elevation. Everywhere rabbits are numerous, and considerable areas occur covered with dead willows and young aspens, barked by these animals in the winter about two feet six inches above the ground. The height of the bank is 80 feet above the valley, denoting a rapid rise in the general level of the country.

On the morning of the 20th we entered the Bad Woods, and followed the road cut by the hunters in 1852. The aspens were much disfigured by countless numbers of caterpillars resembling those of the destructive Palmer worm. In the afternoon we arrived at the Sandy Hills; they consist of rounded knolls covered with scrub oak and aspens. Our latitude to day was ascertained to be 49° 46′ 19″, the height of the prairie 150 feet above the river, the breadth of the valley in which the river flowed 5680 feet, and the variation of the compass 13° E. After passing the point where the foregoing observations were made, the trail again enters the Bad Woods and continues through them until it strikes the Sandy Hills again. These rounded eminences have all the appearance of sand dunes, covered with short grass and very stunted vegetation.

As we emerged from the Bad Woods a noble elk trotted to the top of a hillock and surveyed the surrounding country; a slight breath soon carried our wind as the hunter was endeavouring to approach him, he raised his head, snuffed the air and bounded off. Another terrible thunderstorm came on at sunset, with heavy rain and boisterous wind. The aspect of the country for many miles is that of a plain sloping gently to the west, covered with innumerable mounds or hillocks of sand, scarcely clothed with vegetation; here and there small lakes or ponds are found, fringed with rich verdure, but its general character is that of sterility. From the summit of an imposing sand-hill, formerly a drifting dune, which we ascended on the 21st, the country lay mapped at our feet; as far as the eye could reach, north, east and west, sand hills, sometimes bare, but generally covered with short grass met the eye.

On the afternoon of this day a hail storm of unusual violence caused us to halt. The stones penetrated the bark of our canoes and broke off the gum. The grasshoppers, which were very numerous just before the storm began, suddenly disappeared, but they might be found quietly clinging to the leaves of grass in anticipation of the storm. After it had passed, they re-appeared, apparently in undiminished numbers, although every member of the party, crouching for shelter under the carts and waggon, fully expected the complete annihilation of these destructive and troublesome insects. A singular instinct enables them to seek and find refuge, even from a pitiless hailstorm or a drenching rain. The same evening a thunderstorm again visited us, but the sun set in gorgeous magnificence, with a brilliant rainbow and vivid flashes of lightning in the east. The Cinnamon thrush is not uncommon among the sandy hills, we saw several during the day. The next day we reached the pines, for which we had been anxiously looking, but to our disappointment they proved to be nothing more than balsam spruce in scattered clumps. Another thunderstorm this evening.

On the 23rd we passed for a distance of eight miles through

a country of sand ridges, until we reached Pine Creek. Here the Sand Hills are absolutely bare, and in fact drifting dunes. Sending the main party in advance, Mr. Dickinson and I set out to examine the valley of the Assiniboine where Pine Creek disembogues. The sand dunes were seen reposing on the prairie level, about 150 to 180 feet above the river. In crossing the country to regain the carts, our course lay across a broad area of drifting sand beautifully ripple-marked, with here and there numbers of the bleached bones of buffalo protruding from the west sides of the dunes, memorials probably of former scenes of slaughter in buffalo pounds similar to those which we witnessed some weeks afterwards at the Sandy Hills on the South Branch of the Saskatchewan. The progress of the dunes is very marked, old hillocks partially covered with her bage are gradually drifted by the prevailing westerly wind to form a new one. Sometimes the area of pure sand was a mile across, but generally not more than half that distance. The largest expanse we saw was near the mouth of Pine Creek, it is called by the Indians "the Devil's Hills," and a more dreary, parched-looking region could scarcely be imagined.

We reached the mouth of the Little Souris River on the 24th, and made preparations to cross the Assiniboine at this point. The distance travelled through the Sandy Hills was about 48 miles, their breadth does not exceed ten miles. At the mouth of the Souris the grasshoppers were in countless numbers and so voracious as to attack and destroy every article of clothing left for a few minutes on the grass. Saddles, girths, leather bags, and clothing of any description were eaten without distinction. Ten minutes sufficed them, as our half-breeds found to their cost, to destroy three pair of woollen trousers which had been carelessly thrown on the grass. The only way to protect our property from the depredators was to pile it on the waggon and carts out of reach. There were two distinct broods of grasshoppers, one with wings not yet formed, which had been hatched on the spot, the other full-grown, invaders from the prairies south of the Assiniboine. We noticed here to-day the first flight of these insects which afterwards were witnessed on a scale of alarming magnitude, giving rise in their passage through the air to optical phenomena of very rare and beautiful descriptions. As we cautiously approached the bank of the river opposite the mouth of the Little Souris on the look-out for Sioux Indians, some jumping deer and a female elk were observed gambolling in the river. A shot from a Minie rifle dispersed them and started from their lair two wolves who were watching the deer, patiently waiting for an opportunity to surprise them.

The volume of water in both rivers was carefully measured at the point of junction. The Assiniboine was found to be two hundred and thirty feet broad, with a mean depth of six feet, and a current of one mile and a quarter per hour. The Little Souris was one hundred and twenty-one feet broad, two feet four inches mean depth, and flowing at the rate of half a mile an hour. Observing numbers of fish rising at grasshoppers in the Souris we stretched a gill net across the mouth of the river and succeeded in taking pickerel, goldeyes and suckers, the grey and the red. In a second attempt we caught a tartar, a huge sturgeon got entangled in the meshes of the gill net, and before we could land him he succeeded in breaking away and carrying a portion of the net along with him.

Signs of Sioux Indians in the neighbourhood led to our keeping watch during the night; and on the morning of the 25th we proceeded cautiously up the valley of the river, keeping a sharp look out. On the left bank the Blue Hills of the Souris are visible ten miles from the mouth of the stream, and towards the west the Moose Head Mountain is seen to approach the Grand Rapids of the Assiniboine. The first rock-exposure in the valley was observed about fifteen miles from the mouth of the Souris. It consisted of a very fissile, dark-blue argillaceous shale, holding numerous concretions containing a large per-centage of iron, partly in the state of carbonate and partly as the peroxide. Some very obscure fossils were found, with fragments of a large Inoceramus. The shale weathers ash white. It is exposed in a cliff about ninety feet high. The upper portion of the cliff consists of yellow sand, superimposed by sandy loam holding limestone boulders and pebbles. The exposure of shale is seventy feet thick, in horizontal layers. The country west of the Souris, so far, is an open, treeless, undulating prairie. On the east side the Blue Hills are very picturesque, with their flanks and summits wooded with aspen. Rain as usual; the day closed with a thunder storm.

On the 27th we arrived at the westerly Bend of the Souris in the midst of a very lovely, undulating country; the river is here fifty feet broad, and in its passage through the Blue Hills it has excavated a valley fully four hundred and fifty feet deep. Rock exposures are of frequent occurrence, the dip being 3° south. Fragments and perfect forms, but very fragile, of a large Inoceramus are very common. The feruginous concretions are disposed in regular layers and constitute a marked feature of the Cretaceous rocks of this valley. A continuation of the valley of the Souris extends in a direction nearly south-east towards Pembina River, with which it is said by the half-breeds to interlock. Three lakes visible from our camp were said to be the sources of the Pembina River; a little stream issuing from the most westerly of these is called Back-fat rivulet, it flows into the Souris. Deer are very numerous at this beautiful bend of the river; it appears to be a favorite watering place. The half-breeds of St. Joseph often cross it at this bend when on their hunting expeditions to the Grand Coteau. It is not improbable that it will become a point of importance if ever an emigrant route should be established from Minnesota to the Pacific, *viâ* the South Branch of the Saskatchewan; and from the great distance saved by going through St. Joseph, instead of Fort Garry it is not improbable that this may yet be the case.

On the30th we succeeded in passing the Blue Hills, and enjoyed on the evening of the same day one of the most sublime and grand spectacles of its kind which it is possible to witness. Before leaving the last ridge of the Blue Hills we suddenly came upon the borders of a boundless level prairie, one hundred and fifty feet below us and of a rich, dark-green colour, without a tree or shrub, and with one solitary conical hill in its centre. Here we expected to find Buffalo, but not a sign of any living creature could be detected with the aid of a good glass. The prairie had been burnt last autumn, and the Buffalo had not arrived from the south or west to people this beautiful level waste. What a magnificent spectacle this vast prairie must have furnished when the fire ran over it before the strong west wind!

From beyond the South Branch of the Saskatchewan to Red River all the prairies were burned last autumn, a vast conflagration extending for one thousand miles in length and several hundreds in breadth. The dry season had so withered the grass that the whole country of the Saskatchewan was in flames. The Rev. Henry Budd, a native Missionary at the Nepoween, on the North Branch of the Saskatchewan, told me that in whatever direction he turned in September last the country seemed in a blaze; we traced the fire from the 49th parallel to the 53rd, and from the

98th to the 107th degree of longitude. It extended, no doubt, to the Rocky Mountains.

A few miles west of the Blue Hills, being anxious to ascertain the dip of a very remarkable exposure of shale with bands of ferruginous concretions, Mr. Dickenson levelled with the utmost care an exposure facing the south, and found it to be perfectly horizontal. At the base of the exposure, and on a level with the water's edge we succeeded in finding a layer of rock full of gigantic Inoceramus. One specimen measured 8½ inches in diameter, it was very fragile, but the peculiar prismatic structure of the shell was remarkably well preserved. On attempting to raise it, it separated into thousands of minute prisms so characteristic of this shell.

Vast numbers of pigeons were flying in a north-westerly direction, and our friends the grasshoppers were everywhere abundant. From the Blue Hills to the South Bend of the river, rock exposures possessing the characteristics already noticed occurred at every bend of the river. The first specimen of Lignite was seen near the mouth of Plum Creek, where we camped on the 29th. It was a water-worn rounded boulder. On points of the river valley some fine oak, elm, balsam poplar and aspen are found for the first twenty miles. The guelder rose is common, wild prairie roses abundant, snowberry and two varieties of cherry of frequent occurrence, as well as woodbine, wild convolvulus and hop.

A little beyond Plum or Snake Creek we found numerous pebbles and boulders of Lignite, and with a view to ascertain whether the Lignite existed in situ we made an excavation in the bank of the river and exposed the stratification for a depth of twenty-five feet. The last exposures of the Cretaceous shales were observed about three miles east of the bank where this trial was made. A few hours labour revealed five old beaches, probably of an inland lake. These beaches were composed of sand and boulders of Lignite from the size of a hen's egg to one foot in diameter. No fragment of Lignite was found which did not possess a rounded or spheroidal form and a roughly polished or worn surface. An abundant supply was speedily obtained for a fire which was soon made on the bank; a strong sulphurous odour was emitted from the iron pyrites in the Lignite. The section exposed the stratification shown in the wood-cut.

Some boulders of Lignite when broken open exhibited streaks and small particles of amber.

The low hills about Snake Creek are sand dunes, and on their sides an Opuntia is very common. The prairie on the west of the Souris as well as on the east is treeless, the banks of Snake Creek support a thin belt of small forest trees, such as oak, and ash, with a few ash-leaved maple. The annual fires prevent the willows and aspens from covering the country, which they undoubtedly would do until replaced by other species, if not destroyed to within a few inches of the root every time the fire sweeps over them. The banks of the Souris here are not more than 40 feet high, with level prairies on either hand, a few miles beyond the Snake Hills. Within four miles of the mouth of Snake Creek, Oak Lake, several miles in diameter attracted the hunting portion of our party; they brought back some pelicans and a score of duck. Thunder storms as usual to day and yesterday.

On the 1st July we arrived at the Souris Sand Hills, and made a section of the river bank where a land slip occasioned a fine exposure to the water's edge. The formation consisted of blue clay above the level of the river five feet, supporting four feet of ferruginous sand and gravel, on which reposed twelve feet of sandy loam and sand to the prairie level. The blue clay, capped by the ferruginous sand was traced for a distance of 2½ miles, and showed a dip to the south of two feet in the mile, the clay disappearing beneath the water. No organic remains of any description were found, although a

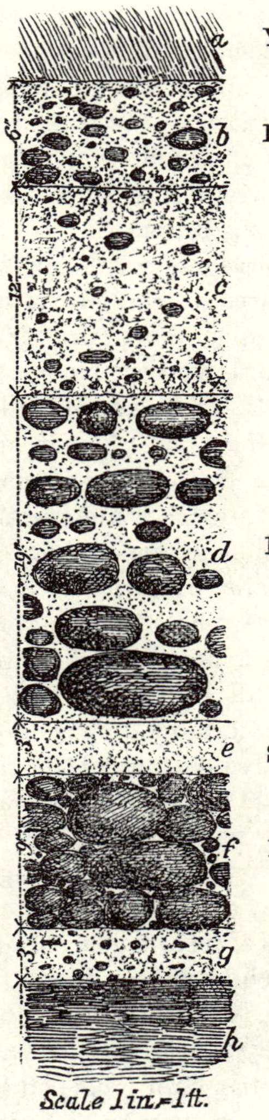

SECTION ON THE LITTLE SOURIS SHOWING ANCIENT BEACHES WITH LIGNITE BOULDERS.

careful search was made. Boulders of lignite from 6 inches to 9 in diameter, were frequently seen in the bed of the river. The eggs of the nighthawk were several times found on the bare ground, with no approach to a nest for the helpless young. The parent birds endeavored to draw us away from their eggs, fluttering as if wounded a short distance from them and uttering cries of distress. The Hudson's Bay Company have a post on the river among the Sand Hills, which is maintained only during the winter; the Sioux in summer and autumn, being altogether opposed to the approaches of civilization in their hunting grounds, and entertaining besides a feeling of deadly hostility to the Red River half-breeds.

Near the Company's house we found on the river bank an extensive deposit of bog iron ore, capped by shell marl, and above the marl drifted sand. The banks of the river are here not more than 25 feet high, and on the east side there is a narrow fringe of fine timber. The Bois de Vache (dried buffalo dung) is distributed very abundantly in the prairie and through the Sand Hills and ranges near to the post. In fact the buffalo were very numerous during the whole of the winter of 1856 and spring of 1857 on the banks of the Souris, but the great fires during the autumn of last year, have driven them south and north-west, and between the two branches of the Saskatchewan.

The country is very low after passing the last sand hills, and over a large extent of prairie south of these, drift timber is found, showing the extraordinary rise in the waters of the river during the floods of 1852.

On the second of July we observed the grasshoppers in full flight towards the north, the air as far as the eye could penetrate appeared to be filled with them. They commenced their flight about nine in the morning, and continued until half-past three or four o'clock in the afternoon. About that hour they settled around us in countless multitudes, and immediately clung to the leaves of grass and rested after their journey. On subsequent days when crossing the great prairie from Red Deer's Head River to Fort Ellice, the hosts of grasshoppers were beyond all calculation; they appeared to be infinite in number. Early in the morning they fed upon the prairie grass, being always found most numerous in low, wet places where the grass was long. As soon as the sun had evaporated the dew, they took short flights, and as the hour of nine approached, cloud after cloud would rise from the prairie and pursue their flight in the direction of the wind, which was generally S.S.W. The number in the air seemed to be greatest about noon, and at times they appeared in such infinite swarms as to lessen perceptibly the light of the sun. The whole horizon wore an unearthly ashen hue from the light reflected by their transparent wings. The air was filled as with flakes of snow, and time after time clouds of these insects forming a dense body casting a glimmering silvery light, flew swiftly towards the north north east, at altitudes varying from 500 to perhaps 1000 feet.

Lying on my back and looking upwards as near to the sun as the light would permit, I saw the sky continually changing colour from blue to silver white, ash grey and lead colour, according to the numbers in the passing clouds of insects. Opposite to the sun the prevailing hue was a silver white, perceptibly flashing. On one occasion the whole heavens, towards the south-east and west appeared to radiate a soft grey-tinted light with a quivering motion, and the day being calm, the hum produced by the vibration of so many millions of wings was quite indescribable, and more resembled the noise popularly termed "a ringing in one's ears," than any other sound. The aspect of the heavens during the greatest flight we observed was singularly striking. It produced a feeling of uneasiness, amazement and awe in our minds, as if some terrible, unforeseen calamity were about to happen. It recalled more vividly than words could express the devastating ravages of the Egyptian scourges, as it seemed to bring us face to face with one of the most striking and wonderful exhibitions of Almighty power in the creation and sustenance of this infinite army of insects.

In the evening, when the grasshoppers were resting from their long journeys, or in the morning, when feeding on the grass leaves, they rose in clouds around us as we marched through the prairie—if a strong wind blew they became very troublesome, flying with force against our faces, in the nostrils and eyes of the horses, and filling every crevice in the carts. But fortunately, comparatively few flew on a windy day, otherwise it would have been almost impossible to make headway against such an infinite host in rapid motion before the wind, although composed individually of such insignificant members.

Those portions of the prairie which had been visited by the grasshoppers wore a curious appearance; the grass was cut uniformly to one inch from the ground, and the whole surface was covered with the small, round, green exuviæ of these destructive invaders.

The valley of the Souris, along which we travelled during the day, varies from one quarter to one mile broad, the river is not more than twenty five feet across and very shallow. It flows through a rich open meadow, 20 to 35 feet below the general level of the prairie, which on either hand is undulating, treeless, light and covered with a short stunted grass, with abundance of last year's bois de vache. The first fresh buffalo tracks were seen to-day, and while taking observations for latitude, tracks of a different character and greater significance were discovered by one of the half-breeds—the fresh print of horses' feet, pronounced to be a few hours old, denoting the presence of Sioux or Assiniboines in our neighbourhood.

Before reaching the 49th parallel, the Souris meanders for several miles through a treeless valley, about a mile broad and 60 feet below the prairie level. Turtle Mountain on the east rises nobly from the great plain, the boundary line between British and American territory cutting it. The country west of the Souris is a treeless desert, in dry seasons destitute of water, and without a shrub or bush thicker than a willow twig. We ascertained the breadth of this arid, woodless tract to be at least 60 miles, north of the Red Deer's Head River on the 49th parallel. Near the boundary line the Souris expands into a series of large ponds and marshes which are called the Souris Lakes. In high water they form a continuous lake of imposing magnitude, extending many miles south of the 49th parallel, consequently far within the United States Territory.

A vast number of boulders are strewed over the hill bank of the Souris, near the 49th parallel, and on a point between a small brook and the Souris, we found a number of conical mounds, and the remains of an intrenchment. Our half-breeds said it was an old Mandan village; the Indians of that tribe having formerly hunted and lived in this part of the Great Prairies. We endeavoured to make an opening into one of the mounds, and penetrated six feet without finding anything to indicate that the mounds were the remains of Mandan lodges. There is a Mandan village near Fort Clark on the Missouri, and in the country unwatered by the Yellowstone the remains of this once fine and powerful tribe are now to be found.

Having reached the 49th parallel and traced the Souris in search of Lignite in position for a distance of 100 miles, we altered our course to a good camping ground on Red Deer's Head River, and made preparations for crossing a treeless, arid, prairie at least 60 miles broad, in a direction nearly due north.

The Little Souris nowhere approaches the Missouri nearer than thirty or forty miles.* Beyond the Souris Lakes it flows in a valley 200 feet below the level of the prairie, with a wooded bottom from one half to two miles wide. The nearest timber in the direction of the proposed Pacific Railroad, near the 49th parallel, east of the Souris, is in the valley of Red River, two hundred miles distant, and with the exception of cotton wood there is no timber west of the Souris for four hundred miles, at the Bear's Paw.† Where Mr. Tinkham crossed the Little Souris,‡ far within the limits of the United States Territory, (Lat. 48.02) he found it on the 21st July to be 120 feet wide, and too deep to ford. The effects of evaporation are plainly seen in the diminished volume of water which flows through the Blue Hills, only a few miles from its junction with the Assiniboine.

* Governor Steven's Explorations and Surveys, page 21.

† *Ibid*, page 41, Report of the Secretary of War. ‡ *Ibid*, page 357.

CHAPTER II.

FROM THE FORTY-NINTH PARALLEL ON THE LITTLE SOURIS TO FORT ELLICE—FROM FORT ELLICE TO THE QU'APPELLE MISSION.

Indian Signs—Smell of fire—The Sioux—Precautions—"Something"—"Souris Lakes"—Red Deer's Head River—The Great Prairie, character of—Mirage—Birds—Grasshoppers—Pipestone Creek—Country changed—Forest disappeared—Cretaceous Rocks—Buffalo Bull—Fort Ellice—McKay—Crees—Hunters—Provision Trading Posts—Pemican—Dried Meat—Thunder Storms—Mammoth Bones—Ojibway Hunter—Half-breeds—En Route for the Qu'Appelle Mission—Grasshoppers—Thunderstorm—Trail—Weed Ridge—Kinni-Kinnik—Mode of Manufacturing—Boulders—White Crane—Magpies—Birds—Dew—Aridity of Great Prairie—Charles Pratt—Chalk Hills—Indian Turnip—Qu'Appelle Lakes—Fresh arrangements—Descent and ascent of the Qu'Appelle—Qu'Appelle Mission—Dimensions of Valley—Character of Lakes—White Fish—Rev. James Settee—Garden of Mission—Grasshoppers—Christian Worship—Baptism—"Praying Father" and "Praying Man"—Rum—Indian wishes.

While engaged in taking observations for latitude at the mouth of Red Deer's Head River, on the night of the 3rd July, John McKay, a Scotch half-breed, observed what he thought to be a wolf approach the brow of a hill, about 200 yards from us, and after apparently gazing at the encampment for a few minutes it retired beyond view. The night was clear and as we were encamped in the valley of the river, close to its junction with that of the Souris, surrounded by steep hills about 150 feet high, an object appearing on the brow of those in our rear could be seen projected against the clear sky. McKay took no further notice of the strange visitor than to mention that he saw it and thought it was a wolf, but before we retired to our tents at 2 a.m. we saw another figure, which he declared to be an Indian, appear near the same spot. Two of the party cautiously approached the foot of the hill, but before they could reach it the figure crouched and slowly retired. The horses were gathered near the carts and a watch set, but this night passed without the re-appearance of the object of our suspicion. On the following morning we endeavoured to discover tracks at the spot where it had appeared, but the hill being composed of gravel, the soil had received no impression which our most sharp-sighted half-breeds could detect.

In the afternoon of the following day having verified our observations on polaris by a solar observation at noon, we started for a new camping ground about twelve miles up Red Deer's Head River, where we proposed to take in a supply of wood for fuel before crossing the great prairie to Fort Ellice. On our way thither the old hunter who had joined us at Prairie Portage said he *smelt fire*; every member of the party strained his olfactory nerves to the utmost but without detecting the smell of fire, nevertheless the old hunter persisted in his statement that he had "smelt it." We camped at sunset close to the river, and when taking supper distinctly heard the distant neigh of a horse; this was considered sufficient warning, and taken in connection with the appearance of the object on the hill in the rear of our camp the night before, was held to be conclusive evidence that we were watched by the Sioux, and that an attempt would be made in the night to steal our horses.

Our fires were put out, the carts placed close together and a watch set; the half-breeds did not anticipate an attempt until the approach of dawn, but the sudden galloping of several horses who were feeding in the valley about 100 yards from us, towards the carts soon after ten, proved that Indians were already near us. On hearing the horses approach the men started up and ran to stop them, which they succeeded in doing before they passed the carts. Each horse was now tethered and the half-breeds crawling through the long grass arranged themselves in a half-circle about 70 yards from the carts, each with his gun loaded with buckshot. The night was dark, and perfect silence was maintained in the camp; towards morning one man came in to report, he stated that he had heard "something" cross the river and crawl through the grass within a few yards of him, he waited a few minutes for more to follow before he fired or gave the alarm, and then cautiously crawled through the long grass in the track of the "something" which had passed near to him. The track led him to within 30 yards of our tents, and then turned towards the river, and evidently crossed it.

Morning soon dawned, and the watchers came in; we examined the tracks described by the half-breed who had first heard the intruders, and they were pronounced to be those of an Indian. Further examination in full daylight showed that we had been surrounded by a band, who, however, perceiving that we were on the alert, and that the horses were tethered, made no attempt to steal them. Had it not been for the old hunter's excellent nose, there is little doubt that we should have lost our horses during the night.

The mouth of Red Deer's Head River is within a few yards of the 49th parallel, Mr. Hime took a photograph of the valley while others of the party made an excursion to the Souris Lakes, within the United States territory, in the hope of finding buffalo to replenish our stores; but although fresh tracks were seen, and skulls and bones in large numbers, the remains of last year's "run", yet no living animal but a 'cabri' was visible.

On the morning of the 4th, having loaded the carts with wood and taken a supply of water from Red Deer's Head River, which is here a rapid, clear stream twelve feet broad, we started on a nearly due north course to cross the Great Prairie. The water marks on the banks of Red Deer's Head River show that it rises 15 feet during spring freshets, almost filling the low, narrow valley in which it flows. The banks are fringed with small elm, balsam poplar and aspens. The prairie for many miles appears to be perfectly horizontal; we always seemed to be in the centre of a very shallow depression, with a uniform and well defined horizon in all directions. In the morning the distant outline meeting the clear sky was best defined; as the day wore on refraction magnified the tufts of grass and small willows into bushes and trees, destroying the continuity of the fine horizontal line where sky and earth seemed to meet. Occasionally the effects of mirage were very delusive, beautiful tranquil lakes suddenly appeared in the distance, and as quickly faded from our view. Fortunately, the almost daily thunderstorms which had occurred replenished the marshes and small ponds, and gave us an abundant supply of water, but in some seasons the buffalo hunters suffer much from the want of water in crossing this vast treeless prairie.

Among the birds noticed during this monotonous journey were turkey buzzards, ravens, barking crows and black terns —we saw some herds of cabri, and McKay succeeded in killing a female. We came to several shallow lakes, which are often dry in the autumn; ducks were plentiful in them, and afforded us a grateful supply of fresh food. The grasshoppers were very abundant, and for four days filled the air like flakes of snow; they rose simultaneously, when about to take their flight, from areas two to twenty acres in extent, first perpendicularly to the height of twelve or fourteen feet, then in a slanting direction, until they had attained an elevation of from two to three hundred feet, after which they pursued a horizontal course before the wind. In a light breeze, the noise produced by their wings was like a gentle wind stirring the leaves of a forest.

Our half-breeds informed us that this great prairie west of the Souris continues treeless and arid for a distance of 60 miles, it is then crossed by a river, probably an arm of the Souris, connecting, as will be afterwards shown, with the Qu'Appelle River at Elbow Bone Creek; beyond the river the prairie continues for 80 miles further, without tree or shrub; and as this was the utmost westerly limit to which any of them had journeyed in their buffalo-hunting expeditions, they could afford us no further information respecting its extent. They were most of them familiar with the country south of the Great Prairie, the Grand Coteau de Missouri, where the buffalo range during the summer in vast herds. On the 6th July we arrived at Pipe Stone Creek, and found the country swarming with a young brood of grasshoppers, with wings about a quarter of an inch long, showing that their progenitors had arrived in the preceding autumn in time to deposit their eggs in the soil. Innumerable hosts of these insects passed overhead during the day, and on looking up through an excellent marine glass, I could see them flying like scud at an immense height. Had it not been for the thunderstorms which daily refreshed and invigorated the herbage, it is probable that our cattle would have suffered seriously from the devastations of these insects.

Pipe-Stone Creek is 20 feet broad at our crossing place, with a swift current, and a depth of water varying from $1\frac{1}{2}$ to 3 feet. Among the trees fringing its banks the ash leaved maple is most numerous. The valley of this river is narrow, but rich and beautiful. On the hills in its neighbourhood, boulders are numerous and the soil barren. We arrived at the Assiniboine near the Two Creeks on the evening of the 7th July. It may here be remarked that our hunter, who had undertaken to guide us in a straight line across the prairie from Red Deer's Head River, confessed that he did not know the country when within ten miles of the Assiniboine; he nevertheless declared his conviction that we should strike the river at the point to which he had promised to lead us. He had not visited it for twenty years, and the timber, consisting of aspens and willows which then covered the country, had nearly all disappeared. The old man was correct, the face of the country had changed, the aspen forest had been burnt and no vestige remained; we struck the Assiniboine within two miles of the spot to which he had been directed to lead us.

Cretaceous rocks were again recognized on the steep hill-sides of the Two Creeks. They had the same lithological aspect as those of the Souris; organic remains were scarce, but in sufficient numbers and variety to establish their position. On the 9th we passed through a fine grazing country, and here saw the first buffalo bull. After a chase of half an hour's duration, we succeeded in killing him. Although very tough and rather strong flavored, he was an acceptable addition to our larder. Three more bulls were seen on the following morning, but being anxious to reach Fort Ellice, and already provided with meat, they were permitted to pass us unmolested. The country in the neighbourhood of Beaver Creek is very beautiful, but the soil is sandy, supporting a short stinted herbage. We arrived at the Fort on the morning of the 9th, and took up our camping ground on the banks of Beaver Creek, close to the beautiful valley of the Assiniboine.

Fort Ellice was at one period a post of considerable importance, being the depot of supplies for the Swan River District, now removed to Fort Pelly. The buildings are of wood, surrounded by a high picket enclosure. Mr. McKay, one of the sub-officers, was in charge at the time of our arrival. Some twenty years ago, before the small pox and constant wars had reduced the Plain Crees to one-sixth or eighth of their former numbers, this post was often the scene of exciting Indian display. Mr. McKay remembers the time when the entire tribe who now hunt on the Qu'Appelle and South Branch would approach the Fort to receive their supplies, to the number of eight hundred warriors, splendidly mounted, and singing their war songs. Twenty years ago the tribe numbered 4000, in five hundred tents, at the present day they do not exceed 120 tents, which represent a population of 960 or 1000 souls. Formerly Fort Ellice used to be visited by the Crees alone, now it numbers many Ojibways among the Indians trading with it. The Ojibways or Saulteaux have been driven from the woods by the scarcity of game, the large animals, such as moose, deer and bear having greatly diminished in numbers. Many of the wood Indians now keep horses and hunt on the Plains.

On the 11th July, a number of hunters attached to Fort Ellice came in with provisions, such as pemican and dried buffalo meat, which they had prepared in the prairies a few days before, about thirty miles from the Post, where the buffalo were numerous. Fort Ellice, the Qu'Appelle Post, and the establishment on the Touchwood Hills being situated on the borders of the great Buffalo Plains, are provision trading posts. They obtain from the Plain Crees, the Assiniboines and the Ojibways, pemican and dried meat to supply the brigades and boats in their expeditions to York Factory on Hudson Bay, and throughout the northern interior. Pemican is made by pounding or chopping buffalo meat into small pieces and then mixing it with an equal quantity of fat. It is packed in bags made of the hide of the animal, in quantities of about ninety pounds each. Dried meat is the flesh of the buffalo cut into long and broad thin pieces about two feet by fifteen inches, it is smoked over a slow fire for a few minutes and then packed into a bale of about 60 pounds. We had many opportunities of seeing the Cree women on the Qu'Appelle, cut, prepare and pack dried meat.

At Fort Ellice, the thunder storms were as violent as on the Souris, not a day passed without lightning, thunder, and generally violent rain of half an hour's duration. The grasshoppers at this Post had destroyed the crops last year, and, at the time of our visit, the young brood were well advanced, their wings being about one third of an inch long. Full grown insects from the south were flying overhead or alighting in clouds around us, so that all hopes of obtaining a crop from the garden or potatoe fields were abandoned for this year. Provisions were very scarce at the Post, and had it not been for the fortunate arrival of the hunters with some pemican and dried meat, we should have been compelled to hunt or kill the ox.

From Mr. McKay, I received a particular account of the "Great Bones" on Shell Creek, which had long been a source of wonder and awe to the Indians hunting on the left bank of the Assiniboine, and whose magnificent descriptions led me to suppose they might belong to a cetacean, and were worth a day's journey out of our track to visit and examine. They were seen many years ago protruding from the bank of Shell Creek, twenty feet below the prairie's level. Mr. McKay instructed some of the hunters attached to the Post to bring them to him. No Indians would touch them, and the Half-breed only brought a tooth and collar bone, which were stated by a medical gentleman to whom they were shown, to belong to a mammoth. Mr. Christie, of Fort Pelly, we were informed, went to Shell Creek with a view to collect more specimens, he obtained some ribs but in a state of crumbling decay; they were sent to Red River Settlement. The Indians had long regarded them as the bones of a Manitou and worthy objects of veneration. An old Indian on Dauphin Lake, to whom reference will be made hereafter, described similar bones in the banks of Valley River leading to Dauphin Lake; but the season was too late when exploring that part of the country to permit of an examination.

On Monday, the 12th, preparations for continuing our journey westward were completed by engaging an Indian to assist in paddling Mr. Dickenson down the Qu'Appelle or Calling River from the Mission to its junction with the Assiniboine. The half of his wages he stipulated to have in advance. Mr. McKay told me he was a bad Indian and not to be trusted, but we could not succeed in getting another. When on the point of starting, a young Ojibway, painted and adorned with feathers, galloped up to the Post, entered the room, drew from beneath his moose skin robe two moose tongues and a mouffle which he quietly handed to Mr. McKay, and, squatting on the floor without speaking a word, lit his pipe. After a few minutes, he informed us that he and his father had killed two moose, thirty miles off, and desired McKay to send for them. Two Half-breed hunters also arrived at this moment, in sad plight, hungry and tired, with worn horses and torn clothes. They had come from Fort Union, on the Missouri, having being hunting on the Grand Coteau, where they met a war party of 60 Blackfeet. They then fled to the fort, the Blackfeet pursuing them and insisted that the Fort Union people should give them up, a request which was promptly refused.

During the night the Fort Union people gave them a small supply of provisions, and leading them out to the prairies, told them to run for it; they did so, and arrived in safety at Fort Ellice after a harrassing journey.

At 4 P. M. on the 12th July, we left Fort Ellice and travelled due west through a pretty country near the banks of the Qu'Appelle or Calling River. We passed one quagmire, and, after breakfast on the following day, arrived at the Cross Woods; they consist of aspen, with a splendid undergrowth. The pasturage is excellent, and the road good. Observed today the grasshoppers descending from a great height perpendicularly, like hail—a sign of approaching rain. On the 12th, we passed through a fair rolling country, the soil consisting of sandy loam with much vegetable matter in the valleys. Aspen groves are numerous, and many little lakes, margined with reeds afford quiet breeding places for duck. The road is good in summer, but wet and soft in the spring.

The grasshoppers, yesterday, were excellent prognosticators, a violent thunder storm in the afternoon commenced in the east, (all preceeding storms had come from the west) and was accompanied by exceedingly heavy rain and a very boisterous wind. The storm continued for several hours. At 9 in the evening, the air was calm and the heavens clear and bright; at 10, the storm returned from the west, and a more terrific and sublime exhibition of elemental warfare none of us had ever before witnessed. Three times the lightning struck the earth so close to us that there was no perceptible interval between the flash and the shock. It was distinctly heard to *hiss* through the air, and, instead of penetrating the ground at once, it seemed to leap from bush to bush for a distance of 60 or 70 yards. So close did one flash approach us that when we had recovered from the shock and our eyes had regained their powers, several of us met each other, groping from cart to cart, to see if any of the party had been struck. It is remarkable that although the wind was blowing violently before and after the two flashes just described occurred, yet, between them, an interval of about three quarters of a minute, there was a dead calm, and a calm of short duration succeeded each flash in our immediate vicinity.

The trail continued through good land for nine miles, with aspen groves on the crown of each undulation, and willow bushes in the hollows. Then came a prairie, three miles across, but of much greater extent longitudinally. Ponds were numerous, abounding with ducks and ducklings. The grey crane was very abundant, as well as a young brood of grasshoppers. Another rain and thunder storm on the evening of this day, the 14th, lasting as usual for about one hour. On the following morning, we reached a treeless prairie marked at its western extremity by a sandy ridge running N. W. by S. E., known among the Indians as the Weed Ridge. It was covered with the bearberry from which the kinni-kinnik, used to mix with tobacco is made. This was the first time we saw this weed since leaving the Sandy Hills of the Assiniboine. The Indians of the prairies generally use the inner bark of the *cornus sericea*, the red barked willow as they term it. We saw them smoke the inner bark of the dogwood, *cornus alternifolia.*

The mode in which these barks are prepared is very simple. A few branches about three quarters of an inch thick and four or five feet long are procured, the outer bark is scraped off, after having been warmed over a fire; a knife is then pressed against the inner bark and drawn upwards, for a space of six or eight inches, until the whole of the inner bark is gathered in curly clusters round the stick, it is then thrust in the ground over the embers and roasted until quite dry, when, mixed with tobacco in equal proportions, it forms the favourite kinni-kinnik of the North-West Indians. I often saw them smoke bark or the leaves of the bear-berry alone, when their supply of tobacco was exhausted. The Indian who accompanied us to the Qu'-

Appelle Mission, complained of weakness and pain in the chest, he suffered much from cough, and was evidently consumptive; he was, however, treacherous and indolent, and, as will be shown hereafter, soon left us in the lurch.

Beyond the Weed Ridge the country is very undulating; boulders of both fossiliferous (silurian limestone) and unfossiliferous (gneiss) rocks were strewed on the flanks and summits of the hills. The white crane was first seen to-day. This beautiful bird is common in the Qu'Appelle Valley and in the Touchwood Hill range. It is a dangerous antagonist when wounded, striking with unerring aim and great force with its powerful bill. When a bird is wounded, the best way to avoid its attacks is to present the muzzle of the gun as it approaches, it will fix its bill in the barrel and may then be destroyed without danger. Instances have been known of this bird driving his bill deep into the bowels of a hunter when not succesful in warding off its blow. Magpies are numerous on the Weed Ridge, and the cat bird is heard in every little wooded dell.

On the 15th we passed two streamlets flowing into the Qu'Appelle. Their banks were fringed with small timber and quite lively with birds. In general, birds are far more abundant here than on the Souris. On all the wooded brooks we saw magpies, cat birds, crows, and, occasionally, the solitary thrush; in the wet prairies, the rice bird, black tern, plover, the golden legged and common, the yellow headed black bird, common meadow lark, chipping sparrow, and grackle; on ponds and in marshes, ducks of many species, bittern and cranes. In the morning after a clear night, we always observed heavy dew; this phenomenon was not so frequently noticed on the Souris under similar circumstances. There can be little doubt that the aridity and barrenness of the Great Prairie between the Qu'Appelle and the 49th parallel is owing to the small quantity of dew and rain, and the occurrence of fires. North of the Qu'Appelle, the country seemed to be far more humid and the vegetation infinitely richer than south of that great valley.

Another prairie eight miles broad succeeding to that last described, and bounded by ridges having a N. W. and S. E. direction, introduced us on the 16th to a hilly country for some miles; the range is called the Indian Head; it contains many beautiful lakes and is well wooded. Here we met with Charles Pratt and party going to Red River. Charles Pratt is a Half-breed catechist of the Church Missionary Society, well acquainted with the habits of Indians and of buffalo, but apparently scarcely sensible of the importance of his duties and the responsibility of his charge. He gave me a good deal of valuable information respecting the country, and, with characteristic generosity, if not christian sympathy, told John McKay to take a young heifer belonging to him when we arrived at the Mission and kill it in honour of our arrival. Pratt showed me some specimens of lignite which he had taken from a bed two feet thick at the Wood Hills about 80 miles south-west of the Hudson's Bay Company's Post He described the hill or range of hills as an island in the Prairie. Probably it was the remains of a Tertiary coal bed, which, like the Stony Mountain near Red River, had escaped denudation.

An old Indian accompanying Charles Pratt, born in this part of the country, told us that he remembered the time when the whole of the prairie through which we had passed since leaving Fort Ellice was one continuous forest, broken only by two or three narrow intervals of barren ground. The view from the Indian Head range is excedingly beautiful; it embraces an extensive area of level prairie to the north, bounded by the Aspen Woods on the borders of the Qu'Appelle Valley. A portion of the old forest alluded to by the Indian still exists on this range. It consists of aspen of large growth and very thickly set. A few cabri (prong horned antelope) were seen in the Indian Head range; they used to abound in the country unwatered by the Qu'Appelle.

On Saturday the 17th we entered a very beautiful and fertile prairie at the foot of the Indian Head Range, our course leading us in a northerly direction to the Qu'Appelle Mission. The common yarrow was very abundant, and with the hare-bell reminded us of other scenes far away. Six miles from the hills we arrived at a subordinate, shallow, broad valley, parallel to that of the Qu'Appelle. The aspect of its boundary suggested the shore of a lake or bank of a large river. The lower prairie consisted of a sandy loam, in which the Indian Turnip was very abundant. We soon came up with a group of squaws and children from the Qu'Appelle Lakes, who were gathering and drying this root, which the Crees call the *Mis-tas-coos-se-ne-na* or big grass root. The French half-breeds call it the pomme de prairie. The Sioux, *Tip-si-nah.* It is an important article of food in these regions. The botanical name is *Psoralea esculenta.* Many bushels had been collected by the squaws and children, and when we came to their tents they were employed in peeling the roots, cutting them into shreds and drying them in the sun. I saw many roots as large as the egg of a goose, and among those brought with me to Canada are some of even larger dimensions. The Crees consume this important vegetable in various ways. They eat it uncooked, or they boil it, or roast it in the embers, or dry it and crush it to powder, and make soup of it. Large quantities are stored in buffalo skin bags for winter use. A sort of pudding made of the flour of the root and the mesaskatomina berry, is very palatable, and a favourite dish among the Plain Crees.

We reached the Qu'Appelle Lakes at 6 p. m., after passing through a magnificent prairie the whole day; in fact the country north of the Indian Head and Chalk Hill ranges is truly beautiful, and will one day become a very important tract. The Chalk Hills are a continuation of the Indian Head range. In the language of the Indians they contain bands of "soft, white earth or mud." The half-breeds call them "Chalk Hills." It is a matter of regret that the time at our disposal did not permit us to make an excursion to them, notwithstanding that no indications of rocks in position were seen on the Indian Head range; they were recorded as composed of drift; which may or may not conceal rocks in position above the general level of the prairie north of them.

Great was our astonishment on arriving at the Qu'Appelle Lakes to find that they were narrow bodies of water, occupying an excavated valley about one mile broad, 250 feet deep, and differing in no important particular from the same valley at its junction with the Assiniboine—one hundred and twenty miles distant by the river, or one hundred and thirty-four by the trail. The importance of the Qu'Appelle valley began to develope itself when the Crees at the Lakes informed us that it continued through to the Saskatchewan without losing its breadth, and maintained, except for a short distance, a great depth below the prairie level. I determined, therefore, to explore the whole valley from the South Branch of the Saskatchewan to the Assiniboine, and ascertain the relation it bore to those rivers. With this view the canoes were put in order, the party and supplies divided, and the arrangements detailed in the following paragraph completed.

Mr. Dickinson with a French Canadian and a Cree half-breed was to descend the Qu'Appelle river from the first Fishing Lake to its mouth. Mr. Fleming and myself were to ascend it from the same starting place to its source, and follow up the valley to the South Branch of the Saskatchewan. Mr. Hime was to explore Long Lake and meet Mr. Dickinson at Fort Pelly. I intended, upon reaching the South Branch, to descend that magnificent river in canoe to the Grand Forks, and then by the main Saskatchewan to Lake Winnipeg and Red River, a distance of about one thousand miles canoe navigation.

The Qu'Appelle Mission is situated between the second and third Fishing Lakes. The situation is beautiful and the country on all sides of a very novel and peculiar description. Here the Qu'Appelle valley is one mile and a quarter broad, and two hundred and fifty feet deep. Both north and south a vast prairie extends, fertile, inviting, but treeless on the south, and dotted with groves of aspen over a light and sometimes gravelly soil on the north. Most beautiful and attractive, however, are the Lakes, four in number, and from the rich store of fish they contain, are well named the Fishing Lakes. A belt of timber fringes their sides at the foot of the steep hills they wash, for they fill the entire breadth of the valley. Ancient elm trees with long and drooping branches bend over the water; the ash leaved maple acquires dimensions not seen since leaving the Red River, and the Me-sas-ka-to-mi-na is no longer a bush, but a tree eighteen to twenty feet high and loaded with the most luscious fruit.

The Qu'Appelle Mission was established last year (1858). For some time past, however, Charles Pratt, the catechist, has resided where the Mission is situated, and has constructed a comfortable log house, fenced in a garden, and now possesses six or seven cows and calves. An old half-breed, whose name is obliterated in my note-book, took up his residence with Pratt; he had been engaged for the better part of his life at different fishing stations belonging to the Hudson's Bay Company throughout Rupert's Land, and he declared that in all his experience he had never seen the white fish (*corregonus albus*,) so large, numerous and well flavored as in the Qu'Appelle Fishing Lakes.

The Rev. James Settee, the missionary, a native of Swampy Cree origin, occupied Pratt's house; he arrived at the Mission last autumn. In the garden, where we found him, Indian corn was growing, as well as potatoes, turnips, beans, and other culinary vegetables. The grasshoppers had not yet visited the Mission, but vast flights had passed over it. They were seen passing the Company's post, 20 miles south, on the 8th of the month. They were then flying to the east. They had missed the Mission in 1857, for they visited the Touchwood Hills, forty to fifty miles north, and deposited their eggs in the ground, and during the present summer the young brood, as I learned a few weeks afterwards, destroyed all garden crops at the Touchwood Hills, and on the 28th July took their flight to the south-east.

On Sunday we attended service in Pratt's house; the Rev. Mr. Settee read the prayers in English with great ease and correctness; he preached in Ojibway, and a hymn was sung in the Cree language. Before the sermon the missionary surprised us by waking up a drowsy Indian who was enjoying a quiet nap in a corner of the room, and leading him to the temporary reading desk, commenced the ceremony of public baptism. My astonishment was not diminished when the reverend gentleman turning to me, without any preliminary notice, said abruptly, Name this man! After a moment's reflection I said, John, and without any unnecessary loss of time or words, John walked to his bench, and was soon apparently lost in noisy slumber to all consciousness of the privileges and blessings of which adult Christian baptism, duly received, had made him the inheritor.

When the Rev. James Settee arrived at the Mission last autumn, the Crees of the Sandy Hills having received intelligence that the bishop had sent a "praying man" to teach them the truths of Christianity, directed messengers to enquire whether "the Great praying father had sent plenty of rum, if so, they would soon become followers of the white man's good Manitou." The messengers returned with the sad intelligence that the great praying father had not only omitted to send rum, but he hoped that the Plain Crees would soon abandon the practice of demanding rum in exchange for their pemican and robes. The messengers were directed to return to the missionary with the announcement, that "if the great praying father did not intend to send any rum, the sooner he took his praying man away from the Qu'Appelle Lakes, the better for him."

There are very few tents about the Mission at present. Mr. Settee speaks English very fluently, and gets through the service without loss of time. The field for his labor is extensive, but not at present promising. When conversing with the Crees of the Sandy Hills, many of them expressed a wish to have their children taught by white men, but they did not appear to like the idea of their being taught by a native of a different origin. The school, however, appears here, as elsewhere among Indian tribes, to be the only sure ground for establishing the true faith among them. "Teach my children for two or three years, but let me follow the ways of my fathers," said the son of the Chief of the Sandy Hills to me. Many expressed a wish that their little ones should know the white man's cunning, and learn to cultivate the soil, but they would stipulate to remain themselves still the wild prairie Indians, hunting the buffalo, and occasionally tasting the savage excitement of war.

On the 20th July we launched our canoes on the Third Fishing Lake, and having seen Mr. Hime *en route* for Long Lake, my carts and horses on the way to the Grand Forks of the Qu'Appelle, Mr. Dickinson started for the mouth of the river, Mr. Fleming and myself with an Ojibway and Cree half-breed, paddled up stream with a view to trace out the valley to its junction with the South Branch of the Saskatchewan. The succeeding chapter contains a narrative of this exploration, which is followed by Mr. Dickinson's description of his canoe voyage to the Assiniboine. We arranged to meet at Fort Ellice forty-three days after our simultaneous departure from the Third Fishing Lake.

CHAPTER III.

FROM THE QU'APPELLE MISSION TO THE SOUTH BRANCH OF THE SASKATCHEWAN.

Depth of Fishing Lakes—Cross-sections — Confervæ—Lower Lakes 66 feet deep — Birds — Vegetation — Water-mark— Third and Fourth Fishing Lakes — Fish — Soundings in Fourth Lake—Fishing Lakes probably once united—Geese— Pelicans—Fourth Lake—Water-mark—Aspect of Valley in 1852—Qu'Appelle River—Prairie—Depth of Valley—White Cranes—Section of Alluvial Flats—Temperature—Character of Prairie —Birds— Shrubs — Antelope — Hare — Roses— Grand Forks—Plain Crees—Temperature of River—Ice marks—Buffalo tracks—Character of stream—Willow bushes —Fetid air—Drift clay—Erratics—Freemen's houses—Prairie—Want of Timber—Thunder-storms—Touchwood Hills— Indians — Tolls — Diplomacy — Indian resolve—The Grand Forks—Long Lake—Souris Forks—Souris of Qu'Appelle and Assiniboine — Dimensions of Valley — The Grand Coteau—Prairie Fires—Indian signs—A Prairie on fire— Buffalo—Consequence of Prairie fires—Reclamation of sterile areas—Indian Telegraph—Scarcity of Wood—Ancient Indian encampment—The Plain Crees—Cree Tents—Provisions— Buffalo Hill Pound Lake—Indians—Shortstick—Aspect of Country — Coteau de Missouri—Last Mountain—Treeless Plain—The Grand Coteau—Character of—Buffalo—Birds Plain Crees, Camp of—The Qu'Appelle Valley—Marrow— Precautions—The Sandy Hills—Crees—Bois de Vache—Salt Lake—Dimensions of Valley—Erratics—Indian hospitality Eye-brow Hill—Source of Qu'Appelle—Buffalo—Character of Qu'Appelle Valley—Water-marks—Sandy Hills—Distribution of Boulders—Section—Rock exposure—Shortstick— Sand Dunes—South Branch—The Qu'Appelle Valley—Cree Camp—Height of Land—Section of Valley—Levels—Buffalo Pound—Camp moving—' Dead men '—Old Buffalo Pound —Horrible spectacle—New Pound—Bringing in Buffalo— Slaughter in Pound—Shortstick—" Talk "—Objections to Half-breeds—To the H. B. Co.—Shortstick's wants—Rock exposure — Boulders in Valley — Character of the South Branch.

Three quarters of a mile from the mouth of the little stream joining the second and third Fishing Lakes, the lead showed 44 feet of water. This great depth surprised us, as we had been paddling since leaving the Mission in shallows not exceeding four and five feet in depth. Cross sections subsequently made showed that the lakes were generally deep on the north and shallow on the south side. An abundant growth of green confervæ covered the surface, which, in its aggregations and general distribution, reminded me of a similar profusion on the Lake of Woods during August, in 1857. The hill sides of the valley are deeply ravined; two excellent photographs, taken near the Mission, of the lakes and hills, display the chief characteristic of the valley with the fidelity which can only be approached by that wonderful art. The ravines are wooded, but the hills they separate bare, and we soon noticed that the north side began to show far less timber than the south, and of more stunted growth. The snow berry was seen in every hollow. Ash, leaved maple and elm were numerous on the south side of the lake.

Soundings near the middle of the lake showed 56 feet, which, when added to 249 feet, the depth of the valley below the prairie as ascertained by trigonometrical measurement, make the total excavation 305 feet. Another sounding 200 yards from the N. W. point, gave 57 feet of water. This was the greatest depth we obtained; but Mr. Dickinson found the lower lakes to be 66 feet deep. The shores of gravel are strewed with blocks of drift limestone and the unfossiliferous rocks. Gulls are numerous about these remote lakes and a pair of eagles have had their eyrie for many years in a fine elm tree, near the west end of the third Fishing Lake. The hop grows very luxuriantly in the thin belt of woods on the south side, and the frost grape hangs in beautiful festoons from the drooping branches of the elm. The water mark shows that this lake rises six to seven feet above its present level.

A low plateau, inundated every spring, separates the third from the fourth lake. It is the delta of two ravines which in the spring and autumn bring down a large quantity of water from the prairie above. Third Fishing Lake is connected with fourth Fishing Lake by a rapid stream flowing through the plateau, about 100 feet broad. At its mouth we saw a large number of fish rising at the grasshoppers which dropped from flights of these insects passing over at the time. In the same stream were many large fish, and among them several individuals of a species to which further reference will be made. Soundings in the fourth lake showed 54 feet; this depth was maintained for a long distance with great regularity. In fact, these lakes appear to be nearly uniformly deep and point to an excavating force, or peculiarity of rock formation deserving of further enquiry. The deltas at the mouth of the ravines coming in from the prairie at right angles to the general course of the valley, give a clue to the mode in which the lakes were separated one from the other. It is very probable that they were once all united.

Geese appeared in large flocks in the Fourth Lake, and at its western end we saw a splendid flock of pelicans numbering thirty-five individuals; as we approached they sailed majestically round and round, but took flight before we arrived within gun shot. Magpies are very numerous in the thin woods fringing the lakes, so also are grackles, the cat bird, and many smaller birds. The Fourth Lake is very shallow at its western extremity, six feet being the greatest depth recorded. The hills on the north side are quite bare, and trees on the south side are found only in the ravines. It is full of weeds and its water emits a very disagreeable odour, but the watermarks show that during spring freshets its level is eight feet higher than in the summer season. This is an important fact when taken in connection with the alleged appearance of the whole valley during wet springs; it is then said to resemble a broad river from a few miles east of the Saskatchewan to the Assiniboine. In 1852, a year memorable in Rupert's Land for the great floods which covered an immense tract of country, the Indians represent the Qu'Appelle Valley as filled with a mighty river throughout its entire length, flowing with a swift current from the lakelets at the height of land, soon to be described, to the Assiniboine, and as a mountain torrent through the short distance of 1? miles which separates them from the South Branch of the Saskatchewan.

After leaving the Fourth Lake and the marshes at its west extremity, we paddled, sailed or tracked up a narrow swift stream, four and five feet deep and seventy feet broad, winding through a low alluvial flat in a valley of undiminished breadth and depth. The hill sides were absolutely bare, not a tree or shrub was to be seen. The prairie on either side is also treeless and arid. On the 21st, after spending a restless night owing to the attacks of multitudes of mosquitoes, we left the canoe in the hands of our Half-breeds to track up the stream, and ascending to the prairie walked for some miles on the brink of this great excavation. We waited five hours for the canoe to reach us, the windings of the stream involving a course three times as long as a straight line up the valley. The hill sides began to acquire a more imposing altitude and probably exceeded three hundred feet. White cranes appeared in flocks of four and seven together; there were very wary and could not be approached.

The river was often seen to draw near to either side of the Great Valley, and it had excavated a channel ten to twelve feet deep in the alluvial flats through which it pursued its tortuous course. Its banks revealed the following section:

6 inches light vegetable mould with sand,
4 inches yellow clay,
10 inches light vegetable mould (former surface),
9 feet yellow clay,
2 to 3 inches hard ferruginous sand to the level of the river.

The last layer was hard, compact, and very coarse-grained. The river is here 60 feet broad and flows at a rate of one mile and a half an hour. The temperature at noon was 71.5° F. At the mouth of Long Creek, an insignificant affluent, the hills are covered with limestone and granite boulders; the north side is treeless like the vast prairie beyond it, the south side has aspens in the ravines and aspen groves in the prairie. The width of the valley remains uniform, never exceeding one mile and a quarter or less than one mile. The pasturage in the flats is superb, the grass long and very thickly set. Robins, magpies, and yellow birds enliven small aspen groves on the south side, or the thickets of cherry, mesaskotomina, dogwood and snowberry, which fill the hollows and ravines; the cat bird is also common and the tyrant fly catcher everywhere. In the river are vast numbers of ducks and geese, the young birds frequently made us an excellent meal, but no four footed animals were seen, with the exception of one prong horned antelope and one prairie hare.

In the afternoon of this day, we made many miles by sailing before a strong east wind; notwithstanding a heavy rain and thunder storm we were glad to push on through this seemingly interminable and now monotonous valley, as the air from the marshes on either side of the river was fetid and oppressive. A scramble to the summit of this steep hill bank, three hundred feet high, though very fatiguing, was amply repaid by the cool, pure and delightful breeze blowing over the desolate prairies around us. Roses of three different varieties, red, white and variegated, were numerous on the upland, and, in the morning, when the dew was on them, or at night when it was falling, the fresh air from above came down in puffs into our deep, hot valley with delicious and invigorating fragrance. On the 4th day after our departure from the lakes we sighted the Grand Forks; leaving the canoe I hastened on to a point where the men with the carts and horses were to await our arrival, and found them safely encamped on a beautiful meadow anxiously looking for us. An empty cart and a couple of horses were despatched for the canoe still some miles below us, and in the evening we were joined by Mr. Fleming and the two voyageurs.

Soon after sunset our camp received an unexpected addition of six Plain Crees, who were on their way to Fort Ellice with dried buffalo meat and pemican. During the day the temperature of the River was found to be 74°. At the mouth of a dry bed of a stream which we called Maple Creek, some very old trees of the ash-leaved species were observed. Many of them showed marks where they had been tapped. The willows which fringed the banks of the Qu'Appelle were barked by ice eight feet above the surface of the water. Numerous buffalo tracks began to appear, and where these animals had crossed the river, they had cut deep roads to the water's edge, and lanes through the willow bushes. The bones of many a young bull and cow were seen sticking out of the banks where they had been mired.

The tortuous character of the stream before we took the canoe out of the water, may be imagined from the fact that eleven hours constant, steady tracking enabled us to progress only five miles in a straight line through the valley. Some little time was lost in crossing from one side to the other in order to avoid the willow bushes, which only grew on the inside of a bend, rarely or never on the outside or longest curve. The breadth of the river where we left it was forty feet, and the speed of its current one mile and a quarter an hour. The fetid air from the marshes made most of the party feel unwell, and I therefore determined to carry the canoe in a cart on the immediate edge of the prairie, keeping the valley in constant view, and occasionally descending into it and crossing it, to ascertain by levelling and measurement its leading dimensions.

No rock exposure has yet been seen. It appears that drift covers the country to a great depth. Where land slips have occurred and exposed an almost perpendicular section, the yellow gravelly clay is alone visible. Some of the limestone erratics strewed over the sides of the ravines resemble those frequently seen on the south-east side of Lake Winipeg. Near our camp are six or seven log-houses, occasionally inhabited by *freemen* (that is, men no longer in the service of the Company,) during the winter months. The prairie above the freemen's houses, slopes gently to the edge of the valley from the distant horizon on both sides. Clumps of aspen vary its monotonous aspect, and though clothed with green herbage, due to the late abundant rains, the soil is light and poor. Some distance back from the valley it is of better quality, the finer particles not having been washed out of it; the grass there is longer and more abundant, but the greatest drawback is the want of timber.

Since we have been on the Qu'Appelle we have frequently noticed thunderstorms towards the north-west and north, in the neighbourhood of the Touchwood Hill range, which did not reach us; the day before yesterday, (22nd July,) a very violent thunderstorm in the Qu'Appelle valley, which delayed us for several hours, did not wet the carts ten miles to the south. Rain clouds appear to follow the Touchwood Hill range; the frequency of storms in that region is proverbial, and the richness of the vegetation proves that an abundant supply of rain falls during the hot summer months. The Indians who visited our camp had been hunting between the two branches of the Saskatchewan—they represented the season as very dry and the buffalo scarce. We passed a quiet and friendly night with them, and on the following morning made them a small present and pursued our way to the Grand Forks.

I happened to be about 100 yards in advance of the carts, after we had travelled for about a quarter of an hour; when hearing a loud clatter of horses' feet behind me, on looking round I found the six Indians galloping up behind. One of them, who had represented himself as a chief, seized my bridle, drew the horse's head round and motioned me to dismount. I replied by jerking my bridle out of the Indian's hand. My people came up at this moment and asked in Cree what this interference meant. We wanted to have a little more talk, said the soi-disant chief. The real state of the case being, however, that they wished to establish a sort of toll of tobacco and tea for permission to pass through their country, threatening that if it were not given they would gather their friends in advance of us, and stop us by force. We knew that we should have to pass through about 100 tents, so there was some little meaning in the threat. The old hunter, however, who knew Indian habits and diplomacy well, at once remarked that we were taking a large present to the chief of the Sandy Hills, and we should not distribute any tobacco or tea, until we had seen him, according to Indian custom. They tried a few more threats, but I closed the parley by unslinging a double barrelled gun from the cart, and instructing the men to show quietly that they had theirs in readiness; wishing the rascals good day, we marched on; they sat on the ground, silently watching us, but made no sign. In the evening one of them passed near us at full gallop, towards some tents which we saw in the distance, as we ascended the hill at the Grand Forks. One rather significant statement they made proved to be correct, namely, that the Plain Crees, in council assembled, had last year "determined that in consequence of promises often made and broken by the white men and half-breeds, and the rapid destruction by them of the buffalo they fed on, they would not permit either white men or half-breeds to hunt in their country or travel through it, except for the purpose of trading for their dried meat, pemican, skins and robes."

We crossed to the north side of the Qu'Appelle when we arrived at the Grand Forks, and ascended the hill bank to the prairie. The Grand Forks consist of the junction of two deep, broad valleys; the south valley being that in which the Qu'Appelle river flows, the other is occupied by Long Lake, or Last Mountain Lake, forty miles in length, and from one-half to two miles broad, being in fact an exact counterpart of the Qu'Appelle Valley; narrow, deep, filled throughout with water, and inosculating with the South Branch of the Saskatchewan some miles below the Elbow. In its general aspect Last Mountain Lake is similar to the Fishing Lakes. A rapid, winding stream, 30 feet broad, runs from it into the Qu'Appelle. Both valleys are of uniform breadth and depth, and very little narrower than when united they form the main valley of the Qu'Appelle. From the Grand Forks to the Souris Forks (Elbow Bone Creek) the country is treeless, slightly undulating and poor. The Indians say that the Souris River of the Qu'Appelle, coming from the Grand Coteau de Missouri, inosculates with an arm of the Souris of the Assiniboine before described, and a canoe in high water might pass from one river to the other without a portage. If this be the case, the diversion of the waters of the South Branch down the Qu'Appelle valley would acquire additional importance, and give value to an immense extent of territory, now comparatively inaccessible, and destitute of water.

A few miles west of the Souris Forks the Qu'Appelle is 19 feet wide and 1½ feet deep, but the great valley is still a mile broad and 200 feet deep. Here on the 25th we caught a glimpse of the blue outline of the Grand Coteau, with a treeless plain between us. After passing these Forks, the country is more undulating, small hills begin to show themselves; the general character of the soil is light and poor, the herbage consists of short tufted buffalo grass, and the plants common in dry arid plains. This afternoon we saw three fires spring up between us and the Grand Coteau. They were Indian signs, but whether they referred to the presence of buffalo, or whether they were designed to intimate to distant bands the arrival of suspicious strangers we could not then tell, and not knowing whether they were Crees, Assiniboines, or Blackfeet, we became cautious. In a few days we ascertained that the fire had been putout* by Crees, to inform their friends that they had found buffalo.

The grandeur of a prairie on fire belongs to itself. It is like a volcano in full activity, you cannot imitate it, because it is impossible to obtain those gigantic elements from which it derives its awful splendour. Fortunately, in the present instance the wind was from the west, and drove the fires in the opposite direction, and being south of us we could contemplate the magnificent spectacle without anxiety. One object in burning the prairie at this time, was to turn the buffalo; they had crossed the Saskatchewan in great numbers near the Elbow and were advancing towards us, and crossing the Qu'Appelle not far from the height of land; by burning the prairie east of their course, they would be diverted to the south, and feed for a time on the Grand Coteau before they pursued their way to the Little Souris, in the country of the Sioux, south of the 49th parallel.

Putting out fire in the prairies is a telegraphic mode of communication frequently resorted to by Indians. Its consequences are seen in the destruction of the forests which once covered an immense area south of the Qu'Appelle and Assiniboine. The aridity of those vast prairies is partly due to this cause. The soil, though light, derives much of its apparent sterility from the annual fires. In low places and in shallow depressions where marshes are formed in spring, the soil is rich, much mixed with vegetable matter, and supports a very luxuriant growth of grass. If willows and aspens were permitted to grow over the prairies, they would soon be converted into humid tracts in which vegetable matter would accumulate, and a soil adapted to forest trees be formed. If a portion of prairie escapes fire for two or three years the result is seen in the growth of willows and aspens,

* A native expression; 'put out fire,' signifies to set the prairie on fire.

first in patches, then in large areas, which in a short time become united and cover the country; thus retarding evaporation and permitting the accumulation of vegetable matter in the soil. A fire comes, destroys the young forest growth and establishes a prairie once more. The reclamation of immense areas is not beyond human power. The extension of the prairies is evidently due to fires, and the fires are caused by Indians, chiefly for the purpose of telegraphic communication, or to divert the buffalo from the course they may be taking. These operations will cease as the Indians and buffalo diminish, events which are taking place with great rapidity.

Wood began to be a great treasure in the prairie after passing the Moose Jaws Forks; we were compelled to go supperless to bed on the night of the 24th, because we had neglected to take a supply at the last aspen grove we passed, thinking that the bois de vache (dried buffalo dung) would be found in abundance, but the fires had burned it also, and not even a fragment was to be procured. No tree or shrub, or even willow twig could be seen in any direction from our camp on the morning of the 26th. Our customary breakfast of tea and buffalo meat was impossible. We had to content ourselves with uncooked pemican and water from a marsh.

Immediately on the banks of the Qu'Appelle Valley here, are the remains of ancient encampments, where the Plain Crees, in the day of their power and pride had erected large skin tents, and strengthened them with rings of stones placed round the base. These circular remains were twenty-five feet in diameter, the stones or boulders being about one foot in circumference. They wore the aspect of great antiquity, being partially covered with soil and grass. When this camp ground was occupied by the Crees, timber no doubt grew in the valley below, or on the prairie and ravines in detached groves, for their permanent camping grounds are always placed near a supply of fuel.

Making an early start in search of wood, we came suddenly upon four Cree tents, whose inmates were still fast asleep; about three hundred yards west of them we found ten more tents, with over 50 or 60 Indians in all. They were preparing to cross the valley in the direction of the Grand Coteau, following the buffalo. Their provisions for trade, such as dried meat and pemican were drawn by dogs, each bag of pemican being supported upon two long poles, which are shaft, body and wheels in one. Buffalo Pound Hill Lake, thirty two miles long, begins near the Moose Jaws Forks, and on the opposite or south side of this long sheet of water, we saw eighteen tents and a large number of horses. The women in those we visited on our side of the valley and lake, had collected a great quantity of the Mesaskotomina berry, which they were drying. They announced the cheering intelligence that the Chief Shortstick, with some thirty tents was at the Sandy Hills impounding buffalo. Leaving the hospitable Crees, after an excellent breakfast on pounded meat and marrow fat, we arrived at Buffalo Pound Hill at noon. The whole country here assumed a different appearance; it now bore resemblance to a stormy sea suddenly become rigid; the hills were of gravel and very abrupt, but none exceeded 100 feet in height. The Coteau de Missouri is clearly seen from Buffalo Pound Hill towards the south, while north-easterly the Last Mountain of the Touchwood Hill Range looms gray or blue in the distance. Between these distant ranges a treeless plain intervenes.

The Grand Coteau runs parallel with the Missouri; its average breadth is 60 to 80 miles, and it rises from 400 to 800 feet above the bed of the great river it flanks, and between 400 to 800 feet above the high plains through which the Shayenne and James River meander to the Red River of the north, and the Missouri.* The vegetation on the Grand Coteau is very scanty, the Indian turnip is common, so also is a species of cactus; no tree or shrub is seen, and it is only in the bottoms and marshes that rank herbage is found.

Ponds and lakes are numerous on the Grand Coteau side, and it is probably on this account that the Buffalo cross the Qu'Appelle valley near the Moose Jaws Fork and west of Buffalo Pound Hill Lake; in the winter they keep towards the Touchwood Hills for the sake of shelter, and the excellent herbage which grows in the beautiful meadows between the aspen clumps. The prairies there too are not so often burned as south of the Qu'Appelle, the valley of that river serving as a great barrier to prevent the onward progress of the devastating fires. We began to find the fresh bones of Buffalo very numerous on the ground, and here and there startled a pack of wolves feeding on a carcass which had been deprived of its tongue and hump only by the careless, thriftless Crees. Pelicans and ducks are seen in vast numbers on the lake, while on the high banks of the valley the remains of ancient encampments in the form of rings of stones to hold down the skin tents being everywhere visible, testify to the former numbers of the Plain Crees, and afford a sad evidence of the ancient power of the people who once held undisputed sway from the Missouri to the Saskatchewan. The remains of a race fast passing away, give more than a transient interest to Buffalo Pound Hill Lake. The largest ancient encampment we saw lies near a shallow Lake in the prairie about a mile from the Qu'Appelle valley. It is surrounded by a few low sandy and gravelly hills, and is quite screened from observation. It may have been a camping ground for centuries, as some circles of stones are partially covered with grass and embedded in the soil.

At noon on the 26th we rested for a few hours opposite to a large camp of Crees on the other side of the lake; our sudden appearance at the edge of the prairie threw them into a state of the greatest excitement as evinced by their haste in collecting their horses and gathering in groups in the valley below. A few of them set out to ride round the head of the lake but in the wrong direction, so that the chance of their overtaking us was highly doubtful, as they would have to make a round of thirty miles in consequence of the intervening lake. This magnificent sheet of water, never less than half a mile broad and thirty-two miles long, shadowed forth what the Qu'Appelle valley might become if a river like the Saskatchewan could be made to flow through it. As we neared the height of land the physical structure of this great valley became a deeply interesting and almost exciting subject of enquiry. So far it had preserved its breadth and depth with astonishing uniformity all the way from the Mission, and we were within forty miles of the South Branch of the Saskatchewan. The hill banks of the river now became wooded again, ash-leaved maple and elm in the ravines, sustained, no doubt, by the presence of so large a body of water as Buffalo Pound Hill Lake.

Towards evening we arrived at another Cree encampment, where we were again hospitably treated to beaten buffalo meat and marrow fat. Birch bark dishes full of that nutritious but not very tempting food was placed on the ground before us and

* Explorations and Surveys for a railroad route from the Mississippi to the Pacific Ocean. Governor Stevens, page 87.

we were requested to partake of it. The Indians took a piece of the pounded meat in their fingers and dipped it into the soft marrow; they were delighted to receive a small present of tea and tobacco, and while we were engaged in the tent with the men, the girls, children and old women came round our carts asking if we had any rum, and snuffed the boxes and bags containing provisions, in search of that odoriferous stimulant. We left our hospitable friends in the evening and camped about three miles from the last Cree tent. The Chief of the band, an old man, expressed very kindly feelings towards us, and hinted that it would be as well to keep a watch over our horses during the night, for there were some young scamps among his band who would think it an honour to steal a white man's horse. Visitors came during the evening, and from their actions we thought it advisable to keep watch and tether the horses; observing these precautions they retired at an early hour after a friendly smoke.

At dawn on the following morning we were *en route* again, and towards noon approached the Sandy Hills, the valley continuing about 140 feet deep and maintaining its width. Two days before our arrival the Indians had been running buffalo, and many carcases of these animals were scattered over the arid, treeless prairie through which our route lay. Several herds of buffalo were visible wending their way in single file to the Grand Coteau de Missouri distinctly looming south of the Qu'Appelle valley. After travelling through a dry, barren region, strewed with erratics, until 2 p.m., we arrived at the Lake of the Sandy Hills, and on the opposite side of the valley saw a number of tents with many horses feeding in the flats. When within a mile of the lake a buffalo bull suddenly appeared upon the brow of a little hill on our right. A finer sight of its kind could hardly be imagined. The animal was in his prime and a magnificent specimen of the buffalo. He gazed at us through the long hair which hung over his eyes in thick profusion, pawed the ground, tossed his head and snorted with proud disdain. He was not more than 50 yards from us, and while we were admiring his splendid proportions he set off at a gallop towards some low hills we had just passed over.

Our appearance on the brink of the valley opposite the tents surprised the Indians, they quickly caught their horses and about twenty galloped across the valley, here quite dry, and in a quarter of an hour were seated in friendly chat with the half-breeds. We kindled a fire with *bois de vache*, of which there was a vast quantity strewn over the plain, but no wood was near at hand. When the men were going to the lake for water to make some tea the Indians told us it was salt, and that the only fresh water within a distance of some miles was close to their camp on the opposite side of the valley. We were therefore constrained to cross to the other side and erect our tents near to the spring. Advantage was taken of our passage across the valley to make an instrumental measurement of its leading dimensions. It was found to be 140 feet deep, estimating from the abrupt edge of the bank, and one mile five chains broad. The depth below the general level of the prairie is considerably greater, for there was a descent of fifty or sixty feet by a gentle slope not included in the foregoing measurement. A vast number of erratics strewed this slope, indeed it was with great difficulty that we steered the carts through the formidable accumulation of boulders which beset our path. The bed of the Qu'Appelle is quite visible in the valley, but on account of the porous nature of the soil the overflow from Sand Hill Lake penetrates it in dry weather, and reappears about half a mile below in the form of a little stream about ten feet broad, issuing from a marshy tract occupying the entire breadth of the valley. In crossing, the carts and horses sank deeply in the soft grassy bottom, already much cut up by the passing of a large number of buffalo during the week preceding our arrival.

Sand Hill Lake is four and a half miles long, very shallow and contains water strongly impregnated with epsom salts and common salt. We made ourselves acceptable to the Indians by making them a present of powder, shot, tea and tobacco, and in return they invited us to partake of pounded meat, marrowfat and berries. The chief of the band assured us that his young men were honest and trustworthy, and in compliance with his instructions property would be perfectly safe. During the night a heavy rain filled the hollows with water and gave us promise of an abundant supply until we arrived at the Sandy Hills where the main body of Plain Crees were encamped. On the following day, the 28th, I rode to the Eye-brow Hill range, a prolongation of the Grand Coteau, and distant from the Qu'Appelle Valley about four miles. It was there that the Indians told us we should find one of the sources of the Qu'Appelle river. After an hour's ride I reached the hills and quickly came upon a deep ravine at the bottom of which bubbled a little stream about three feet broad. I followed its course until it entered the prairie leading to the great valley, and traced it to its junction with the main excavation, through a deep narrow gully.

The Eye-brow Hill range is about 150 feet above the prairie and forms the flank of a table land stretching to the Grand Coteau, of which it is the western extension. The recent tracks of buffalo were countless on the hill sides, and in the distance several herds could be seen feeding on the treeless plateau to the south. In the afternoon we bid farewell to our Cree friends and travelled west on the south side of Sand Hill Lake until we arrived at the gully through which the stream from the Eye-brow hill range entered the Qu'Appelle valley. It was here nine feet broad and three deep, having received accessions in a short course through the prairie from the hills where I had observed it scarcely three feet broad. We camped in the valley and employed the evening in taking levels.

About four miles west of us we saw the Sandy Hills and could discern the Great Valley passing through them, and containing as the Indians had alleged, ponds which sent water both to the South Branch and the Assiniboine. An important physical fact which we afterwards verified instrumentally and by optical proof. We found the streamlet from the Eye-brow Hill range strike the Qu'Appelle Valley eight and a half miles west of Sand Hill Lake, and four miles from the height of land where the ponds lie. The fall between the ponds and our camp was about five feet, and the valley 150 feet deep, and one mile seventy chains broad. The Eye-brow Hill stream had excavated a channel nine feet deep in the bottom of the Great Valley, and was joined by a sluggish brook coming from the ponds a few yards from our camp. Water marks on the hill banks showed that the entire breadth of the valley is flooded during spring.

The Sand Hills commence on the north side about two miles west of Sand Hill Lake as it appears in summer. They are drifting dunes, and many of them present a clear ripple marked surface without any vegetation, not even a blade of grass. They have invaded the Great Valley and materially lessened its depth. One feature in its banks is worthy of special notice. Many boulders or erratics are distributed over the west extremity of small hills or ridges into

which the steep banks are broken, seventy to one hundred and twenty feet above the level of the flats. These ridges have the form of long, narrow islands, their longitudinal axes being parallel to the sides of the valley, and the erratics are deposited and arranged on the top of each ridge and at their western extremities. The form of these ridges is also peculiar, they are sharp at the west end where the erratics lie, and rounded at the east end. The slope is gentle at the west end, abrupt at the east end. This peculiarity is a constant feature of all the ridges seen on the sides of the banks of the valley. They vary in height from 10 to 30 feet, and in length from 60 to 140 feet, and in breadth from 20 to 80 feet. They have evidently some relation to the excavating force which has produced this great valley, and cannot be attributed to the long continued action of a small stream; however competent running water may be to produce deep and long depressions in loose drift, or a soft friable rock. (*See wood-cut on page* 57.)

A section of the bank of the Eyebrow Hill stream, on its course through the flats, showed fine clay brought by recent rains from the hill banks, sand blown from the dunes, and loam produced by the blending of the two. Where it leaves the prairie the little river has exposed a section of a drift hill ten feet above the level of the flats, which reposes upon an ochreous stratified rock, seamed with veins of selenite. It exhibits yellow and red ferruginous clay about 6 feet thick, and below it is a hard greenish sandstone, in which gigantic concretionary masses are numerous. Veins of selenite penetrate the greenish coloured rock, but are most abundant in the ferruginous clay. This is the first rock seen in position above the Mission.

On the morning of Thursday, 29th, we prepared to visit the main body of the Crees at the Sandy Hills, and with a view to secure a favourable reception, sent a messenger to announce our arrival, and to express a wish to see Shortstick, the Chief of the Sandy Hills. Soon after breakfast we crossed the valley and entered the sand dunes; one which we measured was 70 feet high, quite steep on one side, beautifully ripple-marked by the wind, and crescent-shaped. Sand dunes are on both sides of the valley. From the summit we saw the woods and hills beyond the South Branch of the Saskatchewan, and what was more delightful to us traced with the eye the Qu'Appelle valley with undiminished depth and breadth through the Sandy Hills, until it was lost as it dipped towards the South Branch

At 8 o'clock, a.m., we came in sight of the Cree camp, and soon afterwards messengers arrived from Shortstick, in reply to the announcement of our arrival, expressing a hope that we would delay our approach until they had moved their camp half-a-mile further west, where the odour of the putrid buffalo would be less annoying. We employed the time in ascertaining the exact position of the height of land, and soon found a pond from which we observed water flowing to the Saskatchewan and the Assiniboine. The pond was fed by a number of springs and small streams, a foot or two broad, issuing from the Sandy Hills, on both sides, at right angles to the valley. We selected this spot to level across the valley, and found its depth to be 110 feet below the first plateau, its breadth, although partially invaded by sand dunes, 73 chains, or nearly one mile. Here we commenced taking the levels to the South Branch, twelve miles distant from us, an operation which we soon found necessary to close for the present, in consequence of the arrival of about sixty Cree horsemen, many of them naked, with exception of the breech cloth and belt. They were accompanied by the Chief's son, who informed us that in an hour's time they would escort us to the camp. They were about constructing a new pound, having literally filled an old one with buffalo, and being compelled to abandon it on account of the stench which arose from the putrifying bodies. We sat on the ground and smoked, until they thought it time for us to accompany them to their encampment. Shortstick had hurried away to make preparations for bringing in the buffalo; the new pound being nearly ready. He expressed through his son a wish that we should see them entrap the buffalo in this pound, a rare opportunity, few would be willing to lose.

We passed through the camp to a place which the Chief's son pointed out, and there erected our tents. The women were still employed in moving the camp, being assisted in the operation by large numbers of dogs, each dog having two poles harnessed to him, on which his little load of meat, or pemican or camp furniture was laid. After another smoke, the Chief's son asked me, through the interpreter, if I would like to see the old buffalo pound, in which they had been entrapping buffalo during the past week. With a ready compliance I accompanied the guide to a little valley between sand hills, through a lane of branches of trees, which are called 'dead men' to the gate or trap of the pound. A sight most horrible and disgusting broke upon us as we ascended a sand dune overhanging the little dell in which the pound was built. Within a circular fence 120 feet broad, constructed of the trunks of trees, laced with withes together, and braced by outside supports, lay tossed in every conceivable position over two hundred dead buffalo. From old bulls to calves of three months old, animals of every age were huddled together in all the forced attitudes of violent death. Some lay on their backs, with eyes starting from their heads, and tongue thrust out through clotted gore. Others were impaled on the horns of the old and strong bulls. Others again which had been tossed were lying with broken backs two and three deep. One little calf hung suspended on the horns of a bull which had impaled it in the wild race round and round the pound.

The Indians looked upon the dreadful and sickening scene with evident delight, and told how such and such a bull or cow had exhibited feats of wonderful strength in the death struggle. The flesh of many of the cows had been taken from them, and was drying in the sun on stages near the tents. It is needless to say that the stench was overpowering, and millions of large blue flesh flies, humming and buzzing over the putrefying bodies was not the least disgusting part of the spectacle. At my request the Chief's son jumped into the pound, and with a small axe knocked off half a dozen pair of horns, which I wished to preserve in memory of this terrible slaughter. "To-morrow," said my companion, "you shall see us bring in the buffalo to the new pound."

After the first run, ten days before our arrival, the Indians had driven about 200 buffalo into the enclosure, and were still urging on the remainder of the

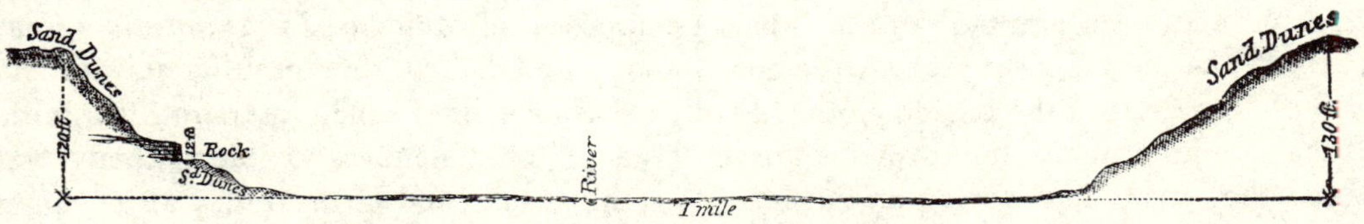

Transverse Section of the Valley of the Qu'Appelle at the Height of Land.

(Horizontal Scale, 16 *chains to an inch; Vertical Scale,* 200 *feet to an inch.)*

herd, when one wary old bull, espying a narrow crevice which had not been closed by the robes of those on the outside, whose duty it was to conceal every orifice, made a dash and broke the fence, the whole body then ran helter skelter through the gap, and dispersing among the sand dunes, escaped, with the exception of eight who were speared or shot with arrows as they passed in their mad career. In all, 240 animals had been killed in the pound, and it was its offensive condition which led the reckless and wasteful savages to construct a new one. This was formed in a pretty dell, between sand hills, about half-a-mile from the first, and leading from it in two diverging rows, the bushes they designate dead men, and which serve to guide the buffalo when at full speed, were arranged. The dead men extended a distance of four miles into the prairie, west of and beyond the Sand Hills. They were placed about 50 feet apart, and between the extremity of the rows might be a distance of from one and a half to two miles.

When the skilled hunters are about to bring in a herd of buffalo from the prairie, they direct the course of the gallop of the alarmed animals by confederates stationed in hollows or small depressions, who when the buffalo appear inclined to take a direction leading from the space marked out by the dead men, show themselves for a moment and wave their robes, immediately however hiding again. This serves to turn the buffalo slightly in another direction; and when the animals having arrived between the rows of dead men, endeavour to pass through them, Indians here and there stationed behind a dead man, go through the same operation, and thus keep the animals within the narrowing limits of the converging lines. At the entrance to the pound there is a strong trunk of a tree placed about one foot from the ground, and on the inner side a shallow excavation is made, sufficiently deep, however, to prevent the buffalo from leaping back when once in the pound. As soon as the animals have taken the fatal spring they begin to gallop round and round the ring fence looking for a chance of escape, but with the utmost silence the women and children on the outside hold their robes before every orifice until the whole herd is brought in, they then climb to the top of the fence, and with the hunters who have followed closely in the rear of the buffalo, spear or shoot with bows and arrows or firearms at the bewildered animals, rapidly becoming mad with rage and terror, within the narrow limits of the pound. It is then that a dreadful scene of confusion and slaughter begins, the oldest and strongest animals crush and toss the weaker; the shouts and screams of the excited Indians rise above the roaring of the bulls, the bellowing of the cows and the piteous moaning of the calves. The dying struggles of so many strong, full grown animals crowded together, furnish a revolting and terrible picture, but with occasional displays of wonderful brute strength and rage; while man in his savage, untutored and heathen state shows both in deed and expression how little he is superior to the noble beasts he so wantonly and cruelly destroys.

Shortstick is about fifty years old, of low stature, but very powerfully built. His arms and breast were deeply marked with scars and gashes, records of grief and mourning for departed friends. His son's body was painted with blue bars across his chest and arms. The only clothing they wore consisted of dressed elk or buffalo hide, and the breech cloth; this robe was often cast off the shoulders and drawn over the knees when in a sitting posture; they wore no covering on the head, their long hair was plaited or tied in knots, or hung loose over their shoulders and back. The forms of some of the young men were faultless, of the older men bony and wiry, and of the aged men, in one instance at least, a living skeleton. I enquired the age of an extremely old fellow who asked me for medicine to cure a pain in his chest; he replied he was a strong man when the two Companies (the Hudson's Bay and the North West) were trading with his tribe very many summers ago. He remembers the time when his people were as numerous as the Buffalo are now, and the buffalo thick as trees in the forest. The half-breeds thought he was more than 100 years old. Shortstick accepted the presents of tea, tobacco, bullets, powder and blankets I made him, with marked satisfaction, and expressed a wish to learn the object of our visit. We held a "talk" in my tent, during which, the chief expressed himself freely on various subjects, and listened with the utmost attention to the speeches of the Indians he had summoned to attend the Council.

All speakers objected strongly to the half-breeds' hunting buffalo during the winter in the Plain Cree country. They had no objection to trade with them or with white people, but they insisted that all strangers should purchase dried meat or pemican and not hunt for themselves.

They urged strong objections against the Hudson's Bay Company encroaching upon the prairies and driving away the buffalo. They would be glad to see them establish as many posts as they choose on the edge of the prairie country, but they did not like to see the plains invaded. During the existence of the two companies, all went well with the Indians, they obtained excellent pay and could sell all their meat and pemican. Since the union of the companies they had not fared half so well, had received bad pay for their provisions, and were growing poorer, and weaker, and more miserable year by year. The buffalo were fast disappearing before the encroachments of the white men, and although they acknowledge the value of firearms they thought they were better off in old times, when they had only bows and spears, and wild animals were numerous. I asked Shortstick to name the articles he would like to have if I came into his country again. He asked for tea, a horse of English breed, a cart, a gun, a supply of powder and ball, knives, tobacco, a medal with a chain, a flag, a suit of fine clothes, and rum. The talk lasted between six and seven hours, the greater portion of the time being taken up in interpreting sentence by sentence, the speeches of each man in turn. They generally commenced with the creation, giving a short history of that event in most general terms, and after a few flourishes about equality of origin, descended suddenly to buffalo, half-breeds, the H. B. company, tobacco and rum.

Early on the morning of the 30th I retraced my steps to examine an exposure of Cretaceous rock, forming part of the bank at the summit level of the Qu'Appelle valley, while Mr. Fleming continued taking the levels of the South Branch. (*See Section*, p. 55.) The rock is a sandstone, dipping very slightly to the south-west. The length of the exposure is about fifty yards, east and west; it is covered with drifting sand. Near the summit the layers are highly fossiliferous, and almost wholly composed of *Avicula Linguiformis* (*Evans and Shumard*); above and below the fossiliferous portion there is a coarse greenish coloured sand, interstratified with brown ferruginous layers. The thickness visible is about twelve feet. The rock occurs at the bend of the valley at its summit level; the exposure is perpendicular, and about 60 feet above the bottom of the valley. Some of the beds, those which are unfossiliferous, are very soft and friable, easily disintegrating, and may, farther west, be the origin of the sand dunes distributed over so wide an area in this part of the country. In descending the slope from the summit level to the Saskatchewan, the boulders on the ridges in the valley were found to be generally deposited upon the west side. The inclination of the boulders was towards the east, those forming the upper stratum were inclined against or superimposed upon the west side of those beneath, leading to the inference that the current which directed the course of ice which bore them, came here, as on the other side of the summit level in the valley, *from the west*.

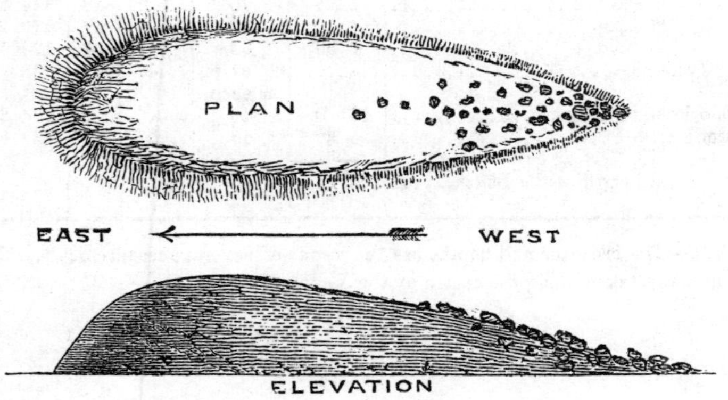

Ridges, with Boulders, on the East and West side of the Height of Land in the Qu'Appelle Valley.

About fourteen miles from the Saskatchewan there is a gigantic erratic of unfossiferous rock on the south side of the valley. It is seventy-nine feet in horizontal circumference, three feet from the ground; and a tape stretched across the exposed portion, from side to side, over the highest point, measured 46 feet. The Indians place on it offerings to Manitou, and at the time of our visit it contained beads, bits of tobacco, fragments of cloth and other trifles.

At noon I bid farewell to Shortstick, and joining the carts we wended our way by the side of "the River that Turns," occupying the continuation of the Qu'Appelle valley, to the South Branch of the Saskatchewan. The carts were accompanied by several Indians who watched with much curiosity the progress of taking the levels, and were very anxious to know what "medicine" I was searching for when sketching the position of the erratics in the valley.

Now and then a fine buffalo bull would appear at the brow of the hill forming the boundary of the prairie, gaze at us for a few minutes and gallop off. The buffalo were crossing the South Branch a few miles below us in great numbers, and at night, by putting the ear to the ground, we could hear them bellowing. Towards evening we all arrived at the South Branch, built a fire, gummed the canoe, which had been sadly damaged by a journey of 700 miles across the prairies, and hastened to make a distribution of the supplies for a canoe voyage down that splendid river. We were not anxious to camp at the mouth of "the River that Turns," in consequence of a war party of Blackfeet who were said to be in the neighbourhood of the Cree camp, watching for an opportuity to steal horses, and if possible to "lift a scalp."

The Indians who had accompanied us hastened to join their friends as soon as they saw the canoe in the water, and just as the sun set, the canoe containing Mr. Fleming and myself, with two half breeds, pushed off from the shore; the rest of the party, with the carts and horses in charge of the old hunter, retired from the river to camp in the open prairie, where they would be able to guard against a surprise by the Blackfeet, or the thieving propensities of treacherous Crees. Great precautions were undoubtedly necessary, as sure signs had been observed within three miles of the Sandy Hills, proving that a war party of Blackfeet were skulking about. The Crees, always accustomed when on the South Branch to their attacks, merely adopted the precaution of posting watches on the highest dunes, about a mile from the camp, and it was owing to the advice of Shortstick that we embarked so late in the evening in our canoe. We drifted a mile or two down the river until we came to a precipitous cliff showing a fine exposure of rock, which proved a temptation too great to be resisted, so we drew the canoe on the bank and camped for the night on the east side of the river, making arrangements to watch in turns.

The first view of the South Branch of the Saskatchewan, fully six hundred miles from the point where the main river disembogues into Lake Winnipeg, filled me with astonishment and admiration. We stood on the banks of a river of the first class, nearly half a mile broad, and flowing with a swift current, not more than three hundred and fifty miles from the Rocky Mountains, where it takes its rise. We had reached this river by traversing either within it or on its banks, for a distance of two hundred and seventy miles, a narrow deep excavation continuous from the valley of one great river to that of another, and exhibiting in many features evidences of an excavating force far greater than the little Qu'Appelle which meandered through it, was at the first blush, thought capable of creating. How were the deep lakes hollowed out? lakes filling the breadth of the valley, but during the lapse of ages not having increased its breadth, preserving too, for many miles, such remarkable depths, and although in some instances far removed from one another, yet maintaining those depths with striking uniformity. What could be the nature of the eroding force which dug out narrow basins fifty-four to sixty-six feet deep at the bottom of a valley already 300 feet below the slightly undulating prairies, and rarely exceeding one mile in breadth? It was easy to understand how a small river like the Qu'Appelle could gradually excavate a valley a mile broad and three hundred feet deep. The vast prairies of the North-West offer many such instances; the Little Souris River, for example, in passing through the Blue Hills; the Assiniboine, for a hundred and fifty miles, flows through a broad deep valley, evidently excavated by its waters; the rivers in western Canada often flow in deep eroded valleys; but in no instance to my knowledge are deep and long lakes known to occupy a river valley where the altitude and character of the rocks preclude the assumption that they may have been occasioned by falls, without having increased its width by the action of their waves on the banks, or without leaving some traces of the force which had excavated them. It was certainly with mingled feelings of anxiety and pleasurable anticipation that we embarked on the broad Saskatchewan, hoping during our long journey down its swift stream to find some clue to the origin of the curious inosculating valley of the Qu'Appelle we had traced from one watershed to another.

Leading dimensions of the Qu'Appelle or Calling River Valley, and of the Lakes which occupy it.

Table showing the length, with the breadth and depth, of the Qu'Appelle Valley at different points.

	Miles.	Chains.
Length of Valley from the South Branch of the Saskatchewan to the Assiniboine	269	0
Breadth of Valley 70 miles from the Assiniboine	0	78*
Do do 177 do do	1	30
Do do 239 do do	1	5
Do do 253 do do	1	70
Do do 258 do do	0	73**

At its junction with the Assiniboine its breadth exceeds one mile.

At its junction with the Saskatchewan its breadth exceeds one mile and a quarter.

	Feet.
Depth of the Valley 70 miles from the Assiniboine........	320
Do do 177 do do	250
Do do 230 do do	220
Do do 253 do do	140
Do do 258 do do	110
At the junction with the Assiniboine the prairie slopes to the Valley of that river, and its depth here is........	240
At its junction with the Saskatchewan the prairie also slopes to the Valley of the Saskatchewan, and its depth was estimated to be ...	140

* One mile less 44 yards. ** One mile less 154 yards.

Table showing the length, breadth, mean depth, greatest depth, and distance from the Assiniboine, of the Lakes in the Qu'Appelle Valley.

Name of Lake.	Length.	Breadth.	Depth.		Distance from Mouth.
			Mean.	Gr.	
	m ch.	Chains.	Feet.	Feet	m. ch.
Round Lake, or Ka-wah-wi-ya-ka-mac..	4 56	60	28	30	41 20
Crooked Lake, or Ka-wa-wa-ki-ka-mac	6 10	60	31	42	56 0
Fishing Lake, No. 1, or Pa-ki-ta-wi-win	6 0	40	52	66	108 0
" " No. 2	3 25	40	32	48	114 20
" " No. 3	4 30	60	41	57	119 20
" " No. 4	8 50	60	37	54	124 12
Long Lake	....	60 as far as seen.	......	....	168 0
Buffalo Pound-hill Lake	16 0	40	......	....	194 20
Sandhill Lake	4 50	45	......	....	239 50
Total length of the Lakes......	53 61				

Note.—The breadths and depths are the *means* of several measurements. The distances are taken along the centre of the Valley.

CHAPTER IV.

FROM THE QU'APPELLE MISSION TO FORT ELLICE, DOWN THE QU'APPELLE RIVER.

The Second Fishing Lake—Depth of—Indian Map—Origin of name Qu'Appelle, or "Who calls River"—The First Lake, or Pakitawiwin—Great depth of First Lake—Fish—Confervæ—Depth of Valley—Width of River—High-water mark—Valley flooded—Affluents—Depth of Valley—Crooked Lake, or Ka-wa-wa-ti-ka-mac—Dimensions of—Effects of fires—Trees in Valley—Boulders—Character of the country—Indian surprise—Indians—Summer berry Creek—Dimensions of Valley—Valley and Prairie scene—Camp scene—Character of Valley—Ka-wah-wi-ya-ka-mac, or Round Lake—Dimensions of—Stony Basin—Granite Boulders—Little Cut-arm Creek—The Scissors Creek—Rock exposure—Grasshoppers—Big Cut-arm Creek—Dimensions of Qu'Appelle—Flooding of Valley—Timber—Undergrowth—Birds—Minks—Deer—Uniformity of Qu'Appelle Valley.

Mr. Dickinson's Narrative.

Dear Sir,—Soon after parting from you on the morning of July 20th, at the Church of England Mission in the Qu'Appelle Valley, my instruments for surveying, with watch, a magnetic compass, a log line and sounding line, all arranged for ready use, and a cargo of kettles, pans, pemican, and blankets stowed away, our little canoe commenced its voyage down the river. In half an hour we reached the lake, which is generally called the second of the Fishing Lakes. Before venturing to go down it we were obliged to stop for the purpose of gumming the canoe, as it was leaking more than was desirable. To save time we took breakfast here. The distance between this lake and the one at the Mission is 1½ miles, while the actual length of the river is upwards of 2 miles. Its width averages 80 feet, and its depth 3 feet; the rate of current, which is nearly uniform throughout its length, is one mile per hour. The difference of level between these two lakes, obtained instrumentally on a previous day, is 1.50 feet. These measurements, not valuable in themselves, are taken for the purposes explained in the "Rules for conducting the Exploring Survey," namely, as the means for calculating approximately the total fall in the river. I may mention, that at every opportunity similar measurements and observations were made, with the assistance of Mr. John Fleming, from which we were able to deduce some general laws for guiding us in estimating the fall in rivers. I may mention also, as it has not often been used before, I think, on similar surveys, that the log line was found to be most invaluable in ascertaining the rate of the canoe on the rivers as well as on the lakes, being a much more accurate way than that of estimating it by the eye.

The canoe being now declared to be sea-worthy, we started on our way again. The lake is 3¼ miles long and three quarters of a mile in breadth, extending between the slopes of the valley, and appearing to be merely an expansion of the river, but on trial found to be something more than that. For some distance out from the mouth of the river it is only from three to four feet deep, but on trying it when we were about half a mile distant with a sounding line 30 feet long, to my great surprise, I could find no bottom; having added more line, the depth proved to be 42 feet. About the middle of the lake the depth is 48 feet.

A stream a quarter of a mile in length, flowing sluggishly through a marsh, connects this lake with the next, the first of the Fishing Lakes, or as it is in Cree, *Pakitawiwin.* All the Indian names of the lakes and tributaries of the Qu'Appelle I got afterwards on my arrival at Fort Ellice, from an old Indian 70 years of age, who had been once upon a time a great hunter and warrior, now in peace and comfort spending his remaining days at the hospitable Fort. With a piece of charred wood he drew on the floor a map of the Qu'Appelle Valley from the Fishing Lakes to the Assiniboine, shewing every little creek so accurately that I easily recognised them. Mr. McKay, who was then in charge of the Fort, kindly acted as interpreter on the occasion. The Cree name of the Qu'Appelle river is *Katapaywie sepi*, and this is the origin of the name as told me by the Indian:—A solitary Indian was coming down the river in his canoe many summers ago, when one day he heard a loud voice calling to him, he stopt and listened and again heard the same voice as before. He shouted in reply, but there was no answer. He searched everywhere around, but could not find the tracks of any one. So from that time forth it was named the "Who Calls River."

Pakitawiwin is six miles long and half a mile wide, and is most wonderfully deep. In one place by means of putting together various pieces of cord, sashes, &c., the sounding line being too short, the depth was found to be about 66 feet. The mean of several depths is 52 feet. It is famous for the quantity and quality of its fish. For three miles we passed through a dense decaying mass of *confervæ*, which an east wind had driven to the upper parts of the lake. The smell of it was most unpleasant; the men pushed through it as hard as they could, no easy matter, as it impeded the progress of the canoe considerably. The valley here is about the same depth as it is at the Mission, but the slopes are not so precipitous; one of them, that on the south side, has been the whole way covered with a dense growth of young aspens, and the other has been bare of trees except in some of its many hollows and ravines.

Leaving the lake we now descend the river at an average speed of four miles an hour, the rate of current being generally about one mile and a quarter per hour. Paddling was easy work, but the steering by no means so, for the bends of the river are innumerable and very sharp, and the waters sweep round them with great velocity; oftentimes, but for the strong and dexterous arm of the steersman, the canoe would have been dashed against the bank, as it was he could not avoid sometimes getting entangled among the overhanging branches of the willows. The width varies from one chain to one and a half, and the depth from four and a half to two feet. The bed for the most part consists of soft mud and is quite free from boulders, as is the case the whole way to the mouth, excepting in one place to be mentioned hereafter. The high water mark, very apparent on the willows growing along the banks, was eight feet over the present level of the water; the whole bottom of the valley, I was told, is often flooded to a depth of three feet.

Nineteen small creeks flow into this portion of the river, two only of them having names, the first and second Pheasant Creeks, called in Cree *Akiskoowi sepisis*, named after a hill which lies to the north some miles away, from near which they both take their rise. I took a cross section of the valley here, and found it to be 320 feet deep and 78 chains wide; it is I think the deepest part of it. At noon, on July 23rd, we reached Crooked Lake, called in Cree *Kawawak-kamac*, the most picturesque of the Qu'Appelle Lakes. Several streams draining the prairies on both sides have excavated deep and wide gorges opening into the main valley, which here sweeps in graceful curves, so that Crooked Lake seems to be embosomed amongst hills, and thus differs from the others which have very much the appearance of a gigantic canal. It is a little more than six miles in length, and its mean width is three quarters of a mile. The greatest depth I found was 36 feet, and the mean of several soundings was 31 feet. The south slope, as before, is clothed with a dense foliage of young aspens, willows, and dogwood; a great contrast to the opposite side, on which only grows short and scanty grass, leaving the granite boulders which lie scattered over it, exposed to view; only in the ravines and the deep hollows are seen patches of young aspens and straggling oaks which have escaped the devastating fires.

For some time I could not understand why one side should be covered with trees and the other quite bare, the soil on both being exactly similar, until I discovered unmistakeable evidences of fire, which may be the cause of it. On enquiry afterwards I found that Indians often travel along the valley on the north of the river, which accounts for the fires being on that side.

Between the gravelly beach and the first of the slopes a fringe of willows runs all round the lake, and several points of low land jut out on both sides, on which grow oak, elm, and ash; not very large trees certainly, but healthy and thriving looking, and giving additional beauty to the landscape.

I ascended a bluff on the north side by a well-worn deer path, on which there were many foot-marks quite fresh, for the purpose of taking some observations connected with the survey and seeing the nature of the surrounding country. A gently undulating prairie, dotted with clumps of small poplars and willows, stretched away on every side, and as far as I could see, the soil was a light sandy and gravelly loam, and in many parts strewed with boulders. I rather think that such is the character of a considerable extent of this section of the country.

As I stood upon the summit of the bluff, looking down upon the glittering lake 300 feet below, and across the boundless plains, no living thing in view, no sound of life anywhere, I thought of the time to come when will be seen passing swiftly along the distant horizon the white cloud of the locomotive on its way from the Atlantic to the Pacific, and when the valley will resound with the merry voices of those who have come from the busy city on the banks of Red River to see the beautiful lakes of the Qu'Appelle. The view down the valley, where the river after issuing from the lake commences again its strange contortions, was doubtless very pretty, but it shewed too the trouble that was before me, that there would be no rest for eye or finger, such as I had when taking long straight courses on the lake.

Again re-seated in the canoe we soon passed out of the lake into the river, the current of which for some distance is very strong and rapid, about $2\frac{1}{2}$ miles per hour according to the log-line, and the width averages 70 feet, and the depth 3 ft. 6 in. A little way down it, as we swiftly and noiselessly glided round a sudden bend, we were borne by the current very close indeed to a group of Indian women who were enjoying the pleasures of a bath, quite as much to our astonishment as to theirs. First a loud chorus of screams arose, and then there was a rushing about for blankets and other apparel, which they adjusted with most wonderful rapidity, and then away they scampered to their wigwams laughing heartily as they went. Presently men and boys came trooping down to us simply arrayed in blankets, some worn in rather a *negligé* fashion, for the day was very hot. The chief man of the party, which consisted of six families, invited me in the most polite and hospitable manner to go to his lodge and have something to eat; but I had to decline as he had told me previously, in answer to a question as to how many days' journey it was to Fort Ellice, that we would have to sleep four or five times before we reached it, and this was now our fourth day from the Mission; and, moreover, I thought that the interior of a wigwam would not be a very agreeable place on such a hot day.

While we were speaking the young ladies, whom we had so unintentionally disturbed, came down one by one to see us. Although their toilets were quite completed, so very modest were they, that they remained behind the bushes and peeped at us through the branches. Having given the men some tobacco, and receiving in return a large supply of Pembina berries (High-bush Cranberries), wewished them good-by and resumed our journey. We went at the average rate of four miles an hour for two hours and a half, and camped before sunset at the foot of a bluff on the south side of the valley, of which I had taken a bearing from the end of the lake, and close to a creek about ten feet wide called Nipimenan sepesis, or summer berry creek.

The valley is here of the same breadth as heretofore, that is, about one mile, and its depth is from 250 to 300 feet. The bottom is covered with willows interspersed with young sugar maples, with here and there an open patch of long luxuriant grass. With some difficulty I made my way to the level of the prairie through a dense and tangled mass of aspens and underwood of willows, dogwood, and rose trees; but the beauty of the glorious sunset, and the cool refreshing breeze that came across the plains, more than repaid the trouble. I need not try to describe the exceeding beauty of the scene, for I could not; I will merely state what the components of the picture were. The sun just merged from behind a bank of crimson clouds reflected in the waters of Crooked Lake. Part of the valley in deep shade and part brightly illuminated. The vivid green of the young poplars on one side, and on the other large granite boulders lying on the bare and rugged surface of the slope. The blue smoke of the wigwams rising up high and straight from the bottom of the valley. The river, with its complicated coils, gliding among the willow bushes. To the south the great prairie, ocean-like, with its many islands of poplars and single trees, looking in the distance, and by twilight, like becalmed ships. As this view just dissolves away, another arises very pleasant to see,—our camp fire is now burning brightly below, and over it swings a kettle, and passing round and about it are my two men, one busily engaged in preparing supper, the other in spreading out the blankets on the ground between the fire and the canoe.

Next morning (24th) we started as soon as it was daylight, glad to escape from our insatiated tormentors the mosquitoes and black flies, that would not let us rest or sleep all night. While at breakfast at 8 o'clock a great thunder storm from the south-west came upon us. Having thrown an end of the tarpaulin over the canoe, and resting the other end on the paddles stuck into the ground, we got beneath it and very soon fell fast asleep, ands lept till 1 o'clock, when I was awoke by the sudden calm, for the storm had apparently only just then ceased.

The valley and river still retain their old character and dimensions till we come to the lowest of the lakes, called *Kawahwiga-kamac*, or Round Lake, which varies from one mile to half a mile in width, and is nearly five miles long. The name is by no means an appropriate one, as it is far from being round. The mean of some soundings I took was 28 feet, the greatest being 30 feet. On the sand banks which are at the head of the lake, were myriads of duck, and large numbers of geese were swimming about in every direction, and a few great northern divers or loons. We camped at a place about two and a half miles down the river, called the Stony Basin, the Cree of which is *Asini-pichigakan*. For about 100 yards in length the river is full of large and small granite boulders, rendering it quite impassible for the smallest canoe when the water is low; at this time the water was just high enough to admit of us passing over it.

Two miles down the river from this spot a little stream brings in its gatherings from the prairies on the south, rejoicing in the name *Isquawistequannak Kaastaki,* which means, 'where the heads of the women lie.' A long time ago two women, one a Cree and the other a Chippeway, were killed by the Mandans on the banks of this stream; their bodies were left unburied, and their skulls are still lying there, from which circumstance the stream derives its name. This was all my informant at Fort Ellice knew of the story. The next creek which is dignified with a name is the "Little Cut-arm," or *Kiskipittonawe sepesis*, the origin of which I could not find out; it flows in from the north.

A few miles further down another creek, ten feet wide and very rapid, joins the Qu'Appelle on the other side; its name is *Pesquanamawe sepesis*, which may be rendered into English, 'the Scissors Creek;' it is not a very literal translation, but is the best that can be given. The incident to which it owes its name exhibits a peculiar habit of the Indian, but is one that cannot be told. Near this spot there is an exposure of rock on the north slope of the valley, which on examination proved to be a shale similar to that on the Little Souris, but so decomposed that the amount or direction of its dip could not be ascertained. There are several extensive patches where the surface of the rock has been re-converted into soft mud, very much cracked, and on which no grass grows. On digging into it I found the mud to be three inches thick, then fragments very small and soft, and gradually increasing in size and hardness to a depth of about two feet, where the rock is perfectly hard but very much shattered. About fifteen miles to the east of this the rock is again to be seen on the south slope of the valley, also much broken.

On the 26th vast clouds of grasshoppers, flying towards the east, passed high over our heads, without intermission, for nearly two hours. It was the last large flight I saw.

Big Cut-arm Creek, or *Kichekiskapettonano sepesis,* the last to be noted, joins the Qu'Appelle about 20 miles from its mouth, and is the largest of its affluents. It is twenty-five feet wide and three feet deep where it issues from a wide ravine on the north side. The Qu'Appelle from thence to its mouth is from eight to twelve feet deep and varies in width from seventy to ninety feet, and the rate of current is one mile and a half per hour.

There is much good land in the valley from the Fishing Lakes to the Assiniboine, but as it is flooded every spring it is questionable whether it will ever be of much importance. For ten miles up it there is an abundance of timber, consisting of aspens, balsam poplars, elm, black ash, oak, birch, and sugar maple. None, however, exceeding 1'.6" in diameter, and few so large. The underwood, which is very beautiful, is chiefly composed of dogwood, roses, cherries, and pembinas, intertwined with convolvuli and vetches. In this wooded part the birds are innumerable. Kingfishers, blue jays, and Canada jays, cat-birds, and American magpies, flitted from tree to tree uttering their discordant notes. Cherry-birds and pigeons were calmly and listlessly perched on the dense trees, having eaten plentifully of their favourite fruits, while the tyrant flycatcher, when alone or with some companions, chased and worried the crows, ravens, hawks, and eagles, who tried in vain to escape from them. The beautiful white-bellied swallow swiftly skimming the surface of the river, helped in addition to enliven the valley. Ducks and geese crowded the river for several miles; there were enough of them, I should think, to supply all the markets in Canada. Minks were perpetually crossing and re-crossing the river in front of the canoe. I was told that deer are sometimes very numerous in the valley, but I was only fortunate enough to see two jumping deer who were coming down to the river to drink, but the moment they got a glimpse of us away they bounded up the slope. The only other animal we saw was a little prairie wolf, Togany as he is called by the Indians, that was standing by the edge of the river, and who was so much astonished at our sudden appearance that he never thought of running away, but stood staring at us incapable of motion.

The wonderful uniformity of the valley, or that part of it which I have described, necessarily causes a great deal of repetition in the description of it; so similar is its character throughout, that my two men, half-breeds, well accustomed to mark any peculiarities in the features of a country, said, that though they might pass up and down it several times they thought they would often be at a loss to know in what part of it they were. The length of the valley from the second Fishing Lakes to its junction with the valley of the Assiniboine is 110 miles, while the river itself is about 270 miles long, which will give an idea of its extraordinary tortuous course. We arrived at its termination on the evening of July 27th, and having hauled up the canoe on the bank, walked across to Fort Ellice, distant about three miles, where I was kindly received by Mr. McKay.

Very truly yours,

J. A. DICKINSON.

Professor H. Y. Hind,
&c. &c. &c.

CHAPTER V.

FROM THE ELBOW OF THE SOUTH BRANCH OF THE SASKATCHEWAN TO THE NEPOWEWIN MISSION, ON THE MAIN SASKATCHEWAN.

Rocks on the South Branch—Cretaceous—Altitude of exposure—Character of—Selenite—Fossils—Concretions—Mesaskatomina berry—Character of river—Drift—Rock exposures—Fibrous Lignite—Treeless prairie—Cree Camp—Mud Flats—Rock exposure—Concretions—Treeless banks and prairie—Low country—Driftwood—Ripple marks—Dimensions of the South Branch—The Moose Woods—Water and Ice marks—Forest Timber—Character of River—Treeless Prairie—Boulders—Soundings—Buffalo—Dimensions of River—Absence of animal life—'The Woods'—Rate of current—Boulders, arrangement of—Artificial pavement—Tiers of Boulders—Temperature—Balsam spruce—Former aspen forest—Good country—Water-marks—Soundings—Absence of animal life—Stratified Mud—Fall of River—Character of River—Colour and temperature of North and South Branch—The North Branch—Absence of Indians—Grizzly bear—Current of North Branch—Coal Falls—Dimensions of North Branch—Boulders—Trees—The Grand Forks—The Main Saskatchewan—Fort à la Corne—Cubic feet of water in North and South Branch and Main Saskatchewan.

The first rock exposure on the South Branch below the Qu'Appelle Valley is a Cretaceous sandstone occupying the river bank, unconcealed by drift for some miles. The altitude of the highest part of the exposure is sixty feet above the level of the river. It is capped by about seven feet of drift, which reposes on twenty feet of soft and easily disintegrated sandstone of a pale yellowish-grey colour, containing a large number of small, bright, pale, yellow, spheroidal bodies, varying from one-tenth of an inch to one inch and a half in diameter, and composed of sand. Below this soft stratum there occurs a layer of sandstone about three feet six inches thick, which is broken into an irregular projecting outline by the protrusion of a series of immense concretions, of a flat spheroidal form, like that of a lemon slightly compressed at its longest diameter. The concretions vary from three feet to six feet in horizontal dimensions. They are very hard in the centre, and show concentric rings for at least six inches from their outer casing, which is a shell of gypsum, often passing into Selenite. Selenite is found in this and lower strata in veins and fragments. Some of the concretions thrust out their rounded forms from the face of the cliff, others have been broken off and show their internal structure. A gray sandstone with a slight tinge of green, soft and friable, then occurs for a space of four feet; it is succeeded by five feet of hard sandstone containing a vast number of obscure cylindrical forms, slightly conical, composed of sandstone, and showing occasionally traces of organization. Below this stratum a layer of sandstone occurs, six feet thick, holding spheroidal forms, which vary in size from six inches to two feet in diameter; they are composed of yellow sand containing a hard central calcareous nucleus often six inches to one foot in diameter, and composed almost altogether of an aggregation of *Avicula Nebrascana*, (Evans and Shumard.) The stratum in which they are imbedded holds *Avicula Linguiformis*, (Evans and Shumard.)

A second layer of huge concretions then occurs, similar in external aspect to those already described. Below them there is a persistent layer of hard calcareous sandstone about four feet thick, containing *Avicula Linguiformis*, (E. and S.)

The lowest stratum exposed is a soft sandstone about six feet above the river, and passing beneath its level. This rock is worn into caves by the action of water. The part of the formation exposed is nearly horizontal, with a slight north-westerly dip. For several miles this rock continues to form the river bank. The concretionary masses are persistent, bold, and prominent; and about three miles in a north-westerly direction from the point where they were first observed, those of the lower stratum are nearly on the same level as the water, thus showing a north-westerly dip of about three feet in the mile.

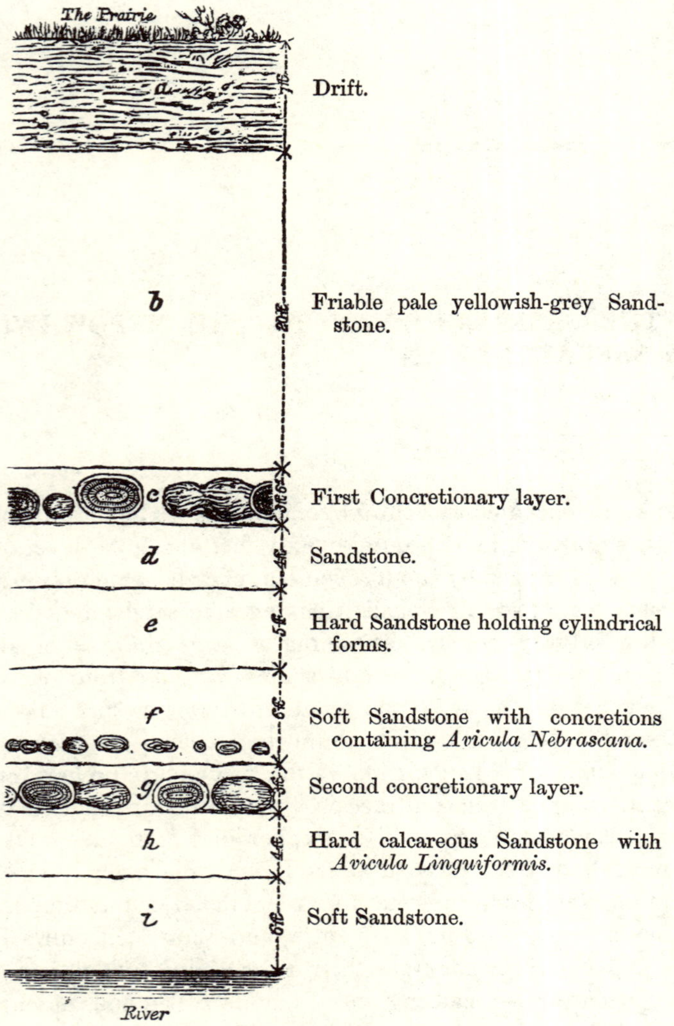

Section on the South Branch of the Saskatchewan, showing Conretionary Layers holding *Avicula Nebrascana* and *Avicula Linguiformis.*

The banks of the river slope gently from the prairie on the south-west side to an altitude of about 250 feet, they then become abrupt. On the north-west side the Sandstone cliff, varying from 30 to 60 feet in altitude, rises abruptly from the river, then follows a hilly slope to the prairie level. Trees, consisting chiefly of aspen and the Mesaskatomina (la Poire), are found in patches on both sides. The river continues about half a mile broad, with numerous sand-bars and low alluvial islands. The drift above the sandstone is gravelly, and many small sand dunes occur on the hill bank sloping to the prairie, and have progressed beyond the prairie to a considerable distance. A treeless prairie, boundless and green, except where the patches of drifting sand occur, is visible on either hand from the top of the bank; below, the river glides with a strong current, two, and two and a half miles an hour, filling the broad trench or valley it has eroded. The Mesaskatomina berry (*Amelanchier Canadensis*) la Poire, is very abundant; shrubs or trees eighteen to twenty feet high, loaded with this fruit perfectly ripe and of excellent flavour, are numerous in every grove; the berries are of the size of large black currants, very juicy and sweet. This shrub is the La Poire of the Red River Voyageurs.

During the morning of this day (31st Aug.) three Crees from a camp on the east bank came to the river, they shouted to us, asking us to land, an ivitation we declined. About twelve miles below the Qu'Appelle the river becomes narrower, being not more than a quarter of a mile broad, but full of mud flats and shoals. The banks are more sloping, and frequently broken into two plateaux, the upper one being the prairie. The lower plateau is dotted with small groves, the intervals consisting of pretty grassy areas, smooth as a lawn.

About fifteen miles from the Qu'Appelle valley the drift is occasionally exposed in cliffs, which disclose its structure twenty to thirty feet above the river. It consists of coarse sand stratified in curves, and often containing beds of gravel; it is also frequently capped by the same material with small boulders. The dip of the rocks to the north-west, and the aspect of the drift appear to indicate a geological depression, which may have been the seat of a large lake during earlier periods.

Some exposures of sandstone appear on the river at intervals lower down, and the drift above them is well stratified with layers of boulders of the same character as the sandstone below, and so regularly placed as to lead, when viewed from a small distance, to the belief that they are part of rock in position. Thirty miles from the Qu'Appelle the rock appears on the south-west side, and consists of a white sandstone, with impressions of fragments of leaves, and some brown, fibrous lignite.

A treeless prairie with a few sand dunes forms the country on either side for a distance of thirty-eight miles, which comprised the extent of our voyage during the day. As evening began to close upon us we came to a camp of Crees just after they had crossed the river. They numbered nineteen tents, and in order to avoid them we drifted several miles further down, and built our fire close to the river at the mouth of a small gully leading from the prairie, 200 feet above us. Mud flats and sandbars continue as before, but the river is not more than a third of a mile broad.

A narrative of a canoe voyage down a river flowing through a prairie country must necessarily involve numerous descriptive repetitions, which will appear perhaps less tedious and more readable in the form in which they were registered at the time in my note book, than if I were to attempt a connected narrative. I shall therefore strictly follow the daily record of what we observed, at the risk of its being nothing more than a dry enumeration of not very interesting facts.

August 1st.—Found a fine exposure of rock on the river bank where we camped last night. There is a change in the aspect of some of the strata. They occur massive, in rusty red and greenish-gray sandstone layers, with the concretionary bands as before described. A belt of sandstone twelve feet from the river level is capped by brown and red argillaceous layers forty feet thick in the aggregate. Drift sand, ten feet thick, to the prairie level succeeds. The upper portion of the drift is hard and reddish coloured; as it approaches the clays below it partakes of an argillaceous character. The upper stratum of the sandstone weathers reddish brown, with bands of deep red and purple. Below this a greenish-gray stratum occurs enveloping more concretions of a reddish-brown colour. The concretions are hard and argillaceous. The greenish-gray matrix is soft when weathered, otherwise hard, and may be split without difficulty into thin layers. The concretions occur in the sandstone in forms easily detached, and often contain abundance of *Avicula Linguiformis*. If the clays above the sandstone are rock in position, the exposure has an altitude of about 60 feet. Fragments of fibrous lignite, dark-brown and sometimes approaching to black in colour, occur in the sandstone. The attitude of the rocks is nearly horizontal. The greenish-gray sandstone is identical with the rocks seen on the south bend of the Qu'Appelle above Sand Hill Lake; the red layers are similar lithologically to those observed at the height of land in the same valley, holding the same species of shells. Sometime layers of grey sandstone occur which are easily split; they contain the impressions and remains of plants. The position of these rocks is about fifty miles from the Qu'-Appelle valley.

The river banks and the whole country is now much lower. This subsidence began about four miles from our camp south of us. The banks at our camp are not more than one hundred feet in altitude, and are getting lower as we proceed north. They are treeless areas, and so is the prairie on either side, with few detached exceptions. The river is about half a mile broad, with a current in the lead fully two miles and a half an hour. Large drifted trees are sometimes seen on the beach, and one pine was noticed this morning. They have probably travelled from the flanks of the Rocky Mountains.

About twelve miles from our camp, or 60 miles from the Elbow, forests of aspen begin to show themselves on the banks, after passing through a low country, which is an expansion of the river valley. Ripple marks are numerous on the fresh mud, the furrows lying parallel to the course of the stream. They are quite recent and similar to those observed on Red River in the spring. The ash-leaved maple begins to show itself, but the aspen is the prevailing tree. The woods are not continuous, and the prairie on either side of the river remains bare; it is fast regaining its former altitude. and hills are visible in the distance from the top of the bank. La Poire is very abundant and fine flavoured. The exposed cliffs consist of reddish loam, and the rock is no longer seen below them. At a point fifty-three miles from the Elbow we made a careful section of the river, and found its breadth to be nearly one-third of a mile (28 chains); its greatest depth was ten feet on the east side, but on the west side there is another channel with nine feet of water.

As we approached the Moose Woods we passed for several hours between a series of low alluvial islands from ten to twelve feet above the water. They sustain some fine elm, balsam-poplar, ash, ash-leaved maple, and a vast profusion of La Poire. The river valley is bounded by low hills leading to the prairie plateau four to eight miles back. The country here furnishes an excellent district for the establishment of a settlement. The spot where we are camped for the night is an extensive, open, undulating meadow, with long rich grass, and on the low elevations rosebushes in bloom grow in the greatest profusion. It is only ten feet from the water, yet it does not appear to be flooded in the spring; water-marks and ice-marks are nowhere seen above four feet from the present level of the broad river.

August 2nd.—The region called the Moose Woods, which we entered last evening, is a dilatation of the Saskatchewan, flowing through an extensive alluvial flat six miles in breadth, and cut into numerous islands by the changing course of the stream. This flat is bounded by sand hills, some of which are nothing more than shifting dunes. The woods are in patches, and in the low land consist of balsam poplar, white wood, and aspen. Small aspen clumps cover the hills, but no living timber of importance has been seen as yet, although many fine dead trunks are visible, probably destroyed by fire. The river continues to flow through a broad alluvial flat for about twenty-five miles. Its water is very turbid, like that of the Mississippi, holding much solid matter in mechanical suspension.

Beyond the Moose Woods the banks close upon the river, and have an altitude not exceeding sixty feet. The breadth of the stream contracts to 250 yards, with a current fully three miles an hour. On the east bank the prairie is occasionally wooded with clumps of aspen, on the west side it is treeless, and shows many sand hills. During the afternoon we landed frequently to survey the surrounding country. Nothing but a treeless, slightly undulating prairie was visible; many large fragments of limestone not much water-worn lie on the hill banks of the river, which is about 100 feet in altitude. The river continues very swift, and maintains a breadth of 250 yards. Frequent soundings during the day showed a depth of ten to twelve feet. A little timber displays itself occasionally on the east bank below the level of the prairie. The dead bodies of buffalo are seen floating down the stream, or lodged on sand-bars in shallow water. The banks expose occasionally yellow drift clay with numerous boulders; the soil of the prairie appears to improve as we progress northwards, and the grass is no longer stunted and withered. Little rapids occur at the bends of the river, but there is always deep water on the other side. A heavy thunder storm compelled us to camp two hours before sunset.

August 3rd.—The river is not more than 200 yards broad, but deep and swift; the volume of water it carries here, ab ut eighty miles from the Grand Forks, is much less than at the Elbow, where it is half a mile broad. No doubt evaporation during its course through arid plains is competent to occasion a large diminution. Recent water-marks show a rise of five and eight feet, but near the top of the lowest bank stranded timber occurs twenty-five feet above the present level of the river. On both sides a treeless prairie is alone visible. There is a remarkable absence of animal life, no deer or bear have been seen, tracks of buffalo are everywhere, but they have already passed to the east. The nights are cold but fine, dew very abundant. The prairie level is not more than eighty feet above the river.

At 8 A. M. we arrived at a part of the river where it showed an increase in breadth, it is now about a quarter of a mile broad, still flowing through a treeless plain, in which only one low hill is visible. This character continues for many miles, the hill banks then begin to increase in altitude, and are about 100 feet high, but the river flows through a dreary treeless plain for 30 miles from our camp, after which "The Woods," as they are termed, begin; they consist of a few clumps of aspen on the hill flanks of the deep valley of the river. The face of the country is changing fast, it is becoming more undulating, and patches of aspen woods appear on the prairie; here and there, however, the remains of a heavier growth are visible in clusters of blackened trunks ten to fourteen inches in diameter. During the afternoon we anchored to measure the rate of the current. The river is 200 yards broad, and it flows three miles and a half an hour. Its average depth is seven and a half feet.

Some remarkable exposures of drift, consisting of clay with long lines of boulders, occur frequently after entering the wooded parts of the South Branch of the Saskatchewan. The drift is exposed in cliffs 50 to 80 feet in altitude at the bends of the river. The fragments of shale, slabs of limestone, and small boulders imbedded in the clay are not arranged according to the position they would take if dropped by floating ice; some of them stand in the drift with their longest axis vertical, others slanting, and some are placed as it were upon their edges. They have the same forced arrangement and position as the shale, &c., in the blue clay at Toronto. (See chap. XI.) Here also are long lines of boulders from ten to twenty feet below the surface, or top of the cliff; they lie horizontally as shown in the woodcut.

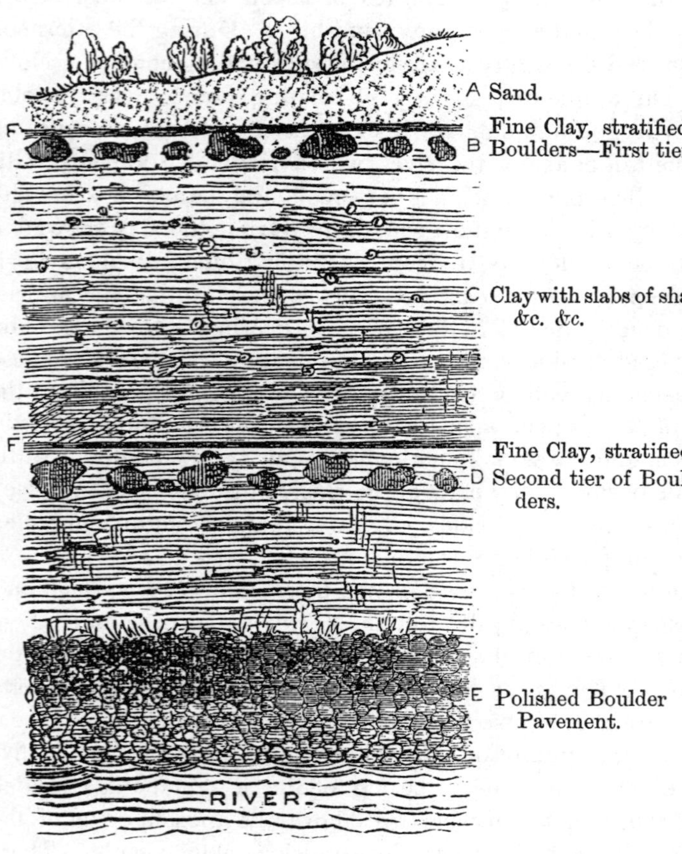

Horizontal Layers of Boulders in Drift on the South Branch, with Polished Boulder Pavement at the edge of the River.

In many places close to the water's edge and rising from it in a slope for a space of 25 to 30 feet, the fallen boulders are packed like stones in an artificial pavement, and often ground down to a uniform level by the action of ice, exhibiting ice grooves and scratches in the direction of the current. This pavement is shown for many miles in aggregate length at the bends of the river. Sometimes it resembles fine mosaic work, at other times it is rugged, where granite boulders have long resisted the wear of the ice and protected those of softer materials lying less exposed.

Two tiers of boulders, separated by an interval of twenty feet, are often seen in the clay cliffs. When first noticed they were about fifteen feet above the stream; as we descend the stream they rise above its level, preserving evidently a nearly horizontal position. The lower tier contains very large fragments of water-worn limestone, granite, and gneissoid boulders, above them is a hard sand containing pebbles, this is superimposed by an extremely fine stratified clay, breaking up into excessively thin layers, which envelope detached particles of sand, small pebbles, and aggregations of particles of sand. Above the fine stratified clay yellow clay and unstratified sand occur. The fine clay must have been deposited in very quiet water. The polished pavement at the foot of the cliff was observed this afternoon inclined at a high angle, so much so, that it was difficult to walk upon it.

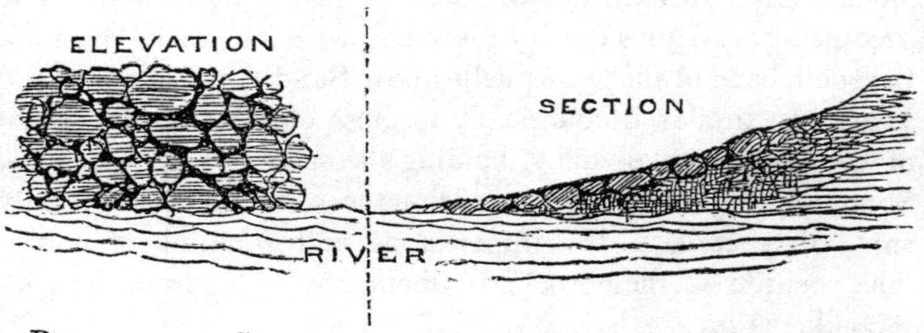

Polished and Grooved Pavement of Boulders on the South Branch.

Towards evening the country began to improve, and the timber to include a few elm and birch. In the prairie are clumps of aspen. On the flats, which occur regularly on the inside of each bend of the river, with steep clay cliffs on the outside of the curve, fine aspens are common, and the herbage is very luxuriant.

August 4th.—Temperature of air at 8 A.M. 61°, of the South Branch 67°. The balsam-spruce begins to appear in groves. The river winds between high wooded banks, with low points and wooded bottoms on one side, high cliffs also wooded with aspen and spruce groves on the opposite bank. The flats are covered with a rich profusion of vetches, grasses, and rose bushes. There are traces everywhere of a former fine aspen forest, with clumps of elm and ash; the dead trunks of these trees, 18 inches in diameter, being frequently concealed by the undergrowth, offer a rude and stubborn obstacle to progress on foot through the tangled mass of vegetation which covers the rich flats. A view obtained from a low hill coming down to the banks of the river, continues to show a deep valley about three quarters of a mile broad, through which the river winds from side to side in magnificent curves. The polished pavement on the banks was frequently seen during the day, with ice furrows and scratches. During the whole afternoon we passed swiftly through a good country, well fitted, as far we could judge from soil and vegetation, for settlement. Islands are numerous in the river, and extensive alluvial flats occur in an expansion of the valley. The water marks are seen seven and nine feet above the present level. The banks of loose clay, when not protected by the pavement before described, are being undermined, and fall bit by bit into the river. A violent thunder-storm at 5 P.M. compelled us to camp.

August 5th.—The early part of the morning was employed in examining the surrounding country, which gave evidence of an excellent soil, and timber sufficient for the first purposes of settlers. Much of the timber, however, has been burnt, and the country is fast becoming open prairie land. Soundings yesterday showed ten to fourteen feet water in the channel; the current maintains its speed of three to three miles and a half an hour. Throughout the entire length of our voyage we have been surprised at the extraordinary absence of animal life. Of quadrupeds, we have seen half a dozen wolves, two or three badgers, several beaver, skunks, minks, foxes, and a number of dead buffalo; of birds, eagles, geese, a few ducks, kingfishers, cliff martins, pigeons, crows, cranes, plover, hawks, and a few of the smaller birds; but no deer, or bear, or live buffalo; and if we had been compelled to depend altogether upon our guns for a supply of provisions, it is probable that our voyage of two hundred and fifty miles down the South Branch would have been attended with some inconvenience and delay. Early in spring and late in the autumn game is more abundant, but during the summer season the smaller rivers in the prairies, the ponds and lakes which abound throughout the country north of the Touchwood Hills, to be afterwards described, are the haunts of vast numbers of aquatic birds and of the larger four-footed animals which now form the small remnant of the earlier representatives of animal life in these wilds, before the fur trade led to their destruction, either for the sake of their flesh or skins.

The stratified layers of fine mud before described, were found again this morning forty feet from the water's edge, above the horizontal layer of boulders which has again made its appearance. The small aggregations of sand are still distributed between the thin layers of fine clay. A great change is coming over the character of the stream; its fall, as ascertained by levelling, is two feet three inches in the mile, with a very rapid current, sometimes six miles an hour. Large boulders are numerous in the bed of the river, but there is always a passage from 50 to 60 yards broad, often however very tumultuous, and for a small heavily ladened canoe rough, and at times hazardous. The hill banks are getting higher as we approach the North Branch. Balsam spruce appears in patches and stripes. The river sweeps in grand curves at the foot of high bluffs, in which fine exposures of the drift may be seen; on the opposite side are low alluvial points covered with aspens, thick and impenetrable. Yellow clay cliffs, 120 feet high, appear at the outside curve of the bends, and where the adjoining flats begin, balsam spruce, two feet in diameter, is not uncommon.

At half past two P.M. we arrived at the North Branch, coming upon it suddenly and finding ourselves in its waters almost before we were aware of its proximity. The temperature of the South Branch was 67°, of the North Branch 62°; an important difference at this season of the year. It is perhaps a fair standard by which to estimate the climatic character of the regions of country through which these rivers flow, in relation to agriculture. The difference in the time of the ripening of fruits on the two Branches has already been noticed. (See page 25, par. 20.) The water of the South Branch is yellowish brown in colour, and turbid; of the North Branch, a shade lighter, and clearer. The one more resembled the waters of the Mississippi, the other those of the St. Lawrence. The South Branch is the larger river of the two at the Grand Forks. After resting for some time at the junction of these mighty rivers, the South Branch being about 180 yards, the North Branch 140 yards broad, their currents meeting one another at the rate of three and a half miles an hour, we turned our canoe up stream and attempted to stem the tide of the North Branch of the Saskatchewan in search of the Coal Falls.

With the exception of the Cree encampment passed during the first and second days of our voyage, we did not meet with a single Indian or Half-breed. Once or twice, smokes, which from their being soon answered in another quarter, we presumed to be signals, and might be raised by Blackfeet in the distant prairies, appeared on the west side of the river. The plan we adopted one night when danger was apprehended, was to cook our supper early in the evening and then drift down the river at sunset for a few miles.

Once only were we disturbed in camp, and this may or may not have been a false alarm. Both of our Half-breeds came into the tent some time after we had retired to rest, and in a low tone whispered 'a grizzly bear,' at the same time seizing a rifle and a double-barrelled gun which were purposely placed at the foot of the tent ready for any unwelcome intruder upon our repose. The night was dark and the fire nearly out. Our men declared they had seen a large animal within ten yards of us, and pronounced it to be a grizzly bear; the alarm they testified was the only proof of the presence of that terrible animal, for the patient watching of the whole party during the greater part of the night, and a careful search for tracks next morning failed to satisfy me that we had been disturbed by this deservedly dreaded monster of the Western Plains.

That the grizzly bear is sometimes found far down the South Branch is a well known fact, and he is such a daring and formidable antagonist that proper precautions are always advisable. A large camp fire often fails to deter this animal from making an attack, and when a large fire might attract the attention of wandering parties of Blackfeet which were known to be following the Crees, who had crossed the river some distance above us, it would not have been wise to have availed ourselves of this doubtful security. Our camp was at the edge of a cliff, we therefore were sure of not being attacked in our rear, and the greater part of the night was passed in quietly watching the open space in front of us. It was the steady determination of the Half-breeds to watch, after a fatiguing day, that led me to suppose they had really seen a grizzly bear, for under ordinary circumstances no people are so unwilling to watch during the night in the prairie as those who have lived the greater part of their lives in them, without they have the best reasons for keeping themselves awake.

During the afternoon of the 6th and morning of the 7th of August we occupied ourselves in dragging the canoe up the North Branch. Paddling was quite out of the question, the current being from six to seven miles an hour a few hundred yards above the Forks, and continuing rapid for a distance of seven miles, that being the furthest limit of our exploration up the North Branch. This rapid current is maintained for eighteen miles above the Grand Forks; the valley of the river, as far as we saw it, resembles in almost all particulars the last ten miles of the South Branch; the river channel is much more obstructed by boulders, and the depth and volume of water considerably less. It is doubtful whether in its present condition a steamer drawing more than two feet of water could ascend it, and in dry seasons the boulders and rapids would probably present an insuperable obstacle. The river was high at the time of our visit, and about 180 yards broad; nevertheless in descending we had a few narrow escapes from striking against huge boulders just concealed by the water. If some of these were removed, the chief difficulties during low summer levels to steamers of shallow draft and great power would vanish.

The character of the Coal Falls, above the point we reached, is described by the people at Fort à la Corne to be similar to the part we saw. The hill banks expose drift in which large masses of Tertiary rock are imbedded containing fish scales. Fragments of Lignite are numerous, but no rock was seen in position. The breadth of the valley is about half a mile and 150 feet deep; the river winds from side to side like the South Branch. The low points are covered with aspen; the hill banks with white spruce, aspen, banksian pine, and poplar. Just below the junction of the two branches, after they unite to form the Main Saskatchewan at the Grand Forks, there is an extensive flat, on which the remains of an old Post of the Company is situated.

The Main Saskatchewan is a noble river, sweeping in magnificent curves through a valley about one mile broad, and from 150 to 200 feet deep. We paddled rapidly round eight points, making a distance of sixteen miles in three hours, and towards evening sighted Fort à la Corne, with the Nepowewin Mission on the opposite or north side of the river. As the description of the Saskatchewan and the valley in which it flows at Fort â la Corne applies equally to the river between it and the Grand Forks, it is unnecessary to incur the risk of needless repetition by enumerating the features of each of the eight points or bends we passed, and the character of the valley through which the river flows. At Fort à la Corne we made measurements of its leading dimensions, a section of the bed of the river, (see sheet of sections,) ascertained its rate of current, examined the cliffs, points, and flats, which are so curiously reproduced at every bend both above and below for many miles, and which will be amply sufficient to illustrate the most interesting and important features of this noble stream between the Grand Forks and a short distance below Fort à la Corne, after which the country begins to assume a different aspect, and will require an independent notice.

An approximate estimate of the number of cubic feet of water passing down the South Branch, North Branch, and Main Saskatchewan, gives the following numbers:—

	Cubic feet per hour.
South Branch	123,425,616
North Branch	91,011,360
Main Saskatchewan, at Fort à la Corne,	214,441,290
Main Saskatchewan near Tearing River,	206,975,000

CHAPTER VI.

FROM FORT À LA CORNE TO FORT ELLICE, AND FORT ELLICE TO THE RED RIVER SETTLEMENTS.

Sandy stripes on the Saskatchewan — Banksian Pine—Fine country—Long Creek — Old forest—Fires, extent of—Extension of the Prairies—Former extent of wooded country—Effect of fires—Long Creek—Hay Ground—Moles—Humidity of climate—Source of Long Creek—The Birch Hills—Flowers—Aspect of country—Carrot River—The Lumpy Hill of the Woods—Lakes—The wooded country—Former extent of—Limits of good land—Raspberries—Mosquitoes—The height of land—Continuation of the Eyebrow Hill range—Valley inosculating with South and North Branch of the Saskatchewan — Grasshoppers—Character of the country—Birds—Destruction of forests—The Big Hill—Boulders—Limit of wooded country—Belts of wood—Great Prairie—Character of the country—Salt Lakes—The Touchwood Hills—Beautiful country — Excellent soil — The Quill Lakes—Flowers—White Cranes—The Heart Hill—The Last Mountain — The Little Touchwood Hills — Lakes numerous—Touchwood Hill Fort—Ka-ou-ta-at-tin-ak—Touchwood Hill Range—Long Lake—Devil's Lake—Garden at the Fort—White Fish in Long Lake—Burnt Forest—Grasshoppers—Winter Forage for horses—White Fish—Buffalo—Medicine man—Climate of Touchwood Hills—Humidity of—Trail to Fort Ellice—Marshes—Little Touchwood Hills—Character of country changes — Depressions — Pheasant Mountain—Character of the country—Heavy dews—Hoar frost—Cut-arm Creek — Willow Prairie — Little Cut-arm Creek—Rolling Prairie — Attractive country — Spy-Hill — Boulders—Aspen groves increasing—Sand hills—The Assiniboine—Dimensions of Valley near Fort Ellice—The Riding Mountain—Rapid River—Character of the country—Well adapted for settlement—Timber of the Riding Mountain—Birds—Cretaceous Shales — Pembina Mountain — White Mud River—Character of the country—Forest Timber—Fish—Luxuriant vegetation—Lake Manitobah—Fishing Station—Red River—Assiniboine Prairies—Arrive at the Settlements.

The trail from Fort à la Corne to the old track leading from Fort Ellice to Carlton House ascends the hills forming the banks of the deep eroded valley of the Saskatchewan in the rear of the Fort. It passes through a thick forest of small aspens until near the summit, when a sandy soil begins, covered with banksian pine and a few small oak. This sandy area occupies a narrow stripe on the banks of the river, varying from half a mile to four miles broad. South of the sandy stripe the soil changes to a rich black mould distributed over a gently undulating country; the pine gives place to aspen and willows in groves, the aspens occupying the crest of the undulations, the willows the lowest portion of the intervening valleys. On the slopes the grass is long and luxuriant, affording fine pasturage. The general aspect of the country is highly favourable for agriculture, the soil deep and uniformly rich, rivalling the low prairies of Red River and the Assiniboine. Our course lay along the banks of Long Creek, which flows in a small depression parallel to the South Branch of the Saskatchewan, and enters the Main river near Fort à la Corne.

August 10th.—During the whole of yesterday afternoon we passed through a good farming country. The remains of aspen forests, in which trees of large growth are numerous, are still to be seen in solitary clumps, or with blackened trunks lie hidden in the long luxuriant herbage until rudely encountered by the carts and horses as we push our way through the rank tangled grass. Raspberries were abundant in patches but not yet ripe; they were fully ripe a fortnight since on the Qu'-Appelle, 200 miles south.

Some of the small aspens near our camp on the 9th have been nipped at the extremities of the branches by frost when in full leaf. The tops of many are black and drooping.

About four miles from Long Creek, and perhaps ten from the South Branch, a low range of hills running north-east and south west, are still covered with an aspen forest of the same age as the blackened poles which stand in clumps on all sides. These poles are from nine to twelve inches thick; the young aspens are from four to six inches in diameter. The fire was here last year. We have now traced the extent of that vast conflagration from Red River to the South Branch, and over four degrees of latitude at least; but the Rev. Henry Budd states, that in the autumn, north, south east, and west of the Mission the country appeared to be in a blaze. The immediate banks of Long Creek, with the exception of a narrow stripe in the prairie South of the Qu'Appelle, is the only part of the country in which we have not recognized traces of last year's fire. The annual extension of the prairie from this cause is very remarkable. The limits of the wooded country is becoming year by year less, and it appears from the almost universal prevalence of small aspen woods that in former times the wooded country extended beyond the Qu'Appelle, or five or six degrees of latitude south of its present limit. It being always borne in mind that the term wooded country is applied to a region in which prairie or grassy areas predominate over the parts occupied by young aspen woods. The South limit of the wooded country is some distance north of the Touchwood Hill range, but there are areas north and south of the Qu'Appelle where the remains of aspen forests of large dimensions exist, and young forests are in rapid process of formation; perhaps, however, soon to be destroyed by fire.

This lamentable destruction of the forest is a great drawback to the country, and a serious obstacle to its future progress. It appears to be beyond human power to arrest the annual conflagrations as long as the Indians hold so vast a prairie region as their hunting grounds. Their pretexts for "putting out fire" are so numerous, and their characteristic indifference to the results which may follow a conflagration in driving away or destroying the wild animals, so thoroughly a part of their nature, that the annual burning of the prairie may be looked for as a matter of course as long as wild Indians live in the country. A fire lit on the South Branch of the Saskatchewan may extend in a few weeks, or even days, to Red River, according to the season and the direction and force of the wind.

Long Creek maintains a breadth of six feet, flows clear and sluggishly through a broad shallow depression, where wild hay is as abundant as if the whole valley were one continuous beaver meadow. The burrows of moles are very numerous; wherever the soil is very rich these little animals are to be found in large numbers; they form excellent indicators of the fertility of a soil; they are never seen where the soil is poor and sterile. Ponds and lakes are very numerous; this extensive distribution of water points to a much more humid climate than is in the country south of the Qu'Appelle.

August 11th.—Still the same excellent soil. The burrows of foxes and badgers have twice shown a light gravelly substratum on low ridges, otherwise the black mould is everywhere distributed. A chain of lakes, lying westerly from our course, give rise to Long Creek. The Lakes are from 200 yards to a third of a mile broad, and form a continuous series connected by a small rivulet for a distance of ten miles. A hill range, called the Birch Hills, whose western flanks we have turned, is said by Indians to extend to the rear of Fort Pelly. A vast profusion of flowers gives remarkable beauty to the large open areas. They generally occur in parterres of several acres in extent occupied by one species, here the yarrow, there the fire weed, then a field of a species of helianthus, followed by *Liatris scariosa*. When viewed from an eminence, the country appeared to be clothed with pink, white, yellow, and blue, in singular contrast to the unform tint which prevails on the great prairies of the Little Souris.

Our course yesterday continued up the valley of Long Creek, which taken as a whole, offers by far the most attractive features for settlement of any part of the country through which we have assed since leaving Prairie Portage. To-day we follow the windings of a shallow brook which runs into the South Branch. It meanders through a fine broad rich valley with hills on its south-eastern side gently sloping towards it, and covered with the, dead standing trunks of burnt aspen. The soil of this valley is good, differing in no respect from that of Long Creek. The flowers are equally numerous and showy, consisting of the same varieties, and distributed in large patches occupied by a single species.

We passed to-day near the source of a river which flows into the Main Saskatchewan at the Pas, about 140 miles distant from us. It is called Carrot River or Root River, and, rising within twelve miles of the South Branch, it drains an extensive area of wooded country, passing also in its course through numerous lakes. The rise of Root River within ten or twelve miles of the South Branch shows that the height of land between the two water-sheds maintains the same distance as on the Qu'Appelle, and at the North Fork of that valley near the Moose Woods. Before us, about four miles distant, is the Lumpy Hill of the Woods, and the range of hills on the north side of which Root River flows becomes better developed. The Birch Hills form the dividing ridge between the water which flows into the Main Saskatchewan and the Assiniboine, or Red Deer and Swan River.

The valley leading to the Lumpy Hill of the Woods is rich in alluvial meadows, ponds and lakes. A view from the Lumpy Hill, which I ascended this evening, is very extensive. The altitude of this eminence is about 400 feet above the general level of the country. From its summit an undulating open country, dotted with lakes and flanked by the Birch Hills is visible towards the east. South and south-west is a lake region, also north and north-east. These lakes are numerous and large, often three miles long and two broad. Seventeen large lakes can be counted from the Lumpy Hill; hill ranges in several directions can also be discerned. The most important of these are the Bloody Hills, the Woody Hills, far in the prairie west of the South Branch, and the chain of Birch Hills running from the Lumpy Hill easterly. The view extends to the borders of the wooded land; beyond is a treeless prairie. The so called wooded land now consists of widely separated groves of small aspens, with willows in the low places. Formerly, the Cree Indian guide we took from the Lake of the Sand Hills states, the woods extended in one unkroken range to the borders of the prairie, which may be 25 miles south-east of the Lumpy Hill. The Moose Woods coming between the prairie and the South Branch to the west.

Much of the soil on the south and east of the Lumpy Hill is sandy and poor, in fact we have reached the limit of the good land, and are about to enter a comparatively sterile country. Low hills and long ridges running north-east by east, and south-west by south, diversify the general level character of the plains, as seen from the Lumpy Hill. This eminence consists of drift sand and clay, with boulders on its summit; the western side is very steep, and partially covered with a burnt forest of birch. Raspberries of very large size abound on the west side, but the mosquitoes start from the bushes in such countless myriads that it is next to impossible to linger five minutes to pick the delicious fruit. I offered the Cree guide a piece of tobacco for a tin cup full of raspberries, he tried to win it, but after a short struggle with these terrible insects he rushed from the hill side and buried his face in the smoke of the fire we had lit to expel the tormentors from the neighbourhood of our camp; the horses became quite frantic under the attacks of their tormentors, holding their heads over the smoke, and crowding together in a vain endeavour to avoid the clouds of insatiable insects which surrounded us. Both man and beast passed a miserable, restless, and sleepless night.

August 12th.—The early part of this morning was spent on the summit of the Lumpy Hill. A strong breeze drove the mosquitoes away, and permitted me to enjoy a quiet view of the country, which lay mapped about 400 feet below. After breakfast, the trail passed nearly due east, over a series of hills and through intervening valleys, constituting a height of land. This range may be from thirteen to fifteen miles from the South Branch. It is a continuation of the Eyebrow Hill range on the Qu'Appelle, before described, and it continues on under the name of the Birch Hill, limiting the valley of the North Saskatchewan, as far as the rear of Fort Pelly. As soon as we passed the crest of this range, and entered the small aspen prairie east of the hills, a valley through the range became apparent to our right. From lakes in this shallow depression water passes to the South Branch and to the North Branch, by a tributary of Carrot River, during spring freshets.

Grasshoppers were seen to-day, flying to the north-east. These are the first that have been noticed since leaving the Mission on the Qu'Appelle. The vegetation still continues luxuriant; lakes are numerous, and flowers abundant. Aspens cluster here and there, and the country presents many attractive features. Wild-fowl are found on all the lakes: cranes, both the brown and white; waders of many species, and a few prairie hens. As we approach the great prairie, the country becomes more undulating, and the soil light-coloured and poor. The aspens, which cap some of the hills, are still large, although many are nothing more than dead trunks. The wooded country through which we are passing is only so called in remembrance of former forest growth. If the devastating fires continue for a few more years, it will become a treeless prairie to the Lumpy Hill; and the aspen and birch woods will then be limited to the country between that eminence and the North and South Branch of the Saskatchewan. A young brood of grasshoppers have been seen to-day, showing that these destroyers reached this part of the country last autumn.

At noon on the 13th, we arrived at the Big Hill, a point of some interest, for south and south-east of it, a boundless, undulating prairie lies before us; the summit of the Big Hill is covered with huge granite or gneissoid and limestone boulders, indeed on all the hills which surround the Big Hill boulders are very numerous. The limit of the so-called 'Wooded Country,' is about seventy miles from the North Branch in an air line, and thirty miles from the South Branch.

August 15th.—In journeying from the Lumpy Hill we crossed three belts of woods before arriving at the great prairie west of the Touchwood Hills. These belts, which consist of groves of small aspen, following a low gravelly ridge about a mile broad, and having a north-east and south-west direction, are separated by prairie valleys which sustain in their lowest parts a good soil, and fine pasturage. Each belt diminished to a point some ten or fifteen miles south-west of our track. We can see the points of these belts from the summit of mounds not more than fifty feet high; beyond them is a treeless prairie, stretching away to the South Branch, north-eastward. The belts of woods become broader in a north-easterly direction until they merge into the wooded country between the Birch Hills and the Saskatchewan. There are many delightful spots in the belts, the herbage is clean as a well shaven lawn, the clumps of aspen are neatly rounded as if by art, and where little lakes alive with waterfowl abound, the scenery is very charming, and appears to be artificial, the result of taste and skill, rather than the natural features of a wild, almost uninhabited country.

In the prairie valleys the ponds are fringed with boulders, and water marks show that during the spring a large area is flooded. The great extent of pond and marsh affords food and shelter to vast numbers of aquatic birds. Grey geese were seen here for the first time; the Canada goose is very abundant; and duck, teal, cranes, and bittern, are numerous. The lakes and marshes all contain salt or brackish water, which we found to our discomfort was not suitable for culinary purposes, or for slaking thirst. Tea made from it had a nauseous taste, and possessed the medicinal effect which might be supposed to result from preparing that beverage with a weak solution of Epsom Salts. The Touchwood Hills seen from the treeless prairie present a bold outline gently rising from the flat country and maintaining a course nearly due east and west for ten or twelve miles, they then assume a more easterly direction; westward they are seen to die away in the prairie.

In the afternoon we began the ascent of a gently rolling slope at the foot of the Touchwood Hills; patches of willow appear here fringing small areas of good pasturage. At 6 p.m. we reached the summit plateau, and then passed through a very beautiful undulating country diversified with many picturesque lakes and aspen groves, possessing land of the best quality and covered with the most luxuriant herbage. From the west side of the summit plateau the Quill Lakes are seen to the north-west; these bodies of water have long been celebrated for the large numbers of goose quills which were occasionally collected there by Indians and brought to the fort for exportation. There is no timber visible on the west side of the range with the exception of small aspen and burnt willow bushes. All the wild flowers so numerous and beautiful in the valley of Long Creek are met with on the summit plateau of the Touchwood Hills, of even larger growth and in greater profusion. Little prairie openings fringed with aspen occur here and there, through which the trail passes; we then come suddenly on to the banks of a romantic lakelet, in which ducks with their young broods are swimming, and flocks of white cranes start from their secluded haunts at so unexpected an intrusion. The breadth of this beautiful plateau is about four miles, its level above the Salt Prairie to the west may be about five hundred feet. Our course lay diagonally across it, so that we had to pass through seven miles of this delightful country. The Heart Hill, with others not seen before, come into view as we approach the eastern limit and begin a descent to Touchwood Hill Fort. The Last Mountain is visible in the west, but blue in the distance; the Little Touchwood Hills lie before us, the trail to Fort Ellice stretching towards their eastern flank. The country between the two ranges is dotted with lakes and groves of aspen. From a small hill near the fort I counted forty-seven lakes.

Touchwood Hill Fort, 16th August.—Arrived at the fort after sunset last evening. It is situated on the south-east flank of the range, and from a hill close behind it an extensive view of the country is obtained. Heart Hill or *Ka-ou-ta-at-tin-ak* is about seven hundred feet above the general level of the plain, and seven miles in an air line N. 12° W. of the post. The general direction of the range is N. 26° E. It appears to consist of a series of Drift Hills, many of which rise in rounded dome shaped forms from the summit plateau. The Last Mountain bears S. 26° W., about 25 miles distant from the post, and the end of Long Lake, as it was pointed out to me by the guide, bears W. 37° S., distant from the fort a good day's journey, or about 30 miles. The Little Touchwood Hills bear south-east, and have a general direction parallel to the main range. At the foot of the Heart Hill and on its northern flank is a lake about five miles long, running east and west close to its foot, and is said to contain white fish. Devil's Lake, which is connected with Last Mountain Lake, lies about 40 miles due west of the post.

The garden or rather the remains of a garden in the rear of the fort, produces every variety of vegetable grown in Canada, but the efforts to culivate it are almost abandoned in consequence of the depredations committed by the Indians from the prairies, when they arrive in autumn with their supplies of provisions. (buffalo meat and pemican). A few of the lakes near the fort are known to contain fish, and it is probable that all of the large fresh water lakes contain them. The officer in temporary charge of the post stated that the people here had only known of the existence of white-fish in the Last Mountain Lake for three years; they are now taken in the fall, and it is probable that the fishery recently established will become of great importance to this part of the country. The Plain Crees are not fishermen like the Ojibways, they did not know how to catch fish when the attention of people at the Touchwood Hill Fort was first directed to the treasures of Last Mountain Lake. Mr. Hoover, the officer in charge at the time of my visit, told me that he had first observed the White-fish under the ice in November of 1854, and since that period they have established a fishery which provides the fort with an ample supply for winter consumption.

The timber on the Touchwood Hills is nearly all small and of recent growth; fires years ago destroyed the valuable forest of aspen which once covered it. The remains of the forest are still seen in the forms of blackened poles either standing erect or lying hidden in the rich covering of herbage which is found everywhere on the south west flank of the range. Last year the grasshoppers visited the Touchwood Hills and deposited their eggs. This year the new brood consumed every green leaf in the garden, and make local ravages in the surrounding country. They took their flight on the 28th July for the south-east, and during the period of my visit but few were to be seen. So rich and abundant is the vegetation here, that horses remain in the open glades all the winter, and always find plenty of forage to keep them in good condition. The cows are supplied with hay; the horses are worked during the winter, either journeying to Fort Pelly or to the Last Mountain Lake to fetch fish. The White-fish weigh on an average 7 lbs., but 10 lbs. each is not uncommon. Buffalo congregate in the beautiful prairie south of the fort every winter, sometimes in vast numbers.

During the greater part of the night we were disturbed by a noted conjuror who was performing his ceremonies over the suffering form of an invalided woman who lay in his medicine tent near to the fort. His drum and song were heard nearly the whole of the night, and his incantations are described in another chapter as well as the remedy for the sickness of the poor squaw, which the conjuror suggested as infallible.

August 17th.—Snow falls on the Touchwood Hills to the depth of two feet and a half in the woods, and in the plain where aspen groves are numerous it is not unfrequently found one foot and a half deep. In the great prairie south, where the herbage is short, the snow is drifted off by winds; the climate of the Touchwood Hill is evidently very humid. Thunder storms appear to travel in the direction of this range and occasion a copious precipitation as they pass over it. Not only are lakes very numerous and well supplied with water, but there are several living streams flowing from the range. Indeed the whole country from the Touchwood Hills to the Riding Mountain, including the country about the head waters of the Assiniboine is dotted with innumerable lakes, annually replenished by summer rains.

A range of hills joins the Greater and Lesser Touchwood Hills, having a course nearly north-west and south-east, or at right angles to those of the main ranges In this subordinate range there are many conical hills, some of them well wooded up to their summits, but the forest trees are small. The trail to Fort Ellice winds round the base of conical hills, past small lakes and aspen bluffs, through luxuriant herbage, and over an excellent soil. About nine miles from the Fort it begins to ascend the eastern flank of the Little Touchwood range and gently winding up it for several miles it finally reaches an extensive marsh which occupies a portion of the summit plateau. The marsh is but the introduction to numerous lakes, which continue to diversify the country in all directions.

On the following day, we entered a region differing in many points from the rich tract we had left. Gravelly hills and areas of coarse drift sand form the surface of the country for a few miles, they are succeeded by a number of curious depressions or hollows, circular or oval in form, and varying from one quarter to one mile in diameter, often with a lake in the centre, but without visible outlet. The land is high in which they occur, and forms a ridge running nearly north-west and south-east, like the general direction of the hill ranges before described, but the country is so undulating that it is difficult to ascertain the true character of the surface until we arrive at the summit plateau. Here boulders are seen; the sand is coarse and mixed with a little clay, so as to resemble a coarse gravelly loam on the ridges and hills, as well as on their flanks, but in the hollows and valleys the soil is excellent and the herbage very luxuriant.

August 19th.—The view this morning from the summit of a mound revealed a rolling treeless prairie, stretching on all sides and bounded only by the horizon. The wooded range of Pheasant Mountain appears low in the south-west, serving only to destroy the uniformity of the general outline. Numerous lakes, ponds and marshes are visible in every direction, covered with wild fowl. The soil in low places is good, supporting long grass which afforded fine pasturage for our cattle. The ridges and mounds are gravelly, and a few boulders of the unfossiliferous rocks are seen here and there. It is remarkable that east of the Touchwood Hills no limestone boulders have been noticed, but limestone gravel is common.

The Pheasant Mountain runs north-east and south-west, and may be twenty miles long. The wet grass reminds me that the dews in the Touchwood Hills are very heavy and abundant at this season of the year. Last night, dew was deposited a few minutes after the setting of the sun, although the sky was cloudy and prevented direct radiation. This phenomenon has been noticed several times; the setting of the sun appears to admit of the cooling of the air sufficiently to allow the dew point to be quickly attained on the surface of vegetables, notwithstanding the screen of clouds which must necessarily obstruct radiation into space, but it would also appear to show that the temperature of the clouds must be very low. With the thermometer at 65° in the air, ten minutes after sunset, and under a cloudy sky, I have observed dew form three times since leaving Fort à la Corne. On clear nights, dew has always been copiously deposited during the summer; so much so at times as to wet the tents. This fact shows not only a certain humidity in the air, but the sudden reduction of the temperature when the sun sinks below the horizon.

On the morning of the 20th, hoar frost on the Buffalo robes reminded us that the season was advancing. We crossed today a rapid stream with a swift current, ten feet broad, and one and a half deep, flowing into the Qu'Appelle. It was thought to be Cut Arm Creek; it meanders through a prairie covered with low willows, and named the Willow Prairie; it embraces an extensive area of excellent land, sustaining fine pasturage. Limestone boulders were seen again to-day. The country preserves a uniform level character, with a few gravelly ridges and mounds; neither lakes nor marshes are numerous, and timber for fuel is very scarce. Little Cut Arm Creek, which we crossed this morning, flows in a ravine about 80 feet deep and 400 broad. Lakes begin to appear again. The prairies are more rolling and are crossed by ridges, which preserve a certain amount of parallelism, generally from north-east to south-west. The aspen replaces the willow in small clumps, and after passing Big Cut Arm Creek, the country is undulating, attractive, and very well watered. Large hills appear near the Big Cut Arm, which flows in a valley 1200 feet broad, and 180 ft. deep, resembling that of the Qu'Appelle, from which we are not now far distant. We camped in the evening near to Spy Hill, called also *Ka-pa-kam-a-ou*, or 'Some one knocked.'

August 22nd.—The Blue Hills across the Assiniboine are visible from Spy Hill, so also are those on the Qu'Appelle. Spy Hill is a gravelly eminence about 120 feet above the prairie. Near it, boulders of the unfossiliferous rocks are very numerous, and of large dimensions. One of gneiss, measured 13 feet in diameter. Our old hunter remarked that the aspen groves were much more numerous west of Spy Hill at the present time, than when he first remembered the country forty-three years ago. After crossing a sandy prairie flanked on our left by numerous bare sand hills, we reached the Assiniboine at the mouth of the Qu'Appelle early in the afternoon, and having crossed that river in preference to the Qu'Appelle, wehad the pleasure on the following day, of meeting Mr. Dickinson within a mile of the Ferry, on his way to Fort Ellice, our place of rendezvous. The distance from Fort a la Corne to Fort Ellice by the route we followed, is three hundred and twenty miles.

We spent two days in the valley of the Assiniboine near Fort Ellice, occupying ourselves in making a section of the valley. We found its breadth to be one mile and thirty chains, and its depth two hundred and forty feet below the level of the prairie on either hand. The river is one hundred and thirty-five feet broad, with a greatest depth of 11.9 feet, a mean depth of 8 feet, and a current flowing at the rate of one mile and three quarters per hour.

On the 26th we set out on our return to the Settlements. Our route lay on the flanks of the Duck and Riding Mountains, and through a country admirably adapted for farming purposes. On the morning of the 27th the herbage was covered with hoar frost, but without any injury to vegetation. Ponds and lakes are very numerous on the flanks of the Riding Mountain, but as far as our opportunities enabled us to judge, the whole country, with the exception of narrow ridges, possesses a rich black fertile mould, supporting very luxuriant herbage, and on the mountain an ample supply of timber, consisting chiefly of aspen of large dimensions. The Riding Mountain consists of a succession of slopes and plateaux on its south-western side; the ascent is almost imperceptible to the thick impenetrable forest which covers the highest plateau.

On Saturday, 28th August, we arrived at the Little Saskatchewan or Rapid River, which Mr. Dickinson had explored for a distance of one hundred miles from its source. The valley of this river is extremely beautiful and fertile until within a few miles of its junction with the Assiniboine; it offers the most attractive and desirable place for settlement in any part of the country we have explored. The stream abounds in fish, the flats in the valley are covered with the richest herbage; timber, consisting of aspen, poplar, and oak, is abundant; the prairies on either side are clothed with the greatest luxuriance of vegetation; the scenery is very attractive, and the river navigable down stream for canoes and batteaux to the Assiniboine. Where the Rapid River enters the Riding Mountain balsam and white spruce appear, and our explorations on the east flank of the range showed that large birch, spruce, poplar, and aspen flourished on the summit plateau.

Fires here as elsewhere have damaged the forest which once covered the country. Vast numbers of young oak and aspen are springing up in all directions on the prairie fringing the river near our trail. Birds are very numerous in this region; every lake contained duck, with their young. The aspen groves and willow clumps were alive with grackle and yellow birds congregating in flocks. Humming-birds were also observed, as well as the American cuckoo and the solitary thrush.

In the marshes, herons, cranes, and bitterns are numerous. Hoar-frost again noticed early this morning.

In a brook emptying into Rapid River, I found an exposure of the Cretaceous shales before described as occurring on the Assiniboine and the Little Souris. The rock was very fragile, and contained a few fossils in an imperfect state of preservation.

On the 29th we reached the south-eastern termination of the Riding Mountain, and obtained a fine view of the successive steps of which it is composed. These were three in number, each step being separated by a gently sloping plateau. The entire mountain appeared to be densely covered with forest trees. The country through which we passed to-day was very wet and swampy in many places. On the ridges the soil is dry and gravelly: we are, in fact, descending the Pembina Mountain, which being here extended over a great breadth, is not easily recognized. In the afternoon we arrived at a beautiful ridge, running N. 12° W. and S. 12° E. One side of this ridge is partly excavated by the White Mud River, and exhibits finely stratified gravel, consisting almost altogether of small limestone pebbles, with a few belonging to the unfossiliferous rocks. The ridge is gently sloping towards the east, and precipitous towards the west,—having on either hand a level country, higher on the west side than on the east. I have no doubt this ridge is either a continuation of the Big Ridge on the Assiniboine and Red River, or of one at a higher level in the rear of those characteristic lake boundaries. Some fine oak grows on the banks of White Mud River near the ridge; and ash-leaved maple begins to show itself again.

August 30th.—Our course to-day lay through the prairies drained by White Mud River. This tract of country is second only in beauty and fertility to the valley of Rapid River. Not only is the herbage of surprising luxuriance, but the trees in the river bottoms are of very large dimensions, and consist of oak, elm, ash, maple, aspen, and poplar. Near the crossing place there is a fish weir, where large quantities of pike, suckers, gold-eyes, and other species, are taken by the people of Prairie Portage, who have established a fishing-station here, as well as one at Lake Manitobah, some miles further east.

The woods fringing the river at the crossing place are very important. The oak and elm are of the largest size; 2 ft. to 2 ft. 6 in. in diameter, with tall, clean trunks. The hop and vine twine around the underbrush, and give a very attractive appearance to the belt of woods which fringe White Mud River.

Wishing to ascertain the character of this stream to its outlet, we gummed the canoe, and once more launched it for a short voyage down the White Mud River, to the fishing-station on Lake Manitobah. Mr. Dickinson proceeded down the river, the carts, with Mr. Hime, journeyed on towards Prairie Portage, while I rode to the fishing-station, in company with a half-breed who was familiar with the history and progress of the station since its commencement

We soon arrived at Rat River, a stream of much interest in connexion with the floods of the Assiniboine. Down its valley the water of that river, during freshets, flows into Lake Manitobah, and by making a very shallow cut, a permanent communication, in time of high water, could always be maintained. The fishing-station at the mouth of White Mud river consists of about half a dozen houses, which are only tenanted during the fishing season. Very large quantities of white fish are caught here, and no doubt when the demand requires it the station at the mouth of White Mud River will become an important source of supply. The Assiniboine prairies extend to the banks of Manitobah Lake, and their elevation as seen here and at Oak Point is not twelve feet above the level of that extensive but shallow sheet of water.

We camped on the banks of Rat River, and the following day made a nearly due south course through a rich but treeless prairie to Prairie Portage on the Assiniboine. In making this traverse we passed the shallow, winding, but dry bed of a brook several times, a tributary of Portage River. In wet seasons this bed is occupied with drainage water from the Bad Woods, while Rat River rises within three miles of the Assiniboine in the same locality. The valley of Rat River and of the dry water-course may yet become of vast importance if it should ever happen that the commercial inducements for effecting a steam communication with the South Branch, by way of the Qu'Appelle valley, should lead to the construction of works for that purpose.

On the 31st of August we arrived at Prairie Portage, and reached the Settlements at Red River on the 4th of September, after an absence of nearly three months. Our course from Prairie Portage lay through the prairies which were described in my Report for 1857.

CHAPTER VII.

FROM FORT À LA CORNE TO THE GRAND RAPID OF THE SASKATCHEWAN.—FROM THE GRAND RAPID TO THE RED RIVER SETTLEMENTS *via* THE WEST COAST OF LAKE WINNIPEG.

Instructions—Equipment—Departure from Fort à la Corne—General direction, current, and breadth of the Saskatchewan; and character of its Valley—Country through which the river flows well adapted for settlement—Sickness and discomforts—Reach Pemican Portage and Cumberland House—Description of Cumberland—The Saskatchewan and surrounding country between Cumberland and the Pas—The Pas—Christ Church—Gradual depression of the country bordering the river—Alluvial flats—Marshes—Delta—Muddy Lake—Rock exposure—Marshes and mud flats—Cedar Lake: its situation and dimensions—Surrounding country—The Saskatchewan between Cedar Lake and Lake Winnipeg—Cross Lake Rapid: its dimensions—Enter Cross Lake—Meet a brigade of boats—Cross Lake: its dimensions and altitude—Surrounding country—The Saskatchewan east of Cross Lake—Rapids: their dimensions—Smooth reach—Drift Clay banks—The Grand Rapid: Portage; running the Rapid; its dimensions; character of its excavated bed; magnificence of the upper portion of the cataract; mode of ascending it; remarks in relation to surmounting this barrier and making the Saskatchewan available for steam navigation—Indian encampment—Lake Winnipeg—Cape Kitchinashi—Storms—Detained on an island—Windbound on mainland — Tempest—Repulsed by the wind—Character of the coast: the sand beaches and swamps—War Path River—Verifying rate of canoe—Tracking—Limestone Point—Encountering a head wind and storm—Lightening canoe—Starving Indians—The Little Saskatchewan—Recapitulation—The prominent features of the coast—Formation of Cape Kitchinashi—Limestone exposures—Tributary streams—General Character of the Country—Indian Chart—Inaccuracy of the Maps of the Lake—Depart from the Little Saskatchewan—Windbound again for three days—Provisions exhausted—Contrary Winds—Driven back and stopped—The Cat Head—Windbound again by a hurricane—Barrier of boulders—Eagle—Stopped by foul winds again at the Wicked Point—Pike Head and River—Opportune supply of Fish—Wide traverse to Grindstone Point—Grassy Narrows—Sandy Bar—Arrive at the Settlements—Conclusion.

Fort à la Corne,
August 9th, 1858.

Dear Sir,—You will start in a canoe from Fort à la Corne and proceed down the Saskatchewan River into Lake Winnipeg, thence by the west coast of that lake to the mouth of Red River, thence to the Settlement.

In your progress down the Saskatchewan you will make as complete a survey of the river as circumstances will permit, ascertaining its course, rate of current, volume of water, fall, and extent and nature of the obstacles to navigation. It is desirable from time to time to make sections of the river and its valley, to level the rapids with precision, ascertain the height to which the water rises and the extent to which it falls in the course of a year.

The west coast of Lake Winnipeg should be attentively examined, and specimens of all rock exposures collected

The object of this exploration is to obtain information respecting the main Saskatchewan, similar to that which has been acquired during the recent exploration of the South Branch from "The River that turns" to the Grand Forks.

I am, dear Sir,
Very sincerely yours,
HENRY Y. HIND.

John Fleming, Esq.,
Assistant Surveyor.

MR. FLEMING'S NARRATIVE.

Dear Sir,—In compliance with your request, I shall endeavour to describe the more prominent topographical features and the general character of the country which came under my observation, while in control of the branch expedition with which you were pleased to entrust me; giving some of the results of the exploratory survey of the Saskatchewan and Lake Winnipeg, conducted according to your instructions dated Fort à la Corne, August 9th, 1858; and such additional information as I was enabled to obtain by instrumental operations and otherwise. In reporting my progress from the time we separated at Fort à la Corne on the 9th of August, until we again met at Selkirk Settlement on the 16th of September, 1858, I shall avail myself of copious extracts from notes daily recorded on the journey.

You are aware that the equipment available for the service to be performed was that with which we surveyed and explored the South Branch of the Saskatchewan; consising of a three fathom birch bark canoe, manned by two voyageurs, (Wigwam, an Ojibway, and James Louis, a Blackfoot Half-breed,) and provided with the necessary instruments for the track survey and for making the requisite observations at intervals. Wigwam continued to act as bowsman, and Louis as steersman; both were expert canoemen, and proved themselves eminently trustworthy throughout, and reliable in time of difficulty and danger. The canoe being leaky, owing to injuries it had sustained in crossing the plains from Red River to the Elbow of the South Branch, was the source of much trouble until we reached Cumberland House, where, through the aid of the letters with which you provided me, and the courtesy of the gentleman in charge, I was enabled to procure a new canoe and some other necessaries.

Before proceeding on our journey, and commencing the continuation of the survey of the Saskatchewan, we were occupied some time near Fort à la Corne in making a transverse section of the river, ascertaining its fall by levelling, and measuring its rate of current by the log, (adopting the mean of a series of observations;) so that it was at a late hour when we made our departure, and we did not accomplish more than 23 miles the first day.

The general direction of the Saskatchewan from Fort à la Corne towards Cumberland House is north-easterly, as will be observed on referring to the plans which have been protracted from my field notes. The current continues strong for a considerable distance below Fort à la Corne, where the average rate was found to be three miles an hour. In some places the mean velocity of the current exceeds this, as I ascertained by repeated trials; and at the points a small rapid is frequently seen, generally caused by a submerged spit or reef of boulders and gravel protruding into the river; but the water is only agitated in its passage over these shoals, which are always on one side of the river; in the bays opposite the points it is quite smooth and deep, averaging in the channel 19 feet.

At Fort à la Corne the breadth of the Saskatchewan (which I obtained by trigonometrical measurement) is 965 feet, and its immediate banks are high; the sides of the valley, which are much higher, being no great distance from the river. The breadth of the river continues very uniform, but its immediate banks become gradually lower, the hill sides of the valley at the same time diverging. About 20 miles below Fort à la Corne the banks of the river are low, and the general character of the adjacent country considerably changed. The high cliffs before seen at the great bends of the river give place to rich alluvial flats, supporting a forest of fair sized balsam-spruce and poplar, and the valley becomes so broad that the high banks are nowhere observed.

The second day of our journey, August 10th, we embarked at 6 a. m., and passed during the day the "Big Birch Islands," and many others; they are all alluvial deposits, and some of them are overflowed in spring. The banks of the river are now quite low, and the country on either side is very flat; but it still continues well adapted for agricultural purposes and settlement; the soil being a rich alluvial loam of a considerable depth, well watered and drained by many fine creeks, and clothed with abundance of timber for fuel, fencing and building. In some places stony points projecting into the river, contract it to a width of five or six chains; stretching out from these points there are shoals over which, as before observed, the current is very strong and rough. Among the islands the river attains a width of from 25 to 30 chains, but where it is broad its depth is diminished in many places by mud flats. We stopped to camp for the night about half past 6 p. m., nearly 53 miles from where we started in the morning.

August 11th.—We left our last night's resting place at daybreak this morning, and passed through an excellent tract of country all day; the soil on both sides of the river consisting of a very rich alluvial deposit, ten feet in thickness, above the surface of the water, well wooded with large poplar, balsam-spruce, and birch; some of the poplars measuring two and a half feet in diameter; and, as far as I was enabled to ascertain, the land continues good for a great distance on either side, but more especially on the south side of the river. In many places the river is studded with large alluvial islands supporting a most luxuriant growth of poplar and willows. Among these islands the channel is sometimes intricate, being occasionally interrupted by sand-bars and snags. We encamped about 6 p. m., having attained a distance of about 47 miles to-day.

On the 12th August we embarked about 4 a.m., although I could only communicate with the men by signs, being unable to speak, owing to a very painful swelling in the throat with which I was seized soon after leaving Fort à la Corne. This distressing malady, from which I fortunately rapidly recovered, was so severe that for nearly three days I could scarcely eat or drink. It was brought on probably by exposure to the frequent rains and lying in wet clothes. The night of the 11th August was to us a sleepless and most uncomfortable one; a terrific thunderstorm came on after dark, and having no tent to protect ourselves from the driving rain, we were drenched to the skin, and had to lie in a pool of water all night. Our constant tormentors, the mosquitoes, were also excessively annoying.

The general character of the country we passed throughout the day, is excellent; the soil being rich and the timber of fair quality. The depth and breadth of the river is variable; in one or two places it is impeded by mud flats and shoals, sometimes holding snags and sawyers. About noon we came to the mouth of a tributary stream 100 feet broad, flowing into the Saskatchewan from the north, which we supposed to lead to Cumberland House, as it corresponded to the description given to us at the Nepoween, but being desirous of keeping the main river, agreeably to your instructions, we went on until reaching an old carrying place, called "Pemican Portage," leading to the fort, where we discharged and hauled up the canoe. I despatched Louis to the fort, and he returned in the evening reporting the road very wet and marshy. We came to-day nearly 29 miles, so that the distance between Fort à la Corne and Cumberland, by the windings of the river, is upwards of 150 miles.

August 13th.—Owing to the thickness of the rushes and the shallowness of the water in many parts of the marsh between the Saskatchewan and Pine Island Lake, we had to go over to Cumberland this morning in the empty canoe; pushing it through the marsh until we reached a strip of dry ground, about half a mile wide behind the fort. Mr. Edward McGillivray, the gentleman in charge *pro tem.*, received us very hospitably. I obtained from him some pemican and flour, and got him to procure for me a new canoe, for which I had to wait, as it was not quite finished. In the forenoon a brigade of boats from the McKenzie River arrived and departed *en route* to York Factory. One of the boats contained Mr. Anderson Chief Factor, who was going direct to Red River and Canada. I mention this, because, although Mr. Anderson left Cumberland three days before us, in a boat of four or five tons burthen, well manned and equipped, and infinitely better fitted for encountering the boisterous gales of Lake Winnipeg, than our little canoe, we reached the mouth of Red River only 24 hours after him.

On Saturday the 14th August, we were aroused at daybreak, by the singing of the voyageurs of another brigade of boats just arriving. It proved to be a detachment from York Factory, bringing J. G. Stewart, Esq., Chief Trader, in charge of Cumberland, with Mrs. Stewart, and Mr. Spencer. Our canoe was not finished till late in the afternoon, when I could have started had I been supplied with a guide for Cedar Lake and the Grand Rapid; but the only man that was competent, and willing to go, being one of Mr. Stewart's boatmen, and they having received their usual holiday and allowance of rum on reaching their destination, no arrangement could be made with him. I was consequently compelled to remain till Monday. During the day, Mr. Stewart, from whom I received the most kind and hospitable attention, opened some packs and enabled me to get one or two articles of clothing, of which I stood greatly in need.

Sunday, August 15th.—A beautiful day. Another brigade from Methy Portage came in and left about noon to-day; bound for York Factory under the pilotage of the veteran guide, L'Espérance.

Cumberland House, the chief depôt or fort of the Cumberland District of the Hon. Hudson's Bay Company is situated on the south shore of Cumberland or Pine Island Lake; in latitude 53° 57′ N., and in longitude 102° 20′ west of Greenwich, (according to Sir John Richardson.) It is about two miles in an air line north of the Saskatchewan, on the north side of what is called "Pine Island," a tract of land of considerable extent between the Saskatchewan and Pine Island Lake, isolated by two branch rivers connecting the lake with the Saskatchewan. The stream we passed before reaching Pemican Portage is the western connection, and bears the name of Big Stone River; it is about six miles long by its windings, and about two chains wide. When the water of the Saskatchewan is high, it passes through this channel or canal into Pine Island Lake, and when low, the water from the lake flows into the Saskatchewan. At the time the accompanying survey was made, (16th August, 1858,) Big Stone River was flowing into the Saskatchewan, at the rate of 1½ miles an hour. The eastern connection is about the same size as Big Stone River and joins the Saskatchewan some distance below Pemican Portage; it is called "Tearing River," and is the route followed by the McKenzie River boats. The Saskatchewan boats go by these rivers when they require to call at Cumberland.

The country around Cumberland is low and flat; the soil in some places is a stiff clay, but in general it consists of a gravelly loam a few feet in thickness, covering an unexposed horizontal bed of white limestone, and supporting a light growth of poplar and birch. Occasional groves of spruce, (the so-called pine of Rupert's Land, from which Pine Island derives its name) are seen here and there. The land being so little raised above the lake and river, a great deal of it is submerged during the spring floods, and some portions upon which the water remains become marshes and swamps; but many of them could be drained and improved without much difficulty.

There is a considerable extent of ground enclosed and under cultivation at Cumberland. I observed a field of barley, and another of potatoes, both looking well, within the fort palings; and there is an excellent garden adjoining the Chief Factor's house; the soil appeared rich and fertile, bearing an exuberant growth of rhubarb, cabbage, peas, carrots, and other vegetables.

Cumberland House being at the junction of two great lines of water communication, one leading from the Pacific, and the other from the Arctic Sea, to the Winnipeg basin, is a place of importance, and was formerly one of the Company's principal depôts. Within the fort there are a number of buildings, one of them (the store-house) is a very large edifice, containing extensive machinery and appliances for pressing and packing furs, and making pemican. Cumberland has been visited by several celebrated Arctic explorers. In the garden there is a sun-dial which was brought from England, and erected by Sir John Richardson, and Sir John Franklin remained here a portion of the winter of 1819, while on his first overland expedition to the Polar Sea viâ the McKenzie River.

August 16th.—We left Cumberland this morning in our new craft, a three-fathom birch bark canoe. Not being so deep nor of the same beam as the old one, our load of baggage, instruments, and provisions, sank it to within a few inches of the gunwale, rendering it rather unsafe in a heavy sea. I succeeded in getting an Indian guide, through the kindness of Mr. Stewart, but could not prevail upon him to accompany us farther than the Grand Rapid; which ultimately proved fortunate for us, as had he continued with our party, the pemican, upon which we had now solely to depend till we reached Red River, would have been exhausted much sooner than it was. We returned to the Saskatchewan viâ Big Stone River; and passed the mouth of Tearing River about 14 miles farther down. Between the mouths of these rivers, the Saskatchewan flows occasionally among low alluvial islands, wooded with small poplar and willows; and in many places its depth is lessened by mud-flats and sand-bars; its banks are here low alluvial flats, only two to three feet above the water, covered with gray willows and sapling poplar. The current in this part of the river is slacker than before, the average rate as measured by the log being two miles an hour. We camped about a quarter to seven, p. m.; but before camping, made a section of the river, which gradually increases in breadth and volume of water; a number of soundings, taken at intervals across the river with the hand lead, shewing a mean depth of 20 feet; and the width of the river at this point as computed from observations made with the sextant being 980 feet. I levelled about ¾ miles along the bank of the river here, to ascertain its fall.

August 17th.—We embarked at 4 a. m., and observed no material change in the general character of the river and adjacent country during the day. The banks of the river are similar to those already described, being low alluvial flats not exceeding two feet above the water, and covered with willows and patches of balsam-poplar. The tract of country back from the river is rather low and wet; and the Indians make portages in one or two places from the river to small lakes north of it. The current is now much slacker than before, being only 1–1½ miles an hour.

About 13 miles below Tearing River, Fishing Weir Creek falls into the Saskatchewan; by which, during high water, boats sometimes go to Cumberland. About 14 miles farther down, at what is called the Big Bend, the general direction of the Saskatchewan changes from a north easterly course, which it has maintained from the Grand Forks, to a south-easterly one. This Big Bend is the most northerly point on the river, being very near the 54th parallel of latitude. The Pas or Cumberland missionary station, where we arrived about sunset is nearly 22 miles below the Big Bend. About three miles above, or west of the Pas, the Saskatchewan makes an abrupt semi-circular curve, (called by the Indians "The Round Turn,") causing eddies and whirlpools, the river being at the same time diminished in width. The depth of the river was here found to be 33 feet, and its breadth about 10 chains. Near the Round Turn, there is a wooded ridge, upwards of 50 feet high, about half a mile from the north bank of the river. About three-quarter miles above the Pas, Root River, a long affluent with a width at its mouth of two chains empties into the Saskatchewan.

The Pas, or Cumberland Station, is a missionary post of the Church of England, situated at the confluence of the Saskatchewan and the Basquia River, a tributary about three chains wide at its mouth. Christ Church, as will be seen in the sketch I made of the Pas, is a neat and rather imposing edifice; and it seemed like getting back to civilization again after all our wayfaring, when, on rounding one of the majestic sweeps of the river, the pretty white church surrounded by farm-houses and fields of waving grain, burst unexpectedly upon our view. It was on a calm summer's evening, and the spire was mirrored in the gliding river and gilt by the last rays of the setting sun.

The Church is situated on the right or south bank of the river; near it is the Parsonage, a large and commodious building, occupied by the Rev. E. A. Watkins, the present incumbent. Adjoining the Church there is a neat school house and several dwelling houses; and on the opposite side of the river I counted seven houses, but they seemed to be uninhabited and in a dilapidated condition; the Indians for whom they were erected disliking a settled life devoted solely to the pursuit of agriculture; and preferring the wandering and precarious life of a hunter in their native wilds. The river banks at the Pas are 10 to 12 feet high, composed of light coloured drift clay holding boulders and pebbles of limestone, and the surface soil is a dark gravelly mould well adapted for cultivation; but the surrounding country is said to be low and swampy with marshy lakes. Barley and other crops growing here looked well, and were just ripening. Mr. Watkins' garden also looked well, and he kindly supplied us with some onions to make our pemican more palatable.

August 18th.—Having to make some observations this morning, and Mr. Watkins wishing to send some letters with me, we did not leave the Pas till about 9 a.m. From the Pas the Saskatchewan flows in a north-easterly direction through a low flat country wooded with scrub poplar and balsam-spruce for about eight miles; when again turning suddenly it resumes its south-easterly course, forming a great bend or elbow. About a mile below the mission, a branch, three chains wide, leaves the Saskatchewan, and cutting across the tongue of land embraced by this elbow, affords a navigable passage about three miles shorter than by the main river; although it is the route generally followed by the boats, had I availed myself of it I must have left a considerable portion of the Saskatchewan proper unsurveyed.

About six miles from where this branch or canal rejoins the Saskatchewan, another branch, leading from Moose Lake and House, falls in; before uniting with the great river it separates into two branches forming a Y, the distance between the mouths being about half a mile. From the Pas to this point the character of the country bordering the river gradually deteriorates, the banks becoming lower and lower, and the timber more scrubby and scanty. The alluvial flats are in many places only one to two feet above the water, and they are at some points covered with driftwood, shewing that they are flooded at certain seasons.

We stopped to cook dinner opposite the Moose Lake branch, where, by ascending a tree, I succeeded in getting a view of the surrounding country. The banks are here three feet above the river, supporting a thin strip of gray willows along the water's edge; and about half a chain back from the river there commences an extensive marsh or swamp with rank reeds and rushes, interspersed with ponds of open water and dotted with clumps or islands of balsam-spruce and willows as far as the eye can reach. From Moose Lake Fork to where we camped, about sixteen miles further down, a slight improvement is observed on the immediate banks of the river; occasional groves of young ash, elm, and ash-leaved sugar maple are seen, but the flats behind are generally very low, and covered only with willows and sapling poplar.

We started on Thursday, August 19th, at break of day with wet baggage and blankets. A thunder-storm with heavy rain came on during the night, and the want of a tent was again severely felt. About four miles below our camping place one or two branches leave the main river and flow to the north into a marshy expanse of water, about one mile broad and two to three miles long, called "Marshy Lake" on the plans returned. Between Marshy Lake and Cedar Lake are seen all the characters of a great alluvial delta. The Saskatchewan ramifies into many different channels, some of them return to the parent stream forming large islands, and several flow into Muddy Lake and other expansions of the main river, before finally emptying into Cedar Lake.

The country bordering the Saskatchewan from Marshy Lake towards Muddy Lake and Cedar Lake, consists of low mud flats not exceeding 18 inches above water, supporting along the river's edge a belt of willows, alder, dogwood, and long rank grass; in the rear is an extensive marsh with occasional islands of small poplar and spruce. These flats, being so little above water, are flooded every spring after the ice breaks up, and no camping place can then be found for a considerable distance up the river. A very rich mud is deposited during these floods, raising and extending the flats every year.

Muddy Lake, near which we were compelled to remain for some time owing to a boisterous head wind, is apparently a dilatation of the Saskatchewan in a northerly direction; it is about two miles wide, and extends to the north for about four miles. We effected a landing on a point of the river four to five feet above the level of the water, where we found an exposure of light coloured limestone in horizontal beds along the waters edge, and several large detached masses adjacent. This was the first outcrop of rock *in situ* we met with on the main Saskatchewan, and I made a very careful search for fossils, but, being unsuccessful, had to content myself with some specimens of the rock. On examining the point it was discovered to be an island eight chains long and four broad, with the river on one side, and on the other a vast reedy marsh interspersed with large ponds. This island is a favorite camping and fishing place of the Swampy Indians, there being on it a clump of good sized poplar, the only timber fit for fuel for miles around; and here they hold their great councils, dog feasts, and medicine dances. Its name in Swampy is *Kash-ke-bu-jes-pu-qua-ne-shing*, signifying, "Tying the mouth of a drum."

Between Muddy Lake and Cedar Lake the Saskatchewan meanders through an immense marsh with tall reeds and rushes. It is now no longer an integral stream but is divided into a maze of reticulating branches. According to our Indian guide, land is being formed here very fast; and what is now marsh and mud flats was, within his recollection, open navigable water for a considerable distance back from where the Saskatchewan at present debouches into Cedar Lake through its numerous mouths In one or two places we saw the trunks and branches of stranded trees sticking above water, where alluvial flats or shoals of mud and drift timber are in course of formation.

The Indians informed me that beyond these extensive alluvial flats and shallow marshes there is not to their knowledge anything but "muskeg" or boggy swamps for a very great distance on either side. I could see no high ground of any kind, and the character of the country bordering the Saskatchewan as above described may be said to continue back from the river for many miles.

Cedar Lake, (so called from the occasional groves of cedar,—a tree rarely seen in Rupert's Land,—growing on its shores, particularly at its western extremity,) is an expanse of water of considerable extent in which the turbid waters of the Saskatchewan are allowed to disseminate and settle before re-uniting into one great river and rushing down the Grand Rapid into Lake Winnipeg. It is situated in about 53° 15′ N. latitude, and 100° W. longitude; and is nearly 30 miles long with a breadth at its widest part of about 25 miles; its coast line embracing an area of water of about 312 square miles. Cedar Lake being more than 60 feet higher than Lake Winnipeg, is consequently upwards of 688 feet above the sea level. The only tributary it has of any size, beside its principal feeder the Saskatchewan, is a branch leading from Moose Lake and House, which enters it from the north. I was unable to obtain soundings of the Lake in consequence of the high winds and stormy weather that prevailed during our voyage through it, but so far as I could learn it has sufficient depth of water for the largest craft, except at the west end, where the Saskatchewan is rapidly filling it up.

We entered Cedar Lake on the morning of the 20th August, and coasted along the north shore till about noon, when we ran into a fine little harbour to eat dinner after making a long traverse. In the afternoon, while crossing a wide and deep bay or sound stretching far to the north (the extremity being below the horizon), a stiff breeze sprang up, soon raising a very heavy sea, in which our canoe became almost unmanageable, pitching tremendously and shipping a great deal of water. On the 21st August we breakfasted at the Rabbit Point, and entered the portion of the Saskatchewan issuing from the east end of the lake about noon.

The northern coast of Cedar Lake is deeply indented and very low, and the country continues flat for a long distance back. At some of the points and on many of the islands along the coast, there are exposures of limestone in horizontal beds, the top of the strata being a few feet above the surface of the lake. It is to be regretted that, owing to the stormy weather and the rate at which we were obliged to travel, no opportunity was afforded for collecting specimens. The main land and islands being well wooded with balsam-spruce, birch, poplar, tamarack, cedar, and banksian pine, could furnish an abundant supply of fuel; thus offering, like the Saskatchewan, facilities to steam navigation; but a considerable portion of the land is reported to be swampy and unavailable for agricultural purposes.

The portion of the Saskatchewan between Cedar Lake and Lake Winnipeg is nearly 20 miles in length, and its general direction is easterly. Through this channel, the great volume of water brought down for many hundred miles by the main river, and its north and south branches, together with that collected by many tributaries through a wide extent of country, is disembogued by one grand mouth into Lake Winnipeg.

Where the Saskatchewan emanates from Cedar Lake the bed of the river is divided for a short distance into two channels, by an island. We entered the smaller or south channel and found it only two or three chains wide, for a distance of about a quarter of a mile. At its narrowest part, near the beginning, the Indians have a fishing station, and white fish and sturgeon are caught there in abundance. Along the side of this water-course there is an outcrop of horizontal limestone, 3–4 feet in thickness, above the water, covered with a thin coating of vegetable mould, supporting small poplar, willow and dogwood. I brought away some specimens of the rock, but could find no fossils. The current in this channel, as in most places where the river is narrower than usual, is strong; measuring 2½–3 miles an hour.

About half a mile below Cedar Lake on the right or west bank of the river, which is now more than half a mile in width, is situated Cedar Lake House, a winter trading post of the Hon. Hudson's Bay Company; lately established, with a view to check or compete with the "Freemen" who come annually from Red River to trade with the Indians.

Between Cedar Lake and Cross Lake Rapid, a little below which the Saskatchewan expands into Cross Lake, the river is very broad and widens here and there into deep bays and funnel shaped indentations. It grows narrower again, a little above the rapid, where a projecting point of limestone, obstructing the current, causes a small smooth rapid on the south side with a fall of about eight inches. The Cross Lake Rapid is occasioned by a band of limestone intersecting the bed of the Saskatchewan nearly at right angles; and this is the first interruption of any magnitude, to the even flow of the river. The Saskatchewan is let down by this rapid about 5½ feet in a short distance. There is a large island near the south side of the river, extending the length of the rapid, and dividing it into two channels. The broadest or northern channel is that which came under my observation. It is about 30 chains wide and is the route followed by the Hon. Hudson's Bay Company's boats. In order to ascend the rapid, the Company's boats of 4–5 tons burden have to be "tracked" or dragged up with half cargo, and the other half of their load has to be carried over the portage, a distance of 230 yards. The fall from the west to the east end of the portage (obtained by levelling,) is 4.08 feet, and from the east end of the portage to the quiet water below, about 1½ feet, making a total fall of 5.58 feet. Loaded boats run the rapid without difficulty, and if the channel were cleared of boulders and improved, it might be ascended by a powerful steamer.

Having spent some time in making observations at Cross Lake Rapid, it was late in the afternoon when we entered Cross Lake; where our Indian guide left us, although he had agreed to pilot us down the Grand Rapid. He expressed himself anxious to return to his family at Moose Lake, and could not be induced to go farther. During the return journey, upon which he set out in a little canoe that he picked up, coming down the river, he would have several days hard paddling against a swift current.

At the east end of Cross Lake, we met Mr. Christie (a gentleman in the service of the Hon. Hudson's Bay Company, who had recently been appointed to the charge of Edmonton House,) in command of a brigade of boats, *en route* from York Factory to Edmonton and the Rocky Mountain District. Mr. Christie's heavily laden boats, (14 in number) were manned by a motley group of Indians, Half-breeds, Orkney-men, Norwegians, and Negroes; they had just made the laborious ascent of the Grand Rapid, and thus far their progress had been very slow. Mr. Christie represented the many difficulties which had to be contended with in a boat voyage; the detentions on the lakes by contrary winds; the strong currents and rapids that had to be encountered in ascending the rivers; and the difficulty of procuring men suitable for the work; (each boat requiring six to eight experienced voyageurs,) and he expressed a hope that the long talked of steamers would soon make their appearance on Lake Winnipeg, to replace the present tedious, toilsome, and expensive mode of conveyance.

In reply as to whether there would be sufficient business to warrant the placing of steam vessels on these north-western waters, (irrespective of the establishment of a continental route to the Pacific, through British Territory;) I was informed that there would be plenty of freight to carry for the present requirements and traffic of Rupert's Land; as during the year (1858) no fewer than 167 freight boats of the largest class, belonging to private traders and merchants, as well as the Hon. Hudson's Bay Company, (many of them loaded with valuable furs,) had passed Norway House, at the northern outlet of Lake Winnipeg, *en route* to York Factory; and returned with heavy cargoes of merchandize brought by sea to York, consisting chiefly of the usual supplies for Selkirk settlement, ammunition, and a variety of goods for the prosecution of the Indian trade both by the Company and "Freemen." The aggregate quantity of freight transported by this fleet of boats from the sea-board to Lake Winnipeg and from thence distributed along its principal feeders would be upwards of 800 tons. It is well known that there are large quantities of goods imported by other lines of communication—chiefly through the United States Territory at present; and as the York Factory route is to be partially abandoned, a large portion of the importations of Rupert's Land will have henceforth to enter the Winnipeg Basin from the south, so that there will doubtless be sufficient commerce in view of the great water facilities afforded by the country, to encourage the initiation of steam navigation.

After remaining at Mr. Christie's encampment about an hour, we set off again in the hope of reaching the Grand Rapid before dark. We soon entered a rapid by which we were lowered about 2½ feet in a distance of 10 chains, followed, after an interval of smooth water by another about a mile long, but with an easy inclination, the descent in that distance not being above 7½ feet; it being nearly dark when the foot of the latter was reached, we camped for the night. (August 21.)

Cross Lake doubtless derives its name from its shape and the peculiar position it bears in relation to the Saskatchewan, of which it is evidently a dilatation. It is an oblong sheet of water, upwards of eight miles in length, having its longitudinal diameter at right angles to the general trend of the river; three miles is its greatest transverse diameter, and this breadth is about the distance between the termination and beginning of the bed of the river on either side of the lake. The altitude of Cross Lake in relation to Cedar Lake and Lake Winnipeg, acquired by levelling the rapids and measuring the currents in the river, would make its approximate elevation above the sea about 680 feet. It is reported to be deeper than Cedar Lake; and its banks on the east and west side are more abrupt and rocky, but its northern and southern shores are very low. Along the coast there are some fine groves of balsam-spruce, and aspen, but the land back from the lake is very flat and poorly wooded, a great portion of the original forest having been destroyed by fire; large tracts of burnt and dead timber are seen here and there; the blackened trunks of poplar and spruce indicating the ridges or dry areas over which the conflagration extended, and the lifeless tamaracks revealing the swamps or flooded land. The lake extends so far to the north, its extremity in that direction is not seen from the traverse line, being below the horizon of the spectator. In the northern arm of the lake there are several wooded islands, but as they were some distance from our track I was unable to ascertain the nature of their formation.

There being two rapids between Cross Lake and the Grand Rapid, the Saskatchewan may be said to descend by four distinct steps from Cedar Lake to Lake Winnipeg; the first one east of Cross Lake, having a length of about 10 chains with an estimated fall of 2½ feet, occurs half a mile below the re-commencement of the channel of the river, and appears to be attributable to a low and nearly level belt of limestone, through which the river has gradually excavated its way by three separate channels. The middle channel, by which we descended the rapid is only 3–4 chains wide and could apparently be ascended by a steamer without difficulty, as it is deep and appears to be free from boulders. The other channels might even be more favourable for steam navigation, being broader as far as could be observed, and containing a greater volume of water; they are however a little out of the direct course and for this reason are not followed by the boats. The smooth portions of the river are really broad here; the width above the two islands formed by these three channels being more than half a mile, and below them upwards of three-quarters of a mile. About a mile below the foot of the first rapid the second one begins. Its length by "dead-reckoning," is fully a mile, and its approximate fall is not more than 7½ feet. It is a long gradual slope with a deep channel of rolling, but comparatively unbroken water in the middle; the water is more turbulent at the sides, where the current is interrupted by points of limestone rock, boulders and débris. The exposures of limestone on the points, are 4–6 feet in thickness above the water, with a horizontal stratification. The loaded boats of the Hon. Hudson's Bay Company descend this rapid easily, and as they are generally "tracked" up with the whole of their lading, a lightened steamer, with powerful engines might surmount it by taking the best channels and other precautions.

It is about four miles from the foot of this last rapid to the beginning or summit of the Grand Rapid. In that distance the river is smooth and deep, but has a very swift current, especially where its bed is contracted. The width of the river in this interval is much diminished, varying from nine chains to a quarter of a mile, and the rate of current is from three to three and a half miles an hour. There are one or two large boulders in the bed of the river here, over and around which the water boils and bubbles like a caldron; and now and then shoals on the north side of the channel are indicated by the rippling water and ground-swell occasioned by the current in passing over them. The land between Cross Lake and the Grand Rapid is generally low and flat, but thickly timbered with balsam-spruce, poplar, tamarack, and birch. At the second rapid east of Cross Lake the banks on the north side of the river are eight to ten feet above the surface of the water, and are composed of a light coloured drift clay. These clay banks gradually increase in height towards the Grand Rapid, where they attain an elevation of upwards of twenty feet; but it is probable that the surface of the country is nearly level, and that it is the descent in the river which causes the apparent rise in its banks.

August 22nd.—This being Sunday, with a view to rest our wearied limbs, we did not proceed on our journey till after breakfast, (about 8 a.m.)

However desirable it might have been, under other circumstances, to have remained inactive on this day; in the position in which we were placed, like a ship at sea, with a limited supply of provisions, and a long and hazardous voyage before us, it would have been altogether out of the question; indeed, the loss of a day or even an hour might have compromised the safety of the whole party.

In about an hour we reached the beginning or west end of the portage at the head of the Grand Rapid, whence my various instrumental observations and measurements in relation to the rapid began. In order to commence operations we disembarked and made the portage, which of course is never done by boats in descending the river. Yet, notwithstanding that boats invariably 'run' the whole of the rapid it would be extremely perilous to descend the upper portion of it in a small heavily laden canoe without a guide.

So much having to be done with so few hands, our little party exhibited a scene of unusual activity and exertion, from the time we landed at the top of the rapid until we camped in the twilight on the coast of Lake Winnipeg. The first thing to be accomplished was the transportation of the canoe and the heavier articles of luggage to the east end of the portage; to effect this, the united energies of the party were required, and owing to the length of the portage it occupied some time. While Wigwam was carrying the remainder of the lading, I was engaged with Louis in making a survey of the portage and rapid, chaining across in one direction and levelling back in another, and so forth.

About 4 p.m. the various observations were completed, and everything had arrived at the east end of the portage. The different operations involved the crossing of the portage (more than a mile in length) many times during the day. While dinner was preparing I occupied myself in making a sketch of the cataract and examining the character of the perpendicular limestone cliffs at its side.

After eating a hasty meal we re-embarked to run the lower portion of the rapid.* The voyageurs wished me to walk through the woods to the foot of the rapid, (probably to lighten the canoe,) but as the day was already far advanced, and being anxious to reach Lake Winnipeg, as well as for other reasons, I deemed it expedient to go down 'in canoe.'

In running the rapid we followed as closely as possible the instructions given to us by our old guide on the Plains, (John Spence,) who had often piloted the old N. W. Co.'s *North* canoes down its entire length. In attempting, according to his directions, to cross from the north to the south side of the rapid in order to get into what was reported to be the best channel for a small canoe, such was the fierceness of the current and the turbulence of the great surges and breakers in the middle that we were nearly engulfed; and although every nerve was strained we were swept down with impetuous velocity, and did not get near the other side till we were about three quarters of a mile below our starting point. We were then impelled with astonishing swiftness along the south side of the torrent, often in dangerous proximity to the rugged wall of rocks bounding the channel, and now and then whizzing past—almost grazing—sharp rocky points jutting out into the river, against which the thundering waters seethed and foamed in their fury. During the descent the voyageurs exerted themselves to the utmost of their strength, and evinced an admirable degree of coolness and dexterity.

* The part here designated as the *lower portion*, although the Grand Rapid is one continuous torrent from beginning to end, is that below the east end of the portage, and is more than one mile and a half in length.

The Grand Rapid is acknowledged by those who have witnessed it, and who have had opportunities of traversing the great river systems of the continent, to be unsurpassed (as a rapid) in magnificence and extent, as well as in volume of water. It is certainly a formidable barrier to the navigation of the Saskatchewan.

The following are the dimensions of some of the leading features of the Grand Rapid:—

1. *Its Length.*—The portage path is nearly straight, with a magnetic course, from the upper to the lower end, of S. 60°E.; it is 87 chains 40 links in length; the distance between its extremes by the river is a little more than this, as the river describes an arc of which the portage is the chord, but as the head of the rapid is a little below the west end of the portage, this distance may be adopted as the length of the upper or most precipitous portion of the rapid. The distance from the east end of the portage to the foot of the rapid by our track is 129 chains. This would make the whole length of the rapid 216 chains 40 links, or nearly $2\frac{3}{4}$ miles.

2. *Its Descent.*—By levelling carefully along the portage path, I ascertained the fall between the smooth water at the head of the rapid to the general level of the water at the east end of the portage to be 28.58 feet; and after observing instrumentally the descent in the lower portion of the rapid as far as the nature of the country would allow, I closed my levels on a bench mark at the surface of a pond of still water fed by an eddy at the lower end of the portage. The fall in the lower portion of the rapid, acquired by levelling and by careful estimation is about 15 feet; this would give about $43\frac{1}{2}$ feet as the total descent of the rapid.

3. *Its Breadth and Depth.*—The width of the river, at the upper end of the portage, is about 20 chains; at the head of the rapid, about seven chains further down, where there is an island in the bed of the river, it is about 30 chains; and at the lower end of the portage, where the rapid emerges from the highest limestone plateau, its width is about 10 chains. From thence it gradually widens towards the foot of the rapid, where it attains a width of 25 chains. I was unable to obtain soundings of the rapid, but, from the depth and volume of water above and below it, where the river is much broader, it is undoubtedly deep.

The Grand Rapid, throughout almost its entire length, washes the bases of perpendicular escarpments of rock. It passes through two plateaux of brittle buff-coloured limestone, with a horizontal stratification; the top of the first, or upper plateau, being nearly on a level with the surface of the water at the head of the rapid, and underlying a stratum of light-coloured clay, 23 feet in thickness, in which are embedded boulders and pebbles of limestone; the whole overlaid by about eight inches of vegetable mould, and clothed by a forest of balsam-spruce, tamarack, and poplar. The surface of this plateau continues nearly level as far as the lower end of the portage, where the top of the rock is 25·36 feet above the surface of the water, and about the same height above the lower plateau. The lower plateau continues some distance further down, but is soon hidden by drift clay banks, which, at the foot of the rapid, have an altitude of 20–30 feet above the water.

It is not improbable that the Grand Rapid is the result of

the eroding influence of the great body of water in the river, upon the rock through which it flows—the limestone being of a friable and yielding nature. At a remote period, the water of the Saskatchewan was perhaps lowered from the top of this rock formation, by a perpendicular cataract; the precipitous leap most probably began at the lower end of the portage, or at the eastern limit of the highest limestone plateau, from whence the river gradually wore away the rock, at the same time diminishing the height of the fall, until it became a foaming rapid from beginning to end.

The upper portion of the Grand Rapid,—of which I succeeded in getting a sketch,—presents a scene that strikes the beholder with wonder and admiration. The great body of water that has been stealing along, swiftly but silently, for many miles, appears to be suddenly imbued with life—the rippling of the river becoming gradually more turbulent, until the surges grow into huge, rolling billows, crested with foam, like waves in a tempestuous sea. The great rollers and breakers seem, to the spectator, to be continually changing in shape and appearance, on account of the lines of surf and the peculiar colour of the water; but although the mighty cataract thus appears to be for ever changing, it really rolls on for ever the same.

The ascent of the Grand Rapid is one of the most laborious duties that has to be performed on a boat voyage from Lake Winnipeg to the Saskatchewan district. The Hon. Hudson's Bay Company's brigades surmount this fearful interruption to the upward navigation of the Saskatchewan in the following way: On arriving at the foot of the rapid, every boat discharges one-half of its cargo of four to five tons. Thus lightened, they are then "tracked" (towed) up to the beginning of the portage—the whole of the crew of six or eight voyageurs, with the exception of the bowsman and steersman who remain in the boat, being engaged in the labour of tracking. Each man is attached to the tracking-line by a leather belt, or portage strap, passing round his body; and harnessed in this manner they drag the boat along, running and scrambling barefooted over the slippery and jagged rocks at the sides of the cataract. When the lower end of the portage is reached, the boat is emptied, and "run" back again to the foot of the rapid, and from thence hauled up as before, with the remainder of its load. The whole of the lading is then carried over the portage, exclusive of 15 pieces, or about 1,350 lbs., which is left in the boat. With this ballast, the boat is pulled across to the south side of the rapid, to be tracked up, as the towing-path is better there than on the north side. In consequence of the rapidity and violence with which the upper portion of the rapid flows, in ascending it, it is necessary to employ the "main line,"—a much thicker and stronger rope than is generally used for tracking. To this line the crews of one or two boats are lashed, and thus they run along the top of the cliffs of limestone,—there being no footing at the bottom of these walls of rock,—hauling the heavy craft up the surging cascades. The utmost strength of the bowsman, with his pole, and the steersman, with his long sweep oar, is required, to prevent the boat from being dashed to pieces among the rocks.

Small brigades, feebly manned, often haul their boats over the portage. The portage road bears evidence of this, as it is deeply scored and furrowed by the keels of boats, from beginning to end.

Although the Grand Rapid is the most serious obstacle that the Company's boats have to encounter, it is not the only difficulty they meet with on the Saskatchewan. The whole ascent of the river is one of labour and fatigue. The current is so swift—as the name of the river is well known to imply—that the voyageurs would track nearly all the way to the Rocky Mountains, if the banks of the river would allow; but where the river passes through marshes and swamps they have no alternative but to pull against the current, however strong it may happen to be.

Before finally determining upon any works or measures for overcoming the Grand Rapid, in order to render the whole of the Saskatchewan navigable for steam vessels from Lake Winnipeg, without interruption; it would be necessary to make a more extensive and elaborate survey; but probably sufficient information and data have been acquired during this *reconnaissance* from which schemes might be devised, and suggestions offered, for surmounting the difficulty. To navigate the Saskatchewan at present, a steamer would evidently have either to be built above the rapid, hauled over the portage, or "warped" up the rapid itself. Seeing that the Company's large batteaux are hauled up the rapid by manual labour, it does not seem impracticable for an empty steamboat, with engines of great power, to ascend it, by the aid of hawsers and guy-ropes stretched from the steamer to the land, using, along with capstans, the motive power of the steamer as far as available. But in any case, unless a canal were constructed, a transhipment of cargo bound upwards would have to take place, whether there were steamers plying above and below the rapid, or whether steamers were forced up the rapid; so that it would be necessary to construct a good road or tramway on the present line of portage. The features of the country in the vicinity of the Grand Rapid are very favourable for a road, and even for a settlement, as the banks of the river are high, with a considerable depth of good soil, from the second rapid east of Cross Lake to near Lake Winnipeg. There is also abundance of timber for fuel and building.

From the foot of the Grand Rapid, the Saskatchewan flows, with a pretty strong current, in a northerly direction till it enters Lake Winnipeg. Its mouth has a width of about 28 chains, and is a little over two miles below the lower end of the rapid. On the coast of Lake Winnipeg, immediately east of the mouth of the Saskatchewan, there are several deep and narrow bays, or estuaries, marshy at their inner extremities, and separated by narrow points or spits of gravel, by which it seems not improbable the Saskatchewan entered the lake at some period of its existence, and that north-easterly gales and shoves of ice have driven up these barriers, and caused the river to excavate new outlets.

We visited an Indian encampment on the north bank of the river, a little below the foot of the rapid, in the expectation of procuring some sturgeon; but were unsuccessful—the fishery carried on here by the Indians having failed this year. This encampment of two lodges was the only one we saw on the Main Saskatchewan. It had been a larger camp, but eight families had just left it, previous to our arrival, for their winter quarters at the Little Saskatchewan. They are Swampy Indians, and generally winter at Fairford, from whence they proceed in summer to the Grand Rapid; where, by assisting in dragging the boats and *portaging*, they get a small recompense in the shape of tea, tobacco, or pemican. They occupy the time between the arrivals of the different brigades of boats, in catching and drying fish, and generally leave after the last fleet has passed up in the autumn.

Reaching Lake Winnipeg about sunset, we proceeded along the coast till it became too dark to continue observa tions, and camped for the night upon a narrow spit of gravel, separated from the wooded shores by a marsh.

August 23rd.—Proceeding on our journey this morning at 4.20 a.m., and being favoured with a light breeze for a few hours we reached the neck of the great promontory, *Cape Kitchinashi*, about noon. From the mouth of the Saskatchewan to this point the coast trends to the south-east, and is indented in a remarkable manner by a series of deep bays of every shape and size. As it would require unlimited time and resources to penetrate into every sinuosity of the coast, we generally steered straight from point to point, although in doing so some long traverses had to be made.

The northern coast line of the promontory being nearly straight with fine sand beaches, affording tolerably good footing, we tracked along the shore for the remainder of the day; although this was hard enough work, the men were glad to avail themselves of it, as a change or relief from paddling. By working 15 hours to day we were enabled to camp at the extreme point of the headland, where, the night being favourable, the magnetic variation of 15° E. was observed. The Ojibways call this cape "Kitchinashi," and the Swampys "Missineo," both names signifying "Big Point." By some it is called "The Détour."

August 24th.—A fine morning, the lake quite calm. After doubling the cape we overtook eight small canoes containing the band of Indians who left the Grand Rapid on Sunday, 22nd. In a short time a light breeze sprang up, and by hoisting a blanket we sailed at a pretty good rate for some hours. About 2 p.m. the wind began to increase in strength and turned suddenly against us, so that we had to run in behind a low point of sand and gravel for shelter. Although the wind still continued high we started again and made a traverse to a small sand island on which we were obliged to remain, being then over two miles from the main land, and the storm having increased in violence. A storm of wind soon raises a very heavy sea on Lake Winnipeg on account of its little depth of water.

The island on which we were detained is one of the Gull-egg group, which, with the point of sand protruding from the main land, form a pretty good harbour on the south side of the neck of the great promontory. The Indians were nearly destitute of provisions, and followed us to the island, where they fortunately got a plentiful supply of eggs and young gulls; but having little ammunition they brought down only a few old ones, although they hovered in countless numbers over the island, screaming at the wholesale destruction of their young brood.

August 25th.—The storm raged all night, and this morning we found ourselves surrounded by a foaming sea on a low island of sand about 100 yards in length, and so narrow that the spray from the breakers dashed completely over it. The gale blew hard from the east till about noon, when it began to subside; I then determined upon starting on our course, but seeing a thunderstorm approaching decided upon taking dinner before making the attempt. It was well that we did so, because just as we were hastily swallowing our meal of pemican, the thunderstorm, accompanied by strong wind and heavy rain, burst upon us with great violence. Some of the Indians were endeavouring to reach the next island in the line of traverse, but had to abandon the attempt and drive before the gale to the main-land, three miles off.

The storm soon abating again, we crossed to the next island and from thence to the main-shore; and after coasting along for some miles encamped on a sandy point, where we found a small bluff of poplar and spruce.

August 26th.—Last night the Northern Lights or Aurora Borealis were unusually brilliant, darting and playing about with extraordinary rapidity in all directions, sometimes extending to the zenith and sometimes to the south of it. The voyageurs said they portended a coming storm, and their prognostications proved correct. The night was clear with a bright moon till about midnight, when a cold north-westerly wind arose, followed in a very short time by a stormy sea. The gale soon veered round to the north increasing to a perfect hurricane, and during the day the lake was white in all directions with breakers and foam. A heavy surf breaking along the coast and tearing away large portions of the bank on which we were camped, warned us to move our canoe and lading back from the shore; yet, notwithstanding every precaution, some of our paddles and poles were swept away during the night. A large marsh being in our rear we could retire but a few yards from the raging lake to wait for the abatement of the storm.

August 27th.—After midnight the wind began to decrease gradually, and by daybreak it had so far subsided as to permit us to continue our voyage. By breakfasting at a point where we witnessed an outcrop of limestone I was enabled to procure some fossils. This, the first rock exposure observed since leaving the Saskatchewan, is apparently the termination of a ridge running at right angles to the coast line, and bounded on either side by marsh and swamp. The top of the rock is ten feet above the surface of the lake, and is covered by a stratum of boulders and drift two feet in thickness, supporting small poplar, tamarack, spruce, birch, and banksian pine; there are only six feet of the limestone exhibited, the remaining four feet being concealed by a talus of boulders and débris. The high water-mark of the lake reaches to the top of the talus.

A contrary wind arising about noon detained us four hours at the mouth of a creek, which we ascended a short distance. The entrance, or where the creek cuts through the sand beach enclosing a marsh, is one chain wide; within the sand beach the creek expands into a deep pond 30 chains in diameter, surrounded by a marsh; this pond is fed by the inner portion of the creek, a broad and sluggish stream five feet deep, meandering through a tamarack swamp. It is reported by the Indians to have its source a long distance inland. As there is but one and a half to two feet of water over the bar this could only be used as a harbour for boats. Its position is delineated on our map about half-way between the Gull islands and War Path River.

We set off again after the wind had moderated a little, but were compelled to camp in an hour and a half in the lee of a point on the weather side of which an adverse wind was blowing hard, driving before it a heavy sea. Being thus repulsed by the wind I directed my attention to the character of the coast in the vicinity of our bivouac. Along the shore there extends a long straight sand-beach 60 feet wide and arched like a roadway, on the inner side of this beach there is a tamarack and black spruce swamp, with a bottom of black muck and moss two feet in thickness, covered with water. This 'muskeg' is said to continue for a great distance back.

By levelling I found the surface of the water in the swamp to be only eight inches higher than the lake; and as the crown of the sand beach is only four and a half feet above the level of the water, and is covered with driftwood, it is evident that the lake washes into the marsh during high water.

Leaving camp at 4.30 a.m., August 28th, we reached the mouth of War Path River at 1 p.m. The Indians say this river rises in lakes, and draining a great extent of swampy country, is very large in spring. There is three feet of water over the bar at its mouth; the channel at the entrance is contracted in summer by the sand to a width of forty feet, with an average depth of four feet; within the entrance there is a basin thirty chains broad, forming a boat harbour of easy access.

An excellent opportunity was afforded to-day for testing the accuracy of the results obtained from observations made with the log line, upon the correctness of which the detail or "filling in" between established points, by track or dead-reckoning survey, in a great measure depends. By chaining three-fourths of a mile along the straight sand beach, near the mouth of War Path River, I was enabled to obtain the rate of the canoe very accurately; the mean of a series of observations registered by the log while making the test, corresponding with the rate computed from the measured distance. The average velocity of our canoe in passing through still water in calm weather was ascertained, by timing it carefully over the standard, to be three and a half miles an hour.

After tracking all afternoon along straight sand beaches, which separate marshes from the lake, we camped nearly opposite Caribou Island on a coast similar to that which we left in the morning. The Indians came up with us, and erected their lodges in our neighbourhood.

August 29th.—Embarking this morning at daylight, we reached Limestone Point about 11 o'clock, after making a traverse of three miles against a strong head wind. On this point there is a very fine exposure of light coloured limestone, containing numerous fossils, some of which I succeeded in procuring. The outcrop on the point is 14 feet in thickness above the lake, in massive horizontal layers, overlaid by 2½ feet of drift and fragments of limestone that have evidently been broken up by ice. This headland is the abrupt termination of a narrow ridge of limestone clothed with aspen, spruce and birch; it is about two miles long, running nearly north and south. On the west side of it, is Portage Bay, so called by the Indians, as they sometimes make a portage from the foot of it, across the neck of the point.

After remaining here about two hours, we proceeded on our journey. On rounding the point we found the wind on the east side of it blowing directly in our teeth, and it required the utmost exertions for two hours to force the canoe against a high gale and stormy sea, until we got into the lee of a small island, it being impossible to land on the main shore. The canoe leaked and shipped so much water during this traverse, that in order to lighten her we were compelled to throw overboard some of the heavier of our geological specimens. It was with great regret I saw one of them, a very large and fine *orthoceratite*, consigned to the deep.

On the island we found part of the Indian band, but the greater portion were hurrying on to the Little Saskatchewan to get fish, as they had nothing to eat. We saw them in the distance, battling against the wind and sea, their little canoes like specks, tossing among the swells and breakers. The Indians remaining on the island were chiefly old men and young children, the more feeble of the party; and being ravenously hungry, they were all in the marshes busily engaged in pulling up and eating the roots of bulrushes. The storm increased towards evening, and we were obliged to camp on the island ourselves.

August 30th.—Although the unfavourable wind had diminished but little this morning, we plied our paddles so well, and made such good headway against it, that we entered the mouth of the Little Saskatchewan or Dauphin River about 11 a. m. We tracked up the river to the Indian encampment, about four miles from its mouth, for the purpose of procuring fish; and found the Indians at the rapids scooping large numbers of excellent white fish from the eddies.

As the west coast of Lake Winnipeg south of the Little Saskatchewan was examined, and will be described and reported upon by yourself, it will be unnecessary for me to do more than give a brief outline of my progress and operations in surveying the coast line from that river to the mouth of Red River where I closed the survey. But before resuming the narrative of my proceedings it would perhaps be as well to give a short recapitulation of the character and general topography of the west coast of the lake between the Main Saskatchewan, and the Little Saskatchewan.

The distance from the mouth of the Main to the mouth of the Little Saskatchewan by our track along the coast, or by the course that canoes or row boats would be likely to pursue, is about 140 miles: but the distance by the coast line embracing every sinuosity of the shore is much greater.

The most prominent feature in the line of coast is the great headland, Cape Kitchinashi. This immense promontory begins to stretch out into the lake in a direction a few degrees north of east, about 15 miles south of the Saskatchewan. Its extreme point is about 24 miles in an air line from the general line of the coast, and its width varies from three to six miles and upwards; its neck is indented by several deep bays, some of which could be used as harbours or roadsteads. The formation of the cape is peculiar; it is very low and flat on the north side, while on its southern boundary the coast is comparatively high and abrupt. Its northern side consists of a series of marshes separated from the lake by a narrow sand-beach; these marshes gradually blending into a tamarack and spruce swamp. Along the south side of the cape there is a continuous escarpment of light coloured clay, 25–40 feet high, yet even on the top of these high banks the character of the land is of the poorest description, being nothing but a "muskeg" or mossy swamp containing a thin growth of very scrubby tamarack and spruce, covered with drooping moss.

The extremity or apex of the promontory is a very low and broad sand beach covered with water worn boulders; the lake is also dotted with boulders a long way out from the shore there being a sand-bar or continuation of the point under water, on which they rest. From the size and position of the cape, and the dangerous shoals extending out from it, if beacons or light-houses are ever required on the lake for the safety and convenience of shipping, no more suitable place could be selected for the erection of one than here.

The coast north-west of the cape as already stated, is very low, and much broken by deep and narrow bays.

From Cape Kitchinashi to the Little Saskatchewan the coast trends generally to the south-east. Between these points limestone is exposed in six places. The exposures are the precipitous extremities of ridges, forming points at intervals along the coast. The stratification in every instance is horizontal, but the escarpments vary in height above the lake; they increase in altitude from 4 to 14 feet towards the south. These ridges are generally wooded with aspen and other deciduous trees; and the swamps intervening are timbered with tamarack and spruce; some of the spruce near the coast are pretty large. Between the ridges low sand-beaches extend along the coast. These beaches separate ponds and open marshes averaging from a quarter to one mile wide, from the lake; in the rear of the marshes is the great tamarack and spruce swamp, or "muskeg."

The tributary streams in this part of the coast are not numerous and they are generally of no great size. The chief are the Gull Egg Rivers or the Two Rivers, the War Path River, Jumping River and one or two others without name; they are not in themselves large, but their estuaries might be available as harbours for boats.

The character of the country exhibited on the coast extends almost an unlimited distance back; indeed the Indians report the whole of the country between Lake Winnipeg and Lake Winnipego-sis as one vast "*muskeg:*" the great moose hunting grounds of the Swampys.

Although the country here described is quite unfit for agricultural purposes, it is not altogether valueless; there are large areas of good timber along the coast, available for fuel, and the limestone cropping out at the various points is well adapted for building.

Being without a guide I got one of the Little Saskatchewan Indians to draw me a map of the lake between Bushkega Islands and Grassy Narrows, shewing the traverses and route to be taken between the islands in order to cross the great arms of the lake, Fisher Bay and Washow Bay. This Indian chart was of great service to us; the best and most recent maps of the lake to which I had access being so incorrect: on them the general contour of the coast north of the Little Saskatchewan is tolerably well delineated, but to the coast north and south of the Dog's Head Straits they bear very little resemblance; the large islands are omitted altogether, and the Great Black Island is represented as forming the extremity of a promontory on the mainland between two bays.

From the beginning our canoe was very weak, the bark being of the poorest description and badly put together; and having now become quite frail I tried to barter with one of the Indians for a new and stronger one; but, taking advantage of our situation, he placed upon it a much greater value than I felt inclined to give. Looking upon our canoe as worthless, he wanted in addition to it £2 stg. and one of my blankets. Considering this an unconscionable price, we determined upon venturing to perform the remainder of the journey with the battered canoe we had.

Having made sections of the river, and examined the country bordering the Little Saskatchewan, we left it on the 31st of August; but were detained the greater part of the day on a point only a few miles from the mouth of the river, by unfavorable wind and in consequence of the sickness of Louis, our steersman; who, being a pretty old man was disabled from over exertion in the storm on Sunday.

On the 1st of September, while sailing with a side wind across the mouth of a deep bay, in which there was rather a heavy sea rolling, a large swell broke over us throwing in a great deal of water; the water got into the compass box, and even my watch in my waistcoat pocket, stopping it at once; and it was some time after effecting a landing and drying it out before I could get it to go again. Having to contend the remainder of the day with opposing winds, we were quite worn out when we camped about sunset.

From the evening of the 1st September until the morning of the 5th we were windbound on a low marshy point on the north east side of the great bay into which the Little Saskatchewan empties. The spot on which we were imprisoned is very much circumscribed; being a narrow sand-beach, about a chain in length, and bounded on three sides by an extensive marsh. During the three days that the storm lasted, the wind blew a hurricane from the N. N. W., raising a tremendous sea on the lake; and the surf beating along the shore, washed away several yards of the sand-beach on which we were encamped. The weather was clear the first day, and I occupied myself in determining the correct time and the variation of the compass. On the second and third days it rained almost incessantly, and it was then for the first time on our voyage that we really felt the privations of hunger; we had no flour from the time we left the main Saskatchewan, and our whole stock of provisions was now reduced to a few pounds of rather mouldy pemican, which I determined to eke out as long as possible, as we were still a great distance from Red River, (upwards of 170 miles by the canoe route;) and with that object in view we made it a rule to eat only one meal a day while we were windbound, unless we were fortunate enough to procure some additional food, in the shape of wild fowl or other animals. We succeeded in getting a gray gull on the second day, on which we made an excellent repast.

On the morning of the 5th, just before we started, an Indian and family from the Dog's Head came to us; they had been windbound seven days on an island not far from where we were; they said they never saw such a continuous succession of winds and storms on the lake before; and informed us that a freeman's boat which passed during the night had been thirty days between Red River and the Saskatchewan, a distance that has been accomplished by a boat, with a favorable wind in three days. After bartering with this Indian for a small *mokok* of fish pemican, (dried fish pounded and mixed with sturgeon oil,) we proceeded on our journey, glad to get away from the dreary spot. Although there was still a heavy retarding ground-swell on the lake, we paddled many miles before halting. On stopping to cook breakfast we were greatly disappointed to find that the fish pemican which I was so thankful to get, was nearly all rotten, there being only a small portion on the top that could be eaten, the remainder had to be thrown away.

A contrary wind freshened up again about noon, but we continued struggling against it, until in attempting to round a point we were completely driven back, and narrowly escaped foundering among the huge swells and breakers that dashed high over the boulders extending out from the beach; we saved the canoe by jumping into the surf and throwing the lading rapidly ashore. As soon as we got everything out of the reach of the waves that were dashing their spray over the dripping shingle beach into the swamp behind, I sent Wigwam off into the marshes to try to procure us some food. Not making his appearance at night-fall I despatched Louis in search of him; they both returned very late, having wandered many miles along the coast, but brought nothing with them. Louis attributed Wigwam's want of success to the fact of his hunting on

Sunday. While they were gone I gathered some green cranberries in the marsh, and with them and a little pemican I made a kind of soup of which we partook and lay down to rest.

Embarking at daylight on the 6th we reached the Cat Head at 2 p. m., after a hard paddle against an adverse wind and rough sea. On the boat voyage upon which I subsequently accompanied you, we passed this bold headland in the night. I may therefore give a description of its leading features. It consists of a perpendicular escarpment of buff-coloured limestone in massive horizontal layers, the top strata overhanging the base; the summit of the rock is 30–35 feet above the lake, and is covered with drift and boulders to the depth of three feet, on which grow scrubby poplar, spruce and tamarack. The water is quite deep up to the foot of the cliff, and as no landing can therefore be effected I was unable to make a minute examination of the rock. There is a series of low, arched caverns in the base of the cliff in which the waves and swells washing to and fro make a singular hollow noise, and for this reason the Indians think it is the abode of a *manitou.*

Some of the Swampys say Cat Head is so named because an Indian hunter was killed there by falling over the precipice while chasing a wild-cat or lynx. The profile of the upper, or over-hanging portion of the cliff, bears a singular resemblance to the " cat-head " of a ship.

The wind becoming more foul we were compelled to camp on a point about a mile and half south-east of the Cat Head, at the extremity of the north-western side of Kinwow (Long) Bay.

During the next day (7th September) the wind blew hard from the east and the waves on the lake rolled mountains high, so that we could not venture out, having a long traverse before us. The narrow point or peninsula upon which we were detained, is of a peculiar character, consisting of a straight barrier or ridge of boulders about three-quarter miles long, running at right angles to the coast, and connecting it with a small area or island of limestone a few feet high ; this barrier resembles very much a railway embankment, or a rip rap breakwater ; although it is 20–25 feet high, the waves wash over it during the great storms on the lake in the fall of the year.

The morning of the 8th dawned, but there still seemed to be little chance of our getting off, and our prospects now began to look cheerless enough; we had but a handful of pemican and one charge of ammunition left ; while deliberating whether to eat the last remnant of our food, a bald headed eagle came wheeling in great circles over us ; he poised himself for an instant as if about to descend upon his prey, when he was fortunately brought down with our last charge of shot. He proved to be a large bird with magnificent plumage ; a Cree or Blackfoot would have given a good horse for his wings or tail. By eating nearly every portion of the animal, except his feathers, we managed to make him serve for two or three meals.

The wind moderated sufficiently at last to permit us to resume our journey, but we had a fatiguing paddle for two hours in crossing Kinwow Bay. The extremity of this long arm of the lake was below our horizon, and the wind came sweeping out of it in great squalls. The wind veered round to the east and stopped us again about noon at the Wicked Point, where we spent the afternoon in drying our clothes and blankets, and gathering sand cherries, on which we supped.

10th September.—The wind fell and allowed us to reach Pike Head yesterday morning. We at once ascended the Pike or Jack Fish River to the "basket" or weir erected across it by the Indians, about half a mile from its mouth ; for the purpose of procuring fish. The basket was much broken, and when we arrived was covered with turkey buzzards waiting to pounce on any fish that might get entangled in its meshes. By repairing the basket and watching it all night we caught an abundance of fish of four species, viz.: gold-eyes, wall-eyed pike, suckers and pike. It rained without intermission during the day and as the wind continued unfavorable we remained at the basket gutting fish to take with us.

We generally boiled our fish, making use of the liquor in which they were cooked as a substitute for tea ; and having succeeded in capturing a small badger by pouring water into his burrow we got sufficient fat or oil to enable us to have fried fish occasionally.

The average width of the Pike River is about a chain, and its depth about five feet, with a moderate current; its banks half a mile from its mouth are of light coloured clay 5–10 feet high, and covered with a rich dark mould supporting a thick growth of aspen, spruce, tamarack, birch and balsam. Near the basket there is an old log house, formerly a missionary station, but now abandoned. When the Indians come to fish here they cut up the flooring and timber of this house for fuel instead of availing themselves of its shelter.

11th September.—Having stowed away as many fish as we could find room for in the canoe, we left the Jack Fish River in the morning, and being favored with a fair wind sailed without stopping till dark, when we camped on a small island in the entrance to Fisher Bay. On Sunday, the 12th, we had to encounter a brisk contrary wind from the south ; but, by working sixteen hours against it and making some wide traverses between the islands we succeeded in reaching the point opposite Dog's Head, at the beginning of the narrows, before night set in.

No opportunity being afforded you for exploring the east coast of the lake while passing through the straits or narrows about ten days after this, I may give some of its characteristics at those points where we landed to examine it. The east coast from the Dog's Head to where we left it to cross to Grindstone Point, consists of a succession of knolls or low domes of granite and gneiss rising generally 8–10 feet above the water, and clothed on their flanks with a scrubby growth of timber, chiefly Banksian pine, spruce and a few aspen ; there are, generally, ponds and swamps between the granite knolls, and the coast line is much broken by deep inlets and small well sheltered bays, forming excellent harbours and coves for boats. The east coast, north and south of the straits is described as being similar to this ; abounding in harbours, and for this reason it is the route by which boats invariably go to York Factory, and generally to the Saskatchewan. Opposite the mouth of Great Washow (Deep) Bay there is an inlet or passage called Loon's Straits, formerly a canoe route of the old North West Company.

By starting at daylight and sailing along the east coast of the lake on the 13th, we got in sight of the Grindstone Point about two p.m., when we set out on a longer and more dangerous traverse than any we had yet accomplished. We had to cross from the east coast of the lake to the Grindstone Point on the west coast, a distance of about twelve miles. From the shape of the lake with its many deep and broad bays this great traverse is unavoidable. When we started from the east side of the lake, the high escarpment of rock forming the point seemed quite low and blue in the distance. By spreading a blanket we were assisted for a while by a side wind; but the wind soon changed and freshened up, so that we had to lower sail and ply our paddles with all our strength until reaching the point nearly four hours from the time we left the east shore. Taking advantage of a little moonlight, which enabled us to coast along a straight shore after dark we did not stop to camp till arriving at the Little Grindstone Point.

By making an early start on the 14th, and creeping along in the shelter of the land, we were enabled to dine at Grassy Narrows. Although our fish had not improved any since leaving Pike River we had always keen appetites and were now by no means fastidious. Sailing from Grassy Narrows across a bay into which White Mud River empties. we arrived at the Sandy Bar a little after dark and camped.

15th September.—The wind and weather being favourable to-day, by working 15½ hours we reached the marsh near the mouth of Red River about dark. We found an Indian encamped on the sand-beach hunting the ducks which are in countless numbers in these marshes at this season. He had killed 100 "stock" ducks during the day, and generously gave us a liberal supply; had it not been for this hospitable Indian we should have been badly off, as we ate our last fish at the Sandy Bar in the morning.

16th September.—Reaching the Stone Fort about dark, and procuring a horse there, I was enabled to join you in the Red River Settlement at 11 p.m., after a canoe voyage of forty-eight days in all; nine of which were occupied in descending from the Elbow of the South Branch of the Saskatchewan to Fort à la Corne, fourteen from thence to the mouth of the Saskatchewan, and twenty-five days in traversing Lake Winnipeg.

The whole distance travelled and explored in canoe is over 940 miles; 600 of which being down the Saskatchewan and 340 miles open lake navigation. In performing this latter part of the journey with a little frail canoe, heavily laden, we were completely windbound for twelve days; and had to contend nearly all the time we were moving with boisterous head winds, foul weather, and a hand to mouth sustenance, frequently without food. This will, in some measure, account for the slow rate of progress we unwillingly made through Lake Winnipeg. I would much regret were it to be supposed that the tardy progress of this expedition was at all owing to the inefficiency of the two men—Wigwam Cullin and James Louis—you were pleased to appoint to accompany me; and must take this opportunity of bearing testimony to their unwearied labour, patient endurance and unflinching devotion under a series of trying circumstances. Their conduct while they were my companions, for nearly two months, was beyond all praise; and they sustained privations, hardships, and risks of no ordinary description without a murmur.

I am, dear sir, most respectfully,

Your obedient servant,

JOHN FLEMING.

To H. Y. Hind, Esq.,
In charge of the Assinniboine and Saskatchewan
Exploring Expedition.

CHAPTER VIII.

RED RIVER SETTLEMENT TO THE MOUTH OF THE LITTLE SASKATCHEWAN, IN A FREIGHTER'S BOAT.

Mouth of Red River—Aurora—Weather Signs—Channel at Mouth of Red River—Storm—Character of the South Coast of Lake Winnipeg—New Land—West Coast—Confervæ—The Willow Islands—Clay Cliffs—Good Land—Drunken River—Aurora—Rock Exposure—Deer Island—Section on —Equivalent of the Chazy formation—Fishing Ground—Miskena—Grindstone Point—Rev. Mr. Brooking—Rocks of Grindstone Point—The Little Grindstone Point—East Coast of Lake Winnipeg—Punk Island—Yellow Ochre—Coast near Dog's Head—Limestone Cave Point—Fissured Rocks—Jack Fish River—Fisher Bay—The Cat Head—Little Saskatchewan Bay—East Coast of Lake Winnipeg—Dimensions of Lake Winnipeg.

A fortnight was occupied at the Settlement in writing reports and making preparations for a voyage through Lake Winnipeg, the Little Saskatchewan River, and Lake Manitobah to the Salt Region, on the shores of Winnipego-sis Lake. Mr. Dickinson prepared for an exploration of the country between the Lake of the Woods and Red River, and between the Assiniboine and the 49th parallel. Both parties were ready by the 18th, and, at noon, started on their respective routes.

Mr. Fleming and I, taking advantage of a fair wind, reached a point about seven miles below the Indian Settlement, where we camped. On the following morning, the temperature of the air at sunrise was 63°, of the river 59°. We reached the mouth of river at 10 a.m., and hastened to avail ourselves of a south east wind just beginning to rise. Last night the aurora was very beautiful, and extended far beyond the zenith, leading the voyageurs to predict a windy day. The notion prevails with them that when the aurora is low the following day will be calm; when high, stormy. The temperature of the mouth of the river was 59°, and of the open lake, 1½ mile from shore, 58½°. Rain commenced as soon as we were fairly in the lake, the wind suddenly chopped round to the North driving a dense fog before it, and in a few minutes enveloped us in a misty shower. The steerman instantly turned about and made for the mouth of the river, there being no harbour nearer than the Willow Islands, at least fifteen miles distant. The breeze rapidly increased to a gale as we regained calm water inside the bar at the mouth of Red River.

The wind subsided about 2 p.m., and a shot heard in a direction due south induced some of the voyageurs to exclaim that the wind would soon come from that direction, according to an impression common among these excellent observers and interpreters of "signs" that a shot heard against the wind is a good omen. But our steersman placed more faith in the aurora, and thought we had not "taken all the wind out of it yet." The sky having a threatening appearance, we determined to camp.

There are four mouths to Red river and the channel we had entered was the main outlet; the breadth of the channel varies from 20 to 28 feet, and on either side shelves rapidly from four to eighteen feet of water. At 3 p.m., when just on the point of starting, one of the voyageurs suggested that we should wait for a few minutes longer as he had observed the water of the lake coming in at the mouth of the river, and thought that the wind would soon blow strong from the north, although at the time the sky was clear and a calm prevailed. In less than half an hour a fresh northerly breeze sprang up, scud appeared drifting before it, and the waters of the lake flowed rapidly up the river into the vast marshes which extend for many miles inland at the southern extremity of Lake Winnipeg. The weather at this season of the year is very changeable, and renders boat navigation of this lake rather hazardous. In anticipation of a storm, we made ourselves as comfortable as circumstances would permit on a low spit of sand, with the lake before us, the river on our left hand, and interminable marshes east and south of us.

Sept. 20th.—Soon after sunset last night, the breeze from the north rose into a gale; the water of the lake ran like a rapid up the river channel into the swamps, and a terrific swell soon set in from the lake, breaking upon the sandy beach with a stunning noise. The water rose to within six inches of the level of the spit on which our tent was pitched and threatened every instant to submerge it. At 10 p.m., the gale was at its height, and as we sat upon a stranded trunk of a tree, looking out upon the lake, a truly magnificent scene lay before us. Huge crested breakers covered the lake as far as we could see through the gloom, lighting up the coast with long glistening streaks of white foam. The noise was so overpowering that we had great difficulty in hearing one another speak; the waves broke over the narrow spit which formed the low bank of the river where our boat was moored and the tent pitched; our camp ground was reduced to a strip of sand eight yards broad and seven inches above the river on one side, with overflowing swamps on the other; if the storm had continued half an hour longer we should have been compelled to take to the boat and drift into the reeds, at the risk of being stranded when the gale subsided and the water retired from the marshes into the lake.

For many miles the south coast of Lake Winnipeg consist of alternate strips of sand sustaining willows, with marshes in the rear running parallel to the coast line. Some of these sand strips show many years of duration when well protected by drift timber, others are of recent origin, clear and bare, enclosing ponds in which rushes are only just beginning to show themselves. They are the records of the progress made by new land in its invasion of the lake at and near the mouth of Red River. A northerly gale throws up a bar or beach about one hundred yards from the main shore. On the new beach, drifted timber accumulates, and in process of time becomes consolidated by the gravel and sand which is washed between the logs. Willows soon grow on the new soil thus formed, and bind the whole into a firm beach with a marsh in the rear. A heavy gale may sweep the new land away or throw up another beach about one hundred yards in advance of it, on which the process of consolidation is renewed. For ages past, this work of construction and destruction has been greatly in favor of the former. Hence it arises that, with the exception of the newly formed spit at the mouth of the river, there is no accessible camping ground for several miles up the stream; marshes surrounding the spits or old beaches on which the willows grow, and extending in all directions as far as the eye can reach.

We employed ourselves during our detention in examining the coast, sounding the river, and in shooting and fishing. Our sporting brought us only six duck, three plover, and three large pike. The flesh of the pike was of a delicate salmon colour, more like that of the salmon trout of the Canadian lakes than of the common pike.

Sept. 21.—We rose at 4 a.m., and in half an hour were *en route*, the morning just beginning to dawn; temperature of the air at sunrise, 51°, of lake 59°. The west coast for a few miles is elevated from five to six feet above the lake, here and there a low beach of limestone gravel, sand, and a few granite boulders, is fringed with a belt of tall aspens which grow within twenty feet of the water's edge. Behind the belt of aspen is a marsh, then another belt of aspens followed by a marsh. This succession continues for a distance of about three miles before good land supporting heavy aspens is to be found in large areas. Near to the spot where we breakfasted, an excellent illustration of the prevailing character of the west coast thus far, occurs. A sandy beach covered with shingle had separated a former bay from the main body of the lake. On this beach which was not twenty feet broad, or more than five above the lake level, willows, dogwood and grasses were growing; a large pond lay inside, fringed with rushes; it was tenanted by hosts of duck. In the rear of this pond a narrow strip of land, clothed with aspen, separated a marsh from it, which had doubtless once been a bay of the lake, then a pond, and finally, a marsh.

At 11 a. m., a vast quantity of confervæ apppeared in clusters on the surface of the lake, resembling in every particular a similar organism noticed in extraordinary profusion on the Lake of the Woods in August, 1857. The sudden appearance of this 'weed,' indicated a calm, according to the experience of our voyageurs. A calm did occur for a short time, soon however, followed by rain in the north, which fortunately did not reach us. Inland ponds cut off from the lake by low beaches appear as far as the Willow Islands, where we arrived in the afternoon; they were found to consist of a few small sandy areas and one long narrow strip of sand and gravel, stretching into the lake in an easterly direction, and separated from the shore by a narrow channel. The Islands are fast wearing away, and in the memory of some of the voyageurs, were covered ten years since with willows, poplar and a few spruce. They have probably afforded much of the material for the formation of the beaches which have cut off portions of the lake on the south-west coast, the materials being drifted along the shore by the long waves which every breeze from the north, or a northerly direction creates. The depth of water near the coast is very small, soundings showed 29 feet water one mile north of Willow Island, the deepest part yet observed.

In the afternoon I landed to examine some cliffs of clay which appear about twenty-three miles from the mouth of the river. They were sixteen feet in altitude, and exposed a clean surface of stratified marl, reposing on a brownish black clay. The stratification was in thin horizontal layers, easily detached one from the other. The brownish black clay showed a very tenacious character, so much so, that it was very difficult to break off with the hand masses larger than ten or twelve cubic inches, in any other direction than that of the plane of stratification. It was worn by the action of the waves into a great variety of forms, and on the beach lay large numbers of rounded, oval, spheroidal or circular forms from one foot in length, and three inches in diameter, to small spherical bodies of the size of peas. They were covered with minute pebbles, or with sand, and when broken, shewed a nucleous of the tough clay which had assumed its regular form by constant rolling on the beach. No organic remains were found, but the impression conveyed by the aspect of the clay and the marl by which it was capped, satisfied me that it was of the same age as the clay and marly substratum of the Red River and Assiniboine Prairies.

The timber in the forest consisted of aspens and birch, with a few oak, elm, and ash. Our steersman, who knew the country well, informed me that good land on which large timber grew, did not extend more than one mile from the lake. It is succeeded by spruce and tamarac marshes, the trees being of dwarfish dimensions. The afternoon was calm and warm, so far verifying the predictions of our voyageurs, which they had based on the sudden appearance of the "weed," in the morning.

Sept. 22nd.—Last night was cold, calm, and beautiful, the thermometer fell to 36° at 10 p. m., and to the freezing point before daybreak; Donati's comet shone a fine celestial object, and with a moon nearly full, and a splendid aurora distinctly visible, notwithstanding the brightness of the moon, the heavens presented a very beautiful spectacle. We camped near the mouth of Drunken River, a small stream which would make an excellent boat harbour, if widened at its outlet. The clay cliffs and marl disappeard before we arrived at our camping place; the shore again consists of a beach, with a swamp or marsh, fringed with small spruce and tamarac in the rear. I aroused the men at 4 a. m. The aurora at that hour was a splendid object, and appeared in the form of sudden flashes of low arcs of light, complete from east to west, rising in vast waves from one constant luminous base, a few degrees above the horizon. The vast waves of pale light followed one another with great rapidity and regularity, for many minutes together.

A strong westerly breeze early this morning, soon enabled us to reach the Sandy Bars, fourteen miles from Drunken River, and then the Grassy Narrows, a distance of seven miles. Both of these points are low, sandy and gravelly peninsulas stretching out into the lake opposite to Big Black Island. The first exposure of limestone was seen on a small island opposite Big Black Island, which we named Guano Island. It dipped very slightly to the south-west; a search for fossils was fruitless, but on Big Black Island, and those adjacent to it, near the Little Grindstone Point, limestone appears in the form of low mural cliffs on the west shores, which were alone seen. This limestone is a continuation of a fine exposure afterwards found on Deer Island, where we arrived at 1 p. m.

The following section occurs on Deer Island.

Lake Level.

Shingle Beach; (Limestone).

No. 1. Four feet of dark green argillo-arenaceous shale, with thin layers of sandstone of uneven thickness—Fucoids very abundant in the sandstone. The weathered sandstone is reddish brown; fresh surfaces are white or gray. White Iron Pyrites assimilating the forms of disks, spheroids and shells occurs in the sandstone.

No. 2. In many respects like the former; the sandstone layers are from one to four inches in thickness and predominate over the shaly portions. Its thickness is six feet. The character of these formations (1 and 2,) is very variable; the green argillaceous portion sometimes predominates, and occasionally the sandstone.

No. 3. Ten feet of sandstone with green bands of a soft argillaceous rock, from one quarter to four inches in thickness. The sandstone often white, but generally red. A persistent green band, a few inches thick, filled with obscure forms, resembling fucoids is very characteristic. The red coloured sandstone is often soft and friable, the white frequently embodied in the red. Both red and white, contain obscure organic forms. The green patches which are found throughout the sandstone contain impressions of fucoids; an *Orthoceratite* was found in the sandstone. In some parts of the exposure on Deer Island the sandstone layers are much harder, although partaking of the characters already described. When thus hard, the white portion is extremely brilliant, of a pure white, and very siliceous; it would form an excellent material for the manufacture of glass. Forms coloured brown, often pervade the white sandstone, and appear to resemble fucoids and corals replaced by brown ochreous sand.

No. 4. Eighteen feet of limestone, perfectly horizontal, very hard, and breaking off the cliff where the soft sandstone has been weathered away in huge rhomboidal slabs, eight to twenty-five feet in diameter, and four to ten inches thick.

The surface of the limestone shows silicified shells and corals, among the shells an *Orthoceras* nine inches in diameter was seen, with fossils belonging to the genera *Rhynchonella* and *Tetradium.* This formation is equivalent to the *Chazy* of New York and Canada, and consequently lies near the base of the Lower Silurian System.

In the Shingle immediately below the cliff, many fine orthoceratites were found, with a large *Maclurea*, and *Catenipora escharoides.**

Limestone appears for some miles on the west coast, south of Big Grindstone Point, where we arrived in the evening. This part of Lake Winnipeg is very beautiful, resembling, in many pleasing particulars, the scenery on Lake Simcoe towards the Narrows, with wooded islands rising from the lake in clusters and rows. Between Grindstone Point and Deer Island, the lead showed sixty feet of water. It is the great fishing-ground of some of the bands of Indians who make this part of the lake their wintering place. White fish are very abundant, and caught by the Indians in large numbers: their flavour is not so fine as those of Lake Manitobah, or of the Qu'Appelle Lakes. Sturgeon are also numerous, and, according to the belief of the miserable natives who fish here during the winter, the deep part of the lake is their great place of resort at that period of the year, where they lie with Mis-

* For an enumeration of the fossils from this and other localities, in the region about Lake Winnipeg, Manitobah, &c., the reader is referred to the Chapter by E. Billings, Esq., Palæontologist to the Canadian Geological Survey.

a, Punk Island.

EXPOSURE ON DEER ISLAND, NEAR GRINDSTONE POINT, LAKE WINNIPEG. (FOR DESCRIPTION, SEE PAGE 86.)

ke-na, the chief of the fishes in the southern portion of Lake Winnipeg.

Longfellow alludes to the same superstition held by Lake Superior Indians, in the song of "Hiawatha," where he makes his hero go—

"Forth upon the Getche Gumee,
On the shining Big-Sea-Water,
With his fishing-line of cedar—
Of the twisted bark of cedar—
Forth to catch the sturgeon Nahma,
Nishe-Nahma, King of Fishes,
In his birch canoe exulting:
All alone went Hiawatha."

We approached Grindstone Point after dark, and observed a camp-fire on the beach, with a freighter's boat close in shore. It belonged to the Rev. Mr. Brooking and his family, who were returning to Rossville from Red River. Mr. Brooking is a Wesleyan missionary, for some years a resident in Rupert's Land, and engaged in the unthankful labour of attemping to christianize the Indians. He had travelled from the head of Lake Winnipeg to Red River Settlement, to obtain medical advice for Mrs. Brooking, who was very unwell. Our interview was short—the voyageurs, in Mr. Brooking's boat, being anxious to take advantage of a fair wind which had just arisen. As soon as supper was ended they embarked, and proceeded by moonlight on their lonely journey. He was twenty days in coming from Norway House to Red River, having been kept back by contrary winds. His prospects of traversing the lake rapidly were now more favourable, as the south wind which prevailed, would soon drive a freighter's boat to Norway House.

September 23.—The rocks at Grindstone Point, about six miles north of Deer Island, are similar to those already described in the previous section. Being further north, the exposure is higher, and the sandstone bands more fully shown. Beneath No. 1 of Deer Island, a hard, yellow, compact sandstone appears, and is exposed for a space of four feet above the level of the water. Strata No. 1 and No. 2 of Deer Island appear in a slightly different form here: the sandstone bands are thicker; the green shaly portion more distinct as a separate band, and two feet thick; while above the hard yellow sandstone, the base of No. 1 appears in the form of a purple band of very soft sandstone, about one foot in thickness, containing a vast number of *stains*, which seem to have been occasioned by fucoids.

At Little Grindstone Point, the Limestone No. 4 of Deer Island comes to the water's edge. The Sandstone No. 3 is just below its level. Little Grindstone Point is a little more than seven miles south-west of Big Grindstone Point, and the altitude of the limestone, where it touches the sandstone at the last-named place, is about 25 feet, which would give an inclination of a sectional exposure in a south-westerly direction of about three feet in the mile. It appeared, however, to have a slight westerly dip, shewing the true dip to be a few degrees more to the west than south-west, as was afterwards ascertained. In the limestone, turbinated shells are numerous, with *orthoceras* of large dimensions. The scenery is attractive, and, in a geological point of view, eminently interesting. The opposite coast is formed of the unfossiliferous rocks belonging to the great Laurentian formation, which extends from Labrador to the Arctic Ocean. Within three miles of Grindstone Point, islands of this important formation occur a short distance in advance of the east coast, which is wholly composed of it.

The depth of Lake Winnipeg immediately opposite Grindstone Point is forty-eight feet. A storm afforded us another opportunity of examining the fossiliferous rocks of this locality, for no sooner had we started in the direction of the "Granite Islands," opposite the point, that the wind turned round to the north, and compelled us to seek shelter in a bay of Punk Island, three miles south-east of the Grindstone Point.

On Punk Island, strata 1, 2, 3, and 4 of Deer Island were recognized in a bay, with some lithological differences. In Nos. 1 and 2 here, which could scarcely be distinguished from one another, a great number of a *Modiolopsis** were found.

On the north-east side of Punk Island, above the purple sandstone, mentioned as occurring at Big Grindstone Point, a thin stratum of buff coloured limestone occurs, possessing some peculiarities. On raising slabs, between each stratum a soft and very pure ochre, of a beautiful yellow colour is found, from one-eighth to half an inch in thickness. The ochre when moist and fresh is easily worked by the fingers, quite destitute of gritty or hard particles, of a uniform, pale yellow, and when burned, of a beautiful cinnabar red. It is used by the Indians in both states as a pigment; the limestone in which it occurs is extremely porous and often honeycombed.

Sept. 24th.—At half-past 2, a.m., the wind being fair, and the sky clear, we prepared to start. There was a sharp frost during the night, and the thermometer registered 28°. We made the traverse of Great Washow Bay, thirteen miles across, and breakfasted at a point half-way between Bull's Head and Dog's Head. The limestone cliffs here were about thirty feet high, and occupy the coast from Bull's Head to Whiteway's Post, opposite the Dog's Head. Where seen at breakfast, the coast is fringed with broken masses, which lie piled one on the other in picturesque confusion. Ascending the cliff, I found large portions detached from the main body, forming deep clefts or cracks. Some of these fissures were twelve feet wide and twenty feet deep, others three feet wide and of greater depth. Sometimes the fissures were roofed with masses which had slipped forward, forming long, narrow caves, lined with moss. One cave was more than sixty feet long, and with the exception of a small aperture, closed at one end and roofed throughout. We named the spot Limestone Cave Point. From the description given by one of the voyageurs who had wintered near this place, and knew the country well, the rock along the coast, from the Bull's Head to Pike Head, is fissured in the manner described. Very roomy walled caverns can be found, which are easily converted into excellent wintering houses for trappers. The sides of the fissures are perpendicular, and the fracture is so even as to form chambers of a rhomboidal shape. The passages between them are beautifully covered with moss, while gracefully drooping overhead the birch and white spruce obstruct the rays of the sun, giving to these lonely cells a gloomy and desolate aspect. The limestone is similar to that which has already been described as No. 4 of Deer Island. At the Narrows, or Dog's Head, the limestone and unfossiliferous rocks are in close proximity. The east side of the strait being composed of the Laurentian formation, on the west side of lower Silurian limestone.

The wind being favourable, we sailed during the whole day, and at 4 p.m., reached the mouth of Jack Fish River, making a traverse across Fisher Bay, a very deep indentation, whose western limit could not be seen from the canoe. In Fisher

* See Chapter by E. Billings, Esq.

Bay, islands are numerous and some of them of large area, such as Great Moose Island, in the mouth of the bay, and Juniper Island, four miles to the north. Due west of the Dog's Head, Black Bear Island contains an excellent boat harbour, a feature worthy of note, as it occurs near the beginning of the great traverse across Fisher Bay. Jack Fish River issues from a marsh separated from the lake by a belt of sand and shingle about 100 yards broad. The river runs in a westerly direction from a series of small lakes and swamps, through a level, low country abounding in fine spruce and tamarac forests, broken by gravelly ridges supporting poplar and birch. The breadth of the river at its mouth is thirty feet, but where it passes through the swamp it is broad and deep, and so continues for some distance into the country. Jack Fish River is a favorite fishing station of a tribe of Ojibways, and was once the seat of a missionary establishment.

It will be mentioned in another chapter, that this tribe were deterred during the winter of 1858 from wintering here, by a threat from a noted conjuror of the Grand Rapid, illustrating the abject position in which superstition frequently places these unfortunate people.

Leaving Jack Fish River, or the Pike Head, as it is also termed, from a promontory bearing that name near to the mouth of the stream, we coasted under sail past Wicked Point across the traverse of Kinwow Bay, rounded Macbeth Point, and camped at Lynx Point, beyond the Cat Head. The coast at the Cat Head is very precipitous; the limestone cliffs rise about thirty-five feet from the water, without any intervening beach, so that boats cannot land, and must necessarily push on until a narrow beach is found a few miles beyond it. Limestone cliffs, similar in all respects to those of the Cave Point occupy the coast at intervals as far as the Cat Head, and probably fringe the Mantagao-seebe Bay, as they are seen near the mouth of the Little Saskatchewan, and on the north point of the great bay which derives its name from that river. Taking advantage of a fair wind and fine night we carried on across Lynx Bay, and camped at half-past eleven, p.m.

At half-past four on the following morning a westerly wind enabled us to round Point Turnagain, pass Bushkega and the Sturgeon Islands, and make the traverse across the Little Saskatchewan Bay to the mouth of the river. In making the traverse we could not see the extremity of this deep indentation in a south easterly direction, where the Mantagao-seebe debouches. The greater portion of the bay was coasted by Mr. Fleming, during his voyage from the Grand Rap d to the mouth of Red River. The temperature of the Little Saskatchewan was found to be 52½°, of Lake Winnipeg one degree higher.

The description of the west coast of Lake Winnipeg from the mouth of the Little Saskatchewan to the Great Saskatchewan is given in Mr. Fleming's Narrative, pp. 79–81. In order to complete a description of the entire coast line of this lake I append the following extract from Sir John Richardson's Journal of a Boat Voyage through Rupert's Land and the Arctic Sea. The south-eastern coast of Lake Winnipeg from the mouth of the Winnipeg to Red River was described in my Report for 1857, page 251.

"The eastern coast-line of Lake Winnipeg is, in general, swampy, with granite knolls rising through the soil, but not to such a height as to render the scenery hilly. The pine-forest skirts the shore at the distance of two or three miles, covering gently rising lands, and the breadth of continuous lake surface seems to be in process of diminution, in the following way. A bank of sand is first drifted up, in the line of a chain of rocks which may happen to lie across the mouth of an inlet or deep bay. Carices, balsam-poplars, and willows speedily take root therein, and the basin which lies behind, cut off from the parent lake, is gradually converted into a marsh by the luxuriant growth of aquatic plants. The sweet gale next appears on its borders, and drift-wood, much of it rotten and comminuted, is thrown up on the exterior bank, together with some roots and stems of larger trees. The first spring storm covers these with sand, and, in a few weeks, the vigorous vegetation of a short but active summer binds the whole together by a network of the roots of bents and willows. Quantities of drift-sand pass before the high winds into the swamp behind, and, weighing down the flags and willow-branches, prepare a fit soil for succeeding crops. During the winter of this climate, all remains fixed as the summer left it, and as the next season is far advanced before the bank thaws, little of it washes back into the water, but, on the contrary, every gale blowing from the lake brings a fresh supply of sand from the shoals which are continually forming along the shore. The floods raised by melted snows cut narrow channels through the frozen beach, by which the ponds behind are drained of their superfluous waters. As the soil gradually acquires depth, the balsam-poplars and aspens overpower the willows, which, however, continue to form a line of demarkation between the lake and the encroaching forest.

"Considerable sheets of water are also cut off on the north-west side of the lake, where the bird's-eye limestone forms the whole of the coast. Very recently this corner was deeply indented by narrow, branching bays, whose outer points were limestone cliffs. Under the action of frost, the thin horizontal beds of this stone split up, crevices are formed perpendicularly, large blocks are detached, and the cliff is rapidly overthrown, soon becoming masked by its own ruins. In a season or two the slabs break into small fragments, which are tossed up by the waves across the neck of the bay into the form of narrow ridge-like beaches, from twenty to thirty feet high. Mud and vegetable matter gradually fill up the pieces of water thus secluded; a willow swamp is formed; and when the ground is somewhat consolidated, the willows are replaced by a grove of aspens.* Near the first and second Rocky Points,† the various stages of this process may be inspected, from the rich alluvial flat covered with trees and bounded by cliffs that once overhung the water, to the pond recently cut off by a naked barrier of limestone, pebbles, and slabs, discharging its spring floods into the lake, by a narrow though rapid stream. In some exposed places the pressure of the ice, or power of the waves in heavy gales, has forced the limestone fragments into the woods, and heaped them round the stems of trees, some of which are dying a lingering death; while others, that have been dead for many years, testify to their former vitality, and the mode in which they have perished, by their upright stems, crowned by the decorticated and lichen-covered branches which protrude from the stony bank. The analogy between the entombment of living trees, in their erect position, to the stems

* The fact of the formation of these detached ponds, marshes, and alluvial flats, points either to a gradual elevation of the district, or to an enlargement of the outlet of the lake, producing a subsidence of its waters.

† The strata at these points contain many gigantic orthoceratites, some of which have been described by Mr. Stokes in the Geological Transactions.

of *sigillariæ*, which rise through different layers in the coal-measures, is obvious."*

The following are the dimensions of Lake Winnipeg:

Area of Lake..................	8500 square miles.
Length, not in'g. Play Green Lake	280 statute miles.
Greatest breadth	57 " "
Length of coast line	930 " "
Approximate height above the sea	628 feet.

This estimate of the altitude of Lake Winnipeg above the sea level, was deduced in 1857, from the levels taken across the portages along the line of the canoe communication between Fort William on Lake Superior, and Fort Alexander on Lake Winnipeg—(*See page* 257, *of the Report for* 1857). The height of the dividing ridge which separates these lakes from one another, is 1485 feet above the level of the sea; and distant, by the canoe route, 104 miles from Fort William and 510 miles from Fort Alexander.

Major Long, in 1823, found the sources of St. Peter and Red River to be 830 feet above the ocean, and Lake Winnipeg 630 feet above the same level—a difference of only two feet in excess of the estimate we made in 1857.

When it is remembered that the St. Peter River is an affluent of the Mississippi flowing into the Gulf of Mexico, and Red River communicates with Lake Winnipeg, which sends its surplus water to Hudson's Bay by Nelson River, the extraordinary lowness of the water-shed becomes apparent.

As it is not improbable that coming events will make all communications between the Mississippi Valley and Lake Winnipeg interesting, if not important, I venture to introduce the subjoined extract from the "Narrative of Major Long's Expedition to the Source of St. Peter's River," performed in 1823:—

* If one of the spruce firs included in the limestone debris, had its top broken off, and a layer of mud were deposited over all, we should have the counterpart of a sketch of Sir Henry de la Beche's Manual (p. 407). The thick and fleshy rhizomata of the *Calla palustris*, marked with the cicatrices of fallen leaves, and which are abundant in these waters, bear no very distant resemblance to *stigmariæ*.

"The St. Peter, in our opinion, probably never can be made a commodious stream; for although it flows over gradations, and not upon a slant, yet as these gradations are accumulated into the upper third of the distance between Big Stone Lake and the mouth of the river, the expense of rendering it navigable, by damming and locking, would far exceed the importance of the object. The plan would doubtless be found very practicable; but the scarcity of water during the greater part of the year would render these works unavailing.

"From considerations upon which it is unnecessary to dwell, and the accuracy of which might be disputed, though they appear to us to lead to correct results, we have estimated the fall in the river, or difference of level between the Lac qui Parle and the mouth of the river, at about fifty or sixty feet. According to this estimate, the average fall does not exceed two or three inches per mile.

"The river having taken a bend to the west, we continued our route in what appeared to have been an old water-course, and, within three miles of the Big Stone Lake, found ourselves on the banks of Lake Travers, which discharges its waters by means of Swan, or Sioux River, into the Red River of Lake Winnipeg, whose waters, as is well known, flow towards Hudson's Bay.

"The space between Lakes Travers and Big Stone is but very little elevated above the level of both these lakes; and the water has been known, in times of flood, to rise and cover the intermediate ground, so as to unite the two lakes. In fact, both these bodies of water are in the same valley; and it is within the recollection of some persons now in the country, that a boat once floated from Lake Travers into the St. Peter. Thus, therefore, this spot offers us one of these interesting phenomena which we have already alluded to, but which are no where, perhaps, so apparent as they are in this place.

"Here we behold the waters of two mighty streams, one of which empties itself into Hudson's Bay, at the 57th parallel of north latitude; and the other into the Gulf of Mexico, in latitude 29°; rising in the same valley, within three miles of each other, and even in some cases offering a direct natural navigation from one into the other."

CHAPTER IX.

THE MOUTH OF THE LITTLE SASKATCHEWAN TO THE SALT SPRINGS ON WINNEPEGO-SIS LAKE.

The Little Saskatchewan—Height of Bank—Country in rear—Tracking—Swamps—Banks of River—Ojibway Camp—White-fish—Character of Country—Canoe Fleet—Spruce—Boulders—Marsh—St. Martin Lake—" Money"—Pounded Fish—Wavys—Fine Land—The Narrows—Boulder Barriers—Sugar Island—Indians—Gneissoid Islands—St. Martin Rocks — Beach Barriers — Depth of St. Martin Lake—Thunder Island—Thunder Storm—Partridge Crop River—Rushes—Old Mission—Low Country—Indian Farmer—Wide Spread Marsh—Fairford—The Character of the Country—The Mission—Evening Service—Rev. Mr. Stagg—The Farm—Hudson's Bay Company's Post—Rum—Lake Manatobah—Progress of the Season—Rocks—Fossils—The Coast—Steep Rock Point—Devonian Rocks—Indian Superstition—Water-hen River—Eagles—Character of Water-hen River—Pelicans—Indians—Wood and Prairie Indians—Barter— Winnipego-sis Lake—Ermine Point—Elms—Salt Spring—Snake Islands—Duck Mountain—Snake Island Fossils—Arrive at Salt Springs.

A few hundred yards above the mouth of the river, horizontal Lower Silurian limestone shows itself on both sides, and it is through this rock that the Little Saskatchewan has excavated its bed. The limestone contains fossils in abundance, but in very bad state of preservation in many of the layers. They are similar to those found on Lake Winnipeg at Cave Point, and in its lithological aspect there is no appreciable difference between the exposures in either locality. The Little Saskatchewan, as its name implies has a very rapid current, varying from one to four miles an hour. The banks are not more than 20 to 25 feet above its level near the mouth, and diminish in altitude in ascending the stream. They are fringed with aspen, poplar, spruce and tamarac. In the rear swamps occur, often covered with deep moss, and sustaining clumps of tamarac and spruce of fair dimensions, but scarcely suitable for any other purposes than those which a limited settlement might occasion.

The river proving too rapid for using the sweeps, we were compelled to track up, a difficult and tedious labour to the men, but offering an excellent opportunity for making traverses into the country, which, however, were never deep, the swamps soon arresting progress inland. The general aspect of the river for the first four miles is very attractive, resembling in many particulars Rainy River. About three miles from the lake the limestone disappears, being covered with drift or alluvial clay. The banks rise gently with the stream, which is rapid and shallow. The yellow autumnal foliage of the aspens contrasts beautifully at this season of the year with the spruce and tamarac, and gives a charming appearance to the river banks. Towards evening we arrived at a camp of Ojibways, containing four tents. They had an abundance of white fish, and told me the river was full of them. Anxious to test the statement I intimated a wish to purchase a score of fresh fish, and offered an Indian some tea and tobacco if he would catch them immediately. He accepted the offer, entered his canoe, crossed over to a well known eddy and in fifteen minutes brought back twenty white fish, weighing on a average three pounds each. We camped close to the Ojibways, as we knew that if we tracked a mile or so up the stream they would follow us, and our party might be increased by others in advance of them. As it was, the guns they fired at our arrival had been heard, so that at sunset several canoes came swiftly down the stream, filled with men and women to " learn the news." The whole body camped close to us, and what with talking, shouting, screaming of children and howling of dogs, we enjoyed no rest until late in the night.

By day-break on the following morning we rose and employed a few hours in examining the country in the rear of the camp. The banks of the river are here about twenty feet above the present level of the river, but the country is very marshy, and clothed with tamarac and spruce behind the belt of aspens which fringe the river banks. After breakfast, the wind being fair, we hoisted sail, and in company with our Ojibway friends proceeded up the river. A little fleet of twenty-three canoes, each with a birch bark sail, glided quickly ahead of us, but the breeze freshening we soon caught and passed them one by one. The banks of the river are not more than ten feet above its present level about nine miles from its mouth, but are rarely flooded. They consist of alluvial clay, and sustain many groves of fine spruce and aspen. At some of the bends there is a large accumulation of boulders consisting chiefly of the unfossiliferous rocks. The colour of the trees is truly beautiful, nearly all the aspens in front are yellow even at this early period, while those in the rear, protected in some measure from the night frost, still retain their green.

About five miles from St. Martin Lake a marsh begins, on the edge of which we camped, our Indian friends soon closing with us. Some of the old men were anxious to show me some specimens of ' Money ' they had carefully folded in bits of cloth or birch bark. The ' Money,' respecting which they have no distinct idea except that it is ' white,' according to information they have obtained from half-breeds, consisted of fragments of selenite, iron pyrites, and silver mica. They profess to know where a large quantity of this ' Money ' is to be found, and demand tea and tobacco for the intelligence. These Indians have been making their autumnal fishing hunt, and have with them large birch bark vessels filled with pounded white-fish, previously dried and smoked, a miserable substitute for pemican. They had also sturgeon bladders filled with white-fish oil. The pounded fish and the oil form part of their winter stores; some samples which were submitted to me for inspection, with a view to barter, were the reverse of inviting.

September 27th.—A stormy uncomfortable night. Wavys (*Anser hyperboreus*) flying to the south early this morning in large flocks, a sure sign, it is said, of approaching winter. The Indians say there is some fine land and large trees in the rear of this part of the river. The river from our camp to St. Martin Lake, about thirteen miles in an air-line from Lake Winnipeg, has marshy banks. St. Martin Lake once reached, small eminences, which in this flat country almost deserve the name of hills, appear on the south side, so also on the north side before entering the Narrows. In general the shores are very low, particularly to the south-east. The Narrows are caused by a remarkable barrier of boulders, chiefly consisting of the unfossiliferous rocks, about six feet above the lake and twenty feet broad. On the west side of the barrier there is an extensive wide-spreading marsh, but the water of the lake is clear, as in most limestone regions.

We arrived at this isolated body of water soon after noon, and camped on a beach or barrier thrown up in the form of semi-circular ridges about half a mile across the arc, and connected in the form of the letter S. In the formation of these ridges granite or gneissoid boulders are first pushed by ice upon a limestone gravel bar, aspens and willows grow on the ridges rapidly formed by sand and gravel washed up in the rear of the boulders, and the space partly enclosed or sheltered by the curve is soon filled with reeds, thus forming extensive marshes at the eastern extremity of St. Martin Lake. Near the channel which separates this maze from the main body of the Lake, a new beach is now in process of formation, and consists at present of a long semi-circular line of stranded boulders, over which the sea washes in easterly and westerly gales. Round about the boulders limestone gravel is accumulating, and thus in this direction at least the lake is slowly diminishing in size, the materials being in great part supplied from the wearing away of islands, and the adjoining coast.

September 28th.—We succeeded in passing the Narrows before breakfast this morning, and made our way into the main lake through a channel varying from three to nine feet in depth, kept open no doubt by the Partridge Crop River, which takes the name of the Little Saskatchewan after it has passed through St. Martin Lake. We breakfasted on Sugar Island, being followed by the little fleet of canoes, whose owners appeared determined to reach Fairford before us, if possible.

On Sugar Island I found what appeared to be partially metamorphosed sandstone rock, tilted at an angle of 50°, with a S. 30° W., and N. 30° E. strike. At one extremity of the island it approached the character of gneiss, at the other extremity it presented the appearance of impure sandstone layers tilted at a high angle. Sugar Island is about a mile from the Narrows, and lies S. 75° E. from three small islands, which upon examination were found to consist of gneiss intersected with quartz veins. The rock on Sugar Island is exposed on one side in the form of a precipitous cliff 20 feet high. On the opposite side it slopes gradually to the water's edge. The Indians in eighteen canoes followed us to the island, and the chief with some ostentation informed me that it belonged to him, but he had no objection to my exploring it. He further stated, that as chief of the band he claimed the whole country from Fisher River, on Lake Winnipeg, to the mouth of Partridge Crop River.

Sugar Island is a favorite camping ground of the Ojibways, who now occupy this part of the country. We found some graves near to a garden in which potatoes were planted. A few pieces of tobacco procured us a small supply of this precious vegetable in these regions. Sugar Island is so named from a grove of the ash-leaved maple, the trees of which bore old marks of tapping.

We went out of our course to visit the gneissoid islands before referred to. The first island bore nearly due east of Sugar Island. It consists of gneiss with rose coloured felspathic veins, pursuing a general direction of S. 40° E. The axis of the island is also S. 40° E, and the gneiss is intersected by fissures nearly at right angles to one another, one set bearing S. 20°—40 E. The surface of the gneiss on the highest point, which may be 23 feet above the lake, is polished and furrowed in a direction S. 55° E. The south-east shore is precipitous, the opposite sloping.

The second island consists of gneiss with large quartz veins meandering through it. It is dome-shaped. The third island, within a few yards of the first and second, shows far less metamorphic action, and with a strike S. 15° W., has a dip 75° from the vertical. It is precipitous to the N. W. and slopes to the S. E.

Proceeding along the south-west coast we found a barrier of beaches along the shore about 300 yards distant from it, on which boulders of the partially metamorphosed sandstone and gneiss were piled up, farther on were worn and large unworn fragments of a silicious limestone, which, however, was nowhere found in position. The occurrence of these gneissoid islands in a flat limestone country is very interesting; the metamorphosed sandstone shows that the epoch of their elevation must have been before the deposition of the limestone found on Thunder Island, to which we next proceeded, and after the deposition of the sandstone on Sugar Island. The three gneissoid islands, having no name, we called St. Martin's Rocks. It is not improbable that the epoch of their elevation was simultaneous with outbursts which have been observed in other parts of the continent. At noon we arrived at a semi-circular island of beaches similar to those at the east end of the Lake. They are due to the great shallowness of St. Martin Lake, which, with an area of over three hundred square miles, was nowhere found to be more than eighteen feet deep and often only five and six feet for long distances.

In the afternoon we landed on an island on which stratified limestone, in horizontal layers, was exposed. The limestone possessed some singular peculiarities. Numerous cup-shaped forms, of very large dimensions, were visible in projecting masses over the whole of the surface exposed. Many of these cups were fully thirteen inches in diameter at the surface, and would hold at least one quart of water. They consisted of concentric rings, or cups, regularly arranged, and from ten to fifty or more in number. The thickness of each cup varied from one-tenth to one-quarter of an inch. A single specimen resembled a gigantic onion which had been cut in half, with a few of the inner layers extracted, leaving a cavity or depression. Many square yards of surface were varigated with this structure. The colour of the limestone is a buff-yellow, its fracture is uneven and masses are difficult to separate. It is extremely hard and silicious. The height of the exposure is sixteen feet, and so nearly horizontal, that no inclination could be detected. The island having no name, and being remarkable for its rock formation, it was thought worthy of some designation: we therefore called it "Thunder Island," in memory of a storm of hail and rain, accompanied by lightning and thunder of more than ordinary violence, which made us very uncomfortable for the rest of the day and during the ensuing night. It was the last of twenty thunder storms which we had encountered since entering the prairies on the 14th of June, and was only second to one in violence and sublimity.

Anxious to get on, we pulled at the sweeps until after dusk, having reached an island about four miles from Thunder Island. We found a sheltered cove, and all slept in the boat, there being no spot on the boulder-beach, or barrier on which we could discover six feet of level ground.

September 29th.—When morning dawned, which it did in a drenching, cold rain, we found we were attached to one of the stony barriers which protect certain aspects of the islands, or main shore. The ever-present marsh lay between us and the timber we so much needed for fuel; but the wind now rising to a gale, we were compelled to content ourselves with an exploration of our boulder barrier to its utmost limits. It was about one hundred yards broad, two to three miles long, and consisted of waterworn masses of limestone and gneiss, with limestone gravel between them. The marsh which separated it from the island was full of weeds, and harboured wild fowl, some of which we succeeded in killing.

We found great difficulty in discovering the mouth of Partridge Crop River, or St. Martin River as it is also called. A maze of rushes inland, extending as far as the eye can see, hides it from view. Half a mile up the stream we saw the houses of the Mission, established, but afterwards abandoned, by the Rev. Mr. Cowley. All the houses were in ruins, and tenantless. The country is very low, and liable to be flooded in the autumn and spring. There are but a few hundred acres of land fit for agricultural purposes, four or five feet above the river. The spot was one, however, of great resort among the Indians of this part of the country, and hence the probable reason why a selection of this site was made for the establishment of a Mission. On landing, we found one Indian family who are determined to continue the cultivation of the little fields which have been cleared and enclosed. They had accumulated three small stacks of hay, were possessed of a yoke of oxen, and were living in one of the least dilapidated houses.

We took to our boat at the beginning of Partridge Crop River, having secured a guide from the fleet of canoes in the rear, to take us through a narrow passage between beds of rushes which cover many square miles, and constitute the 'Crop,' so called by the Indians on account of the resemblance which the outline of this reedy expanse bears to the 'crop' of a partridge. We threaded our way through the mazes of a marsh supporting rushes so tall that, without climbing the mast of the boat, it was impossible to see beyond the masses which enclosed us. The rushes measured from ten to twelve feet in length, and grew so thickly together that they formed a compact green wall, past which the current flowed as if they were formed of solid, stable materials. Through little openings, which were now and then disclosed, we saw tranquil ponds, with a scarcely perceptible stream. Here revelled hosts of ducks of many species.

We arrived at Fairford at three, p.m., having occupied about two hours in passing through the Crop.

Fairford is very prettily situated on the banks of Partridge Crop River (a continuation of the Little Saskatchewan), about two miles from Lake Manitobah. The banks are here about twenty feet high, and show alluvial clay with boulders; but the limestone approaches the surface a short distance in the rear of the river. It is covered with eight to ten inches of vegetable mould; and although the appearance of the country is attractive, the shallowness of the soil would not permit of extensive agricultural operations. The dip of the rock is towards the south-west, but at so small an angle as to be imperceptible, except when a surface of several square yards is exposed. Fossils are few in number, and obscure: the limestone breaks up into thin slabs, being very compact and hard.

We attended evening prayers in an excellent school-house, which serves the purpose of a chapel. There were forty persons present, consisting chiefly of Half-breeds. The service consisted of a hymn and a chapter from the New Testament, respectively sung and read in the Ojibway language; an exposition of the chapter by means of an interpreter, and a concluding prayer: the Lord's Prayer was repeated aloud in Ojibway by the whole congregation.

There are one hundred and twenty Christians, adults and children, at this Mission. The houses are fifteen in number, neat, comfortable, and in excellent order. Several new dwellings are in process of erection. The appearance of this Mission is very promising, and in every way most creditable to the unceasing labours of the zealous missionary, the Rev. Mr. Stagg. We were supplied with potatoes, onions, turnips, fresh bread, and butter, and otherwise most hospitably entertained by Mr. and Mrs. Stagg. A young lady from Nottingham, England, Miss Thompson, is residing at the mission, and devotes herself with exemplary industry, in connection with Mrs. Stagg, to the education and care of Indian and half-breed children. The farm is in capital order, and although the area adapted for cultivation is not likely to induce the establishment of a large settlement, yet Fairford will become an important centre.

The Hon. Hudson's Bay Company have a post at this mission, but it is matter of deep regret that the heathen Indians who come to barter their furs here should be permitted to have access to rum. The little fleet of canoes before spoken of arrived during the evening, and at nightfall the sounds of drunken revelry told how terribly the debasing influence of this traffic must operate against the Christian and humanizing influence of the missionary. The post had been but recently established, and the distribution of intoxicating liquors to the Indians appeared to be a subject of deep anxiety and trouble to the Rev. Mr. Stagg.

We reached the mouth of the river at noon on the last day of September, and entered Lake Manitobah with a head wind, which soon compelled a retreat to a low sheltered beach. The exposed aspens are now quite yellow, but a tint of green remains on groves at some distance from the lake shore. Large boulders are piled up high upon the beach, and behind them is the unfailing marsh. In bays limestone gravel forms a sloping beach to the water's edge, but here again in the rear is a marsh. It is only at the headlands that rock in position, or firm of soil appears as yet.

In the afternoon we set sail, and arrived at Flat Rock Bay,

where limestone of Devonian age is seen on the south side. Some of the layers are highly fossiliferous, and hold numbers of *Atrypa reticularis* and *A. aspera*. The stems of crinoids are common, but the species are very few. The rock is nearly horizontal, and the general dip south-west, at a very small angle, but many slight undulations occur, giving an inclination of equal extent in an opposite direction. The exposure in the bay is ten feet high, worn into caves. The colour is a pale-buff, with some reddish-brown layers. Fucoids are abundant, and become when weathered, yellowish-buff. Small oak are scattered near the spot where we camped, interspersed with aspen. In the rear tamarac and spruce swamps prevented an examination of the country for more than a few hundred yards from the shore. Where rock in position does not form the beach, the marginal barrier of boulders is found with a beach, marsh or swamp in the rear.

October 1st.—Collected fossils, breakfasted, and pulled to Steep Rock Point. Here the limestone (Devonian) is 20 feet high, quite abrupt, with 6 feet of water at the base of the cliff. The layers are more massive and compact than before noticed; they occur from one to three feet in thickness, are very hard, and hold many organic forms replaced by crystalline carbonate of lime. Three and a half fathoms water were found within one hundred yards of Steep Rock Point. A number of swans were seen sailing in a little bay to the south of this land-mark in Lake Manitobah, which, by the way, the Indians who hunt in this part of the country do not visit, being persuaded that "little men" live in the caves and holes into which the rock has been worn by the action of the waves. We ran on before the wind, past Cherry Islands and Point Pao-nan, until dark, and then made for the shore, soon finding a small sheltered bay in the inside of a boulder beach in process of formation, about two hundred yards from land. Temperature of the lake, 53 deg.; greatest depth of water recorded, 22 feet.

A fair wind on the 2nd started us at dawn. We steered for the mouth of the Water-hen River, leaving on our left Crane River and Bay, where salt springs are found, and then passed through a narrow channel in a reef of boulders, which stretched from east to west, as far as we could see. The wind being fair, we pressed on, notwithstanding a heavy rain, and landed, rather late in the day for breakfast, on an island near the mouth of Water-hen River, which connects Lake Manitobah with Water-hen and Winnipego-sis Lake. Here we found a pair of white-headed eagles engaged in fishing; and as we came suddenly upon them after rounding a point, one of them dropped a fine white fish he had just caught, which was immediately seized and appropriated by our men for their own breakfast.

We entered one of the many mouths of the river at 2 p.m., and pulled up a broad channel through a vast marsh, whose limits are well defined by a belt of aspens on either hand. Having reached an attractive camping-place, where the woods came down to the edge of the river, we landed with a view to make a short traverse into the country. The river is swift, very broad, and prettily varied with well-wooded islands. At our camp the trees consisted of white spruce, 1 ft. 6 in. in diameter; poplar, aspen, birch, and tamarac. The land is low, not ten feet above the water. In the rear we found a tamarac swamp, with belts of white spruce. The channel through which our course lay was about three hundred feet broad and three feet deep, with a flat limestone bottom. The water was clear and brilliant, fish very numerous, and water-fowl abundant.

October 3rd.—*En route* at nine, a.m.,—the early part of the morning being employed in drying clothes after the rain of yesterday. We commenced pulling up Waterhen River, which here appears to contain many low islands, and its aggregate breadth must be several hundred yards near our camp. Signs of the approach of cold weather began to thicken around us; a large flock of pelicans, wheeling in circles far above, suddenly formed into an arrow-headed figure, and struck straight to the south. Yellow leaves drifting in the air before a cold north wind, promised us, as the Half-breeds say, by the beautiful aurora of last night. (*See auroras.*) Islands, low and reedy, continue to appear until we arrive at the Great Bend, where a band of Indians have their winter quarters. The Indians are Roman Catholics, originally from Oxford House. I persuaded one of them to act as guide up Moss River to Dauphin Lake, after we had visited the salt works. Their tents were dirty and excessively odorous. In general, the Indians of Lake Winnipeg and Manitobah, in point of cleanliness, cannot bear comparison with the Prairie Indians.

We met here, also, a freighter-boat, in charge of a French half-breed, who, with his family, were returning from the Salt-Springs to Oak Point with about twelve bushels of salt. We exchanged a little tea and tobacco for ducks and fish; and on the following morning started by the Middle Branch of Waterhen River for Winnipegoo-sis Lake, leaving Waterhen Lake to the north. The river is broad, shallow and reedy; a low belt of aspens, a mile off, on either side, shows the only land visible.

A fair wind drove us swiftly on, and at noon we stopped at Ermine Point, on Winnipego-sis Lake. This is a low beach, with a marsh behind, and is remarkable for some fine old elms, crooked and gnarled, still flourishing on the spit, near to a salt spring. At four we reached Snake Island, where we camped early, for the purpose ef examining an exposure of rock, and to collect the fossils which a glance showed it contained in abundance. The Duck Mountain loomed a grand object in the north-west.

The rock exposures on Snake Island are very interesting, not only on account of the fossils they contain, but in consequence of the evidence they afford of a slight upheaval, so rare in the present disposition of the rocks of this region.

The exposure at its highest point does not exceed twenty feet, but it is the centre of a low, narrow anticlinal, running north and south nearly. The dip on the east side is S. 75, E. $\angle$ 18°; and on the west, W. 20, S. $\angle$ 5°. The limestone is highly fossiliferous, beautifully stratified, very hard, and bituminous. It holds abundance of *Atrypa reticularis*; *Tellina ovata*; with fossils belonging to the genera *Favosites*; *Euomphalus*; *Productus*; *Gomphoceras*; *Orthoceras*; *Lituites*; together with *Trilobites*; *Crinoids*, &c. Mr. Billings thinks this locality unquestionably Devonian.

On the morning of the 4th of October, we set sail from Snake Island, and arrived at the Salt Works and Springs at Noon.

CHAPTER X.

THE SALT-SPRINGS, ON WINNIPEGO-SIS LAKE, TO THE SUMMIT OF THE RIDING MOUNTAIN—THE SUMMIT OF THE RIDING MOUNTAIN TO MANITOBAH HOUSE.

Character of the Country—The Duck Mountain—The Salt Springs—The Wells—The Manufacture of Salt—Salt Springs and Lagoons—Moss River—Rapids—Character of River—Valley or Dauphin River—The Riding Mountain—Lake Ridge—Hay Ground—Dauphin Lake—Pike—Snow Birds—Journey to the Summit of the Riding Mountain—Marshes—Ridges—Character of the Country—Whiskey Jack—Quaking Bog—Pitching Track—Rabbits—Foot of Mountain—Cretaceous Rocks—Plateaux—Conical Hills—White Spruce—Brown-nosed Bear—Summit of the Riding Mountain—Character of the Country—Former Character of the Riding Mountain—Denudation—Table Land—Snow Storm—Source of the Rapid River—Indian Superstition—Descent of Riding Mountain—Character of the Mountain—Fish—Sickness—Cupping—Ta-wa-pit—Great Bones—Grasshoppers—Journey from Dauphin Lake to Lake Manitobah—Character of the Country—Bogs—Aspen Ridges—Ridge Pitching Track—Ebb and Flow Lake—Indian Tent—Interior of—Supper—Sleep—Buffalo Runner—Manitobah House.

The surface of the country where the Salt Springs are found is only a few feet above the level of Winnipego-sis Lake, and apparently nearly horizontal for many miles inland, in a north-west course. The barren area occupied by the Springs and wells is about ten acres in extent; but the open country, with points of surrounding forest converging towards the Springs, may include several hundred acres. The trees in the vicinity consist of spruce, aspen, willow, birch, and a few stunted oak. The wells are five feet deep, and the water in them was 2 ft. 5 in. above the level of the lake on the 5th of October, as ascertained instrumentally. The wells are found upon a slight elevation, probably mechanically raised by the ascending brine, to about two feet above the country in the rear, which, in a southerly direction, gently inclines and blends with a vast marsh connected with Moss River. The woods fringing this marsh approach within a mile of the Springs, west and north-west.

The level country extends across the peninsula from Red Deer's Point, about three miles in breadth, to a deep indentation of Lake Winnipego-sis, about five or six miles broad; after which it continues low and marshy, with tamarac, aspen, and white spruce woods to the foot of the Duck Mountain, a distance of sixteen to eighteen miles. From Snake Island, and even from the level of Winnipego-sis Lake, a few miles from shore, the country between the foot of Duck Mountain and the Lake, does not present a single eminence to break the level from which the Duck Mountain rises. It resembles, in every important physical feature, the level tract at the base of the Riding Mountain. These observations apply only to that part visible from Snake Island and the Lower portion of Winnipego-sis Lake.

The soil at the Salt Springs is a very retentive yellowish-white clay, containing small limestone boulders and pebbles, with boulders of the unfossiliferous rocks. The wells, for obtaining a supply of brine, are sunk wherever a small bubbling spring is observed to issue from this retentive clay. The springs are constantly changing their position, and as the wells become exhausted from time to time, a fresh excavation is made where a new spring is observed to issue. No doubt boring, or deeper wells, would prevent these changes, and not only secure a larger flow of brine, but ensure its permanency. The wells at present are twenty-five in number; but some of them appear to have been lately abandoned, and others have long since ceased to yield brine. They are situated four hundred yards from the lake shore, and were first worked forty years since, by James Monkman. This enterprizing individual struggled for many years against the importation of English salt, which was sold in the settlements at a cheaper rate than he could afford to manufacture salt on Lake Winnipego-sis. He has made salt at Swan River and Duck River. The manufacture is now carried on with profit for the Hudson's Bay Company, at Swan River, and at Winnipego-sis Lake by Monkman's sons.

At the "Works" there are two small log-houses and three evaporating furnaces. The kettles, of English construction, are well-made rectangular vessels of iron, five feet long, two feet broad, and one foot deep. They are laid upon two rough stone walls, about twenty inches apart, which form the furnace. At one extremity is a low chimney. The whole construction is of the rudest description; and at the close of the season the kettles are removed, turned over, and the furnace permitted to go to ruin, to be rebuilt in the following spring.

The process of making salt is as follows: When a spring is found, a well, five feet broad and five feet deep, is excavated, and near to it an evaporating furnace erected. The brine from the wells is ladled into the kettles, and the salt scooped out as it forms, and allowed to remain for a short time to drain, before it is packed in birch bark roggins for transportation to Red River, where it commands twelve shillings sterling a bushel, or one hundred weight of flour, or a corresponding quantity of fish, pemican, or buffalo meat, according to circumstances.

The brine is very strong. From one kettle two bushels of salt can be made in one day in dry weather. There are nine kettles at the "Works," seven being in constant use during the summer season. The Half-breeds engaged in the manufacture complained of the want of fuel—in other words, of the labour and trouble of cutting down the spruce and poplar near at hand, and the difficulty of hauling it to the furnaces. An objection of no moment, but characteristic of some of the people, who are generally unaccustomed to long-continued manual labour. Unfortunately, I had no instrument with me for ascertaining the specific gravity of the brine, and a supply which I took to Red River for that purpose, as well as with a view to its analysis, still remains in the settlements. It will be seen that the processes employed in the manufacture of salt are of the rudest description, so that without any outlay beyond a few days' labour, the quantity might be largely increased. I spoke to John Monkman, who now makes salt here, of pumps and solar evaporation. Of a pump he knew absolutely nothing. He had heard that such an apparatus had been contrived, but had never seen one. He readily comprehended the advantage to be derived from pumping the water into shallow troughs, dug in the retentive clay near the springs, and strengthening the brine by solar evaporation. An Indian guide, who accompanied us up the Moss River, assured me that all along the west coast of Winnipego-sis and Manitobah Lakes, there are salt lagoons and springs. The Indians we met on the Dauphin Lake make the same acknowledgment, but declined to give precise information, alleging that the manufacture of salt drove away the game, and spoil their hunting.

The extent, character, and importance of the Salt Region in Rupert's Land, will be discussed at length in another chapter.

October 6th.—Left the Salt Springs, and sailed before a stiff breeze to the mouth of Moss River. We found four feet of water on the bar, and nine feet at the mouth of the river. A low exposure of limestone occurs near the entrance, and another one mile and a half up the stream. The dip is very irregular. The fossils are few in number, and obscure. In its lithological aspect, it resembles the exposure on Snake Island, seven miles distant, in a north-east direction. The rock is curved, and fractured in places, showing in an exposure 120 yards long and nine feet high, inclinations varying from 20° to 40° east, with short horizontal intervals. Some of the layers are extremely hard, others fissile, others crystalline, with crystals of calc spar between the layers and in the fractures.

We arrived at the first rapids on Moss River during the afternoon. They have a fall of two feet, and consist of an accumulation of boulders resting on rock. The second rapids are formed by similar obstructions. The river is here 120 feet broad, and very shallow. The bank, ten feet above the water, sustains fine aspens, with a very thick undergrowth. The soil is clay, and evidently fertile near the river, but in the rear, the country passes into muskeg. In ascending the second rapids, the boat had to be lightened, and hauled up by the men walking in the middle of the stream. The temperature for such work was not conducive to comfort or health, and two of the men caught severe colds, with cramps and pain in their limbs.

October 7th.—A sharp frost during the night. Ice formed on the oars in the morning. Temperature of air, at eight, a.m., 30°; of the river, 42°. The thermometer, during the night, fell to 26°. All the leaves are now fallen, and the country presents a very dreary appearance. The whole of the day was spent in rowing or tracking up Moss River. The bank continues from twelve to fifteen feet high, and sustains some very fine aspens, twelve to fifteen inches through, with a dense growth of young trees springing up in the place of a former fine aspen forest, of which the large trees are the remains. The river continues very shallow, and contains many boulders of the unfossiliferous rocks.

There is a large area of good land on the west side, near to Dauphin Lake, which secluded sheet of water we entered at four, p.m., and came at once in sight of the Riding Mountain in front and the Duck Mountain on our right. Both are very imposing ranges from this point of view, presenting similar aspects; both rising from a level country, a few feet above Dauphin Lake; and, as far as the eye could judge, both maintaining the same elevation, and presenting abrupt wooded escarpments towards the east. They are separated by Valley River, and it is apparent that they were, at one epoch, united, forming part of the great table-land which stretches south-westwards, toward the Grand Coteau de Missouri.—(*See Section along the 51st parallel, and in the direction of the dotted line on the Geological Map.*)

In the evening we passed the mouth of the Valley, or Dauphin River, or *Te-wa-te-now-seebe*, and camped on a beautiful beach, with a few fine oaks growing upon a ridge close to the water's edge.

On the following morning we started at daylight for a part of the coast nearest to the Riding Mountain. At a distance, this magnificent range appears to be clothed with forest, and to rise from a level plain to the height of about 800 feet above the level of Dauphin Lake.

As soon as we arrived opposite to what appeared to be the highest part of the range, we landed, and despatched an Indian to explore the country, and report on the nature of the swamps we should have to pass in attempting to reach the summit. During his absence we set nets, and levelled across the ridge which separates the lake from an extensive fertile meadow which lies between it and the mountain. The ridge is 8·96 feet above the present level of Dauphin Lake; the meadow, 5·70 feet; and so continues for a distance of one mile, with an almost imperceptible rise until a second low ridge is reached. The meadow is covered with long, luxuriant grass; a few clumps of poplar and thickets of willow vary its uniformity. There are, no doubt, many thousand acres of excellent hay ground on the banks of Dauphin Lake, but the breadth of the tract did not appear to exceed two miles. The oak, on the ridges, occurs in patches, and the trees are from twelve to fifteen inches in diameter.

Dauphin Lake is twenty-one miles long, has a greatest breadth of twelve miles, and an area of one hundred and seventy square miles. Its approximate elevation above the level of the sea is 700 feet, or 72 feet above Lake Winnipeg. It is very shallow.

Our nets produced five splendid pike, weighing about fifteen pounds each.

At the close of the day the Indian returned. He had advanced to the first great ridge, about nine miles distant, and reported eighteen inches of water in the swamps, with ice a quarter of an inch thick.

Snow-birds were seen for the first time during the afternoon. They came about our camp in large flocks, but they did not appear to have quite assumed their winter dress. The evening and part of the night were spent in making arrangements for an ascent of the Riding Mountain. We took provisions for four days, a blanket for each man, with a good supply of guns and ammunition. Three of the men were left in charge of the boat, with instructions to cure all the fish they could take, as the danger of being arrested by ice in Lake Manitobah was not improbable. That large body of water has been known to freeze as early as the 25th of October.

As sunrise on the morning of the 9th of October, we set out for the ascent of the Riding Mountain. Each man had a pack weighing about thirty pounds. My share consisted of a sledge-hammer for geological purposes, which proved to be an extremely inconvenient article to carry across swamps or through bushes. Once, indeed, when crossing a quaking-bog, with the hammer on my shoulder, I received a severe blow on the back of the head as I broke through the covering of moss over which we were pushing our steps, and endeavoured to fall flat on the springy surface.

For the first mile and a half the country is quite level and dry, with the exception of three narrow shallow marshes. The soil is excellent, and the hay abundant; but no doubt in spring this extensive flat must be very wet, and, probably, to a considerable extent under water. In its present condition, the pasturage it affords is very luxuriant and abundant. We soon arrived at a low ridge which marks the limit of the good land, not averaging more than two miles from Dauphin Lake. To the ridge succeeded marshes and willow brakes. These were bounded by low gravelly ridges, clothed with aspen, which were again succeeded by marshes.

Finding it quite impossible to outflank the marshes, which appeared to stretch from river to river descending from the mountains, and to be co-extensive with the shores of the lake, we determined to push through to the highest peak, which was in reality the nearest point of the mountain to us—its greater altitude being only apparent on account of its proximity, as we afterwards ascertained. In an hour we arrived at a white spruce swamp, in which many fine trees, fully eighteen inches in diameter, were observed. Beyond the white spruce swamp we came to an old lake ridge, about fifteen feet above the general level, rounded, and composed of limestone gravel, with many boulders of the unfossiliferous rocks on the south, or land side.

This ridge resembled the Big Ridge of the Assiniboine in most particulars. Our Indian guide told us that it extended for many days' journey north and south of Dauphin Lake. It forms the Indian pitching track, at the foot of the Riding Mountain.

The term "pitching track" is applied to an Indian trail from one part of the country to another. West of Lake Manitobah, Dauphin Lake, and Winnipego-sis Lake, the "pitching track" follows the ridge described in the text. It is, in fact, the main and only dry road in this region. On the crest of the ridge there is a narrow well-worn path, which, for many generations probably, has bëen the highway of the Indians passing from Lake Manitobah to the Assiniboine, through the valley of *Te-wa-te-now-seebe*, or "The River that divides the hills." This pitching track is marked on the map as "Scrub Oak and Aspen Ridge." It is connected with "The Ridge pitching track," between Ebb and Flow Lake and Dauphin Lake.

The Whisky Jack is numerous on the Scrub Oak Ridge, and in the Spruce Swamp were several ravens. Formidable marshes succeed the ridge. We waded knee deep for half a mile, and then rested for a short time on a small island, on which stunted aspens grew. We continued to pass through marshes, aspen islands, and over low ridges clothed with willow, until a bog of such a quaking character, and of such great breadth, presented itself, that the men demanded a smoke, before attempting to cross. Our Ojibway half-breed, Wigwam, insisted upon carrying the sledge hammer in addition to his pack, declaring that he was accustomed to quaking bogs, and we should have enough to do to get across, without carrying anything that might impede our progress. The surface of the bog consisted of a thick elastic covering of moss, sufficiently tough to bear our weight when passing quickly over it, but if we stopped for more than half a minute, the moss slowly sank, and a pool of water collected around us. We marched or rather trotted in single file, about ten yards apart. The Indian who took the lead, passed nimbly over the surface, so also did the half-breeds; Mr. Fleming and I, however, two or three times broke through with one leg, but succeeded in withdrawing the unfortunate member without further damage than immersion in water and mud, emitting a very unpleasant odour. Occasionally, we rested on a narrow strip where the tangled roots of a few willows afforded a firm footing. The breadth of this bog was about one mile where we crossed ; it was succeeded by a belt of tall reeds, growing in water one foot deep ; plunging and wading through this, we arrived at a gently sloping ridge, about eighteen feet in altitude. On the other side of this ridge, a narrow deep swamp separated us from the foot of the mountain ; wading through it, we ascended a hill about forty feet high, and found ourselves upon a dry plateau, on which we determined to camp, having accomplished a distance of about eleven miles. The men soon dispersed to hunt rabbits ; a dozen were killed in a few minutes, skinned and placed on sticks before the fire to roast.

In passing through the swamps, we saw many fresh Moose tracks,—hunters' signs, which aroused the Indians and half-breeds to a high pitch of excitement, and caused them to steal swiftly on their wet and treacherous path with a speed which we found it very difficult to maintain, although they entertained no hope of coming within shot of such noble game, in consequence of an unfavourable wind, even if we had been able to preserve the necessary silence in passing through the haunts of this wary animal.

The timber on the hill at the foot of the mountain consists of aspen, with a few small oak. The soil on the plateau is of excellent quality and the underbrush very luxuriant.

The night promising to be very cold, ice forming on the kettles within a few yards of the camp, we built two large fires and slept between them, having previously dried our wet clothes as far as circumstances would permit. At 8 p. m., the sky was quite free from clouds ; the comet shone with brilliant lustre, a flashing aurora gradually spread over the northern sky, the stars shone like diamonds in the south, and the whole heavens assumed that aspect of silent beauty which renders night in the wilderness so impressive and sublime.

October 10th.—Soon after breakfast, we arrived at a steep embankment about 70 feet high, which formed the termination of a plateau about a mile broad, covered with small aspens, and threaded with Moose paths. The plateau ascends very gradually and is abruptly bounded by a hill bank, from which a broken hilly tract rises towards the escarpment, which forms the eastern limit of the Riding Mountain. This broken tract is covered with aspens and spruce of large size, especially in the hollows. We crossed the beds of two or three streams, which flowed through deep gullies to the plain below. So far, the soil consisted of drift clay with many large boulders in the beds of the rivulets; but at an altitude of about 400 feet above Dauphin Lake we arrived at a cliff-like exposure of Cretaceous rocks, through which a stream had cut a channel 70 to 90 feet deep. These rocks seemed to form the boundary of a third plateau, on which were numerous conical hills consisting of gravel and boulders of the unfossiliferous rocks. The stratification appeared to be nearly horizontal, with a very slight dip to the south-west. Although a careful search was made for organic remains, very few were discovered. These were identical with those found on the Little Souris, and in every particular, except the occurrence of bands holding *Inoceramus*, the rocks on the Riding Mountain resembled the exposures on the Little Souris. The layers containing feruginous concretions were found, as well as a soft thin band from which the Indians make their pipes. The total thickness of the exposure exceeded 100 feet.

We now followed a Moose path until we arrived at a high conical hill, which promised a fair view of the surrounding country. Having reached the summit, the relation of the conical hills and plateaux became evident. A wide deep valley separated us from the table-land of the Riding Mountain, about one mile distant in an air line, and perhaps 200 feet above us. Three plateaux were distinctly visible below us; a range of conical hills, the result of atmospheric agencies, lay at the foot of the precipitous escarpment of the Mountain, and followed its general direction. Limestone and unfossiliferous boulders were strewn on the summits and flanks of the weather-worn hills, while in the hollows between them, small lakes lay half concealed by a fine forest of white spruce and aspens. From the brow of the hill where we stopped to dine, the Indian shot a large brown-nosed bear, which suddenly appeared on the plateau below us about 70 yards from our camp fire. His skin was in good condition, and remarkably handsome; the animal might weigh 350 lbs., although not yet fat. Leaving three men to cut up and prepare the meat, we commenced the last ascent, and arrived at the summit of the Riding Mountain at three in the afternoon. The last ascent was very abrqpt; it consisted of a steep escarpment of drift clay with boulders, covered with a fine white spruce, birch, and aspen forest. At the foot of the escarpment were ponds, or small lakes, which fed the mountain streams we had crossed.

The view from the summit was superb, enabling the eye to take in the whole of Dauphin Lake and the intervening country, together with part of Winnipego-sis Lake. The outline of the Duck Mountain rose clear and blue in the north-east, and from our point of view the Riding and Duck Mountains appeared continuous, and preserved a uniform, bold, precipitous outline, rising abruptly from a level country lying from 800 to 1,000 feet below them. The swamps through which we had passed, were mapped in narrow strips far below; they showed by their connection with the ridges, and their parallelism to Dauphin Lake, that they had been formed by its retreating waters. The ancient beach before mentioned, as extending far to the north and south, could be traced with a glass, by the trees it sustained, until lost in the distance; it followed the contour of the lake, whose form was again determined by the escarpment of the Riding Mountain. It required no effort of the imagination to recall the time when the whole of the flat country below us, towards the Laurentides on the east side of Lake Winnipeg, was occupied with the continuation of the Riding and Duck Mountain ranges, and when the Cretaceous series, superimposed in patches by Tertiary rocks, extended to the basin of Lake Winnipeg as it now is. The whole of this immense denuded tract of country, is a splendid instance of the power of water and ice to remove many thousand cubic miles of rock.

It seems very probable that before the Boulder Drift period, the chain of mountains beginning with Turtle Mountain, near the 49th parallel, and terminating with the Porcupine and Basquia Ranges, including the Riding and Duck Mountains, were part of a high table land, composed of Cretaceous and Tertiary rocks, which extended from the Grand Coteau de Missouri to the Laurentides. The areas most affected by denudation are now occupied by Lakes Winnipeg, Manitobah, Winnipegosis, and the vallies of their tributaries; the precipitous eastern escarpment of the Mountains shew the action of oceanic agencies to which they would be directly exposed, if the country were submerged to more than 1,000 feet, and from the distribution of boulders, there can be no question that a submergence to a far greater extent has taken place since the Tertiary epoch. The connection of these ranges will be best seen by an inspection of the map.

The occurrence of drift and boulders of the unfossiliferous rocks on the summit of the Riding Mountain, proves that this portion of the country was submerged to an extent exceeding 1,700 feet, that being the average altitude of the range above the ocean.

The summit of the Riding Mountain is a vast table land declining in steps to the Assiniboine. The forest which covers the upper plateau, consists of very fine white spruce, birch, poplar, and aspen; the dimensions of some of the trees about our camp are given on page 14. Soon after our arrival at the summit, clouds began to gather in from the north-west, and towards evening, a snow storm set in, which continued during the greater portion of the night. Beneath the shelter of the overhanging branches of a spruce we made an excellent camp, and having built a roaring fire, soon found ourselves comfortably supping on bear steaks as we reclined on a couch of spruce boughs, under a roof impenetrable to snow, constructed of the same excellent material.

October 11th.—When morning dawned we found the country covered with a mantle of snow, six inches deep. This did not prevent us from making a traverse in the direction of the lakes from which the Rapid River takes its rise. The course we had taken led us, as was afterwards ascertained, to within a few miles of the spot reached by Mr. Dickinson when he ascended the valley of Rapid River, a few weeks before. This was precisely the result I was anxious to attain. An inspection of the map will show that our explorations when combined, passed through a comparatively unknown country, nearly along the 100th degree of longitude west of Greenwich, and stretching from the 52nd to the 49th parallel of latitude; thus embracing part of Winnipego-sis Lake, Moss River, Dauphin Lake, the Riding Mountain, the Little Saskatchewan or Rapid River, and the Little Souris, to the 49th parallel.

Our progress to the south was soon arrested by a lake, and the lateness of the season made it advisable not to linger too long in this region, lest we should be arrested by ice forming in the great lakes below. Anxious to kill a Moose, I endeavoured to persuade the Indian to follow a fresh track, but he declared that the Mountain was full of devils, and that the grizzly bear was not unfrequently met with, so that no persuasion could induce him to follow the track unless a half-breed accompanied him. The sky and air becoming quite free from clouds and mist before 10 a. m., we were enabled to take bearings of different prominent points. After measuring a number of trees in the neighbourhood of our camp, we commenced to retrace our steps at noon. The course followed was a little to the left of our track on the preceding day, and the following rough estimate of the ascents, descents and distances were carefully noted.

The first descent from the summit is about 250 feet deep, and very precipitous; where the snow had not lodged, boulders were seen reposing on unstratified clay and gravel; a narrow gully is then crossed, and an ascent of forty feet made to a terrace sloping towards the east; on this terrace are the conical hills, before alluded to. The descent continues for a further depth of 150 feet in a distance of half a mile, this brought us to the edge of a ravine seventy feet deep. At the bottom flows a small stream over gravel and boulders. A rise of thirty feet, led us to the top of the opposite bank, along which we travelled, until we came to its termination at the beginning of a second terrace about eighty feet below us. This narrow table land is consequently 480 feet below the summit, and on it we found the second range of conical hills. A gradual descent for a quarter of a mile lowered us about thirty feet; we then ascended a bank about twenty feet high, and found ourselves on the edge of a precipitous descent, 150 feet deep, which brought us to the third terrace, and to the edge of a ravine sixty feet deep. On the sides of the ravine and far above it, exposures of Cretaceous rocks were seen, the highest spot where the rock was observed, in position, is probably between four and five hundred feet below the summit, or about 500 above Dauphin Lake. A sudden descent of 120 feet then occurred, which brought us to a fourth terrace, bounded by a steep bank, to which succeeded a gentle slope, and then a low ridge, where we had formed our camp on the 9th. We arrived there wet, cold, and uncomfortable; the temperature was much higher than on the mountain, and during the day the snow of the previous night had entirely disappeared as we descended, but a drenching rain instead, promised a very disagreeable night, as we could find no friendly spruce near at hand to afford shelter and protection.

Tuesday, October 12th.—The greater part of this day was spent in retracing our steps to Dauphin Lake. The walk through the marshes and bogs was found to be more fatiguing than during our ascent, in consequence of rain and the ice-cold water in the swamps. Two of the men complained of rheumatic pains, and were incapable of doing any work upon their arrival at the camp in the afternoon. During our absence, the men left at Dauphin Lake had set the nets, and caught some fine pike. The precipitation which had occurred on the Riding Mountain in the form of snow, was here a drizzling rain, which again commenced, soon after our arrival, and continued throughout the night. On the following morning one of our best half-breeds was seriously ill, he complained of excruciating pains in the head and limbs, he found, however, great relief from cupping, which the Indian performed with a flint and bowl of a tobacco pipe. At noon, we started in the boat for an Indian encampment at the west end of the lake, about six miles distant. Here we found Ta-wa-pit, an old Ojibway, with two sons, and their wives and children. Having made arrangements with Ta-wa-pit for the hire of two horses and a guide to cross the country to Manitobah House, Mr. Fleming took charge of the boat, to return by Moss River, while I remained with one half-breed to make the land journey round the south side of Dauphin Lake to the Company's post on Lake Manitobah, which was to be our rendezvous.

October 14th.—Ta-wa-pit stayed during the greater part of the night by our camp-fire, talking with the half-breed, smoking and drinking tea. He pointed out the spot near to us, where he was accustomed to take salt from the edges of a spring during the summer months. He described also at length the appearance and virtues of some gigantic bones exposed in the bank of Valley River near where it cuts through the old Lake Ridge. Ta-wa-pit calls these bones a great medicine, and, contrary to the usual custom of the Indians, he now and then takes small fragments, bruises them to powder, and uses them as a medicinal preparation. From his description I infer that the bones are those of a mammoth; his rough drawing in the sand of the ribs and teeth, corresponded in point of dimensions, with those of that gigantic animal.

Ta-wa-pit and family live a very retired life on the shores of Dauphin Lake. The old man is evidently of a misanthropic turn of mind; he does not associate with other Indians who hunt and live on Moss River and the northern part of the lake. His potatoes, of which he planted a small patch in the spring, were completely destroyed by grasshoppers; affording another proof of the immense range and devastating progress of these insects in Rupert's Land, during the past two or three years. Ta-wa-pit showed me a knife he had made out of an old file, and some pipes he was making from a soft shale, procured in the Riding Mountain some miles south-west of his tent. The shale was similar in all respects to a band I had noticed on the little Souris, and in the exposure just described as occurring on the flank of the Mountain, and from which the half-breeds had taken small blocks to make pipes. A couple of pounds of buck shot which I divided among the old man and his sons, delighted them beyond measure; in return for this welcome present, Ta-wa-pit presented me with a new pipe, and the moufle of a Moose.

The hop grows in great luxuriance and abundance at the south end of Dauphin Lake; there is fine pasturage as far as the old lake ridge, but the narrow strips of marsh and quaking bog almost on a level with the dry portion show that these extensive flats are liable to be submerged in the spring.

Our course to-day followed for a few miles the shore of the lake until we came to Turtle River; having crossed this affluent from the Riding Mountain in a small canoe we took an easterly direction and entered a dreary region of swamp, ridge and quaking bog. During the whole of the afternoon our course lay through marshes and bogs, separated by low ridges covered with aspen. The horses were quite useless, and frequently stuck fast; when this occurred we were compelled to carry the bedding and provisions to the nearest ridge and help the wretched animals through the deep bogs into which they sank at every step, breaking through the elastic covering of moss which was generally of sufficient strength to support a man running lightly over it, but not tenacious enough to bear the weight of a horse. Just as night closed in we arrived at a dry gravelly ridge where there was a plentiful supply of dead aspen, from which we made a roaring fire and soon dried our wet clothes and blankets. The night was bitterly cold and the exertion of wading for many hours together through ice-cold water caused every limb to ache; the Indian guide thought nothing of it, and immediately after supper lay down before the fire and was soon sound asleep. Two or three times in the night I rose to replenish the fire and found the Indian without any covering but the wet skin clothes he had worn during the day, curled up on the bare ground and enjoying profound slumber.

Early on the following morning we arrived at the Ridge Pitching track, which we continued to follow for a few miles, and then again descended into a region of swamps and quaking bogs. In no respect does the Ridge Pitching track between Dauphin Lake and Ebb and Flow Lake differ from the Big Ridge of the Assiniboine except in altitude. It is about one hundred yards across, evenly rounded, composed of gravel and covered to a great extent with the bearberry. On either side are small oaks and aspens, succeeded by marshes. Its altitude above the marsh is about fifteen feet. The guide said it formed an extension of the ridge on White Mud River described in chapter VII, and if this be the case no better means of communication by land with this part of the country could be found than the Ridge Pitching track.

Soon after leaving this excellent road we stuck fast in a quaking bog about one mile broad. The horses were mired, and it was only by dint of the greatest exertion and much cruel beating that the Indian and half-breed succeeded in getting them on to dry land. In the afternoon we arrived at Crow Creek, and the country becoming drier we were enabled to make better progress. After passing Sucker Creek, which, with the streamlet before named, flows sluggishly in a trench about 10 feet deep, we arrived at a small open prairie surrounded with tall aspen woods and covered with a splendid crop of wild hay. Here we met an Indian who was setting traps, the hunting season having already commenced. He invited us to his tent, which was placed on the shores of Ebb and Flow Lake, not more than twelve or fourteen miles from Manitobah House. It turned out that the half-breed with me knew the Indian well by reputation; he is one of the most successful and industrious hunters in this part of the lake region; his tent was well supplied with Indian luxuries, such as tea, tobacco and coarse clothing. In the small prairies near us were several fine buffalo runners, and if Indian habits and customs would permit of the accumulation of wealth, our host might soon become a rich man.

His tent was of birch bark, roomy and clean. Thirteen persons including children squatted round the fire in the centre. On the floor some excellent matting was laid upon spruce boughs for the strangers; the squaws squatted on the bare ground, the father of the family on an old buffalo robe. Attached to the poles of the tent were a gun, bows and arrows, a spear, and some mink skins. Suspended on cross pieces over the fire were fishing nets and floats, clothes, and a bunch of the bearberry to mix with tobacco for the manufacture of kinni-kinnik.

Soon after we entered the squaw began to prepare supper, which was done by boiling white fish and potatoes together; when cooked the whole was poured into a large tin dish and handed to me, together with a cup of tea. Helping myself I passed the dish to the Indian, but he laid it at his feet; as soon as I had finished my supper, the Indian helped himself and the half-breed, and then passed the dish to his squaw, who divided the remainder among the other inmates of the tent. These consisted of an old, watchful, restless Indian woman, the mother of the mistress of the tent; a newly married couple related to our host; the Indian guide from Dauphin Lake, and five children. After supper I spread my blanket and lay down, quite overcome with the long continued exertion of wading through swamps and quaking bogs, but too tired to sleep. The half-breed and Indians sat talking for many hours before they turned their feet to the fire, rolled themselves up in a blanket and seemingly at once "found sleep."

On the following morning I rose with a few aches and pains, which the succeeding events of the morning rapidly dispelled. After breakfast my Indian host offered me a favourite buffalo runner to ride to Manitobah House. The exertion required to manage this animal soon removed all unpleasant rheumatic symptoms. Her extraordinary sagacity is elsewhere related (Chapter XII), but her mouth was evidently formed of sole leather and not amenable to persuasions administered through an Indian bridle. The country on the shore of Ebb and Flow Lake is low but well fitted for a limited settlement. There is an abundant supply of aspen timber with a few oak and birch. I arrived at Manitobah House soon after noon, and was cordially received and hospitably entertained by Mr. Mackenzie the gentleman in charge.

CHAPTER XI.

MANITOBAH HOUSE TO MANITOBAH ISLAND—MANITOBAH ISLAND TO OAK POINT—OAK POINT TO THE SETTLEMENTS ON RED RIVER.

Mr. and Mrs. Mackenzie—Manitobah House—Messenger—Missionary privations—Want of supplies—Communication with St. Paul—Future supplies more constant—Snow storm—Indian Summer—Snow birds—Manitobah House—Rock—John Campbell—Whitefish—Importance of—Aspect of country—The Narrows—Manitobah Island—Dimensions of—Oak — Rock formation — Fossils — Indian superstitions—Fairies—Signals—Arrival of Boat at Manitobah Island—Coast of Lake Manitobah—Old Mission Station—Unfitness of this part of the coast of the Lake for Settlement—Indian liberality—Monkman's Point—Cause of the formation of Marshes—H. B. Co. Breeding establishment—Oak Point—Dimensions of Lake Manitobah — Prairie bordering the Lake—Shoal Lake—Character of the country—Big Ridge—Little Ridge—Arrive at the Settlements.

I remained one week at Manitobah House, waiting for Mr. Fleming, who was detained by contrary winds. To Mr. and Mrs. Mackenzie I am indebted for much generous hospitality, and have great pleasure in possessing this opportunity of acknowledging their kindness and the endeavour they made to give me all the assistance and information in their power. Manitobah House is in a very dilapidated condition, but Mr. Mackenzie has erected another dwelling, which was nearly completed during my stay. A Roman Catholic Mission formerly existed here, but having been abandoned, the buildings were sold to the H. B. Co., and in the year following the transfer, they were accidentally destroyed by fire.

On Saturday, the 16th October, a messenger arrived from Fairford on his way to Oak Point, whither he was journeying for the supplies of the Mission in charge of the Rev. Mr. Stagg. Those who have not experienced the privations resulting to Missionaries in distant out-posts from the non-arrival of their supplies by the customary route and at the expected season, can form but a feeble conception of the troubles and anxieties which chequer the life of a zealous Missionary in Rupert's Land. It is not mere personal inconvenience which causes him care and embarrassment; it is the impossibility of taking advantage of many opportunities for inducing wandering Indians to settle around the Mission, of clothing and feeding the children entrusted to his charge, and of securing, by aid judiciously applied, the respect and affection of those he is endeavouring to christianize or educate, or seeking to draw from their faith in strange and imaginary gods.

The Indian generally, from his habits and precarious mode of subsistence, requires something tangible in the first instance to arrest his attention, and practical encouragement, often repeated, to secure his good-will, before an impression can be made on his heart. If the Missionary is cut off from his supplies in the infancy of a Mission much of his work has to be done over again. Indian wants are few and simple, but they must be supplied without fail at new stations; hence the importance, if success is to be secured, of effecting and sustaining a tolerably regular communication once or twice a year with the Settlements at Red River.

Mr. Stagg has suffered much inconvenience from being disappointed in obtaining supplies of clothing and other indispensable articles for the children and adults, now Christian members of his Mission, and the messenger who arrived at Manitobah House was despatched at his expense to bring the necessaries which had been brought from York Factory to Red River, but not forwarded to the Mission at the usual time by the H. B. Co.'s brigade.

It has sometimes happened that this is not convenient or perhaps quite impossible; it is natural to suppose that when, from missing a season or from other causes, the supplies for the service of the different posts of the Company are in arrears, and the brigade of boats can take only a certain quantity of goods, those for the purposes of the trade will first receive attention. It has happened two or three times that one year's supplies for the whole Settlement of many very important necessaries have been unavoidably left at York Factory, causing no little inconvenience and trouble to the settlers as well as the Missionaries. In the Settlements at Red River their wants can be in part supplied from Fort Garry, but at the Missionary out-posts such relief can not be looked for.

Now that communication may be said to be established between Fort Garry and St. Paul by steamboat and stage coach, there will always be an abundant supply of necessaries at the Settlements, which was not the case when the chief means of communication with the outer world lay through York Factory. Opportunities may now be embraced for supplying distant outposts, which did not exist before Fort Abercrombie or the mouth of the Shayenne was connected by steam with Fort Garry.

In the afternoon of this day a snow storm commenced which continued all night, and covered the ground with nine inches of snow. The thermometer was at the freezing point, but Mr. Mackenzie stated his conviction that the 'Indian Summer' not having yet occurred, the snow would soon disappear and we might have fine weather for ten days or a fortnight; a prediction borne out by the rapid disappearance of the snow on the following day, and the occurrence of beautiful weather with frosty nights to near the end of October.

On Monday, the 18th, snow birds were flying about the Post in large flocks; ducks wending their way to the south, and all appearances and signs of approaching winter rapidly following one another.

Tuesday and Wednesday were occupied in writing letters and making up my journal. On Thursday, the 21st, the boat not having arrived, I proceeded to examine the surrounding country. The day was warm and fine, with much smoke from the south-west, coming no doubt from the burning prairies.

Manitobah House is very prettily situated near the Narrows of the lake. Immediately before it is a cluster of low islands, on which some fine ash-leaved maple and elm grow; they are the favourite camping grounds of the Indians who hunt and fish in the country about Lake Manitobah. The land in the rear of the House is stony, but good, and there is an area of many thousand acres in extent, well adapted for a settlement. The timber consisting almost altogether of aspen on the main land, is of fair dimensions, trees from twelve inches to fifteen inches in diameter being common. Near the Post, but on the opposite side of the lake, there is a considerable quantity of balsam, spruce and tamarac. There are no rock exposures visible near the Post, but in making an excavation for a cellar under the New House, the workmen came upon limestone rock, four feet below the surface. It was apparently horizontal, but in the fragments procured no organic remains were visible; its lithological aspect was similar to the rock on Manitobah Island, to be hereafter described. When the surface of the exposed rock was cleaned with a bucket or two of water, well preserved ice groves were visible, their direction was N. 10° W.—S. 10° E.

I visited the house of a freeman named John Campbell a few hundred yards south of the Post, and found there two comfortable log shanties, a potatoe field, two or three haystacks and some cattle. Campbell's son informed me that it was much easier to live here than at the Settlements. Some of his cattle were permitted to remain in the woods and swamps all winter, but they became very poor towards spring. White fish are abundant; the fishing season having already begun, Campbell had caught 500 white fish, but he wanted 4,000 for his winter supply. As soon as the fish are caught in the gill nets, and brought to shore, a slit is made above the tail, through which a pointed stick is pushed. Ten fish are placed on each stick, and the sticks are staged in the open air about nine feet from the ground, beyond the reach of dogs. No curing, cleaning, or any preservative process is employed, the dry air and frost preserve them until they are needed. The importance of the white fish in this region may be imagined when it is known, that not only does it form the chief food of the Indians in the lake region for a great portion of the year, but *three white-fish* per diem, constitute the sole daily allowance of the half-breeds attached to this Post; absolutely nothing more. Flour, tea, sugar, &c., are luxuries, which, if they wish to indulge in, must be purchased at high prices, nevertheless, they are healthy, happy, and according to their notions comfortable.

The white fish I saw staged at Campbell's might average three to four pounds each. They are considered to be superior to those caught in Lake Winnipeg. This important source of food in these regions is well named At-ik-um-aig, or the Reindeer of the Water, by the Ojibways. It forms a principal article of diet during a large portion of the year, not only of the Indians, but also of the settlers at Red River. The price the frozen fish fetch in the Settlement is five for a shilling, or 100 for a pound sterling. During our winter journey to Canada, we purchased them at Red Lake, in Minnesota State, at the rate of $6 (£1 9s.) per hundred, to feed the dogs; each dog was allowed one white-fish, and a morsel of pemican, as long as the pemican lasted.

Indian summer began to-day, October 21st. The weather is warm, smoky, but very delightful. No boat being yet within sight, I visited the Islands opposite Manitobah House, the marshes at the mouth of Ebb and Flow Lake, and the country in the rear of the Post. Its extraordinary flatness is shown by the great expanse of marsh about the islands, and along the coast north of the Hudson's Bay Company's Post. The level of the lake was three feet below high water mark, and about two feet above the lowest point to which it has been known to fall for many years. The boat not arriving on the evening of the 22nd, I determined to take a small supply of provisions and go with Whiteway the half-breed, who had accompanied me from Dauphin Lake, as far as Manitobah Island, about twelve miles in a direction due north, and there await its arrival. This part of Lake Manitobah is not more than from three to four miles across, studded with low islands, and on the east side the coast is indented with deep bays. The straight is shallow, twenty-one feet of water close to the Narrows, being the greatest depth recorded.

Manitobah Island, from which the lake derives its name, is about 600 yards long, and 200 yards broad; on its north side, there is a perpendicular limestone cliff fifteen feet high; a few yards from its edge, a well defined ancient lake beach crosses the island, resembling in most particulars the Ridge Pitching track, or the Big Ridge of the Assiniboine. The part that remains in a good state of preservation is not more than 150 yards long, the breadth of the Island being here about 220 yards. From this ancient beach, the land slopes gradually in a southerly direction to the present beach, with its fringe of rushes at the south extremity of the Island. The timber consists of oak and birch; many of the first named tree have been cut by the people of Fairford and Manitobah House.

The native carpenter employed to build Mr. Mackenzie's new residence accompanied us to the Island, and although very anxious to make the traverse across the lake after passing the Narrows, he remained for a few hours to cut a couple of oak logs which he proposed to take with him to Fairford, to mend the old freighter's boat which had formed his pay for six week's labor. He embraced this opportunity, on account of the difficulty of procuring oak timber near the Mission. Although oak was seen several times on the shores of Lake Manitobah, north of the Narrows, yet nowhere was it found of such serviceable thickness (fifteen inches) and length, as on Manitobah Island. In the rear of the marshes which border the lake it is known to exist in small quantities.

Among the Devonian fossils procured on the island were *Atrypa reticularis: Atrypa aspera*, two species of *Chonetes*, a small *Productus*, an *Orthoceras*, and fragments of a large fish. (Mr. Billings.)

I remained on this island with Whiteway for three days; we shot a mink, a few duck, and saw a red fox, but although the island was so small, we found it impossible to kill him. Indians appeared occasionally in their canoes on the northeast coast of the Lake, but although they heard our guns and fired in return, yet they would not venture near us. They have all a great aversion to caves and overhanging rocks, conceiving that such places are the abode of fairies, or Manitou. The origin of this superstition in relation to Manitobah Island is due to the sounds produced by the waves as they beat upon the beach at the foot of the low cliffs at its northern extremity.

During the night time, when a gentle breeze is blowing from the north, the various sounds heard on the island are quite sufficient to strike awe into the minds of superstitious Indians. These sounds frequently resemble the ringing of distant church bells; so close indeed, is this resemblance, that several times during the night I woke with the impression that I was listening to chimes. When the breeze subsided, and the waves played gently on the beach, a low wailing sound would be heard from our camping place, about 300 yards from the cliffs where the noise was produced. At night it was peculiarly impressive, and as we lay on the moss-covered rock, it was very easy to comprehend the objection which uneducated Indians, naturally of a fanciful and superstitious turn of mind, should have to land or remain on this 'fairy' island.

On the night of Monday, the 25th October, we built as usual a large fire on the beach to serve as a beacon light to Mr. Fleming, and at nine lay down to sleep. Whiteway was telling me about the adventure of Sho-Shons, (Long-ears,) whose tent was within a few miles of us, and who was tossed by a buffalo bull during the past summer, when at 10, P. M., three shots were heard, apparently about three miles north of the island. We sprang up and replied with three shots, and proceeded at once to supply the beacon fire with dry wood. Whiteway put his ear to the water's edge, and after a short pause declared that he heard oars. After a few minutes we fired three more shots, and waited the result; in half an hour the boat loomed through the gloom, and before eleven o'clock Mr. Fleming and the crew were on Manitobah Island.

They had been detained by contrary winds, but had plenty of sport, killing prairie hen,* duck and plover in the upper part of the lake, near Crane Bay. At sunset, Mr. Fleming touched a low point a few miles north-east of our island, where a few Indians were encamped; they told him that they had repeatedly heard shots from the Narrows, but did not care to know who had fired them in that quarter. A quiet admission that the terrors of Manitobah Island were sufficient to check the curiosity, even of an Ojibway Indian.

It was past noon on the morning of the 26th, when we reached Manitobah House; we remained there for an hour to partake of the hospitality of Mr. and Mrs. Mackenzie, and procure a supply of white-fish and potatoes. In the afternoon we pulled towards McKay's Point, passed between Sugar Island and Birch Island, both low and marshy areas, and camped at sunset on a circular sandy beach enclosing an extensive marsh, in which duck still remained in considerable numbers. The lake near the coast is shallow, the greatest depth recorded being thirteen feet.

The whole of the coast as far as Swan Creek is very low, and bordered by beaches enclosing marshes. Here and there wooded points ten to twelve feet above the lake level separate the marshes from one another; on one of these points we observed some very fine elm, but the prevailing timber consists of aspen. A mission was established some years since at Elm Point by the Reverend Mr. Cowley, but abandoned soon after. An attempt was made to open a cart track from this Mission to the prairies near Oak Point, but it was thought that the Indians who professed to guide Mr. Cowley through the driest part of the country, took him through the most swampy portion. The Indians now say that dry ridges exist, with few intervening marshes, over and through which a cart track could be established without difficulty; but it is evident that the character of the country on this part of Lake Manitobah is not fitted for farming purposes. Isolated areas like Elm Point are, doubtless to be found, but not sufficiently extensive to give to this region any value in an agricultural point of view.

We met an Indian in a canoe near Elm Point, and Whiteway, at my request, told him we were starving. I wished to ascertain the truth of the statement so often made respecting the liberality of these Indians in cases of necessity. The answer was a happy one; approaching our boat in his canoe the Indian said, "Look, if you see any thing to eat, take it." In his canoe were sixty fine white fish and a few pike. I gave him some potatoes, tobacco, and tea, and accepted a dozen white-fish which he pressed us to take.

The shore continues low as far as Sandy Point; it is bounded by beaches fringed with fine aspen forests in the rear of marshes filled with rushes, which occupy part of every sheltered cove and bay open to the lake. We camped at Monkman's Point, where one of the family has a fishing station. They were catching their winter supply of white fish. Monkman* pointed out a marsh in the rear of our camp which he said was once dry ground and afforded splendid pasturage for horses. It is separated from the lake by a gravelly beach. This probably occurred during a period of low water. A fall in the level of the lake to the extent of two feet would not only drain and dry this marsh, but many thousand acres of marshy tracts formed under similar circumstances, and at the same period. Mr. Mackenzie of Manitobah House, told me that former residents at that Post had seen the lake for a long period of time two feet lower than at present. In fact before the floods of 1852, the lake was at its lowest level, and the swamps and marshes fringing its low north-eastern coast were then dry areas covered with rank grass. In the course of a few years this will again take place, and for a long period, perhaps, settlers may enjoy fine pasture lands, destined again to revert to an intermittent condition of swamp or marsh. Monkman informed me that many years since the Hudson's Bay Company had a breeding establishment near this Point; and he remembered the time when 120 horses were pastured in the neighbourhood of Swan Creek, about twelve miles from Oak Point.

On the 28th we passed through an immense expanse of reeds called Marshy Point, threading our way through an intricate channel in which large numbers of duck still lingered. About one o'clock we arrived at Oak Point, where we found John Monkman and a number of settlers from Red River catching their winter supply of white fish in gill nets.

Lake Manitobah is one hundred and twenty miles long by twenty-four broad in its widest part, from headland to headland, but if estimated from Oak Point to the mouth of White Mud River on the west side, the breadth does not fall far short of thirty miles The area of the lake is about 1900 square miles, and its approximate altitude above the sea six hundred and seventy feet, or forty-two feet above Lake Winnipeg. An inspection of the map will show that in the parts sounded, which were sometimes twelve to fifteen miles broad, the depth never exceeded twenty-three feet; this occurred half way between Cherry Island and Sandy Point in the upper portion of the lake. In the two traverses between Manitobah Island and

* The Prairie Hen or Pinnated Grouse, *Tetrao Cupido*, is not often found so far north as lat. 52° in the wooded country.

* The brother of John Monkman of Oak Point—a celebrated character at Selkirk Settlement—more will be said of this individual in a future chapter.

Cherry Island not more than twenty-one feet was recorded, while within four miles of the coast in the southern or larger portion of the lake, eighteen feet was the greatest depth found. The soundings are shown on the map.

The effects of winds on the large surfaces of water exposed by the great lakes of the Winnipeg Basin, is very remarkably seen at the Narrows, near Manitobah Island, the Dog's Head (Lake Winnipeg), Waterhen River, and the mouths of the Winnipeg and Red Rivers. The currents produced by the pressure of the wind changing the level of the lake has probably exercised an important influence in connecting different parts of the same lake basins.

At the Narrows, Lake Manitobah, a northerly wind will cause a strong current to flow through the straits into the lower or southern half of the lake; while a south wind produces a corresponding effect in the northern portion, and perceptibly increases the volume of water in the Little Saskatchewan. At the Dog's Head the current sometimes approaches the force of a rapid when the wind blows from the north; the great depth of Lake Winnipeg at this point, which I was assured by half-breeds and Indians who fish there during the winter, exceeds one hundred and twenty feet, is doubtless the result.

At first sight it appears strange that the limestone cliffs should not have been gradually broken away and the communication between the upper and lower portion of Lake Winnipeg enlarged. But running water exercises comparatively little effect in excavating a deep channel through a rocky barrier, or in widening a water-course; ice, beyond all question, is the main instrument in abrading, denuding, and excavating. At the Dog's Head the ice has little force on account of the proximity of islands, either when acting with a thrust or bearing away masses of rock frozen to its substance. By far the greater portion of the ice formed on this part of the coast is so protected by the islands as to melt before it can be moved by winds with its rocky burdens to distant parts of the lake.

At Manitobah House I observed the water rise fully eighteen inches before a storm. Canoes left in calm weather on a beach high and dry are not unfrequently washed away when a strong south or north wind sets in; and it often happens that even before the approach of a change in the direction of the wind is indicated by clouds, the water of the lakes show by rising the operation of a distant pressure which has not yet manifested itself at the point of observation. The Indians and half-breeds in the fall of the year, when winds are variable, frequently notice the mouths of streams or rude registers, such as a stone set up by themselves on the beach, to see if any indications are afforded of a change in the wind, not appreciable by any other means.

In 1823 Mr. Keating in his narrative of Major Long's Expedition to the sources of St. Peter's River, described the effects of winds on the waters of Lake Winnipeg taking place at the mouth of the Winnipeg River as follows:—"A question which has been much discussed by travellers, is that of the supposed periodical rises in the lakes; we do not propose to take part in the discussion at present, but we may state that we observed at Fort Alexander an appearance, such as has probably more than once been mistaken for the effect of a tide.

On our arrival we pitched our tents upon a sort of wharf projecting into the river, and elevated about two feet above the level of the water. In the afternoon a very high wind blew from the lake and accumulated the waters in the bay, so as to cause them to overflow the wharf and oblige us to remove our tents. The next morning the waters had subsided to their former level."

The splendid prairies bordering on the southern shores of Lake Manitobah may be said to begin at Oak Point Their boundary is an imaginary line extending south-easterly towards the Indian Settlement on Red River on the one hand, and to the old lake Ridge, where it is cut by White Mud River on the other; a distance in an air line of one hundred and ten miles. North of this line the country is in general marshy, full of reticulating lakes and low aspen covered ridges.

The Settlement at Oak Point contains about a dozen houses, their appearance does not give a stranger a favourable impression of the industry and energy of their occupants. No advantage appears to be taken of the splendid country by which they are surrounded, and with the exception of John Monkman, who at times is a marvel of energy injudiciously directed, they do not seem to have made any progress in improving their dwellings or in enclosing a farm since they first established themselves at Lake Manitobah. About ten miles in a south-westerly direction from Oak Point a number of French Half-breeds have formed a settlement on the shores of the lake. They enjoy the advantage of having a resident Missionary (R. C.) among them.

On the 29th we made preparations for a journey on horseback to the Settlements, striking diagonally across the prairie region just described. The country in the neighbourhood of Oak Point is very attractive; its general level is about ten feet above the lake; it resembles in every respect the region about White Mud River. Our road, for a few miles, lay across a very rich and fertile tract, until an almost imperceptible ascent introduced us to a low gravelly ridge upon which aspen woods grow in narrow strips; the forest preserving a uniform outline as far as the eye could reach, in a direction corresponding to the present form of Lake Maintobah, indicated without glancing at the soil, the direction and extent of the subaqueous ridges, afterwards a low coast line, which were formed over the floor of Lake Manitobah at a higher level. Succeeding this low flat ridge is a broad plateau slightly undulating and studded with straggling clumps of young poplar and small oak, with willows in the shallow depressions. The soil becomes rich in vegetable mould again as we approach Shoal Lake, an extensive sheet of water, shallow, reedy, connected with numerous lakes lying to the north, and a favourite haunt of aquatic birds.

The south shore of Shoal Lake is particularly attractive. Ridges supporting heavy oak fringe the shore, beautiful meadows bordered with aspen and oak woods, reveal themselves in making a short traverse to the south. Although the shores of the lake are marshy yet the oak ridges, some few hundred yards south of it are high and dry. For a grazing establishment on the largest scale, Shoal Lake is admirably fitted. Wild hay in any desirable quantity exists around its marshy shores, and in the beautiful prairies lying south of it timber of excellent quality for building purposes and fuel may be procured in abundance; in the spring and autumn the lake is covered with wild fowl of every variety. Shoal Lake is a favorite sporting ground of the gentlemen of Fort Garry and the half-breeds of the Settlement. It is on the main road to Lake Manitobah, and is probably destined to become a place of some note as a grazing station in the course of time.

On the 30th October, I set out with Whiteway in advance of the carts in the hope of being able to reach the Settlements before nightfall. We passed through an excellent prairie country studded with aspen groves, and occasionaly relieved by a broad shallow ridge probably of subaqueous origin like those already described. The Big Ridge of the Assiniboine is not well defined where we descended it, about eight miles west of Stony Mountain. It appears to be divided into two portions, part expanding into an undulating tract of country a few hundred yards broad, the other preserving the outline and character of the Big Ridge, but named in consequence of its diminished altitude the Little Ridge. The level country at the base of either is everywhere beautiful, fertile, and admirably adapted for settlement. We descended the Little Ridge, a step of the Big Ridge, at about four in the afternoon, and in the distance could see the twin steeples of St. Boniface with their tinned roofs glancing brilliantly in the south-east about 15 miles off. We then passed through the magnificent prairies lying between Stony Mountain and Red River, reaching the edge of the Big Swamp just before sunset, and arrived at our temporary quarters in the Settlement half an hour after dark.

It has been stated in a preceding chapter that the Ridges of Red River and the Assiniboine mark the limits of land of the first quality in these valleys, north of the 49th parallel and east of the Sandy Hills, near Prairie Portage. But it must not be supposed that the country between Oak Point and Stony Mountain is of greatly inferior quality; in many parts no difference in the rank luxuriance of the grass on these prairies and those south of the Big Ridge could be distinguished, but the area of light or gravelly soil, covered with short stunted grass is far greater, and thus diminishes the available extent of soil adapted for agriculture. It is doubtful whether this drawback is not counterbalanced by the proximity of the country north of the Big Ridge to the forest-covered tract between the great lakes, and to the haunts of vast numbers of wild fowl which breed on the borders of the small sheets of water so numerous in this region. On the map this tract, south of the probable limit of the forest has been recorded as a "vast level prairie adapted for agriculture," the groves and strips of aspen and oak only serving to break a vast level expanse into a series of very attractive plains, apparently bounded by forests, which are found as the traveller penetrates them to be but narrow belts separating one beautiful prairie from another.

TABLE SHEWING THE LEADING DIMENSIONS AND APPROXIMATE HEIGHT ABOVE THE SEA OF THE LAKES IN THE GREAT BASIN OF LAKE WINNIPEG.

LAKE WINNIPEG.		
Area	8,500	sq. miles.
Length	280	st. "
Greatest breadth	57	st. "
Length of coast line	930	st. "
Approximate height above the Sea	628	feet.
LAKE MANITOBAH.		
Area	1,900	sq. miles.
Length	122	st. "
Greatest breadth	24	st. "
Height above the Sea	670	feet.
LAKE WINNIPEGO-SIS.		
Area	1,936	sq. miles.
Length	120	st. "
Greatest breadth	27	st. "
Height above the Sea	692	feet.
ST. MARTIN LAKE.		
Area	316	sq. miles.
Length	30	st. "
Greatest breadth	16	st. "
Height above the Sea	655	feet.
CEDAR LAKE.		
Area	312	sq. miles.
Length	30	st. "
Greatest breadth	25	st. "
Height above the Sea	688	feet.
DAUPHIN LAKE.		
Area	170	sq. miles.
Length	21	st. "
Greatest breadth	12	st. "
Height above the Sea	700	feet.
Total water area in the Great Basin of Lake Winnipeg	13,134	sq. miles.

TABLE SHOWING THE AREAS AND ELEVATION ABOVE THE SEA OF THE GREAT CANADIAN LAKES.

Names of Lakes.	Area in Square Miles.	Elevation above the Sea.
Lake Superior	32,000	600
Green Bay	2,000	578
Lake Michigan	22,400	578
Lake Huron	19,200	578
Lake St. Clair	360	570
Lake Erie	9,600	565
Lake Ontario	6,300	232
Total area	91,860	

CHAPTER XII.

INDIAN WEALTH—THE BUFFALO—THE HORSE AND THE DOG.

The Bison or Buffalo—Its value—Two kinds of Buffalo reported to exist by Half-breeds—The Plain Buffalo and the Wood Buffalo—Characters of—Former range of the Buffalo—Existed on the Atlantic Coast—Throughout the United States Territory, not including all the New England States—Modern range of—The Red River bands—The Saskatchewan bands—Wintering quarters of the North-western bands of Buffalo—Summer ranges—Systematic Migration of—Buffalo Hunt—Census of Red River Half-breed Hunt—Blind Buffalo—Crossing of Buffalo with Domesticated Cattle—Character of Mixed Breeds--The Horse—Training of Horses—Docility of—Illustrations—Attachment of Indians to their Horses—Hoppings—Smokes—The Dog—Its uses—The Midnight Howl—Dog Feasts—Dogs at the H. B. Posts—Voracity of—Cross with the Wolf—Sacrifice of Dogs.

The bison or buffalo, the horse and the dog are to Prairie Indians what domesticated animals and the productions of the farm and the forest are to civilized races. During the greater part of the year the Prairie Indians follow the buffalo, and not only subsist upon the flesh of this animal, but from its skin and sinews they make their tents, clothing, saddles, bowstrings and dog harness. The hide cut into strips serves them for cordage, the sinews split into threads for twine. The dried dung is often their only fuel for weeks together in the treeless plains between the Assiniboine and the Grand Coteau, and on the South Branch of the Saskatchewan. Dried meat, pemican, marrow, soft fat, sinews, dressed skins and robes, all from the buffalo, form their articles of commerce, in exchange for which they demand tea, which is now becoming a most coveted luxury, tobacco, powder and shot, and if possible, rum. It may truly be said that they exist on the buffalo, and their knowledge of the habits of this animal is consequently essential to their existence.

That there are two kinds of buffalo appears to be still a matter of doubt; they are stated to be the prairie buffalo and the buffalo of the woods. Many old hunters with whom I have conversed on this subject, aver that the wood buffalo is a distinct species, and although they are not able to offer scientific proofs, yet the difference in size, colour, hair, and horns, are enumerated as the evidence upon which they base their statement. Men from their youth familiar with these animals in the Great Plains, and the varieties which are frequently met with in large herds still cling to this opinion. The plain buffalo are not always of the dark and rich bright brown which forms their characteristic colour. They are frequently seen from white to almost black. A grey buffalo is not at all uncommon. Buffalo emasculated by wolves, the Half-breeds say, are often found in the prairies, they grow to an immense size; the skin of a buffalo ox is recognized by the shortness of the wool and by its large dimensions. The skin of the so-called wood buffalo, of which I saw two at Red River, is much larger than that of the common animal, the hair is very short, mane or hair about the neck, short and soft, and altogether destitute of curl, which is the common feature in the hair or wool of the prairie animal.

The wood buffalo is said to be very scarce, and only found north of the Saskatchewan, and on the flanks of the Rocky Mountains. It never ventures into the open plains; the prairie buffalo on the contrary generally avoids the woods and keeps to the open country, but in winter they are frequently found in the woods of the Little Souris, the Saskatchewan, the Touchwood Hills, and the aspen groves on the Qu'Appelle. There is no doubt that formerly the prairie buffalo ranged through open woods almost as much as he now does through the prairies.

Great Slave Lake is the northern limit of the buffalo, and the country between that large body of water and the Saskatchewan is partially wooded. The buffalo are now found in considerable numbers on the east flank of the Rocky Mountains. The former limits of the wanderings of these animals are carefully recorded in the narrative of Major Long's Expedition, from which the following extracts are taken: "The buffalo was formerly found throughout the whole territory of the United States, with the exception of that part which lies east of Hudson River and Lake Champlain, and of narrow strips of coast on the Atlantic and the Gulf of Mexico. These were swampy, and had probably low thick woods. That it did not exist on the Atlantic coast is rendered probable from the circumstance that all the early writers whom Mr. Colhoun has consulted ou the subject, and they are numerous, do not mention them as existing there, but further back. There can be no doubt that the animal approached the Gulf of Mexico, near the Bay of St. Bernard, for Alvar Nunez, about the year 1535, saw them not far from the coast, and Jontel, one hundred and fifty years afterwards, saw them at the Bay of St. Bernard. It is probable that this bay is the lowest point of latitude at which this animal has been found east of the Rocky Mountains. There can be no doubt of their existence west of these mountains, though Father Venegas does not include them among the animals of California, and although they were not seen west of the mountains by Lewis and Clarke, nor mentioned by Harmon or Mackenzie as existing in New Caledonia, a country of immense extent, which is included detween the Pacific Ocean, the Rocky Mountains, the territory of the United States, and the Russian possessions on the north-west coast of America. Yet its existence at present on the Columbia appears to be well ascertained, and we are told that there is a tradition among the natives, that shortly before the visit of our enterprising explorers, destructive fires had raged over the prairies, and driven the buffalo east of the mountains. At present it is scarcely seen east of the Mississippi, and south of the St Lawrence. Governor Cass' party found, in 1819, buffaloes on the east side of the Mississippi, above the falls of St. Anthony. Every year this animal's rovings are restricted. In 1822, the limit of its wanderings down the St. Peter was Great Swan Lake (near Camp Crescent). In 1823, the gentlemen of the Columbia Fur Company were obliged to travel five days in the north-west direction from Lake Travers, before they fell in with the game, but they then succeeded in killing sixty animals. There can be no doubt but this constant subtraction from his roamings must affect his numbers; certainly more than the practice of killing only the cows and leaving the bulls; a custom which has probably prevailed among the Indians for a long while, and which we cannot therefore consider as the source of the great modern diminution in their numbers."

The ranges of the buffalo in the north-western prairies are still maintained with great exactness, and old hunters, if the plains have not been burnt, can generally tell the direction in which herds will be found at certain seasons of the year. If the plains have been extensively burned in the autumn, the search for the main herds during the following spring must depend on the course the fires have taken.

Red River hunters recognize two grand divisions of buffalo, those of the Grand Coteau and Red River, and those of the Saskatchewan. Other ranges of immense herds exist further to the south, as far as Texas and Mexico. The north-western buffalo ranges are as follows, and first with respect to the Red River Range: The animals winter on the Little Souris, and south-easterly towards and beyond Devil's Lake, and thence on to Red River and the Shayenne. Here too, they are found in the spring. Their course then lies west towards the Grand Coteau de Missouri, until the month of June, when they come north, and revisit the Little Souris from the west, turning round the west flank of Turtle Mountain to Devil's Lake, and by the main river (Red River,) to the Shayenne again. In the memory of many Red River hunters, the buffalo used to visit the prairies of the Assiniboine as far north as Lake Manitobah, where in fact their skulls and bones are now to be seen; their skulls are also seen on the east side of the Red River of the North, in Minnesota, but the living animal is very rarely to be met with. A few years ago they were accustomed to pass on the east side of Turtle Mountain through the Blue Hills of the Souris, but of late years their wanderings in this direction have ceased; experience teaching them that their enemies the Half-breeds have approached too near their haunts in that direction.

The country about the west side of Turtle Mountain in June last was scored with their tracks at one of their crossing places on the Little Souris, as if deep parallel ruts had been artificially cut down the hill-sides. These ruts, often one foot deep and sixteen inches broad, would converge from the prairie for many miles to a favorite crossing or drinking place; and they are often seen in regions in which the buffalo is no longer a visitor.

The great western herds winter between the South and the North Branches of the Saskatchewan, and south of the Touchwood Hills; they cross the South Branch in June and July, visit the prairies on the south side of the Touchwood Hill range, and cross the Qu'Appelle valley anywhere between the Elbow of the South Branch and a few miles west of Fort Ellice on the Assiniboine. They then strike for the Grand Coteau de Missouri, and their eastern flank often approaches the Red River herds coming north from the Grand Coteau. They then proceed across the Missouri up the Yellow Stone, and return to the Saskatchewan as winter approaches, by the flanks of the Rocky Mountains. We saw many small herds, belonging to the western bands, cross the Qu'Appelle Valley, and proceed in single file towards the Grand Coteau in July last. The eastern bands which we had expected to find on the Little Souris were on the main river, (Red River is so termed by the Half-breeds hunting in this quarter). They had proceeded early thither, far to the south of their usual track, in consequence of the devastating fires which swept the plains from the Rocky Mountains to Red River in the autumn of 1857. We met bulls all moving south, when approaching Fort Ellice; they had come from their winter quarters near the Touchwood Hill range. As a general rule the Saskatchewan bands of buffalo go north during the autumn, and south during the summer. The Little Souris and main river bands, (Red River) go north-west in summer and south-east in autumn. It is almost needless to remark again that fires interfere with this systematic migration, but there are no other impediments which will divert the buffalo from their course. The Half-breeds state that no slaughter by large parties of hunters or Indians can turn large herds from the general direction they have taken when on the march; want of food is alone able to make them deviate from the course they have taken. The approach of numerous herds can be recognized by a low rumbling sound they occasion,—best perceived by applying the ear to a badger hole,—fully twenty miles before they arrive, if the weather be calm. During the rutting season they can be heard bellowing for a great distance on a still night; when we arrived at the Sandy Hills on the South Branch, the Crees, on being asked if the buffalo were numerous near at hand, answered, 'listen to-night, and you will hear them.'

In my report for 1857, I introduced a description of the buffalo hunters of Red River in the field, and described the arrangements and regulations of the hunt from information given me by Mr. G. Flett.* The start is usually made from the Settlements about the 15th of June for the summer hunt, the hunters remaining in the prairie until the 20th August or 1st of September. One division (the White Horse Plain) goes by the Assiniboine River to the Rapids crossing, and then proceed in a south-westerly direction. The other, or Red River division pass on to Pembina, and then take a southerly direction. The two divisions sometimes meet, but not intentionally. In Mr. Flett's division in 1849 there were according to a census taken near the Chiefs' Mountain, not far from the Shayenne River, Dacotah Territory, six hundred and three carts, seven hundred half-breeds, two hundred Indians, six hundred horses, two hundred oxen, four hundred dogs and one cat.

Mr. Ross† gives the following census of the number of carts assembled in camp for the buffalo hunt at five different periods:—

In 1820.	Number of	carts assembled	for the first	trip.	540
In 1825.	"	"	"	"	680
In 1830.	"	"	"	"	820
In 1835.	"	"	"	"	970
In 1840.	"	"	"	"	1210

The mode in which the Crees impound buffalo is described in Chapter III, page 56.

*See page 356, Red River Report for 1857.

†The Red River Settlement, its rise progress and present state. London 1856.

Blind buffalo are frequently found accompanying herds, and sometimes they are met with alone. There eyes have been destroyed by prairie fires; but their quickened sense of hearing and smell, and their increased alertness enable them to guard against danger, and makes it more difficult to approach them in quiet weather than those possessing sight. The hunters think that blind buffalo frequently give the alarm when they are stealthily approaching a herd in an undulating country. When galloping over stony ground blind buffalo frequently fall, but when quietly feeding they avoid the stones and boulders with wonderful skill. The domestication of the buffalo is a subject of much interest to the future population of Red River, and the following information on that subject may be implicitly relied on.

Humboldt in his "Aspects of Nature" says that Albert Gallatin, who, before he appeared in Europe as a distinguished diplomatist, had obtained by personal inspection great knowledge of the uncultivated part of the United States, assures us that "the mixed breed was quite common fifty years ago in some of the north-western counties of Virginia; and the cows, the issue of that mixture propagated like all others." "The favorite food of the buffalo is *Tripsacum dactyloides* (buffalo grass) and an undescribed species of clover nearly allied to *Trifolium-repens*, and designated by Barton as *Trifolium bisonicum*. According to the statement of Gomara, there was still living in the north-west of Mexico, in latitude 40°, an Indian tribe whose principal riches consisted in herds of tame bisons or buffalo. But notwithstanding the possibility of taming the bison, notwithstanding the quantity of milk it yields, and notwithstanding the herds of lamas in the Cordilleras of Peru, no pastoral life or pastoral people were found when America was discovered, and there is no historical evidence of this intermediate stage in the life of nations ever having existed there." (page 66.)

In a description of domesticated herds of buffalo, and the results of crossing with the common cow, from the Patent Office Reports, it is stated that the mixed breeds are of various colours; striped with black on a grey ground, like the zebra; some others brindled red; some pure red, with white faces; and others red, without any markings of white. The mixed bloods have not only produced from the tame and buffalo bull, but it is known that the half-bloods re-produce, viz: those that were the product of the common cow and wild buffalo bull. At the first settlement of the country, cows that were considered the best for milking, were the half-blood down to the quarter, and even eighth, of the buffalo blood. But the writer's experiments have not satisfied him that the half buffalo bull will produce again. That the half-breed heifer will be productive from either race, he has tested beyond the possibility of doubt.

"The domesticated buffalo retains the same haughty bearing that distinguishes him in his natural state. He will, however, feed or fatten on whatever suits the tame cow, and requires about the same amount of food. I have never milked either the full-blood or mixed breed, but have no doubt they might be made good milkers, although their bags or udders are less than those of the common cow; yet, from the strength of the calf, the dam must yield as much, or even more milk, than the common cow."†

Next to tho buffalo the horse is the mainstay of the prairie Indians. Good horses are not very common among the Crees; they are, however, very intelligent and well trained. A good buffalo runner is invaluable to them, for although it does not require a fast horse to catch a bull, the cows possessing greater speed often outstrip them. A good Indian horse possesses some excellent characteristics, the result of training, which it may be interesting to enumerate, for the purpose of exhibiting how admirably this animal serves his rude and savage masters. When galloping after a buffalo, an Indian horse watches the animal as intently as his rider, always swerving when he observes the buffalo's tail begin to vibrate, and breaking into short gallop at his utmost speed when he sees the tail erect, a sure indication of an immediate charge. The rider may with safety entrust himself to his horse if mounted on a trained buffalo runner; he will be carried within three yards of the flanks of the animal, and safely withdrawn when danger is threatened. If the horse stumbles and throws his rider, the sagacious animal stops instantly and waits for him to mount again. A happy instance happened to myself when riding a fiery grey mare an Ojibway Indian lent me to gallop from his tent to Manitobah house, a distance of ten miles. "She is my favorite buffalo runner, said the Indian, and will not need the thong." She ran away with me, however, as soon as we reached a grassy opening about a mile across, and in the midst of her gallop the belly band broke, and the little Indian saddle slipping round, I was thrown at once on the soft turf. The mare stopped immediately, turned round and stood by my side, waiting until I had risen and adjusted the saddle. As soon as I mounted she started off again, as if my sudden and unexpected descent had been intentional. At another time, when driving a small cariole over the frozen waters of Red River last winter, the horse, an Indian one, not being roughshod, slipped and fell, but without an effort to rise remained perfectly quiet until I had loosened the harness, when he scrambled up, gained a rough portion of ice, and quietly waited to be harnessed afresh.

Indian horses are excellent watchers by night; our half breeds were accustomed to note with care the aspect of the horses before retiring to rest; if they showed the least signs of uneasiness, such as staring about them instead of feeding quietly, or, when feeding with the "bite" in their mouth, stopping to listen, or snuffing the air, or approaching the fires when the flies were not troublesome, they would look for the cause and sometimes set watchers. When during the night, however dark, the horses suddenly approached the carts, the Half-breed would go to them, caress them, and watch the direction in which they fed or looked, knowing that their heads would be turned towards the danger, whether of Indians, or bears, or wolves.

One more instance will suffice to show the docility and training of Indian horses. I was riding a small horse which we had procured from the Crees on the Qu'Appelle, in company with a Black-foot Half-breed, some distance before the carts, in the valley of Long Creek.* As we ascended a small hill we saw a bear 250 yards before us. My companion could speak but few words of English, so with signs he motioned me to dismount, and, having satisfied himself that the horses saw the bear, he led them a few yards aside behind a clump of willows, and tying their bridles together he patted them on the neck and pointed to the bear, caressed them again and afterwards motioned me to follow him. The horses,

* See Red River Report, 1857, Appendix.

† U. S. Patent Office Report.

* Long Creek flows into the Main Saskatchewan, near Fort à la Corne.

with pricked ears, followed with their eyes every movement of the bear now slowly moving from us, but occasionally stopping to crop the twigs of willow. We crawled to leeward, and got within 70 yards of the bear, he then perceived us, I fired and sent a ball through his lungs. We waited to see if he would rise again. Finding that he lay struggling on his back, we approached and dispatched him; on looking round for the horses we saw them standing in the same place intently watching us. My companion called them, they came slowly up and stopped within 40 yards, eyeing the bear all the time. Finding that we approached it and handled it, they began to feed, evidently being satisfied that it was harmless.

Prairie Indians become very much attached to their horses, if they succeed in getting possession of a valuable animal. They often keep him in a tent, when in the neighbourhood of an enemy's country or among noted thieves of their own tribe. During the day time, when the camp is well supplied with meat and the buffalo are near, they tether him in the prairie, and indolently stretching themselves at full length on the grass, patiently watch him feed—removing the stake to a fresh spot as soon as he has cropped the best portion of the area limited by his tether. At night, when it was not thought necessary to tether our horses, we always hopped them, that is tied their fore feet together with dressed buffalo hide. Iron hoppings are in great request among Half-breeds, on their hunting expeditions. They can then more safely allow their horses to feed some distance from the camp, but instances have been known of Indians who have succeeded in approaching and catching a horse furnished with iron hoppings, in revenge for their disappointment at not being able to gallop away with their prize, sending an arrow through the animal or otherwise seriously injuring him. During the fly season, smokes are made every night for the horses, and if this precaution is neglected they will remind their masters of their want of care by surrounding the camp fire and pushing their nose into the smoke. It is this habit of crowding round the smoke of a fire to avoid the torment of the flies which makes Indian horses so difficult to drive from a prairie on fire. Many are burned every year on account of their being unable to comprehend the danger which threatens them. The buffalo are more wary, the smell of fire is often sufficient to drive them from pastures where they have been quietly feeding.

Next to the horse the dog is the Prairie Indians's most valuable friend. The dog is the great stand-by of the Squaws, who have to attend to all the duties of the camp, the men employing themselves solely in hunting and fighting. The dogs drag on poles the camp furniture, the provisions, the little children, and all the valuables of the family. It is a very amusing sight to witness several hundred dogs solemnly engaged in moving a large camp. They look wistfully at passers by, and take advantage of the least want of attention on the part of their mistresses to lie down, or snarl and snap at their companions in the work. They nevertheless obey the word of command with alacrity and willingness if not fatigued.

The midnight howl of three or four hundred dogs is an awful and appalling sound. It rises suddenly from a low prolonged whine to a deep melancholy howl, caught up again and again to the distraction of tired travellers anxious to take rest in sleep. When any great event takes place, a dog feast is proclaimed, and it is sufficiently disgusting to see the men handle and feel the unfortunate animals as if they were sheep, with a view to select the fattest, so powerful are early habits and associations in directing our feelings and tastes. Although some of the Indian dogs we saw among the Crees of the Sandy Hills are large and ferocious looking animals, we never found them vicious or inclined to attack us; they were always deterred from approaching by the sight of a stick or a feint at picking up a stone.

Although I made many inquiries, the Indians could give no information respecting the occurrence of hyprophobia among their dogs, and the same observation, as far as I could discover, applies to the dogs so numerous at Red River, and at the different Posts of the Hudson Bay Company. Large numbers of dogs are kept at the Company's Posts to haul sleds during winter; in summer time, they are fed on fish at fishing stations; in the prairie, they feed upon the offal of buffalo. Dogs will go for a week without food, and yet get into condition for travelling if well fed, in a fortnight or eighteen days. At Manitobah House, I saw them devour large pike alive, which were thrown to them as they were taken from the nets. Indian dogs are terrible thieves, especially those originating from a cross with the wolf. It was necessary to place out of reach or under cover every article bearing the least resemblance to leather when we were among the Crees. A careless Half-breed would wake in the morning and find his harness eaten, or his whip devoured; and it sometimes happened that the long tether of buffalo hide would be found partly consumed by dogs if their appetite had not been lately appeased. The wolves have this trick also when food is scarce, especially when the tether is allowed to trail loosely from the horses neck without being attached to a stake, thus leaving him at liberty to wander some distance from the camp, during the night. The voracity of dogs during the winter when travelling is astonishing, several curious instances occurred during our homeward journey which will be found at the close of this narrative.

With Crees, Ojibways, Swampys, and Sioux, the dog is supposed to be the most acceptable sacrifice to offended deities; five dogs is the common number for this propitiatory offering. In the following chapter some instances are given of their superstition in this respect.

CHAPTER XIII.

INDIAN ANTIQUITIES—SUPERSTITIONS AND CUSTOMS—RELATIONSHIP AND KINDRED*—NUMBERS AND DISTRIBUTION.

Rarity of Indian Antiquities in the Valley of the Saskatchewan—Mandan Houses—Tumuli on Rainy River—Pottery—Mr. Schoolcraft's views—Inter-marriage of Tribes—Country of the Ojibway—Scalp Dances—Wood and Prairie Indians—Indian punishment—Treatment of Prisoners—Conjurors—Conjuror's Song — Incantations — The happy Hunting Grounds — Influence of the Conjurors — The Badger—Haunted Holes and Caves — Sacrifices and Offerings — Treatment of Wives—Decorations—Ties of Kindred and Relationship ; Illustrations—Relationship among the Iroquois.—Census of Indians—No. of Indians frequenting H. B. Co's. Posts—Indians of the Saskatchewan Valley—Blackfeet Tribes—Assiniboines—Crees—Sioux or Dakotah's—Tribes of—Conjurors—Weeks—Language—Common and Sacred—Character of Language—Blackfeet Indians—Census of—Tribes of—Country Inhabited by Blackfeet.

Indian antiquities are rarely found in the Valley of the Saskatchewan south of the North Branch. The customs of wandering tribes inhabiting a prairie country are generally opposed to the rude arts which exist among barbarous races preserving a fixed abode. Not even at the fishing stations on the lakes and rivers, where different tribes have congregated at certain seasons of the year, probably for centuries, do we find any lasting memorial of individual handicraft or combined labour.

Antiquities to be ascribed to different races than those which now occupy the country exist here and there. Such are the underground houses on Rainy River,† the Mandan houses with their entrenchments on the Little Souris; but with these exceptions no other ancient monuments were seen during the exploration.

The rings of stones marking the site of Cree encampments on the Qu'Appelle‡ are of comparatively modern date, and belong doubtless to the ancestors of the present races now in possession of the country.

Rude pottery and arrow heads have been found at Red River settlements, about two feet below the surface of the soil. The fragments resemble those so common in many parts of Canada, and from their numbers lead to the inference that at a remote period the banks of this stream were peopled by races familiar with the art of making vessels from clay.

The underground houses at the Second Rapids in the Valley of Rainy River, one of which is forty feet high and about one hundred broad at the base, and the Mandan houses and fort on the Little Souris, give wider limits to the mound builders than Mr. Schoolcroft suggests in the History, Condition and Prospects of the Indian Tribes of the United States. Part 6th, page 603 :

"The whole field of antiquarian research, as represented in the Mississippi Valley monuments, may be regarded as the local nucleus and highest point of development of arts and industry attained by the Red Race, after their segregation from the Nomadic Toltec Stocks.

"These monuments are widely scattered, but they assume the same mixed sepulchral and civic character which is apparent in those found along the Alleghany branch of the Ohio in Western New York, and in other parts of the Union. The largest mound in the Union, and those which are truncated or terraced, bear the closest resemblance to the Mexican teocalli. They occupy the most southern portions of the Mississippi Valley and Florida. They become less in size as we progress north, and *cease entirely after reaching* the latitude of Lake Pepin on the Upper Mississippi, the head waters of the Wisconsin, and the mining excavations of Lake Superior."

One result of the active pursuit of the fur trade for upwards of a century in the valley of the Saskatchewan, is seen in blending of the different tribes by intermarriage. The Crees of the Plains and the Ojibways and Swampys of the Woods, although speaking different languages, are often found hunting the buffalo in company, and not unfrequently form family connections. The Ojibways of Lake Winnipeg may now be discovered, summer and winter, near the Grand Forks of the Saskatchewan, having emigrated four hundred miles west of Red River, where they have permanently established themselves. All the Ojibways now found west of the Lake of the Woods, and the east coast of Lake Winnipeg are invaders of the country. The real home of the Ojibway is the region about the south, west, and north of Lake Superior. Their habits of life have changed with the character of the country the emigrants or invaders now occupy. They are no longer dependant upon the forest for their supply of food and clothing ; but many of them, on the banks of the Assiniboine, Red River, Lake Manitobah, and Dauphin Lake, and on the west flank of the Riding and Duck Mountains, possess horses, and join the half-breeds in their annual spring and fall hunts.

* A portion of this chapter has been published in the *Canadian Journal* for July.

† At the second rapids on an extensive area denuded of trees, having a very beautiful appearance, are two immense mounds which appeared to be tumuli. We forced our way to them, through a dense growth of grasses, nettles, and helianthus, twisted together by the wild convolvulus. Our path to the mounds passed through a neglected Indian garden, and near the decaying lodge poles of an extensive encampment. The mound ascended was about forty feet high, and one hundred broad at the base. It was composed of a rich black sandy loam, containing a large quantity of vegetable matter, and on digging a foot deep no change in the character of the soil was observable.—*Red River Report*, 1857.

‡ Chapter III. page 53.

Nothwithstanding this intercourse and blending of different nations, most of the superstitions and customs peculiar to each, are still maintained and practised.

Nearly one hundred years ago (1770), Mr. Hutchins of the Hudson Bay Company's service, framed an enumeration of the tribes between Lake Winnipeg, and within one hundred miles of James' Bay, speaking the Ojibway tongue. The names of the tribes will be found in Sir John Richardson's journal, page 265, American edition. The tribes enumerated have evidently derived their names as in the present day, from their hunting and fishing stations.

It is often asked whether the thrilling descriptions of savage life, as given in Cooper's delightful romances, are imaginary or real; and, if real, whether they exist now among the tribes which have long been familiar with civilized man, such as the Plain Crees, the Sioux, the Swampys, and the Ojibways. It is enough to visit the secluded Ojibway graves, on the banks of Red River, and behold there, Sioux scalps decorated with beads, bits of cloth, coloured ribbons, and strips of leather suspended at the extremity of a long slender stick, near the head of the grave, to feel satisfied that one barbarous custom still prevails. But, to be an eye witness of a scalp dance, or a skull dance is more than enough to press home the conviction that the fiendish passions, so faithfully described by Cooper, still find expression in violent gesture, loud vociferation, triumphant song, and barbarous feasting, with undiminished strength and bitterness, even after a century's intercourse with civilized man.

In the following paragraphs I shall endeavour to describe some incidents which will show how far old superstitions and customs prevail among the Indians occupying the country between Red River and the South Branch of the Saskatchewan.

Early last spring, the warlike bands of Ojibways, called the Lac la Pluie Indians, were thrown into a state of savage excitement by the arrival of messengers from their friends on Red River, with tidings that two Sioux had been killed and scalped in the Plains. In testimony of this triumph, they brought with them two fingers severed from the hands of the unfortunate Sioux. The announcement of the intelligence that the scalps would be sent after their Red River brethren had celebrated war dances over them, was received with wild clamour and shouting. After the scalps had been carried from hand to hand and the victory that won them triumphed over with dancing, singing, and feasting, they would be returned to the warriors who took them, and finally suspended over the graves of relatives or friends mourning the loss of any of their kindred by the hands of the Sioux.

The wood Indians assemble in the spring to celebrate their medicine feasts and other ceremonies. During the summer they separate into families or small bands, and hunt, fish or go to the Plains in search of buffalo. At the approach of winter, they " take debt " or otherwise obtain supplies at the different posts of the Company, and retire to their winter quarters to trap the fur-bearing animals. The Plain or Prairie Indians follow the buffalo, and vary the monotony of their existence by forming war parties against their enemies, such as the Plain Crees against the Sioux and the Blackfeet, the Ojibways against the Sioux.

When on the south branch of the Saskatchewan last August, we found the Plain Crees hastening from the west to the east bank of the river, at the Elbow, with a strong war party of Black feet in pursuit. The chief Shortstick, pointed out some of his band who had penetrated through the Blackfeet country to the Rocky Mountains two years ago, and returned with several scalps, grizzly bear claws, necklaces, pipes, and other trophies of success. He also related with much feeling how twenty-five young warriors had gone on a similar excursion the summer before last, but none had yet returned. Last July (1858) a large body of the Plain Crees met a portion of the Blackfeet tribe, at the Eagle Hills, on the North Branch of the Saskatchewan, to arrange terms of peace. All matters went on smoothly, and the representatives of the two nations separated as friends. Some of the Crees, however, incapable of resisting the opportunity, stole some horses from the Blackfeet. They were pursued, and three of them taken. One was killed instantly, the others were led back in triumph to the camp of the Blackfeet. They were stripped, their hands were tied behind their backs, a hole was bored through both wrists and a stick passed through them and so tightly fastened that it could not be removed without assistance; the captives were then separated and dismissed singly to find their way to their friends. One only reached his tribe and was lying in a tent which we passed on the banks of the Qu'Appelle near the South Branch.

Short-stick, when relating these adventures, held up the pipe he had in his hand and exclaimed, " this is what my Blackfoot friend gave me one day, the next he killed my young men; he is now my enemy again." I expressed a wish to purchase the pipe; the chief's reply was " take it," handing it to me with a gloomy frown, and silently extending his hand for the common " clay " which I was smoking at the time. The great chief of the Plain Crees is styled " the Fox ;" he is held in high estimation by all the Plain Indians with whom he comes in contact, either in peace or war. He is dreaded by the Sioux, the Blackfeet, the Bloodies, the Fall Indians, the Assiniboines, and all the tribes who occasionally hunt on the Grand Coteau de Missouri and the South Branch of the Saskatchewan.

The cruel, barbarous treatment of prisoners so often described in narratives of Indian warfare, is common even now in the prairies south of the Qu'Appelle and the Assiniboine. Not a year passes without two or more of the Red River half-breeds being scalped by Sioux; sometimes, as was the case last year, quite close to the settlement of St. Joseph, near the boundary line, about thirty miles west of Red River. When a prisoner is taken, the Sioux sometimes adopt a terrible mode of death, during the summer season. They have been known to strip a half-breed, tie him to a stake on the borders of a marsh in the prairie, and leave him exposed to the attacks of millions of mosquitoes, without being able to move any part of his body; and when the agony of fever and the torment of thirst come upon him, they leave him to die a dreadful, lingering death, with water at his feet, and buzzards hovering and circling around him in loathsome expectation.

By way of illustrating the character of the medicine or conjuring ceremonies, which may be witnessed during all seasons of the year, when several families are encamped together, I shall describe a scene of which I was an eye-witness last summer near the Hudson Bay Company's post in the Touchwood Hills, between the South Branch of the Saskatchewan and the Assiniboine. The conversation was carried on in Cree, but, I believe, faithfully interpreted to me by the officer then in charge of the post, who was present. The interpretation was pronounced exact by one of the Cree half-breeds attached to my party.

At the time of my arrival at this Post, a conjuror of some celebrity was endeavoring to cure an invalided woman by the exercise of his cunning. The sick woman was lying in a buffalo skin tent: the conjuror, painted and decorated, employed himself in beating a medicine drum within a few feet of her, and singing at intervals the following words, first uttered slowly, with a pause between each word, then as in ordinary conversation; lastly, with energy and rapidity:

"Great—is—the—man—who—walks—
In—the—middle—of—the—Earth,—
He—is—the—only—true—Lord."

The word "Lord" is not employed in the sense of supreme master, but is rather intended to convey an idea of independence and individual power; and is better expressed in English, as the half-breeds informed me, by the word "gentleman."

The conjuror occasionally came out of the tent; and whenever the supposed Manitou or Fairy who was the alleged cause of the woman's illness approached, a little bell, suspended from the poles supporting the tent, tinkled, and gave the alarm; the conjuror immediately seized his drum, commenced his song and, by his incantations, succeeded in pacifying the Manitou. These proceedings continued for two nights; at the close of the second night, after a prolonged ringing of the little bell, violent shaking of the tent poles, loud beating of the drum, and chaunting of the words before quoted, the conjuror announced that he had discovered the reasons of the Manitou's anger, and the means to appease it.

You had a dream, said the conjouror, and when you rose in in the morning you promised to make an offering to the Manitou, you have forgotten your pledge, and you are sick.

The woman demanded what she had dreamt, and what she had promised, avowing her ignorance of both dream and promise. The conjuror told her, that when the buffalo were around her tent last winter, and no fear of starvation before her eyes, she had dreamed that the buffalo would always surround her, that famine and sorrow were always to be strangers to her, and, in gratitude, had vowed to make a sacrifice of her best robe. The woman, wearied no doubt with the conjuror's unceasing drum and song, probably too, believing that a false confession was the lesser evil, as it might bring the promised relief, acknowledged that the conjuror was in the right. The penalty she was told to pay consisted of the sacrifice of throwing away two robes, or double the amount of the promise she had made; after which her health was to be restored.

Scenes similar to the one just described may be witnessed whenever several families are camping together; but the sacrifices required to be made depend upon the ability of the deluded creatures to satisfy the demands of the conjuror.

"The Happy Hunting Grounds," the Heaven of Indians, so often spoken of by writers of fiction, are an actual reality in the imaginations of Crees and Ojibways, as well as of other north-western tribes. A Plain Cree on the Qu'Appelle gravely informed one of my men that he had been dead once and visited the spirit world. His narrative was to the following effect:—"I was sick, and fell asleep. I awoke on the bank of a deep river, whose waters were flowing swiftly and black from a great mist on the south to a great mist on the north. Many other Indians sat on the banks of the river, gazing at its waters, and on the gloomy shore which lay wrapped in mist on the other side. Time after time the mist before us would roll away and reveal the mouth of another great river pouring its flood into the one on whose banks I was sitting. The country to the south of this river was bright and glorious, to the north dark and gloomy. On the one side were the happy hunting grounds, on the other the hunting grounds of the bad Indians. Time after time my companions tried to cross the swift stream before us, in order to reach the happy hunting grounds; some arrived in safety, others reached the north bank, and disappeared in the mist which overhung the bad country. I tried at last, but the current was too strong for me, the recollection of bad deeds prevented me from stemming the current, and I was swept on to the north shore of the opposite river. I scrambled up the bank, and spent many moons in hunting in that dreary land; always on the point of starving, or being hurt by enemies, or wet and cold and miserable. At length I came upon a river like the one I had crossed, with mists and a great stream opposite, breaking clouds revealing happy hunting grounds on one side, and a more gloomy and terrible country on the other side. Other Indians were there before me, looking at the river and trying to cross; many succeeded, a few were swept to the bad country, these were very wicked Indians. I tried to cross. I knew I had been a good Indian in this dreary hunting ground. I took courage, and swam strong against the stream. I reach the happy hunting grounds; all my sorrow disappeared as I climbed to the top of the bank and saw before me Indians numerous as grass leaves, buffalo on the distant plains thick as rain drops in summer, a cloudless sky above, and a warm, fresh, scented, happy breeze blowing in my face. I sank to sleep, and woke alone in my tent in these prairies again."

Whatever faith the Indian medicine men possess in the efficacy of their charms, it is certain that they entertain great respect for the white man's medicine. A laughable incident occurred at the Touchwood Hills. The conjuror of whom mention has just been made, entered the room at the post where I was sitting with Mr. and Mrs. H., who were temporarily in charge. The Indian and a companion seated themselves upon one of my boxes which contained a small medicine chest. Mrs. H. asked me to give her some sticking plaster. I crossed the room to open the medicine chest, when Mrs. H. (a halfbreed) said to her husband, in the Cree language: "Will his medicines do me any harm if I stop here while he opens them?" Mr. H. answered jestingly, "Yes, you had better go into the other room." On motioning the Indians to move, they rose, and I opened the chest. The moment they saw the bottles, they hurried out of the room, hastened to the summit of a neighboring hill, and, divesting themselves of every article of clothing, shook their garments repeatedly, and, after hanging them on bushes in the sun, squatted on their haunches to await the deodorizing influence of the breeze.

In the valley of the Qu'Appelle River, we frequently found offerings to Manitou or Fairies suspended on branches of trees; they consisted of fragments of cloth, strings of beads, shreds of painted buffalo hide, bears' teeth and claws, and other trifles. Our half-breeds always regarded them with respect, and never molested or liked to see us molest these offerings to Manitou. This custom prevails everywhere in the valley of Lake Winnipeg, and it may truly be said that the medicine drum is heard far more frequently in some parishes of Selkirk Settlement than the sound of church bells.

A conjuror celebrated for the potency of his charms will often exercise a very injurious influence over an entire band consisting of ten or twelve families, in deterring them from frequenting particular hunting or fishing grounds if they offend him. Out of numerous instances of this dangerous influence, I select the following. It occured on the Dauphin River. When ascending that stream, we came upon a large camp of Ojibways who were on their way to the Hudson Bay Company's Post at Fairford. Their usual wintering place was at the Pike's Head near the mouth of Jack-fish river, an excellent fishing station on Lake Winnipeg, but they had abandoned the intention of wintering there, in consequence of a threat which had been conveyed to them from a noted conjuror styled " the Badger," of the Grand Rapid of the Saskatchewan, to the effect that if the band ventured to winter at the Pike's Head, " He would do something." This ambiguous threat was quite sufficient to deter them from visiting their old haunts, and would probably be instrumental in producing much suffering, if not actual want, to many of the band.

There are many places on Lake Winnipeg and Manitobah which the Indians who hunt and live on the shores of those great lakes, dare not visit. There is scarcely a cave or headland which has not some legend attached to it, familiar to all the wanderers on these coasts.

On the west side of Lake Winnipeg, in the long, dark, and gloomy chambers formed by fissures in the limestone, bad spirits are supposed to dwell, according to the belief of the Indians who hunt on the coast; and he would be a powerful charmer who could induce a heathen Indian to approach, much less enter, the abodes of these imaginary Manitous.

Near Limestone Cave Point are several of these supposed fairy dwellings. When an Indian approaches them in his canoe, he either leaves an offering or cautiously gives them a wide berth.

On Lake Manitobah, Steep Rock Point is a noted dwelling-place for the " Little Men." This locality is described in chapter ix. Some of the traditions connected with these places are very absurd, and appear to have little meaning to civilized men; nevertheless, among the barbarous tribes of those regions, they are associated with their past history, or with the history of the race that preceded them. Manitobah Lake, a body of water of very imposing dimensions, having an area of 1,900 square miles, derives its name from one of these superstitions. I stayed for three days on this dreaded island, where a Manitou dwells, but although Indians passed and repassed, heard and answered our shots, yet they could not be persuaded to land. The only evidence of fairy presence which I met with, was the " fairy-like music " of the waves of Lake Manitobah, beating upon the hard limestone shingle on the beach, and producing a very beautiful and melancholy resemblance to distant church-bells. All night long this ringing musical sound was heard, and would, no doubt, in the active imagination of Indians, suggest the existence of those Manitous with which they people the air, the water, the forests, and the caves of the earth.

Sacrifices and offerings are of very frequent occurrence among the Indians of the Saskatchewan Valley. The customary offering consists of two, three, and sometimes five dogs. At the mouth of the Qu'Appelle River, an Indian, in June last, set his net and caught a large fish of a kind different to any with which he was familiar. He immediately pronounced it to be a Manitou, and, carefully restoring it to the water again, at once sacrificed five valuable dogs to appease the anger of the supposed fairy. On approaching Long Lake, an arm of the Qu'Appelle River Valley, the Crees warned us not to visit the Lake by night, as it was full of devils. They told me very extraordinary tales of the dimensions and power of these devils, and appeared to live in awe and terror of them.

Like most heathens and barbarous races, Indians suffer much from their superstitious fears. When the weather is fine, and their tents are well supplied with provisions, they are an independent and joyous people. Full of frolic, and fond of relating anecdotes, they laugh immoderately at any trifling joke or absurdity, and seem thoroughly to enjoy existance.

When visiting the Crees of the Sandy Hills, on the South Branch, and passing the door of the tent belonging to Short-stick's eldest son (*see chap.* v.), who accompanied me, I observed a young squaw outside, leaning upon sticks, evidently in great trouble, and weeping bitterly. The moment she saw us she hobbled into the tent, with a low cry of pain, and closed the entrance. I asked the interpreter what this meant. After some conversation with her husband, he said that the woman was suffering from a beating he had given her for a violation of her faith during his absence in the spring on a war excursion. " I would have killed her," muttered the husband, " but I thought it a pity to kill two at once. She had her choice whether she would have her hair, her nose, or her ear cut off, or whether she would have a beating. She chose what she has got; and I would have killed her had I not known I should regret having killed both." It is needless to add that the woman soon expected to become a mother.

Smearing the skin with different coloured pigments is a universal custom among tho wood and prairie Indians. Sometimes the operation is very tastefully performed. Warriors on the " war-path " often paint the figure of the hand over the mouth, as used in sounding the war-whoop: this is a distinctive sign that the Indian so decorated has been recently, or is still, engaged in the pursuit of his enemies. Vermillion is the most coveted colour. The Ojibways are very fond of decorating their faces with this brilliant pigment. The Plain Crees are partial to white and green; and not only paint the face, but also the chest and arms. The Plain Crees cut and gash the skin and flesh on the arms, sides, chest, and legs, as a token of grief for any deceased friend or relation. My friend Short-stick's body was dreadfully disfigured by scars from wounds made by himself in manifestation of his grief.

The origin of the aborigines on this continent still remains enveloped in thick darkness. Many of their manners, superstitions, and customs correspond to those of Orientals, and it is not improbable that modern Ethnologists may be on the right track in their efforts to solve this deeply interesting question.

Humboldt tells us, in his "*Aspects of Nature*," that he " regards the existence of ancient connections between the inhabitants of Western America and Eastern Asia, as more than probable; but by what routes, or with what Asiatic nations, the communications took place, cannot at present be decided. A small number of individuals of the educated priestly caste might, perhaps, be sufficient to bring about great alterations in the civil and social state of Western America.

" The stories formerly narrated of Chinese expeditions to the New Continent, really apply only to voyages to Fusang or Japan. On the other hand, Japanese and Sian-Pi, from the Corea, may have been driven by storms to the American coast, and landed there. We know, as a matter of history, that Bonzes, and other adventurers, sailed over the eastern Chinese Seas in search of some medicine which should entirely prevent death. Under Tschin-schi-kuang-ti, 209 years before our era, three hundred young couples (young men and young women) were sent to Japan, and instead of returning to China they settled at Nipon. May not similar expeditions have been driven by storms or other accidents to the Aleutian Islands, to Alashka, or to New California? As the western coasts of the American Continent trend from north-west to south-east, and the eastern coasts of Asia in the opposite direction, or from the north-east to the south-west, the distance between the two continents in 45 deg. of latitude, or in the Temperate Zone, which is most favourable to mental development, is too considerable to admit of the probability of such an accidental settlement taking place in that latitude. We must, then, assume the first landing to have been made in the inhospitable climate of from 55 deg. to 65 deg., and that the civilization thus introduced, like the general movement of population in America, has proceeded by successive stations from north to south. The remains of ships from Cathay, *i. e.*, from Japan or China, were supposed to have been found on the coasts of Northern Dorado (called Quivira and Cibora), at the beginning of the sixteenth century. Our knowledge of the languages of America is still too limited, considering their great variety, for us as yet entirely to relinquish the hope of some day discovering an idiom which may have been spoken, with certain modifications, at once in the interior of South America, and in that of Asia; or which may at least indicate an ancient affinity. Such a discovery would certainly be one of the most brilliant which can be expected in reference to the history of mankind. But analogies of language only deserve confidence when the inquirer, not resting in or dwelling on resemblances of sound in the roots, traces the analogies into the organic structure, the grammatical forms, and into all which in languages shows itself as the product of the human intellect and character."

In order to understand the character and nature of wild Indians, they must be seen in their tents when well supplied with provisions, and disposed to be cheerful and merry. In the prairies when on horseback, they are often quiet and watchful, always on the look out, and if twenty or thirty are in a band they generally manage to see a suspicious object in the distance at the same moment, so that a simultaneous note of exclamation is uttered by most or all of the party. In hunting the buffalo they are wild with excitement, but no scene or incident seems to have such a maddening effect upon them as when the buffalo are successfully driven into a pound. Until the herd is brought in by the skilled hunters all is silence around the fence of the pound, men, women and children with pent up feelings, holding their robes so as to close every orifice through which the terrified animals might endeavour to effect an escape. The herd once in the pound a scene of diabolical butchery and excitement begins; men, women and children climb on the fence and shoot arrows or thrust spears at the bewildered buffalo, with shouts, screams and yells horrible to hear. But when the young men, and even women jump into the arena amidst the dying and the dead, smear themselves with blood, thrust their arms up to the shoulders into the reeking bodies of their victims, the savage barbarity of the wild prairie Indian shows itself in its true colours. Not even a scalp dance over many fallen foes, affords such a terrible picture of degraded humanity as a large band of prairie Indians, some hundreds in number, during and after the slaughter of buffalo in the pound.

The condition of the Indians of the Saskatchewan Valley at the present day is very different to what it used to be half a century since. Not only have imported diseases greatly diminished their numbers, but game of different kinds has become so scarce that during some seasons starvation is no fiction.

In sickness prairie Indians are much depressed, and often seek consolation in the monotonous drum of the medicine man and his heathenish incantations, an infliction which the grossest and most debased superstition alone would tolerate, submitted to with hope and confidence, however, by men who are anxious and timid during the roll of thunder, invoking the Great Bird by whose flapping wings they suppose it to be produced, or crouching from the blink of his all penetrating eye, which they allege is the lightning's flash.

The ties of kindred and relationship are of a very complex character among the Ojibways; in more than one instance a singular exemplification of cross-relationship occurred during our voyage on lakes Winnipeg and Manitobah which is perhaps worthy of being recorded, as it may serve to show the permanency of ancient customs and traditions among families now dwelling nearly a thousand miles west of the hunting grounds of their ancestors. Near the mouth of the Little Saskatchewan, we met an Indian family in small canoes journeying towards the mouth of Red River. The family consisted of a young Indian, his wife and two little children. The father was born on the shores of Lake Winnipeg, and had never travelled east of the lake. After a few words had passed between him and a Half-breed Ojibway from Lake Superior, (Wigwam,) they shook hands and proclaimed themselves related to one another. Each belonged, as I was informed, to the tribe which bore the name of the "Bear," and having by some means which Wigwam could not or would not explain, ascertained this fact, they spoke to one another as brothers. A similar relationship was established between Wigwam and another Ojibway on Moss River, solely, as he informed me, because his own and his newly found friend belonged to a tribe whose distinctive name was the "Bear." The Cree half-breeds told me that in their communication with the Ojibways of Lake Winnipeg, and, further west, this recognition of relationship not unfrequently took place between individuals who met for the first time and who were born and lived in districts far apart. In connection with this singular kind of relationship and the bearing it may possibly have upon the origin of the Indian races, I append the following extract from an ethnological paper, read at the Montreal Meeting of the American Association for the advancement of science, by Lewis H. Morgan, Esq., of Rochester, N. Y.*

It has occurred to me, after a careful examination of the system of consanguinity and descent of the Iroquois, that we may yet be able, by means of it, to solve the question whether our Indian races are of Asiatic origin. Language changes its vocabulary not only, but also modifies its grammatical structure in the progress of ages; thus eluding the enquiries which philologists have pressed it to answer; but a system of consanguinity once matured and brought into working operation, is, in the nature of things, more unchangeable than language;—not in the names employed as a vocabulary of relationship, but in the ideas which underlie the system itself. The Indo-European nations have one system, identical in its principal features, with an antiquity of thirty-five centuries, as a fact of actual record. That of the Iroquois is original, clearly defined, and the reverse of the former. It is, at least, to be presumed that it has an antiquity coeval with the race. That of the Chippewa is the same as the Iroquois, with slight modifications; thus establishing the fact of its existence in two of the principal generic stocks. Besides this, there are traces of the same system among the Aztecs, Mohaves, Creeks, Dhcotaas, Delawares, Winnebagoes, and other races, all tending to show that the system has been, and now is, universal upon this continent.

* See page 225 *Canadian Journal*, 1858.

Should this last fact be established, the antiquity of the system, as coeval with the Indian race upon the continent, will also become established. Upon the basis of these two facts, and assuming that these races are of Asiatic origin, we may predict the existence of the same system in Asia, at the present moment, among the descendants of their common ancestors, if any remain.

A brief explanation of the principal features of the system of the Iroquois is annexed, which will assist in working out every other, particularly if they are founded upon the same ideas.

The institutions of the Iroquois were founded upon the family relationships; n fact, their celebrated league was but an elaboration of these relationships into a complex system of civil polity. At the base of this were their laws of descent. They were unlike both the civil and the canon laws; but yet were original and well defined. The chief differences were two: first descent among the Iroquois followed the female line, or passed through the mother; while in each of the former systems it follows the male, or passes through the father. In the second place the collateral lines, with the Iroquois, were finally brought into or merged in the lineal; while, in the other cases, every remove from the common ancestor separated the collateral lines from the lineal, until after a few generations actual relationship ceased among collaterals.

To bring out distinctly this code of descent, it will be necessary to give a brief explanation of the division of the Iroquois into tribes, the union of the several tribes into one nation, and of the several nations into one league. Without a reference to their civil organization, it would be impossible to present it in an understandable form.

In each of the five nations who composed the original league, there were eight tribes, named: Wolf, Bear, Beaver, and Turtle; Deer, Snipe, Heron, and Hawk. The Onondaga nation, therefore, was a counterpart of the Cayuga, each having the same number of tribes, and of the same name; so also, interchangeably, of the Oneida, the Mohawk, and the Seneca nations. In effect, the Wolf tribe was divided into five parts, and one-fifth part of it placed in each of the five nations. The remaining tribes were subjected to the same division and distribution. Between the individual members of the Wolf or other tribe thus divided, or, in other words, between the separated parts of each tribe, there existed the tie of consanguinity. The Mohawk of the Turtle tribe recognized the Seneca of the Turtle tribe as a relative, and between them existed the bond of kindred blood. In like manner the Oneida of the Hawk tribe received the Onondaga or the Cayuga of the same tribe as a relative, not in an ideal or conventional sense, but as actually connected with him by the ties of consanguinity. Herein we discover an element of union between the five nations, of remarkable vitality and power. A cross-relationship existed between the several tribes of each nation and the tribes of corresponding name in each of the other nations, which bound them together in the league with indissoluble bonds. If either of the nations had wished to cast off the alliance, it would have broken this eight-fold bond of consanguinity. Had the nations fallen into collision with each other, it would have brought Hawk tribe against Hawk tribe—in a word, brother against brother. The history of the Iroquois exhibits the wisdom of these organic provisions; for, during the long period through which the league subsisted, they never fell into anarchy, nor even approximated to a dissolution from internal disorders.

At no time in the history of the Iroquois could a man marry a woman of his own tribe, even in another nation. All the members of a tribe were within the prohibited degrees of consanguinity; and to this day, among the descendants of the Iroquois, this law is religiously observed. Husband and wife, therefore, were in every case of different tribes. The children were of the tribe of the mother. Here, then, we discover one the central ideas of their laws of descent: to place the father and mother in different tribes, and to assign the children to the tribe of the mother. Several important results followed, of which the most remarkable was, the perpetual disinheritance of the male line. As all titles, as well as property, descended in the female line, and were hereditary in the tribe, the son could never succeed to his father's title of sachem, nor inherit even his tomahawk.

A tribe of the Iroquois, it thus appears, was not, like the Grecian and Roman tribes, a circle or group of families, for two tribes were necessarily represented in every family; neither, like the Jewish, was it constituted of the lineal descendants of a common father; on the contrary, it involved the idea of descent from a common mother; nor has it any resemblance to the Scottish clan, or to the canton of the Switzer. It approaches, however, nearer to the Jewish. Denying geographical boundaries, a tribe of the Iroquois was composed of a part of a multitude of families, as wide spread as the territories of the race, but yet united together by a common tribal bond. The mother, her children, and the descendants of her daughters, in the female line, would, in perpetuity, be linked with the fortunes of her own tribe; while the father, his brothers and sisters, and the descendants in the female line of his sisters would be united to another tribe, and held by its affinities. No circumstances could work a translation from one tribe to another, or even suspend the nationality of the individual. If a Cayuga woman of the Hawk tribe married a Seneca, her children were of the Hawk tribe and Cayugas and her descendants in the female line, to the latest posterity, continued to be Cayugas and of the Hawk tribe, although they resided with the Senecas, and by successive intermarriage with them had lost nearly every particle of Cayuga blood. Neither could intermarriage with one of a foreign nation confer the Iroquois nationality upon the wife or children of the marriage, and the same *vice versa*. If a Mohawk married a Delaware woman, she and her children were not only Delaware still, but ever continued aliens, unless naturalized as Mohawks, with the forms and ceremonies prescribed in case of adoption."

The difficulty of obtaining reliable information respecting the Indian population has been acknowledged by all who have given attention to this subject. I am convinced that the number of Indians inhabiting Rupert's Land has been considerably overrated. The estimates published in the Appendix to the Report from the Select Committee on the Hudson Bay Company furnish the following result.

Thickwood Indians on the east side of the Rocky Mountains	35,000
The Plain Tribes (Blackfeet, &c.)	25,000
	60,000

The Indian population of Rupert's Land is estimated at 42,870. Over the plain or prairie tribes the H. B. Company profess to have no control, and they are returned as numbering 25,000 souls. It will appear further on that excellent authorities quoted in the text, do not assign more than half that number to the most numerous tribes of prairie Indians, who hunt on the Saskatchewan and Missouri, with their tributaries, and who occasionally trade on both sides of the international boundary.

The Plain Crees and Thickwood Indians are under the control of the Company, but I think that their numbers are also over estimated, and the grounds on which this opinion is advanced are as follows.

The basis of the census for the Thickwood Indians and the Plain Crees is the number frequenting the establishments of the Hudson's Bay Company in 1856, and the following enumeration at certain posts chiefly visited by the Plain Crees is given:

Post.	No. of Indians frequenting it.
Fort Ellice	500
Qu'Appelle Lakes	250
Touchwood Hills	300
Fort à la Corne	300
	1350 Indians.

Upon perusal of the foregoing table the reader would infer that thirteen hundred and fifty Indians visited the posts named. It happens, however, that many Indians trade with two or even more posts, although every effort is made to limit them to one particular station. Their names, however, appear on the books at different establishments, and in the enumeration of the Indians inhabiting certain districts, some of them are counted twice and even three times. I ascertained beyond doubt, that this practice existed to an extent which would affect the census in a marked degree. The custom of giving credit to the Indians encourages this system, while a natural desire to attach additional hunters to a post, on the part of the traders, induces less caution than would otherwise be exercised. As the result of very careful enquiries wherever opportunities offered of obtaining exact information, I am inclined to think that the estimate of 42,870 is about one-fourth too high.

The estimated number of Indians frequenting certain establishments of the Hudson Bay Company in 1856 are given in the following table.

The posts enumerated are included within the area embraced

by the map which accompanies this Report; not including the east side of Lake Winipego-sis, the Half-breed Settlements and Red River.

Locality.	Number.
Fort à la Corne	300
Cumberland House	250
The Pas	300
Fort Pelly	800
Fort Ellice	500
Qu'Appelle Lakes	250
Shoal River	150
Touchwood Hills	300
Egg Lake	200
Manitobah House	200

On the North Branch of the Saskatchewan, where the Prairie Indians assemble, the following enumeration is given in the Blue Book.

Locality.	No. of Indians.
Edmonton	7,500
Carlton	5,000
Fort Pitt	7,000
Rocky Mountain House	6,000

This census is probably over estimated; although it may approximate to the actual number of Indians visiting a particular post, yet there is strong reason to suppose that the same individuals are to a large extent enumerated twice if not thrice.

The Plain or Prairie Indians belong to the following principal tribes:

Blackfeet,	Crees,
Bloodies,	Assiniboines,
Fall Indians, or Gros Ventres,	Sioux,
Piegans,	Ojibways.

The Wood Indians of the Saskatchewan valley belong to the great family of Crees and Ojibways. The Sioux and Blackfeet are Dakotahs.

Mr. Harriet, a chief factor of the Hudson Bay Company who had passed his life among the Blackfeet, estimated the six or seven tribes going by that general name as mustering 1,600 to 1,700 tents, at eight per tent, 13,000.*

Mr. Rowand, one of the oldest resident traders, estimates the Blackfeet tribes as follows:

Blackfeet proper	300
Piegans	400
Bloods	250
Gros Ventres or Fall Indians	400
Circes	45
Cotones, Small Robes, } Mountain Tribes	250
At 8 persons per tent, 13,1 00.	1645 tents.

The Assiniboines are divided into Strongwood and Plain Assiniboines, or Stonys.

Mr. Harriet, in 1842 estimated the		
Strongwood Assiniboines	at 80 tents, =	640
Mr. Rowand, the Plain Assiniboines	300	2,400
	380 tents, =	3,020

* Colonel Lefroy, R. A.

The Strongwood Crees about Edmonton Mr. Rowand estimated at	400 tents, at 10 per tent,	4,000
Crees of the Plains	200 " " "	2,000
		6,000

Colonel Lefroy* states that the aggregate of the tribes inhabiting the Plains on British Territory was estimated in 1843 at not more than 23,400. Since that period they have diminished in numbers, and some of the Blackfeet bands stationed themselves permanently on the Missouri. In succeeding pages, recent estimates of the Blackfeet tribes, and the limits of the hunting grounds are given.

The Sioux and the Blackfeet being the most warlike tribes of the North-west, and retaining their ancient customs to the fullest extent, the following brief notices of these formidable native races are introduced. The Plain and Wood Crees and the Ojibways are almost altogether amenable to the influence of the Hudson Bay Company, and are in fact the hunters upon whom they rely for their supply of furs and provisions.

The Sioux or Dakotah Indians.

The nation of the Sioux Indians or Dakotahs† are composed of seven principal bands. Their aggregate number probably does not exceed twenty-five thousand. Their hunting grounds extend from the Mississipi River on the East to the Black hills in Nebraska on the west, and from the mouth of the Big Sioux River on the south to Devil's Lake in the north. The area ascribed to this nation by the authors of the map attached to the Report of the Special Committee of the House of Com mons on the Hudson's Bay Company, comprehends a larger extent of territory than that included within these limits. Although the Sioux have no dealings with the Half-breeds of Red River, or with the Hudson Bay Company, yet they often cross the 49th parallel in pursuit of the buffalo, and more frequently in search of a scalp from their hereditary enemies, the Ojibways and Crees. As the most dreaded invaders of the prairies north of the boundary line, this powerful nation deserves a special notice.

The name Dakotah signifies the 'Allied,' and they speak of themselves as the 'Oceti sakowin' or 'Seven Council Fires.' The following enumeration of the principal bands which compose the nation by the members of the American Dakotah Mission, will be found at length in the Grammar and Dictionary prepared with so much care, labour and zeal, under the editorial management of the Rev. S. R. Riggs, A.M., Missionary of the American Board of Commissioners for Foreign Missions.

1. The Mdewakantonwans, *Village of the Spirit Lake.* The name is derived from Mdewakan (Spirit or Sacred Lake,) Mille Lacs, (Minnesota,) in the country now claimed by the Ojibways. This band numbers about two thousand.
2. The Wahpekutes, *Leaf Shooters*, five hundred.
3. The Wahpetonwans, *Village in the Leaves*, twelve hundred.
4. The Sisitonwans, *Village of the Marsh*, two thousand five hundred. Their hunting ground is about the Coteau des Prairies, and they subsist on the buffalo.
5. The Ihanktonwanna, *the End Village Band*—four thousand. Their country is on the north east of the Missouri, as

* See an article upon the Native Indian population of British America, by Colonel Lefroy, R.A. *Canadian Journal*, Vol. 1, Old Series.

†See introduction to a Grammar and Dictionary of the Dakotah language, published by the Smithsonian Institution,

far as Devil's Lake. These are the great enemies of the Red River Half-breeds.

6. The Ihanktonwans, *the Village at the End*—two thousand four hundred. Their country is west of the Missouri. They are frequently termed Yanctons.

7. The Titonwans, *the Village of the Prairie* — twelve thousand five hundred. Their hunting ground is west of the Missouri. They are divided into seven bands: the Sicaugu, *Burnt-Thighs;* the Itazipco, *Bow-pith;* the Sihasapa, *Black-feet;* the Minikanye wozupi, *Those who plant by the water;* the Oohenoupa, *Two-boilings;* and the Oglala and Hunkpapa.

The conjurors believe that their dreams are revelations from Spirit World, and they aver that their prophetic visions are the mental revival of occurrences in a former state of existence. Years with them are enumerated by winters, a distance is estimated by the number of nights a man will sleep on the way. The Ojibways have the same method of expressing time and distance. They divide the year into moons, but weeks are unknown to them. The Dakotahs of the valley of the Minnesota have the following months in the year: *

1. Wi-tehi, January; the Hard Moon.
2. Wicata-wi, February; the Racoon Moon.
3. Istawicayazan-wi, March; the Sore (eye) Moon.
4. Magaokada-wi, April; the moon in which the geese lays eggs.
5. Wozupi-wi, May; the planting moon.
6. Wazustecasa-wi, June; the moon when the strawberries are red.
7. Canpasapa-wi, July; the moon when the choke cherries are ripe.
8. Wasutou-wi, August; the harvest moon.
9. Psinhnaketu-wi, September; the moon when rice is laid up to dry.
10. Wi-wazupi, October; the drying rice moon.
11. Takiyuha-wi, November; the deer rutting moon.
12. Tahecapsun-wi, December; the moon when the deer shed their horns.

The Dakotahs have a common and a sacred language. The conjuror, the war prophet, and the dreamer employ a language in which words are borrowed from other Indian tongues and dialects; they make much use of descriptive expressions, and use words apart from the ordinary signification. The Ojibways abreviate their sentences and employ many elliptical forms of expression, so much so that Half-breeds, quite familiar with the colloquial language, fail to comprehend a medicine man when in the full flow of excited oratory.

The American Missionaries, in their admirable written Dakotah language, employ five vowels, and twenty-four consonants, among which are two c's, two g's, two h's, two k's, two n's, two s's, two t's, and two z's. The repetition of the same letter is used to denote a guttural, an aspirate, an emphatic, or a nasal sound. Thus c is both an aspirate and an emphatic letter; g like the English g and guttural; h like the English h and guttural; k as in English and emphatic; n as in English and nasal; p as in English and emphatic; s as in English and aspirate; t as in English and emphatic; z as in English and aspirate.

All syllables are enunciated plainly and fully, but accentuation often determine the meaning of a word. There are three numbers: singulars, dual and plural; the dual including the person speaking and the person spoken to. The proper names of the Dakotahs are words, simple and compounded, which are in common use in the language. The son of a chief when he succeeds his father usually takes the name of his father or grandfather. As with the Ojibways and Swampys, their proper names consist of a single noun or a noun and adjective. The Ojibway have, however, distinct family or clan names which they employ when speaking of their ancestors; as I am of the family of the Bear; the Eagle, the Thunder-cloud, &c. The Dakotah have no sur-names, the children of a family have particular names which belong to them, in the order of their birth up to the fifth child. In counting they use their fingers, bending them as they enumerate until they reach ten. They then bend down a little finger to record one ten and begin again; when the second ten is counted they put down a second finger, and so on.

* See Grammar and Dictionary before referred to.

Dakotah verbs have only two forms of tense, the indefinite and the future; the other tenses are expressed by the help of adverbs, and the context. Words in a sentence are thus placed, first the noun, second the adjective, third the verb, thus:—

Ateunyanpi mahpiya ekta nanke chin
Father-we-have heaven in thou-art the;

Nichaze kin wakandapi kte;
Thy-name the holy-regarded shall;

Nitokichonze kin u kte;
Thy-kingdom the come shall; *

The Blackfeet.

Mr. James Doty, who resided for many years in the country of the Blackfeet and who is acquainted with a large portion of this nation, gave the following boundaries of their country and estimate of the numbers of the people to Governor Stevens in 1853.† The country in which they reside and hunt is bounded as follows: "By a line beginning on the north, where the 50th parallel crosses the Rocky Mountains, thence east on said parallel to the 106th meridian, thence south to the headwaters of the Milk River, down said river to the Missouri, up the Missouri to the mouth of the Judith, thence up the Judith to its source in the Rocky Mountains, and north along their base to the place of beginning."

The country between the Missouri and the headwaters of the Yellowstone is unoccupied. It is the great road of the Blackfeet war parties to and from the Crows, Flatheads and Snakes. It may also be considered as a transient hunting ground of the Flatheads, as they hunt buffalo there for a short time in the fall.

The Blackfeet nation is divided into four distinct tribes or bands, whose names, numbers and localities‡ are as follows:

The Blackfeet...	250	lodges;	1750	population;	625	warriors.
The Bloods‖	350	do.	2450	do.	875	do.
The Piegans	350	do.	2450	do.	975	do.
The Gros Ventres	360	do.	2520	do.	900	do.
Total	1310		9170		3375	

The Bloods and Blackfeet occupy the country between Milk and Marias Rivers, to the 50th parallel of latitude.

The Piegans occupy the country between the Milk and Marias Rivers, and between the Teton and the Missouri.

* See a grammar and dictionary of the Dakotah language, published by the Smithsonian Institution.

† Explorations and Surveys for a Railroad Route from the Mississippi to the Pacific, page 443.

‡ The country occupied by these tribes is evidently more extensive than supposed by Mr. Doty; their permanent lodges were found far beyond the limits given in the text.

‖ Called by the Half-breeds, "Bloodies."

The Gros Ventres occupy the country bordering upon Milk River from its mouth to the territory of the Piegans. The Bloods, Piegans and Blackfeet speak the same language; the Gros Ventres, the Arapahoe language; they were adopted by the Blackfeet about thirty years since, having seceded from their own nation. On the Upper Missouri, near the great bend, the Gros Ventres have a large village of mud houses. Some of the lodges are capable of supporting 100 persons. One part is appropriated to their horses, dogs, cattle, and chickens, another to their sleeping apartments. The lodges are built entirely by women. The Gros Ventres formerly hunted on the Asiniboine. Mr. J. M. Stanley, the artist of Governor Stevens' Exploration, states that the Blackfeet proper are divided into three distinct bands: the Blood band, 400 lodges; the Piegan band, 430 lodges; and the Blackfeet band, 500 lodges, averaging ten to a lodge, and amounting in all to 13,300 souls. The Piegans and Bloods hunt, trade, and winter on American soil, while the Blackfeet extend their hunts as far north as the Saskatchewan, and trade as frequently with the British as with the American Posts.*

The following census of the Indian tribes of the United States, inhabiting the States and Territories adjoining the 49th parallel, is abstracted from the statistics of the tribes as reported to the Bureau of Indian Affairs.†

Name of tribe.	Numbers.	
Assiniboine	8900	Extending from the Missouri into Rupert's Land.
Blackfeet	9530	Nebraska.
Bloods	1612	Upper Missouri.
Crees	800	Upper Missouri.
Sioux (Ihanktonwanna)	4000	Dakotah Territory.
Gros Ventres	2500	Between the Missouri and the Saskatchewan.

CHAPTER XIV.

ON THE ORIGIN OF THE VALLEY OF THE QU'APPELLE, AND ON THE DISPOSITION OF SOME OF THE DRIFT ON THE SOUTH BRANCH OF THE SASKATCHEWAN.

Depression of the Country in the Region of the Moose Woods—Erosion of the Qu'Appelle Valley—Streams enter the Qu'Appelle Valley at right angles—Breadth of the Valley throughout—Ridges with Boulders—Origin of—Depth of the Fishing Lakes—Ancient Lake—Erosion of Main Saskatchewan—Peculiarity in the Lakes of the Qu'Appelle — Back-Fat Creek—Possible Origin of Qu'Appelle Valley—Former Bed of a River, before the last submergence of a continent—Ancient River Valleys—Dr. Hitchcock's Enumeration of—Illustrations—The St. Lawrence—The Ottawa—Boulders in the Drift of the Saskatchewan—Boulders in the Blue Clay of Toronto—Forced Arrangement of—Mode in which Surfaces in the Blue Clay were exposed—Position of the Blue Clay—Lower and Upper Blue Clay—Disposition of the Boulders and Fragments of Shale— Illustrations of—Drift in Canada—Section of Drift—Discussion of the Mode in which the Boulders and Shale acquired a forced Arrangement—Thrown down a Subaqueous Bank—Objections to—Sorting of Materials—Agency of Ice.

THE ORIGIN OF THE QU'APPELLE VALLEY.

There are many features in the Qu'Appelle Valley which furnish materials for discussion. Some of these have been noticed in the preceding chapters, there are others, however, which deserve enumeration before venturing to express an opinion respecting the erosion of this long and deep excavation.

Our voyage down the South Branch has shown that in the region about the Moose Woods, the whole country is much lower than either north or south of that expansion of the trough in which the South Branch flows. It appears to have been the seat of a former dilatation of the River, if not of an extensive, wide-spreading lake, which existed at the time when the Qu'Appelle Valley began to be eroded by its overflowing waters. The prolongation of the Eye-brow Hill in the form of a low dividing ridge, as far as Lumpy Hill, at the base of which, for a distance of two hundred miles, the South Branch flows in a northerly direction, leads to the inference that the ridge marks the coast-line of a former lake, although no resemblance to beaches or terraces was seen near the South Branch. Yet these might occur at a distance of eight or ten miles on the east side, and not be visible from the high banks of the river.

It does not appear probable that a little streamlet like the 'River that Turns,' or the drainage of the Sandy Hills still in process of formation, or of the Eyebrow Hill range, could have worn away a hard rock at the height of land, and excavated a valley half a mile broad, and even now one hundred and ten feet deep, notwithstanding the sand-drifts, which have certainly diminished its depth by many feet. It has also to be borne in mind, that the Qu'Appelle itself, issuing from the Eye-brow

* Explorations and Surveys, page 449.

† *Vide* History, Condition and Prospects of the Indian tribes of the United States, by H. R. Schoolcraft, LL.D.

Hill range enters the great valley at right angles to its course, and a few hundred yards before joining it, flows through a narrow gully, not two hundred feet wide. The "River that Turns" and all the little streams coming from the Sandy Hills enter at right angles and flow down the bank of the great valley into the ponds which occupy it at the summit level. There is no evidence of any eroding agency besides these streamlets now existing, and no range of mountain or high table land from which streams draining into the valley might be supplied. It pursues a nearly straight course to the South Branch of the Saskatchewan, and maintains its breadth throughout. Were it not for the invasion of sand dunes, its outline would be exactly preserved from the Lake of the Sand Hills to the South Branch.

The plan of the Track Survey of the Qu'Appelle Valley, from Sand Hill Lake westward, showing its junction with the Saskatchewan, at the close of this report, exhibits in detail its most important features. The little streamlet from the Eyebrow Hill ridge, is the real source of the Qu'Appelle. The ponds at the Height of Land are the drainage of the Sand Hills and dunes which stretch far and wide, in a north-easterly and south-westerly direction.

There are three ranges of Sand Hills: one is shown a few miles west of Sand Hill Lake; the other, and most prominent, at the Dividing Ridge; and the third on the western slope, invading that part of the great valley through which the 'River that Turns' flows.

The ridges with boulders *on their western extremities*, occurring on the banks of the valley, on each side of the height of land, as described on pages 56 and 57 are curious illustrations of a force proceeding *from the west*. That force must have been water in motion, and although the forms of the ridges on the west side of the watershed in the valley are not so well defined as those on the east, yet they retain the distinguishing figure which is given to ridges shaped under the action of running water, while the disposition of the boulders on the west flanks appears to show that the direction of the current which bore the ice conveying them was from the west. The impression produced at the time when these ridges were examined, was strongly in favour of the supposition that many or all of them were formed at one and the same period, and by a current bearing ice, such as that of a great river like the St. Lawrence or the Main Saskatchewan, flowing easterly.

It will be observed, from an inspection of the table of the depth of the lakes in the Qu'Appelle valley, page 58, that the deepest Fishing Lake, as far as our soundings show, is the first and most easterly of the four; the smaller depth of the other Fishing Lakes may be explained by the occurrence of streams entering the valley from the prairies, and bringing down with them during spring freshets, solid matter mechanically suspended, which would tend to diminish their depths in proportion to their proximity to the source of supply.

The existence of an ancient lake, of great extent, lying west of the prolongation of the Eyebrow Hill range to the Lumpy Hill of the Woods, is shown by the long horizontal lines of boulders which appear in the clay cliffs of the river below the Moose Woods. Above these parallel lines of boulders, fine stratified mud is seen in layers, together with stratified sand and gravel. These horizontal tiers of boulders are described in Chapter V. page 64.

Conditions similar to those which would be required to produce this arrangement, exist at the present day in Lakes Manitobah and St. Martin. The boulders stranded on the extensive shoals in those shallow bodies of water, as described in Chapter IX. are probably modern illustrations of the mode in which this distribution in long horizontal lines was effected.

I conceive that the South Branch, during the existence of this supposed lake, flowed into it, and that its waters, or part of them, were discharged by the valley of the Qu'Appelle, and during that period the ridges were moulded, and the boulders distributed on their western extremities. The deep Fishing Lakes and the other lakes which now occupy a considerable portion of the valley, are the remains of the excavation. At that period Pembina Mountain, the Blue Hills of the Souris, and the flanks of the Riding Mountain probably formed the limit of Lake Winipeg. These boundaries are more fully described in the chapter on the surface geology of the country explored. During the drainage of this region, and after the ancient lake, whose centre would be near the Moose Woods, had excavated a sufficient outlet for its waters down the present valley of the Main Saskatchewan, the Qu'Appelle valley would no longer contribute to its drainage, but receive only the drainage of the country which it now unwaters. A part of its valley would slowly undergo the process of filling up, either by drifting dunes, as at the height of land, or by washings from the prairie at the mouth of streams coming from the north and south.

Long Lake affords another instance of an ancient river valley, and it does not appear improbable that future observations will establish its connection with the same supposed ancient lake before alluded to. The Back-fat Lakes and Creek, inosculating with Pembina River, were probably the valley of a stream debouching into Lake Winipeg when it washed Pembina Mountain.

The remarkable depth of the Fishing Lakes, and those lying further to the east, considered in connection with other well-known phenomena, may suggest another explanation of their origin. It has been stated in the narrative, that north of the Moose Woods, there are to be seen large blocks of limestone, containing many thousand cubic feet; these repose on the surface of the prairie, and doubtless they now occupy the position they assumed when brought thither by icebergs during the last period when that portion of the continent was under the waters of the ocean. The huge unfossiliferous boulder, seventy eight feet in circumference, which lies in the valley of the Qu'Appelle, was probably slowly sunk to its present position by the wearing away of its foundation as the valley was in process of formation, or it may have rolled from the prairie bank as it became undermined. It is not impossible, however, that it now occupies the spot where it was originally dropped from the ice-floe which bore it from the north. This would involve the assumption that the Qu'Appelle valley dates the epoch of its erosion anterior to the last submergence of the continent, affording an illustration of a river valley before the epoch of the boulder drift. The physical aspect of the country is by no means opposed to this view; although there are other reasons which may be urged in opposition to it.

The occurrence of ancient river valleys on this continent has already attracted attention. In his Illustrations of Surface Geology, Dr. Hitchcock says: "Some of the erosions that have been described in this paper are clearly the beds of antediluvial rivers; that is of rivers existing upon this continent before its last submergence beneath the ocean; which beds were deserted when the surrounding surface emerged from the water, although essentially the same rivers as existed previously, must have been the result of drainage.

"The grounds on which I refer the cases mentioned below, and described in detail in this paper, to the latest of former continents are the following:

"1. The occurrence of pot-holes in the walls of gorges, which are either dry or the bed of a brook too small to have produced them.

2. The outlet of such gorges in one direction into valleys now containing streams large enough to have formed the gorges, and in the other direction, into valleys leading at a gentle descent to some rivers.

These two facts make it certain that the gorges were once the beds of rivers.

3. An accumulation of water-worn and perhaps sorted materials, viz: gravel and sand to a considerable depth. This accumulation appears to me to have been made during the last submergence of the land, and to be the cause that prevented the ancient rivers from occupying their old channels upon the drainage of the country, and compelled them, at least for a considerable distance to find a new channel. I consider the following as examples of the phenomenon, most of them very decided; that is, of these antediluvial river beds."

Here follows an enumeration of ten ancient river beds in Canada, (Niagara) New England and the State of New York.

It is however in the bed of the St. Lawrence and the Ottawa that we find the most striking illustration of ancient river valleys and the most convincing proof that the form of the continent anterior to its last submergence was similar to its present outline. The rivers of a former continent had excavated channels through rock formations extending from the Tertiary to the lower Silurian; during the period of submergence the river valleys were partially filled up by drift, and when the continent rose again, or the sea lowered its level, the new rivers, draining regions differing but slightly from the old physical outline of the former continent, sought out their ancient channels, and if not filled with drift, occupied them at once; or, if obstructed by drift, re-excavated part of their former channels, and pursued their old courses to the sea.

Subjoined is an illustration from the valley of the St. Lawrence, taken from Sir Charles Lyell's *Manual of Elementary Geology:*

"I described, in 1839, the fossil shells collected by Captain Bayfield from strata of drift at Beauport near Quebec, in lat. 47°, and drew from them the inference that they indicated a more northern climate, the shells agreeing in great part with those of Uddevalla, in Sweden.* The shelly beds attain at Beauport and the neighbourhood a height of 200, 300, and sometimes 400 feet above the sea, and dispersed through some of them are large boulders of granite, which could not have been propelled by a violent current, because the accompanying fragile shells are almost all entire. 'They seem, therefore,' said Captain Bayfield, writing in 1838, 'to have been dropped from melting ice, like similar stones which are now annually deposited in the St. Lawrence.'† I visited this locality in 1842, and made the annexed section, which will give an idea of the general position of the drift in Canada and the United States. I imagine that the whole of the valley (B) was once filled up with the beds *b*, *c*, *d*, *e*, *f*, which were deposited during a period of subsidence, and that subsequently the higher country (*h*) was submerged and overspread with drift. The partial re-excavation of B took place when this region was again uplifted above the sea to its present height."

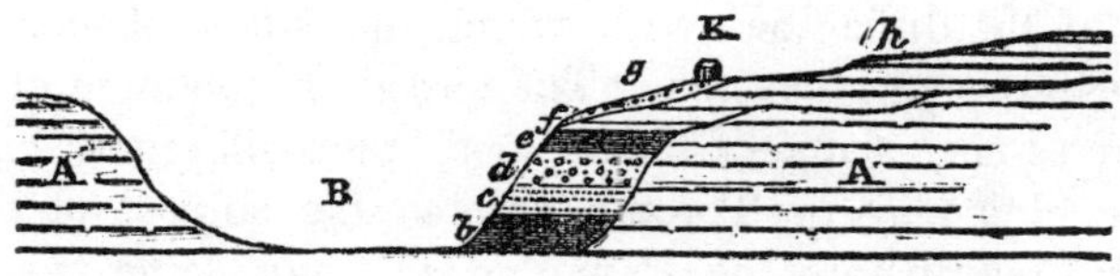

K. Mr. Ryland's House.
h. Clay and sand of higher grounds, with *Saxicava*, &c.
g. Gravel, with boulders.
f. Mass of *Saxicava rugosa*, 12 feet thick.
e. Sand aud loam, with *Mya truncata*, *Scalaria Grœnlandica*, &c.
d. Drift, with boulders of Syenite, &c.
c. Yellow sand.
b. Laminated clay, 25 feet thick.
A. Horizontal Lower Silurian strata.
B. Valley re-excavated.

La Grande Coulée, in the Blue Hills of the Assiniboine, described by Mr. Dickinson, page 18, offers another illustration of an old river valley, but probably of more recent origin than that of the Qu'Appelle, nevertheless a curious and instructive example of surface geology in this part of Rupert's Land.

"We crossed another of these valleys here so numerous, called "La Grande Coulée de la Grosse Butte," deriving its name from a large conical hill about two hundred feet high. The valley varies in width from twenty to thirty chains, and is about eighty feet deep, but appearing much deeper in many places, by reason of the hills adjoining it. The sides are very precipitous and the bottom is quite level and covered with beautiful grass; there is no creek flowing through it, or even the appearance of any recent one. Two miles up in it towards the north there is a small lake, and another valley branching off from it, which we crossed four miles further on; in it there is a small creek six feet wide and one foot six inches deep. The track turning to the north soon comes close to "La Grande Coulée de la Grosse Butte," and continues along it for nine miles. The scenery is now very wild and beautiful; the valley, the bottom of which is eighty feet below the general level of the country, *cuts through ranges of hills, many of them one hundred and fifty feet high, and winds round the base of others,* some bare and rugged and some covered with poplars."

The section of the Qu'Appelle Valley from the South Branch to the Assiniboine, with cross-sections at the several points marked A, B, C, D, &c. (*see lithograph at the end of the volume*), considered with reference to the general features of the country and its geological structure, will be amply sufficient to prove that the deep lakes could not have been occasioned by falls or rapids. Nor can we assume that the strata at these points was of such a soft and yielding nature as to admit of its being eroded into the form of long, deep, and narrow basins at wide intervals apart. The weight of evidence seems to be in favour of the view that the South Branch of the Saskatchewan, at a remote period, flowed down the valley of the Qu'Appelle, and debouched into the low country bounded by the Pembina Mountain, or its continuation northwards.

THE DISPOSITION OF SOME OF THE DRIFT ON THE SOUTH BRANCH.

It has been stated in Chapter V. page 64, that boulders and small masses of shale in the drift cliffs which occur at the bends of the river below the Moose Woods, do not occupy the position they would assume if they had followed the law of

* Geol. Trans., 2nd series, Vol. vi. p. 135.
† Proceedings of Geol. Soc., No. 63.

gravity, supposing them to have been dropped by icebergs or ice floes. Every fact relating to the Drift, whether belonging to the boulder period or of more recent origin, is of interest, and may assist in the elucidation of that stupendous phenomenon and its subsequent changes, as well as tend to remove some of the difficulties with which the whole phenomena of the Drift are still invested. The forced arrangement of blocks of limestone, slabs of shale and unfossiliferous boulders in the blue clay of Toronto, formed the subject of a paper which I read before the Canadian Institute some years ago. As the opportunities for making observations upon this peculiar arrangement were very favourable at that time, I shall here introduce an abstract of the paper, with a view to explain more clearly than would otherwise be possible, the manner in which slabs and boulders are found arranged in the Drift on the South Branch.

The extensive excavations which were made three and four years ago in the clay deposits on which the City of Toronto is built, during the construction of various public works, such as the Esplanade and the Grand Trunk Railway, presented a very favourable opportunity for examining some peculiarities in the arrangement of the materials of which the Toronto blue clay consists. In the construction of the Esplanade, the plan pursued of removing the blue clay was well adapted to show a perfect sectional view of its components, without the risk of changing in the least degree their relative positions. The clay was cut away until a perpendicular wall was left, varying from 10 to 20 feet in height, according to the locality. Wedges were then inserted at the top of the artificial cliff, about two feet from its edge, and driven into the clay until a mass, frequently two feet broad, 15 or 20 feet long, and 12 or 18 feet deep, separated and fell. The fresh surface thus exposed was necessarily quite natural in every respect, not having been touched by the tool of the workman or changed by exposure to the weather.

During the years 1855 and 1856, a large area of sectional surface was exposed in this way on the Bay shore, and frequent examinations of the continually renewed surfaces led me to study the disposition of the materials composing the blue clay. Two varieties of blue clay exist in the neighbourhood of Toronto, forming deposits quite distinct from one another; it is, therefore, desirable to fix at once the position of the blue clay to which reference is now made. The deposit in question overlies the rocks of the Hudson River group, which are exposed in many localities on the lake shore and on the banks of the rivers near the city. Its position was well seen during the working of a quarry opposite the Parliament Buildings; it was there observed to rest upon an argillaceous shale of the same hue, and easily recognized as constituting, in fragments of different sizes, a large proportion of the substance of the blue clay. It can also be seen resting on the rocks of the same formation, a little beyond the new garrison, a few feet above the lake level, where it is not obscured by the debris of the cliff of which it forms the base—the upper portion of which is composed of yellow clay.

The thickness of this deposit of blue clay varies from 10 to 25 feet; its upper surface is irregular and undulating: upon it reposes sometimes stratified sand and yellow clay, sometimes unstratified yellow clay. Resting on the sand or yellow clay, we find another kind of blue clay differing, however, essentially from the blue clay which lies at the base of the whole. This upper blue clay is well seen along the Scarboro' cliffs, where it is best exposed, and it is also recognized in many other localities near and in Toronto. The lower or inferior blue clay contains quartz sand and small rolled pebbles of granitic rocks, a considerable proportion of blue shale containing fossils belonging to the Hudson River group, and frequently large fragments of the last named rock, together with more or less rolled or worn masses of granite, gneiss, &c.

The fragments from the Hudson River group frequently preserve their edges sharp and well defined, showing that they have not been water worn or removed far from the rock from which they originated. They are found not only a few inches from the surface of the parent rock, but in numerous instances as far as 15 to 20 feet above it, imbedded in a peculiar manner in the blue clay. Some of the larger fragments are scratched and grooved.

A cursory inspection of the artificial cliffs, as they existed during the construction of the Esplanade, was sufficient to show that a considerable number of the pebbles and imbedded masses of rock did not occupy the position they would assume if they had not been subjected to some other force besides that of gravity or water in motion. The inclination of the subjacent rock is so slight (30 feet in the mile) that for all purposes of the present inquiry it may be considered horizontal. And it may be further remarked, that there is no reason to suppose that any material change in position has occurred since or during the accumulation of the blue clay. A large number of the fragments of rock seen in the blue clay are symmetrically inclined at an angle of 60, 70, and 80 degrees to the horizon, and frequently lean towards the east and north-east. Whenever favourable opportunities offered, I made measurements of some of the most striking of these rock fragments, and rough sketches of their position as they were revealed by the falling masses of the cliff loosened in the manner already described.

The following brief notes will serve to illustrate this peculiarity better than a more lengthened description.

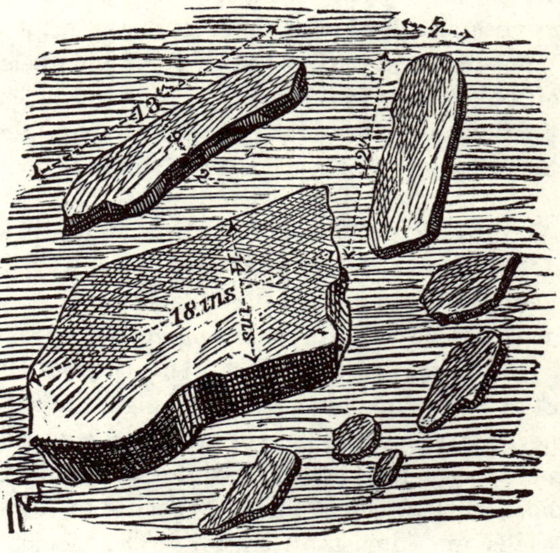

(*Fig.* 1.)

1. A mass of shale (Fig. 1) imbedded in the blue clay about 2 feet from its surface, and 18 from the solid rock. Largest diameter, 18 inches; breadth, 14; thickness, 7; inclined at an angle of about 50 degrees, and leaning towards the north-east. The greater number of water-worn stones and unworn fragments of shale appear to have the same inclination in this spot. Locality near the Water Works.

2. The general inclination of the fragments of shale a few

hundred yards from the last named place, is at an angle of 60° and toward the east. (Fig. 2.)

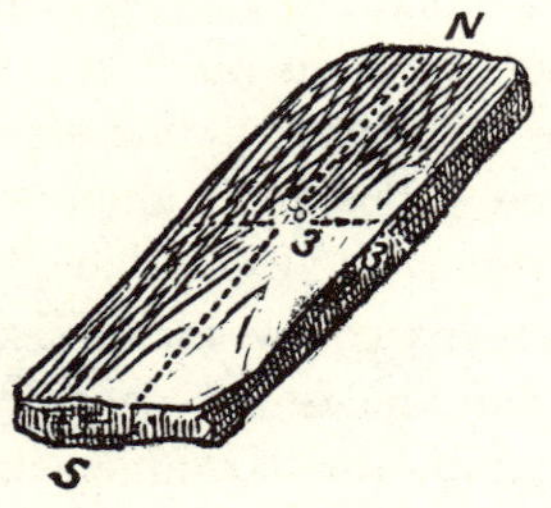

(*Fig.* 2.)

3. Boulders of gneiss distinctly seen in the blue clay, associated with perpendicular fragments of shale.

4. A slab from the Hudson River group 18 inches long, 15 broad, 3 and 4 thick, very little water-worn, 7 feet from the top of the blue clay, and 10 feet from the solid rock, inclined at a high angle towards the north-east. A boulder of gneiss near this block, not much worn, and slightly inclined in the same direction. Around the slab, numerous smaller fragments of rock present the same inclination. (Fig. 3.)

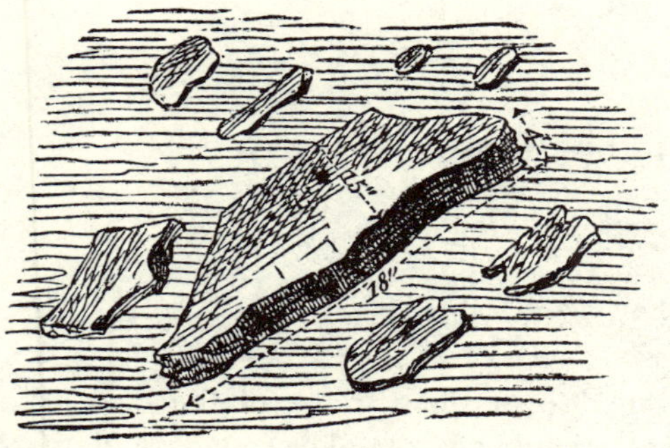

(*Fig.* 3.)

These illustrations represent the general character of the position of rock fragments in the blue clay for several miles along the lake shore. What force has thus symmetrically arranged these fragments of shale, &c.? That they now preserve the position into which they were forced by pressure, or that they were brought from a distance and left in that position, is sufficiently evident, as we cannot entertain the opinion that the rock on which the boulder drift rests has materially changed its inclination since or during the Drift epoch.

The materials composing the blue clay are of two descriptions—foreign and local. The same may be said of Drift generally. It has been observed by Mr. Murray that the coarser fragments reposing upon each successive formation in the order in which they occur in Canada, is made up with the addition of whatever is of primary origin, of material derived from the formation itself, or of the ruins of some lower deposits whose outcrop is to the north.

The granitic fragments present in the blue clay of Toronto are evidently derived from the north or north east, and must have travelled at least one hundred miles before they were lodged in the place where they are now found. There can be no doubt that a very large portion of the drift of Canada has been rearranged since it was first deposited. The inferior layer of blue clay is, however, essentially different from the upper layer which is frequently separated from it by a few feet of sand, and in some instances may even directly overlie it and consist of a re-arrangement of its materials. The superior blue clay, together with the sand and yellow clay, frequently give evidence of stratification, and thus explain at once the nature of the force to which they have been subjected. (Fig. 4.)

FIG. 4

The position of the rock fragments in the inferior blue clay shows that it cannot have been subjected to the action of water, otherwise they would not preserve the forced arrangement which distinguishes them. The fragments of shale, as represented in Fig. 4, if submitted to gravity alone, would not have assumed the position in which they were found, had they dropped through water in motion or water at rest, into soft mud. It is well known that shingle, sand, gravel, and clay, either separately or combined, when thrown down an incline, as in the construction of a railway embankment or as in a land slip, will assume a position upon the surface of the embankment, which, if constructed of sand, is generally inclined about 45°; if of harder or coarser materials, at a higher angle If the embankment or incline be formed under water, like the deltas at the mouths of rivers, this inclination is much less, and is dependent upon the specific gravity of the materials; but under no circumstances is it so high as 45° when the bank is formed under water If, now we conceive a current sufficiently powerful to move masses of shale and boulders of the unfossiliferous rocks, it is not to be supposed that they would be found deposited upon the slope of a bank at so high an angle as the shale and boulders in the blue clay of Toronto; neither is it in the least degree probable that the current which could transport these heavy materials would admit of the mixture of clay, sand, shale and boulders, such as constitutes the blue clay. The materials would be sorted by the current and deposited in the order of their specific gravity. The *sorting* of *materials* is one of the most positive proofs of the action of currents; and where no trace of *sorting* can be discovered, when fine sand, coarse sand, pebbles, and boulders are present, we may reasonably infer that no current assisted in distributing them.

Among the foreign materials entering into the composition of the blue clay, we find granitic masses which have been brought from the outskirts of the fossiliferous rocks in Canada, a distance of at least one hundred miles from their present position; throughout the blue clay we discover also the magnetic oxide of iron, which is found in such abundance in the washed sand of the Peninsula of Toronto Harbour, and in tenfold greater quantity on the Peninsula of the Rondeau in Lake Erie, at treble the distance from its northern source. The materials of local origin exist in great abundance in the form of fragments and masses of shale, limestone, and clay derived from the underlying shales, &c. The nature of the agent which transported the foreign materials from so great a distance is almost universally acknowledged to have been water and floating ice. The finer materials may have been conveyed by water, the coarser drift and erratics would require floating or moving ice. There can be little doubt that both water and floating ice (ice bergs and floes) have been instrumental in bearing from northern fossiliferous and unfossiliferous rocks a considerable proportion of the numberless erratics which strew the surface of a large part of this continent, as well as much of the clayey deposits which we see everywhere around us. But the symmetrical arrangement of some of the slabs, pebbles and boulders in the blue clay at Toronto, in the clay cliffs of the South Branch of the Saskatchewan, and in other localities where the same disposition may be witnessed, points also to the action of *glacial* or *stranded* ice. The phenomena may be explained by coast ice, or the dirt bands of glacial ice, but the entire absence of a sorting of fine and coarse materials, seems to destroy the hypothesis which introduces the agency of currents of water, as the forced but symmetrical arrangement does that of floating ice. May not the plastic and irresistable agent which picked up the materials composing the blue clay, and then melting, left them in their present position, have been largely instrumental in excavating the basins of the great Canadian Lakes?

CHAPTER XV.

CLIMATE OF A PORTION OF RUPERT'S LAND.

Climate of the Laurentides and the Prairies—Frozen Lakes—Mean Annual Temperature — Arid and Humid Region—Sources of Humidity—Cause of Aridity West of the 98th Meridian—Influence of the Gulf of Mexico—Rocky Mountain System—Mississippi Valley — Arid Region of the United States—Humid Region of the Valley of Lake Winnipeg—Causes of—Elevation of the Country—Humid Pacific Winds —North-easterly Current—The Arid Region — Prevailing Winds—Source of the Humidity—Hail Storms—Thunder Storms in 1858—Progress of Dunes—Summer Surface Wind — Rocky Mountain Plateau—Depression in—Table of Elevation of Plateau and Passes—Importance of Capt. Palliser's Discoveries — Seasons of the Valley of Lake Winnipeg — Meteorology of Red River—Winter Temperatures—Winter Temperatures at Montreal—Cold Terms—Quebec Temperatures —Climate of the South Branch of the Saskatchewan—Limit of Permanently Frozen Soil —Growth of Forests—Tail of the Prairies — Prairies Converted into Forest Land in Missouri—Seasons on the Main Saskatchewan—At Fort à la Corne—At Cumberland House—At Carlton House—At Red River—Character of the Great Plains in the United States—Major Emory's Statement—Auroras—Oct. 2nd—Oct. 27th—Colonel Lefroy's Observations—Altitude of Auroras—Connection with the Atmosphere—Sound—Sir John Richardson's Observations—The Twilight Bow.

The climates of Canada and Rupert's Land under the same parallels of latitude, vary to a considerable extent with the rock formations of the country. Throughout the undulating region of the Laurentides the proportion of water to dry land is about one to two, not collected into one large water area, but distributed over the surface of the country in the form of countless thousands of lakes, ponds and marshes. The intense cold of winter is sufficient to solidify the deepest lakes for a depth of several feet, and the thawing of so much ice in spring has the effect of absorbing and rendering latent the heat which would be otherwise expended in warming the soil and advancing vegetation.

Lakes Winnipeg, Manitobah, and Winnipego-sis, together with the smaller lakes belonging to the Winnipeg basin, are deeply frozen every winter, and ice often remains in their northern extremities until the beginning of June, greatly retarding the progress of vegetation on their immediate shores. Hence one reason that north of the 47th or 48th parallel the mildness of the seasons increases rapidly as we advance towards the west, after leaving Red River. The improvement arises not only from greater longitude but also from the character of the rock formations by which the country is underlaid and surrounded. The soil of the Prairies is in general dry and is rapidly warmed by the rays of the sun in spring. The Prairies enjoy too, north of the 58th parallel, the genial, warm and comparatively humid winds from the Pacific, which are felt as far north as the latitude of Fort Simpson.*

The mean annual temperature of 40°, as determined by the Smithsonian Institution, passes through Canada and Lake Superior, curves northward and leaves the United States for British America at about the 103rd meridian, crossing the South Branch of the Saskatchewan north of the Elbow.

The country embraced within the limits of this exploration may be divided into two regions in relation to climate; the arid

* Colonel Lefroy—Meteorological observations at Lake Athabasca and Fort Simpson, p. 139.

and the humid region. The vast treeless prairie west of the Little Souris lies within that part of the area which receives comparatively a small annual rain-fall. Its northern limit is roughly shown by the Qu'Appelle Valley, or more accurately by an imaginary line drawn from the Fishing Lakes to the Moose Woods. North and east of this area the precipitation is considerably greater, and supplies the valley of the Main Saskatchewan, the Touchwood Hill Range, and the valley of the Assiniboine with an abundance of moisture, which is protected and treasured by forests.

The valley of Red River east of the Little Souris, or the 101st degree of longitude, receives much humidity from the moist winds coming from the Gulf of Mexico up the valley of the Mississippi, and over the low Height of Land which separates the waters of Red River from those of the St. Peter.

The Touchwood Hill Range and the country generally north of the Qu'Appelle valley, and in an easterly direction towards and beyond Lake Winnipeg, are made humid by the south-west Pacific wind, in concurrence with the prevailing east wind of this region. These phenomena are referred to in detail in succeeding paragraphs.

The cause of the aridity and unfitness for settlement of fully one-third of the United States has been ably discussed by distinguished meteorologists. The physical geography of that vast region has been very admirably described by Dr. Joseph Henry.* I avail myself of a few extracts from Dr. Henry's paper to illustrate the causes which produce the aridity of a large portion of the valley of Lake Winnipeg, and the probable explanation of the humidity of the region properly belonging to the subordinate valley of the Assiniboine.

"The climate of a district is materially affected by the position and physical geography of the country to which it belongs. Indeed, when the latitude, longitude, and height of a place above the sea are given, and its position relative to mountain ranges and the ocean is known, an approximate estimate may be formed as to its climate.

"At the southern extremity of the United States is the great elliptical basin containing the perpetually heated waters of the Gulf of Mexico, an enormous steaming cauldron continually giving off an immense amount of vapour, which, borne northward by the wind of the south-west, gives geniality of climate and abundant fertility to the eastern portion of our domain. On the western side of the continent the coast presents, as a whole, an outline of double curvature, principally convex to the west in that part which is occupied by the United States, and concave further north. These bends of the coast-line and of the adjacent parallel mountain ridges affect the direction of the winds in this quarter, and consequently of the ocean currents. The Gulf of California at the south, between the high mountains of the peninsula of that name and those of the main land, must also modify materially the direction of the wind in that region.

"The continent of North America is traversed in a northerly and southerly direction by two extensive ranges of mountains—the Alleghany system on the east and the Rocky Mountain system on the west. We give the latter name to the whole upheaved plateau and all the ridges which are based upon it. These two systems separate from each other more widely as we pass northward, and between them is the broad interval which, within the territory of the United States, is denominated the valley of the Mississippi; but in reality the depression continues northward to Hudson's Bay, and even to the Arctic Ocean, giving free scope to the winds which may descend from that inhospitable region. It, however, may be divided into two great basins, one sloping towards the south, comprising the basin of the Mississippi, and the other sloping to the north, including the basins of Mackenzie's river and of Hudson's Bay, the dividing swell which may be traced along the heads of the streams having an elevation of about 1,200 feet.

"The general character of the soil between the Mississippi river and the Atlantic is that of great fertility, and as a whole, in its natural condition, with some exceptions at the west, is well supplied with timber. The portion also on the western side of the Mississippi, as far as the 98th meridian, including the States of Texas, Louisiana, Arkansas, Missouri, Iowa, and Minnesota, and portions of the Territory of Kansas and Nebraska, are fertile, though abounding in prairies and subject occasionally to droughts. But the whole space to the west, between the 98th meridian and the Rocky Mountains, denominated the Great American Plains, is a barren waste, over which the eye may roam to the extent of the visible horizon with scarcely an object to break the monotony. From the Rocky Mountains to the Pacific, with the exception of the rich but narrow belt along the ocean, the country may also be considered, in comparison with other portions of the United States, a wilderness unfitted for the uses of the husbandman; although in some of the mountain valleys, as at Salt Lake, by means of irrigation, a precarious supply of food may be obtained sufficient to sustain a considerable population, provided they can be induced to submit to privations from which American citizens generally would shrink. The portions of the mountain system further south are equally inhospitable, though they have been represented to be of a different character. In traversing this region, whole days are frequently passed without meeting a rivulet or spring of water to slake the thirst of the weary traveller.

"We have stated that the entire region west of the 98th degree of west longitude, with the exception of a small portion of western Texas and the narrow border along the Pacific, is a country of comparatively little value to the agriculturist; and, perhaps, it will astonish the reader if we direct his attention to the fact that this line, which passes southward from Lake Winnipeg to the Gulf of Mexico, will divide the whole surface of the United States into two nearly equal parts. This statement, when fully appreciated, will serve to dissipate some of the dreams which have been considered as realities as to the destiny of the western part of the North American continent. Truth, however, transcends even the laudable feelings of pride of country; and, in order properly to direct the policy of this great confederacy, it is necessary to be well acquainted with the theatre on which its future history is to be enacted and by whose character it will mainly be shaped."

HUMID REGION OF THE VALLEY OF LAKE WINNIPEG.

Prominent among the causes which tend to give humidity, together with an elevated spring and summer temperature, to a part of the valley of Lake Winnipeg, there may be noticed:—First, the comparatively low elevation of the country above the sea level. The prairies of Red River within British Territory are not more than 730 feet above the ocean. Those on

* Meteorology in its connection with Agriculture, by Prof. Joseph Henry, Secretary of the Smithsonian Institution.

the South Branch of the Saskatchewan, at the Elbow, do not exceed 1600 feet, and the mean elevation of the country between the South Branch and the Riding Mountain is only 1200 feet above the same level.

Second, the influence of the warm westerly winds from the Pacific Ocean, in connection with the prevailing northeast wind, which is one of the established physical phenomena of this part of British America. It would appear, at first sight, that the snow-capped ridges of the Cascade, Blue, and Rocky Mountains would abstract so much heat from the warm westerly winds coming from the Pacific Ocean, as to neutralize their influence upon the winter and spring temperature of a large part of the country drained by the Saskatchewan. Such, however, is not the case; and happily for the purpose of practically substantiating this apparent anomaly, we have indisputable testimony.

In the magnetical and meteorological observations at Lake Athabasca and Fort Simpson, by Col. Lefroy, R.A., we find the following important observation, in relation to the phenomena of Pacific winds affecting the climate of the Northern regions :—

" A local phenomena of interest was observed several times at Fort Simpson, (Lat. 61° 51′ 7″ N.; Long. 8h. 5′ 40″ W.; 460 miles from Sitka, (1800 geo. miles from Toronto,) in the rapid rise of the temperature of the air, when the wind changed to the south-west from an easterly direction. It appeared as if the warmer air of the Pacific ocean were transferred across the neighbouring ridges of the Rocky Mountains with little loss of its temperature."

Much of the precipitation in the humid region is due to the Pacific winds, which are not so completely deprived of their moisture in traversing the Rocky Mountain ranges as in lower latitudes, where the average altitude of these ranges is much higher and the eastern slope of the mountain of a *far greater mean annual temperature.*

The prevalent winds at Toronto and Lake Athabasca belong, as shown by Colonel Lefroy,* to different and nearly opposite systems. A *north-westerly* current preponderates in the lower latitudes (43° 39′), a *north-easterly* current, inclined at an angle of about 117° prevails in the higher one (58° 45′). Between these latitudes is a region of calm or of variable winds; and there can be no doubt that the north-easterly current materially affects the humidity of the climate of Rupert's Land north of the 50th parallel. The prevalence of north and north-easterly winds during the winter months occasions a great precipitation of snow throughout the humid region. In the Touchwood Hill range snow not unfrequently accumulates in the woods, where it is undisturbed by winds, to the depth of two feet; on the Riding and Duck Mountain the precipitation is also large, and throughout the humid region very much in excess of the precipitation in lower latitudes.†

Forty-eight inches of rain and thirty-nine inches of snow were registered by Mr. Gunn near the Stone Fort, Red River, between June 1st, 1855, and May 31st, 1856. The precipitation at Toronto during the same period was thirty inches of rain and seventy-two of snow, giving an excess of humidity to the climate of Selkirk Settlement, as compared with Toronto for that period, represented by fourteen inches, a quantity exceeding the annual precipitation over the greater portion of the eastern flank of the Rocky Mountains south of the great Missouri bend.

The arid region, or Great Plain, west of the 101st degree of longitude receives a very small amount of precipitation from the humid south winds coming up the valley of the Mississippi from the Gulf of Mexico. It is too far south to be much affected by north-east winds, or the westerly winds from the Pacific. This vast treeless prairie forms in fact the northern limit of the great arid region of the eastern flank of the Rocky Mountains; but still its humidity is greater than the plains south of the Missouri, in consequence of its high northern latitude.*

* Magnetical and Meteorological Observations at Lake Athabasca.

† Compare Lorin Blodget's Rain Charts.

*From its mouth to the Great Bend, the Missouri admits of almost continuous settlement on its *immediate* banks; thence, to Fort Union, only about one-fourth could be cultivated; and above Fort Union many extensive but detached bottoms show their adaptation for small independent agricultural areas.

The general westward progress of settlement a few miles west of the Upper Missouri River is rendered impossible by the conditions of climate and soil which prevail there. The progress of settlement must necessarily be up the valley of the Mississippi, on the immediate banks of the Missouri, and through the valley of the Red River of the north, to the cultivable areas in the Basin of Lake Winnipeg. The exploration for the Pacific Railroad and the meteorological investigations carried on under the direction of the Surgeon General of the U. S. army show conclusively that no settlement of any importance can be established over a vast extent of country. many hundred miles broad, on the eastern flank of the Rocky Mountains, and south of the Great Bend of the Missouri. Owing to the absence of rain, the apparently great rivers, the Platte, the Canadian, the Arkansas, &c., are often converted into long detached reaches or ponds during the summer months, and forbid extensive settlements even on their immediate banks. This great and important physical fact is contrary to popular opinion, which is mainly based upon an inspection of a map, and guided by the glowing but utterly erroneous descriptions which are periodically circulated respecting the wonderful fertility of the Far West, and its capability of sustaining a dense population.

The arid districts of the Upper Missouri are barren tracts, wholly uncultivable. from various causes.† The arid plains between the Platte and Canadian Rivers are in great part sand-deserts. The " Sage-plains," or dry districts, with little vegetable growth except varieties of Artemisia, begin on the western border of the plains of the eastern Rocky Mountain slope, and cover much the larger portion of the whole country westward.‡ The sterile region on the eastern slope of the Rocky Mountains begins about 500 or 600 miles west of the Mississippi, and its breadth varies from 200 to 400 miles; and it is then succeeded by the Rocky Mountain range, which, rising from an altitude of 5,200 in lat. 32°, reaches 10,000 feet in lat. 38°, and declines to 7,490 feet in lat. 42° 24′, and about 6,000 in lat. 47° Along this range isolated peaks and ridges rise into the limits of perpetual snow, in some instances attaining an elevation of 17,000 feet. The breadth of the Rocky Mountain range varies from 500 to 900 miles. The soil of the greater part of the Sterile Region is necessarily so from its composition, and where well constituted for fertility, from the absence of rain at certain seasons. The general character of extreme sterility likewise belongs to the country embraced in the mountain region.§ The table subjoined is capable of conveying a very good idea of the great barrier to the westward progress of settlement which lies between the Mississippi valley and the Pacific slope of the Rocky Mountains. It is extracted from a table shewing the lengths, sums of ascents and descents, &c. &c., of the several routes surveyed for a railroad from the Mississippi to the Pacific, and published in the " Explorations and Surveys " before quoted,—page 31.

	Length of Railway.	No. of miles of Route through Arable Land.	No. of miles of Route through lands generally uncultivable, arable soil being found in small areas.	No. of sq. miles of sums of areas of largest bodies of arable land in uncultivable regions.
	Miles.			
Route near the 47th and 49th parallel	1864	374	1490	1000
" " 41st and 42nd "	2032	632	1400	1100
" " 38th and 39th "	2080	620	1460	1100
" " 35th "	1892	416	1476	2300
" " 32nd "	1618	408	1210	2300

This table shows that the least distance of uncultivable land through which a railway from the Mississippi to the Pacific must pass, in the United States terri-

† Page 684, Army Meteorological Register, U. S.

‡ Ibid.

§ Exploration and Surveys for a Railroad Route from the Mississippi River to the Pacific Ocean—page 40.

PREVAILING WINDS.

All the thunder-storms we encountered in 1858 in the valley of Lake Winnipeg, came from the west, south-west, or north-west, with one exception. I do not find a single record of thunder-storms with heavy rain coming from the south. This may have been an exceptional year, but the warmth and dryness, often oppressive, of the south wind, west of the 100th degree of longitude, contrasted strongly with the humidity and coolness of winds from the west. This phenomenon is directly opposed to those which prevail in lower latitudes, and may probably be explained as follows :

Warm air from the Pacific, loaded with moisture, passes at certain periods of the year over the whole range of the Rocky Mountains in British America and in the United States. These Pacific winds occasion but a very small precipitation of rain or snow on the eastern flank of the Rocky Mountains, south of the Great Missouri Bend. Similar winds from the Pacific do occasion a considerable precipitation in the northern part of the Saskatchewan valley. Whence, then, this apparent anomaly? It probably arises from the difference in the temperature of the two regions, the direction of the prevailing winds, and the lowness and comparatively small breadth of the Rocky Mountain ranges in that latitude. In spring and summer, warm westerly winds ladened with moisture, in passing over the mountain range south of, say the 46th parallel, are cooled to a certain temperature, and precipitate the greater portion of their moisture, in the form of rain or snow, upon the mountain ridges. On arriving at the eastern flank of the Rocky Mountains, their temperature rises to that of the region over which they pass, being elevated bythe deposition of their moisture in the form of rain or snow, and continually increasing density as they descend ; but the capacity of air for moisture is well known to be dependent upon its temperature, within certain limits, hence the westerly Pacific winds become more warm and more dry as they descend the Eastern Rocky Mountain slope, until they meet the moist winds from the Gulf of Mexico passing up the valley of the Mississippi, towards and through the region of the Great Canadian Lakes and over the low height of land separating the waters flowing into Lake Winnipeg from the Mississippi valley.*

In the latitude of the valley of the Saskatchewan, however, the moist south-west winds from the Pacific, find a broad depression in the Rocky Mountain range, and losing less humidity than those passing over the higher ranges to the south, meet with a prevailing north-easterly wind as they begin to descend their eastern flank, their temperature is consequently lessened instead of being elevated, and their capacity for moisture diminished, hence precipitation in the form of rain and hail takes place as they descend the slope towards Lake Winnipeg.

Hail-storms are not unfrequent during the summer months, and the prairies sometimes retain the records of their occurrence for many weeks. On the Grand Coteau de Missouri hail-storms are so violent that the stones have been known to penetrate the buffalo skin tents of the Indians who hunt on that elevated plateau. The thunder-storms of 1858 are given in the annexed table.

TABLE SHOWING THE NUMBER OF DAYS ON WHICH RAIN FELL, WITH THE CHARACTER OF THE THUNDER-STORMS, DURING THE SUMMER OF 1858, IN THE VALLEY OF THE ASSINIBOINE AND SASKATCHEWAN:

DATE.	TIME.	CHARACTER OF STORM.	LOCALITY.
June 18	2–4 p. m.	Heavy rain, Thunder.............	Prairie Portage.
" 19	6 a. m.	Slight rain	Do do
" 20	Sunset.	A terrific thunderstorm, heavy rain, high wind	Bad Woods.
" 21	11 a.m. 1 p.m.	Tremendous thunderstorm, hail-stones 1–1½ in. in diameter	Bear's Head Hill.
" 21	6 p.m.	Thunderstorm, heavy rain..........	Do do
" 22	3.45 p.m. 6 p.m.	Terrific thunderstorm, continued roar of thunder without intermission for 1½ hours..............	Sandy Hills.
" 25	8–10p.m.	Violent thunderstorm, heavy rain	Little Souris.
" 26–27	Night.	Thunder and rain......................	Do do
" 29–30	Night.	Thunder and rain......................	Do do
" 30	6 a.m.	Heavy rain with rolling thunder, without intermission for 1 hour	Do do
July 4	11 a.m.	Rain..	Great Prairie.
" 5	10 p.m.	Lightning in the east, no rain, thermometer in shade 92°, at noon	Do do
" 9	9 a.m.	Rain..	Assiniboine.
" 11	3 p.m.	Thunderstorm, hail, and heavy rain	Fort Ellice.
" 13	7 p.m. 10 p.m.	Thunderstorm of unusual violence and sublimity—see narrative...	Qu'Appelle Valley.
" 14	2.30 p.m. to 4-30	Thunder and rain.......................	Do do
" 14–15	Night.	Rain all last night....	Do do
" 15	....	Rain North of Qu'Appelle, temp. at 6 a.m. 45°..........................	
" 18	11 p.m.	Heavy rain and thunder............	Qu'Appelle.
" 19	4 p.m.	Rain North of Qu'Appelle Valley	
" 22	Noon.	Violent thunderstorm with heavy rain and hail..........................	Do
" 28	1 p.m.	Rain in torrents.......................	Do
August 2	4 p.m.	Heavy thunderstorm with rain . . .	South Branch.
" 4	4.30 p.m.	Thunderstorm, heavy rain...... ...	Do do
" 5	....	Heavy rain.............................	Do do
" 11	6 p.m.	Violent thunderstorm................	Long Creek & Main Saskatchewan.
" 18	6 p.m.	Thunderstorm, rain and high wind	Main Saskatchewan.
" 25	Noon.	Violent thunderstorm and rain....	Do do

The progress of Dunes affords a very excellent indication of the direction and force of prevailing winds. The Devils Hills and the sand dunes surrounding that dreary waste on the Assiniboine, in long. 99° 40′ W., showed a bare advancing surface towards the north-east, being pushed in that direction by the prevailing south-west wind. The sand dunes at the Height of Land in the Qu'Appelle Valley, in long. 106 W. lat. 51 N., were advancing in an easterly direction ; their clean surfaces were facing the east. Had they progressed under a prevailing south-west wind, they would long since have invaded and filled up the Valley of the Qu'Appelle. These existing records of prevailing winds during the period when the dunes are not frozen, show that while the south-west is the most effective as a summer surface wind in Ruperts Land under the

tory, exceeds 1,200 miles in length,—a barrier sufficient to arrest the general progress of settlement, for very many years to come, in a course due west of the Mississippi.

The only direction which remains for extensive free soil settlement in and near the United States is northwards, partially along the immediate banks of the Missouri, about the head-waters of the Mississippi, and towards the vallies of the Red River and the Assiniboine and the main Saskatchewan. The popular impression that immense areas of land available for the purposes of agriculture, lie between the Missouri and the Rocky Mountain chain, has, as before stated, been completely refuted by the explorations and surveys for the Pacific Railroad. The now well ascertained aridity of climate and its natural consequence, sterility of soil, both combine to confirm the title of "The Great American Desert," given by the early explorers of the eastern flank of the Rocky Mountains to that extensive region of country. This important fact cannot fail to exercise a powerful influence upon the occupation of British Territory north of the 49th parallel of latitude, and on the sources from which that occupation will flow. [From a paper by the author of this Report, on the Great North West, published in 1858.]

* See Meteorology in its connection with Agriculture by Professor Joseph Henry.

99th meridian, on the South Branch of the Saskatchewan, seven degrees further west, westerly winds prevail.

There is no doubt that the south-west Pacific winds, passing through the broad depression in the Rocky Mountains near the 49th parallel without losing the whole of their moisture, give humidity to the large portion of Rupert's Land over which they traverse.

The great plateau on which the Rocky Mountain ranges rest, has an average elevation of 4,000 feet near the 32nd parallel of latitude, the lowest pass in the most easterly range being there 5,717 feet above the ocean. Along the 35th parallel the vertical section across the mountain system is of greater width and elevation. The mean height above the ocean is about 5,500 feet, and the lowest pass 7,750 feet. Between the 38th and 40th parallel the section has an elevation of 7,500 feet, and the lowest pass is 10,032 feet above the level of the sea. Beneath the parallel of 47° the base of the plateau is narrow, and has an average altitude of 2,500 feet, the lowest pass being 6,044 feet above the ocean.* Within British Territory north of the 49th parallel the passes in the eastern range are still lower. The recent measurements by Captain Palliser's Expedition show that the height of the Kutanie Pass in latitude 49° 30′ is nearly 6,000 feet above the sea level; the Kananaski Pass 5985 feet, and the Vermillion Pass, traversed by Dr. Hector, in latitude 51° 10′ only 4,944 feet above the ocean.

The following table exhibits the elevation of the Rocky Mountain plateau, and the height of the lowest Passes above the ocean:

Table showing the elevation and breadth of the Plateau on which the Rocky Mountain ranges rest, and the height above the Ocean of the lowest Passes, from the 32nd parallel to the 51st parallel, North Latitude.

BREADTH AND ELEVATION OF PLATEAU BETWEEN THE 32ND AND 49TH PARALLELS.

	3000 and 4000 feet.	4000 and 5000 feet.	5000 and 6000 feet.	6000 and 7000 feet.	7000 and 8000 feet.	8000 and 9000 feet.	9000 and 10000 feet.
32nd Parallel	170 miles	503 miles	60 miles	....	....	....	....
35th "	185 "	160 "	303 "	235 miles	95 miles	....	....
38th and 39th Parallel........	143 "	725 "	284 "	110 "	155 "	80 miles	20 miles.
41st and 42nd "	160 "	580 "	285 "	270 "	107 "	20 "	....
47th and 49th† "	130 "	97 "	28 "	....	....	....	....

Summit of the lowest passes above the Ocean from the 32nd to the 51st parallel, North Latitude:

	Feet.	
32nd Parallel	5,717	
35th "	7,472	
38th and 39th Parallel	10,032	
41st and 42nd "	8,372	
47th and 49th "	6,044	
Kutanie Pass, lat. 49° 30′	6,000 nearly	Passes discovered by Captain Palliser's Expedition.‡
Kananaskis Pass, north of 49th par........	5,985	
Vermillion Pass, lat. 51° 10′........	4,944	

Not only has the depression in the Rocky Mountain range, north of the 47th parallel of latitude, a remarkable effect upon the climate of the Valley of the Saskatchewan, but its bearing upon means of communication between the Atlantic and Pacific slopes of the Rocky Mountain ranges is of the greatest importance.

SEASONS OF THE VALLEY OF LAKE WINNIPEG.

The natural division of the seasons in the Lake Winnipeg Valley is as follows:

Spring—April and May.

Summer—June, July, August and part of September.

Autumn—Part of September and October.

Winter—November, December, January, February and March.

The natural division of the Seasons is strikingly represented by the early and rapid advance of temperature in May in the valley and prairies of the Saskatchewan; and it is also indicated in a very marked degree by the extension northwards to the same valley, between the 95° and 105° of longitude, of numerous plants, whose geographical distribution, east and west of those limits, has a much more southern climatic boundary. The limits of trees rise with the isothermal lines, and these attain a much higher elevation in the interior of British America than on the Atlantic coast.§

In relation to agriculture, the intensity of winter cold is of comparatively little moment. The elevated spring and summer temperature, combined with the humidity of the humid region

* Dr. J. Henry, Meteorology in its connection with Agriculture.

† Pacific Railroad explorations.

‡ Proceedings of the Royal Geographical Society, Vol. III., No. 5.

§ Gray.

in the Valley of Lake Winnipeg, enable Indian corn and the melon to ripen with certainty, if ordinary care is taken in selecting soil and in planting seed.

The following table of the meteorology of Red River was published in my Report for 1857. In the absence of other information on this subject it is inserted here, but it must be borne in mind that the results of one year's comparison are not of much value in estimating the relative climatic adaptation of regions far apart; nor do they afford sufficient data for a fair estimate of the climate of the locality where the observations were made. It is doubtful in the present case whether the instrument was quite reliable at low temperatures.

Comparison of the meteorology of Red River Settlement with Toronto, Canada West, with refercnce to mean temperature, depth of rain and snow, from corresponding observations at both stations, from June, 1855, to May, 1856, inclusive.

Month.	Mean Temperature.		Rain in inches.		Snow in inches.		Temperature: Rain and Snow at Red River + or — of Toronto.
	Red R.	Toronto	Red R.	Toronto	Red R.	Toronto	
1855.	°	°	°	°	°	°	*Summer.*
June......	69.10	59.93	6.0	4.07	0.0	0.0	Temperature + 3.78.
July	71.16	67.95	12.0	3.24	0.0	0.0	Rain + 21.74 inches.
August ...	63.03	64.06	12.5	1.45	0.0	0.0	Snow 0.0.
Summer ..	67.76	63.98	30.5	8.76	0.0	0.0	
							Autumn.
September.	59.26	59.49	5.0	5.59	0.0	0.0	Temperature — 6.94.
October ...	42.20	45.39	0.0	2.48	2.0	0.8	Rain — 5.16 inches.
November..	21.19	38.58	2.5	4.59	7.0	3.0	Snow + 5.2 inches.
Autumn...	40.88	47.82	7.5	12.66	9.0	3.8	
1856.							*Winter.*
December..	–8.31	26.98	0.0	1.85	8.0	29.5	Temperature — 26.42.
January...	–10.55	16.02	0.0	0.00	5.0	13.6	Rain — 1.85 inches.
February..	–1.71	15.69	0.0	0.00	6.0	9.7	Snow — 38.8 inches.
Winter ...	–6.85	19.57	0.0	1.85	19.0	52.8	
							Spring.
March	9.09	23 06	0.0	0.00	6.5	16.2	Temperature — 2.83.
April.....	39.83	42.27	6.5	2.78	3.0	0.1	Rain + 3.14 inches.
May......	58.46	50.52	4.0	4.58	2.0	Inap.	Snow — 4.8 inches.
Spring....	35.79	38.62	10.5	7.36	11.5	16.3	
Annual ...	34.38	42.50	48 5	30.63	39.5	72.9	

ANNUAL.

Colder mean temperature..............................	8°12
More rain..	17.85 inches.
Less snow..	33.4 "
More moisture and most probably less evaporation..	14.53 "

At Quebec the difference between the mean temperature of summer and winter is 53°93; at Fort Snelling 56°81; and at Red River Settlement 74°61.

The summer temperature of Red River, and the absence of frosts during that season, determine its fitness for agricultural purposes. The followiug table exhibits a comparison, based upon one year's observation only, between the summer temperature of the Settlement and various other well known places in Canada:--

Summer temperature at Red River Settlement	67.76
Montreal, Canada..	66.62
Quebec..	62.91
Toronto...	63.98

The extraordinary cold of the winter of 1855 and 1856 at Red River, is shown by the tables for December, January, and February, (Mr. Gunn's observations,) which give a mean of —6°.85 for the mean temperature of that season; but if we turn to the records for 1857 and 1858* we find the mean temperature of that winter to have been 2°.87, showing a difference of eight degrees in favour of the winter of 1857–'58. The temperatures recorded were as follows:—

	1855–56.		1857–58.
December................	–8.31	......	9.11
January	–10.55	......	6.2
February	–1.71	......	–6.68
Winter Mean	–6.85	......	2.87

The thermometers supplied to the Red River Expedition in 1857 were made by Negretti and Zambra, and were of the best construction.* They had been compared with a standard at the Provincial Observatory, and their errors recorded, but it does not appear that the proper corrections were made after each observation. Mr. Gunn's thermometer was an ordinary instrument, and, like many of its class, liable to errors at low temperatures. It is, therefore, probable that his winter temperatures are too low, and that the true mean of the winter months at Red River, and consequently the annual mean, may be considerably higher than it appears from his observations.

The cold of February 1858 was exceptional. At Montreal it was the coldest February on record, being 14°.05 below the mean temperature of February, 1857.† But it cannot be denied that the winter cold of Red River is excessive, and the temperature sometimes falls so low as to freeze mercury in a few minutes.

The Meteorological Register kept at the Stone Fort, Lower Settlement, in 1847, under the Superintendence of Captain Moody,‡ from which extracts were permitted to be made by Dr. Owen, furnish trustworthy evidence respecting the severity of the climate in winter.

The mean temperature for January, 1847, was —12° 5. Observations being taken at 9 a.m., 3 p.m. and 9 p.m. During twenty-two consecutive days of this period, from the 5th to the 26th inclusive, the thermometer never once rose to zero. The lowest temperature reached was —48°, the highest 30°, giving a range of 78°. On the coldest day, the 20th, when the thermometer showed —48° at the Stone Fort, and — 47° at Fort Garry, mercury froze in fifteen to twenty minutes when exposed in bullet moulds.

Although there is no record of cold terms in Canada approaching the extreme low temperature and extending over so long a period as those instanced above, yet cold terms of great intensity are not uncommon in Lower Canada. In the excellent observatory of Dr. Smallwood's at Isle Jesus, nine miles west of Montreal, the following records of cold terms have been preserved:—§

* Meteorological tables recorded by Mr. Dawson's party.

* These thermometers, together with other meteorological apparatus, were furnished to the different members of the Red River Expedition in 1857, by the permission of the Rev. Dr. Ryerson, Chief Superintendant of Schools, from the stock of instruments provided by the Chief Superintendent for observatories attached to the Grammar Schools throughout the Province. The thermometers were compared and their errors determined and tabulated at the Provincial Observatory. A table of errors was attached to each instrument.

† Records of St. Martin's Observatory, Isle Jesus.

‡ See Dr. Owen's Geological Survey of Wisconsin, Iowa, and Minnesota,—page 181.

§ *Canadian Journal* for 1855. and *The Canadian Naturalist* for April, 1859.

Low Temperatures at Isle Jesus, C. E.

On the 22nd and 23rd Dec., 1854		*On the 9th, 10th, and 11th Jan., 1859.*	
22nd.—8 A. M.,	31.6 below zero.	9th.—6 A. M.,	29.9 below zero.
9 "	27.0 "	7 "	29.0 "
10 "	19.1 "	9 "	28.4 "
11 "	17.8 "	12 "	23.8 "
12 "	16.6 "	2 P. M.,	21.5 "
1 P. M.,	12.1 "	9 "	33.9 "
2 "	11.8 "	10 "	34.2 "
3 "	8.1 "	12 "	36.0 "
7 "	23.4 "	10th.—6 A. M.,	43.6 "
8 "	26.9 "	7 "	43.1 "
10 "	28.2 "	9 "	41.6 "
12 "	34.8 "	12 "	20.1 "
23rd.—6 A. M.,	36.2 "	2 P. M.,	14.3 "
7 "	36.0 "	9 "	28.8 "
8 "	34.3 "	10 "	29.2 "
10 "	24.1 "	12 "	31.6 "
12 "	13.4 "	11th.—6 A. M.,	37.1 "
2 P. M.,	12.6 "	7 "	36.9 "
4 "	9.1 "	12 "	24.8 "
6 "	12.6 "	2 P. M.,	19.9 "
10 "	9.1 "	9 "	21.0 "
		10 "	21.6 "
		12 "	18.1 "

In January, 1859, the thermometer did not rise above zero during a period of 124 hours 30 minutes, or more than five days. Mercury froze in the open air. The mean temperature on the 9th was—27° 8; 10th,—29° 0; 11th,—28° 2. Dr. Smallwood says that this cold term was felt generally throughout Canada and the Eastern States, and seems to have travelled from the west.

The following minimum temperatures were observed at different places :—

Rochester	10.0	below zero.
Brooklyn, (New York,)	9.0	"
Boston	14.0	"
Toronto	38.0	"
Quebec	40.1	"
Huntingdon	44.0	"

For the purpose of comparing the Monthly Mean at Quebec (Lat. 46° 49′ 2″, Long. 71° 16′) with those of Red River, the following table is inserted :*

	Mean.	Maximum.	Minimum.
1854.			
October	46.05	64.7	+ 31.6
November	31.83	59.8	+ 10.0
December	13.08	36.6	− 19.2
1855.			
January	16.70	46.0	− 14.0
February	10.55	36.8	− 29.5
March	21.06	47.3	2.4
April	34.14	59.8	5.9
May	49.03	83.0	32.0
June	58.34	88.0	43.2
July	68.86	90.3	51.9
August	61.54	85.0	38.3
September	55.15	81.3	34.7
October	45.43	60.4	28.4
November	28.75	34.3	21.81
December	18.09	40.1	− 19.2
1856.			
January	8.19	27.0	− 16.5
February	11.99	31.9	− 18.0
March	17.60	39.0	− 11.0
April	36.90	55.3	6.4

In the absence of instrumental observations, the progress of vegetation affords the best indication of climate, apart from latitude and elevation above the sea. It has been observed elsewhere that there exists an extraordinary difference between the characteristic fruit tree of the South Branch, the Misaskatomina, (*Amelanchier Canadensis*,) and the same tree on the North Branch of the Saskatchewan. On the South Branch at the Elbow, and for forty miles down the river, this shrub attains an altitude of twenty feet, with a stem fully three and three and a half inches through; the fruit is large and very juicy; the size of the berry there is equal to the largest black currant, resembling a small grape more than any other fruit.

The period of flowering and fruiting is about three weeks earlier in latitude 51° than between the 53rd and 54th parallels west of the 100th degree of longitude. The prairies of the Assiniboine, of the Qu'Appelle, and of the South Branch of the Elbow, are decorated with brilliant spring flowers, and covered with luxuriant herbage, at a time when the ice still lingers at the head of Lake Winnipeg, or chills the air and arrests vegetation in Cedar and Cross Lakes on the Main Saskatchewan. Two and a half degrees north of Cumberland, the soil is permanently frozen three feet below the surface. Sir John Richardson relates that in 1851 he did not disengage his canoes from the ice at the upper end of Lake Winnipeg until the 9th of June. At the Touchwood Hills horses are allowed to remain in the open air all the winter, finding sufficient pasture under the snow to keep them in good condition. (See page 69 for a short description of the winter climate at the Touchwood Hills.)

The growth of forests is very intimately connected with the climate of a large extent of country. That forests once covered a vast area in Rupert's Land there is no reason to doubt. Not only do the traditions of the natives refer to former forests, but the remains of many still exist as detached groves in secluded valleys, or on the crests of hills, or in the form of blackened prostrated trunks covered with rich grass and sometimes with vegetable mould or drifted sand. The agent which has caused the destruction of the forests which once covered many parts of the prairies in Rupert's Land is undoubtedly fire, and the same swift and effectual destroyer prevents the new growth from acquiring dimensions which would enable it to check their annual progress. Nearly everywhere, with the exception of the treeless, arid prairie west of the Souris, and west of Long Lake on the north side of the Qu'Appelle, young willows and aspens were showing themselves where fire had not been on the previous year. South of the Assiniboine and Qu'Appelle few plains had escaped the conflagration in 1857, and the blackened shoots of willow were visible as bushes, clumps, or wide spreading thickets where the fire had passed.

The end or tail of the prairies is at Fort Liard, a short distance to the south of Fort Simpson, (lat. 61° 51′ 7″ N.). There is a long high belt of prairie land which runs as far as the neighbourhood of that locality, at the foot of the Rocky Mountains.*

In the State of Missouri forests have sprung up with wonderful rapidity on the prairies as the country becomes settled so as to resist and subdue the encroachment of the annual fires from the west. Missouri lies within the limit of the humid south-west wind coming up the Valley of the Mississippi, and enjoys a greater rainfall than the region west of the 100th degree of longitude.

GENERAL CHARACTER OF THE SEASONS ON THE MAIN SASKATCHEWAN, EAST OF CARLTON HOUSE.

The following tables will serve to show the general character of the seasons at important points in the Valley of Lake Winnipeg:

* See *Canadian Journal*—Old Series.

* Col. Lefroy.—Evidence before the Select Committee of the House of Commons.

Extracts from a Journal kept at "Fort à la Corne," on the Main Saskatchewan, Lat. 55.30, *Long.* 104.25.

1856.

April 1.—No frost last night, but thick mist this morning. The weather has been warm although cloudy.
" 2.—Hard frost last night, but mild during the day.
" 4.—Slight frost last night, day very mild. Snow dissolved a great deal during the day. Water making its appearance on edge of river.
" 7.—Froze hard last night, and has been cold most of the day.
" 8.—Do. do no thaw during the day. River rising very much, and boat frozen in.
" 9.—Ice made a start previous to moving.
" 17.—Weather warm, ice drifting down river.
" 19.—Weather fine. Annual Goose Dance of McLeod took place to-day.
" 21.—Rain with N. W. wind.
" 23.—Had good fall of snow during night. Continued snowing without intermission the whole day. Nets set for first time. One sturgeon, ten suckers, and one gold-eye caught.
" 25.—Hard frost last night.
" 26.—Weather fine, considerable quantity of ice in river, but melting fast.
May 1.—Weather warm. Change perceived on trees, they are getting a little *green*,
" 2.—Working in garden; put down peas, onions, radish, and a few greens. Net produced two sturgeon.
" 6.—Weather warm.
" 10.—Storm of snow and rain during last night, with a strong north wind, which continued at intervals during the day.
" 12.—Planted north field with potatoes, and ploughed south field.
" 13.—Cloudy, rain, with N. W. wind. Planted potatoes in south garden.
" 14.—Cold north wind. Sowed four beds of swedish turnips.
" 21.—Thunder and lightning most of last night. Rain poured down in torrents River rose considerably to-day.
" 30.—Saskatchewan Brigade arrived this afternoon. Started same evening.
June 1.—Clear and beautiful to-day.
Sept. 16.—Raining all day, wind east.
" 17.—Clear but rather cold. Slight frost last night. Wind N. E. (light.)
" 20.—Mild and warm during day. Slight frost last night.
Oct. 2.—Raining all morning; wind W. Cleared up in the afternoon. Men in morning cleared all the *potatoe* stalks out of north garden, and in afternoon commenced again the potatoes in south garden.
" 13.—S. W. wind. Fall boats started this morning for Carlton.
" 17.—Fine weather, men employed in garden.
" 18.—Do do putting dung in garden.
" 22.—Very hard frost over night.
" 23.—Severe frost last night.
" 26.—Snowed during night, but thawed as it fell. Blowing very hard.
Nov. 11.—River full of ice.
" 16.—Weather fine. One cow calved.
" 31.—Slight fall of snow last night, but day remarkably fine.
Dec. 1.—Weather fine, not in the least cold. Have had no cold weather as yet, compared to last year.
" 2.—Weather colder than of late.
" 5.—Slight fall of snow during night.
" 11.—Very cold.
" 12.—Cold, very severe.
" 31.—Snowing most of the day.

1857.

Jan. 2.—Cold, and snowing at intervals.
" 3.—Very cold.
Mar. 29.—Hard frost last night.
" 30.—Very warm, snow melting about the fort.
" 31.—Raining during the night. Slight rain during the day.
April 2.—North wind and cold. No thaw these three days back.
" 3.—North wind and very cold.
" 4.—North wind.
" 5.—Weather milder, a slight thaw.
" 6.—South wind, thawing a great deal.
" 8.—Hard frost last night, cold all day. North-west wind accompanied with snow, which continued most of the day.
" 9.—Snowed last night. Cold during day. Water appearing on edges of river.

At this time last year ice started in river. What a difference this year. We cannot go anywhere at present without snow shoes, our cattle are nearly starved, they cannot go about as the snow is so hard.

" 10.—Weather still cold, wind variable.
April 11.—Storm of snow and wind.
" 12.—Hard frost last night. Cold all day. No thaw.
" 13.—Blowing hard, accompanied with snow. Day fine and snow dissolving. Turned very stormy in afternoon. North wind with snow.
" 14.—Still cold. North wind. No thaw.
" 15.—Southerly wind, but still cold. River still rising at edges. Little or no thaw during day.
" 16.—Weather clear, but still cold. Little or no thaw. Notwithstanding the late cold weather the ice went off this day.
" 17.—Weather same. Very little ice drifting down river.
" 18.—Weather still cold. North wind.
" 19.—Fine during day. Sun shining bright. Snow melted a good deal.
" 20.—Day fine, but weather turned cold towards evening. Had a slight fall of snow last night.
" 21.—Beautiful day. Snow dissolving fast. Little or no ice drifting.
" 22.—Cloudy and variable, very little thaw.
" 23.—Strong South wind. Thawing very much.
" 24.—Snowed without intermission the whole day. Wind variable and blowing hard.
" 25.—Beautiful day. Warmest we have had this season.
" 27.—Cloudy and cold, with slight snow.
" 28.—Weather fine and warm.
" 29.—Cold and cloudy. Slight snow.
" 30.—Beautiful day, but blowing hard.
May 1.—Weather and wind from same quarter. Snow dissolving fast.
" 3.—Ice drifting all last night, but not much to-day.
" 5.—Disagreeable day. Snowing without intermission with a cold north wind. River full of ice.
" 8.—Stormy northerly wind, and very cold.
" 12.—Weather warm. Yesterday planted potatoes and onions in south garden, and to-day sowed cabbages in boxes.
" 15.—Mild, wind south.
" 18.—Boisterous weather.
" 20.—Beautiful day. All hands employed planting potatoes. Sowed turnips, carrots, beans, &c. Nets caught three sturgeon and nine suckers.
" 21.—Very warm. Annual Goose Dance came off.
June 2.—Hard frost last night. Froze my beans, and the hops were affected also.
" 7.—Rained hard all last night, and continued without intermission all day.
" 9.—Fine weather, river still rising.
" 15.—Very warm and clear this afternoon.
" 30.—Beautiful day. Bull dogs so numerous that horses had to be put in stable and grass cut for them. Starvation is staring the people in the face. Have caught no sturgeon for some time back. Our nets produced nothing to-day.

1858.

April 20.—Warm and clear, south wind.
" 21.—Ice drifting in river. Large quantity of ice on banks.
" 22.—Cold north wind.
" 24.—Slight fall of snow in morning. Rain towards sunset. Still cold, wind south-west.
" 25.—Warm and fine to-day.
" 27.—South wind. Warmest day this spring.
" 28.—North wind. Cold and blowing hard.
May 1.—South wind. Warm. Sky overcast with smoke. Large fire close to fort. Clearing up north garden.
" 7.—Set 4 men to dig potatoe ground in south garden. Caught 1 sturgeon, first this spring.
" 11.—Cold north wind. Cut the potatoes for planting.
" 12.—Planted potatoes in south field and commenced to dig the north field for sowing. Sowed beetroot, radish and lettuce.
" 15.—South wind. Weather cold. Planted north garden with potatoes.
" 17.—Still cold. Slight fall of snow in night.
" 18.—Wind from north and cold. Think we are going to have a second winter.
" 19.—Continues cold. Wind north.
" 20.—Weather improving, wind south-west.
" 21.—Warm and mild. South wind.
" 22.—Warm and fine.
" 23.—Warm in morning. Thunder and rain towards sunset.
" 24.—Warm. Wind south. Clearing up garden. River muddy and water rising fast.
June 1.—Wind south, and weather warm.
" 8.—Wind north, and appearance of cold. Think we are going to have a cold summer. Garden herbs slow in making their appearance above ground.
" 15.—Weather continues warm.

July 1.—Boisterous weather. Wind north.
" 10.—Very warm to-day. Bull dogs so numerous, horses and cattle had to be kept in stable all day. Men hoeing south garden.
" 21.—A very fine day.

SEASONS AT CUMBERLAND HOUSE.

In the following Table of Phenomena, indicating the *Progress of the Seasons at Cumberland House*, are combined the observations of Sir J. Richardson, in the Spring of 1820, with those of chief factor John Lee Lewis, in 1839 and 1840, distinguishing the remarks by the years. The supposed altitude of Cumberland House above the sea is 900 feet, according to Colonel Lefroy's calculations.*

March 4.—Water collecting in pools round the establishment. 1840.
" 7.—Much bare ground visible.
" 8.—The snow, which covered the ground to the depth of three feet, was observed to moisten in the sun for the first time this season. 1820.
" 12.—Temperature in the shade rose for the first time to +30° F. The melting snow began to drop from the eaves of the houses.
" 21.—Patches of earth became visible, the season being in respect to the melting of the snow fourteen days later than that of 1840. The River Saskatchewan broke up partially, the melting snow covered with *Poduræ*, as it is also frequently in the autumn.
" 24.—A white-headed eagle was seen, this being almost always the first of the summer birds which arrives; it comes as soon as it can obtain fish. In 1840, the first eagle was seen on the 26th.
April 2.—The river Saskatchewan froze over again, after some very cold days.
" 7.—Barking crows (*Corvus Americanus*) seen. They were not observed till the 19th in 1840.
" 8.—First snow bunting seen (*Emberiza nivalis*). 1840.
" 9.—A merganser seen. 1820.
" 10.—Willow catkins beginning to burst.
" 12—Geese and swans seen in 1820. In 1840 they were not seen till the 20th; and pelicans and ducks were observed that year on the 21st.
" 13.—Buds of *Populus balsamifera* bursting. 1820.
" 17.—Plovers, grakles, and orioles seen, and, on the following day, Canadian jays and fly-catchers. Frogs croaking.
" 20.—Coltsfoot, *Nardosmia palmata*, flowering.
" 26.—Alder flowering. The sugar harvest, which is collected in this district from the *Negundo fraxinifolium*, commenced in 1820, on the 20th of this month, and lasted till the 10th of May. The flow of the sap is greatly influenced by the direct action of the sun, and is greatest when a smart night's frost is succeeded by a warm sun-shining day. The flow ceases in a cold night.
" 28.—The Saskatchewan thoroughly broken up. The ice on Pine Island Lake did not disappear until nearly a month afterward. Wahlenberg observes that the mean temperature of the air in Lapland must rise to 40° F. before the rivers are completely free. The Saskatchewan opens in this district before the mean heat for ten days rises so high; but its upper part flows from a more southerly and warmer, though a more elevated country.
" 30.—Commenced ploughing. 1840.
May 1.—*Anemone patens*, or wind flower, in blossom, its leaves not yet expanded. 1820.
" 2.—A fall of snow to the depth of two feet. 1840.
" 13.—Planting potatoes.
" 14.—Sowing barley. 1820. *Negundo fraxinifolium* and gooseberry bushes in flower.
" 17.—Willows, gooseberries, aspens (*Populus tremuloides*) in leaf. Various *Drabæ* in flower. 1820. In 1840 the trees were bursting their buds at this time.
" 17.—Wheat sown on the 8th of this month, above ground to-day, having germinated in nine days. 1840.
" 21.—Barley sown on the 14th above ground, having taken seven days to germinate.
" 22.—Leaves of the trees expanding rapidly.
" 24.—*Ulmus Americana* flowered. 1820.
" 25.—Pine Island Lake clear of ice, 28th. *Prunus pennsylvanica*, *P. virginiana*, and Amelanchier in flower. 30th. From the 23rd to the 30th of this month, in 1840, the temperature in the shade at 2 p.m. varied between 78° F. and 93° F. On the 30th, potatoes planted on the 13th appeared above the ground. 1840.
June 12.—All the forest trees in full leaf. 1820.
Aug. 1.—Commenced reaping barley. On the 15th, 18th, 19th, and September 1, the thermometer at noon ranged between 80° and 90°, being the hottest days in the month. There was much thunder and hail on these days. 1839.

Sept. 2.—Flocks of water-fowl beinning to arrive from the north. 3rd. The first fall of snow this autumn. 4th. Vast numbers of water-fowl flying southward. A severe fall of snow and frost in the north causes these birds to hurry to the south. 11th. First hoar-frost. Birch and aspen leaves turning yellow. 14th. Wild-fowl numerous. 20th Snow; 21st, ditto very heavy. 24th. Thunder and lightning.
Oct. 1.—Taking up potatoes. 5th. Leaves all fallen from the deciduous trees. On the 11th, thermometer at 2 p.m., in the shade, 68° F., being unusually high.
" 14.—Water-fowl passing southward in large flocks, 1839.
" 15.—Bays of the lake frozen over. 16th. The ground frozen hard. 17th. Last water-fowl seen this season. 18th. Lake entirely frozen over. In 1839 the Little River was frozen over on the 24th of this month, but broke up again in part, and remained partially open all the winter.
" 31.—Waveys (*Anas hyperborea*) passing. Lake partially open.

SEASONS AT CARLTON HOUSE.*

The following are the *Phenomena of the Spring of* 1827 *at Carlton House*, in lat. 52° 51′ N., long. 106° 13′ W., on the eastern limits of the Saskatchewan prairie lands, and at an elevation above the sea of about 1100 feet.

Feb. 15.—Snow thawing in the sunshine, and on the 17th many sandy hummocks on the plains were bare. This is at least three weeks earlier than the thaw commences in an early season at Cumberland House, which is a degree further north, but is 200 feet lower.
March 6.—Trees thawed in fine days, and on the 8th the black earth on the immediate banks of the river was softened to the depth of two inches by the power of the sun's rays. At this place the westerly winds bring mild weather, and the easterly ones are attended by fog and snow.
" 13.—Sparrow-hawks (*Falco sparverius*) arrived from the south, and on the 17th several migratory small birds were noticed.
„ 29.—Large flocks of snow-birds (*Emberiza nivalis*) came about the establishment; and, by the 31st, steep banks, which had a southern aspect, were clear of snow.
April 1.—Many *Fringillidæ* (birds of the sparrow tribe) were seen. On the 2d, swans arrived, and, by the 3rd, much snow had disappeared from the plains.
" 4.—The snow at this time was melting in the shade, and the sap of the maple trees (*Negundo fraxinifolium*) began to flow.
" 6.—Geese arrived. Stormy weather, about the middle of the month, retarded the arrival of the summer birds; but the plants continued to grow fast. On the 20th, the Telltale plover (*Charadrius vociferus*) and several small birds came.
" 22.—*Turdus migratorius*, *Pyrrhula ludoviciana*, and *Lanius excubitor* were seen, and the flowers of *Anemone patens* expanded.
" 27.—Ice in the River Saskatchewan gave way. Frogs began to croak.
" 28.—Canada cranes (*Grus Canadensis*) arrived.
May 1.—*Sturnus ludovicianus* arrived, and the last flocks of *Emberiza nivalis* departed for the north.
" 2.—On this day, *Icterus phœniceus* and *Scolecophagus ferrugineus* were seen, and most of the water-fowl had by this time arrived. On the 4th, *Phlox hoodii* flowered.
" 5.—*Ranunculus rhomboideus*, *Viola debilis*, *Nardosmia palmata*, and several carices flowered.
" 6.—*Hirundo viridis* and many gulls arrived.
" 7.—On this day the sap of the ash-leaved maple, which had flowed scantily for ten days, ceased to run altogether, and the sugar harvest closed. *Avocetta americana* arrived. *Populus tremuloides* in flower.
" 9.—Crow-blackbirds were first seen. *Corydalis aurea*, *Corylus americana* and *rostrata*, *Hippophae canadensis*, *Thermopsis rhombifolia*, *Vesicaria arctica*, and *Alnus viridis* flowered. 12th. *Potentilla concinna*, *Townsendia sericea* flowered. 14th. Gooseberry bushes coming into leaf. Ash-leaved maple flowering, seven days after the sap had ceased to flow from wounds in the stem. 16th. The *Picus varius* arrived in considerable numbers, and on the 19th the *Viola nuttalliana* flowered.

The average antecedence of spring phenomena at Carlton House to their occurrence at Cumberland House is between a fortnight and three weeks. The difference of latitude, which is only one degree, is nearly counterbalanced by 200 feet of greater altitude; but the dry sandy soil of the plains, which are early denuded of snow, gives the spring there a great superiority over that of the lower country, where the ground is almost submerged, and the greater part of it ice-bound for a month after the river is open.

* Arctic Searching Expedition. Sir John Richardson.

* Arctic Searching Expedition. Sir John Richardson.

SEASONS AT RED RIVER.

On the progress of the Seasons and state of the Weather at Red River Settlement, from 1st June, 1855, to 31st May, 1856.

1855. June 5th was the coldest day in the month. Thermometer, 7 a.m., 58; 2 p.m, 63; 9 p.m., 56. The 14th was the hottest day. Thermometer, 7 a.m., 72; 2 p.m., 88; 9 p.m., 71. Three inches of rain fell on the 17th, one on the 19th and six on the 25th.

July 2nd was the coldest. Thermometer, 7 a.m., 56; 2 p.m., 78; 9 p.m., 68; light rain. The 25th was the hottest day. 7 a.m., 87; 2 p.m., 92; 9 p.m., 82. 7th, rain 3⅝ inches. 10th, rain ⅝ inches. Thunderstorm on the 17th, rain 3 inches. 26th, 1 inch rain; 29th, 3 inches rain; 30th, 2 inches; total, 14⅝ inches. Wheat out of the ear. On the 12th hay-cutting commenced. Tabani and mosquitoes very numerous and troublesome.

August:—Coldest day, 29th. Thermometer, 7 a.m., 44; 1 p.m., 68; 9 p.m., 56. The hottest day was the 5th. 7 a.m., 67; 2 p.m, 86; 9 p.m., 76. On the 8th, 5 inches of rain fell; 11th, 5¼ inches fell; 14th, 2 inches; 27th, ¼ inch: total, 12½. inches. Barley harvest commenced about the 1st; wheat harvest on the 15th Slight frost on the 30th.

September:—The coldest day was the 30th. Thermometer average +48. The hottest day was the 5th; thermometer, 7 a.m., 70; 2 p.m., 80; 9 p.m., 70. Total of rain during the month, 6½ inches. Finished storing wheat on the 8th. A few leaves falling. 26th, grey geese flying to the south.

October:—The warmest day was the first. Thermometer, 7 a.m., 56; 2 p.m., 70; 9 p.m., 58. Some snow fell on the 4th. Taking up potatoes on the 8th. White geese flying to the south, and continued to do so up to the 20th, and a few flocks later than that: all the larger kind of ducks leave about the same time. The deciduous trees are bare of leaves, except the oak, and some of the hardier kinds.

November:—The 2nd was the warmest day. Thermometer, 7 a.m., 32; 2 p.m., 38; 9 p.m., 36; 2½ inches rain fell on the 3rd; 5 inches of snow fell on the 11th; 12th, river covered over with ice. The coldest day of the month was the 21st, thermometer, 7 a.m., —12; 2 p.m., +8; 9 p.m., +6. Warm weather from the 21st to the end of the month. 7 inches of snow fell during the month. Flocks of snow birds have made their appearance from the north, and all the summer birds are gone.

December:—The warmest day was the 6th. Thermometer, 7 a.m., +22; 2 p.m., +26; 9 p.m., +30. The coldest day was the 24th; thermometer, 7 a.m., —48; 2 p.m., —30; 9 p.m., —40. We had six days of very cold weather, including the 23rd and 28th. The wind blew from the north during three days before the severe cold began; during its continuance there was very little wind and for two of the coldest days it was at the south. 8 inches of snow fell.

1856, January:—The warmest day was the 17th. Thermometer, 7 a.m., +10; 2 p.m., +22; 9 p.m., +16. The coldest was the 7th; thermometer, 7 a.m., —36; 2 p.m., 28; 9 p.m., —36. 5 inches of snow fell. The average cold for this month has not been great; very little wind.

February:—Coldest day the 2nd. Thermometer, 7 a.m., —36; 2 p.m., —20; 9 p.m., —34. The warmest day was the 20th; thermometer, 7 a.m., +26; 2 p.m., +35; 9 p.m., +24. 6 inches of snow fell. After the 12th, spirits of wine in the glass stood with few exceptions above zero, and the weather has been pleasant.

March:—The coldest day was the 8th; 7 a.m., —32; 2 p.m., 24; 9 p.m., —26. The warmest day was on the 22nd. Thermometer, 7 a.m., +28; 2 p.m., +38: 9 p.m., +34. The thermometer fell during the night a few degrees below zero; but on the whole the weather was pleasant; 6½ inches of snow fell. Much of the snow melted during the month. Barking crows made their appearance about the 20th.

April:—Geese made their appearance on the 2nd, and the snow birds left us for the north. The 12th was the coldest day this month. Thermometer, 7 a.m., +16; 2 p.m., +30; 9 p.m., +24. Warmest day, 23rd: thermometer, 7 a.m., +46; 2 p.m., +66; 9 p.m., +44. About six inches of snow, and 5 of rain fell. On the 16th the rain began to throw off its winter coat; clear of ice on the 20th. Sturgeon taken in the river in great numbers; the snow all away. Wild fowl to be seen in every direction on the 29th, and sowing wheat commenced.

May:—The coldest day, 11th. Thermometer, 7 a.m., +34; 2 p.m., +43; 9 p.m., +30. The warmest day was the 18th, 7 a.m., +75; 2 p.m., +84; 9 p.m.; +56; four inches rain fell on the 26th. On the 4th whip-poor-will began his serenades. The wheat sown on the 29th has germinated, and given a green appearance to the field. On the 9th wild geese abundant in the plains; maple in leaf, gooseberry bushes the same; finished sowing wheat on the 10th.

1856. Wheat sown in the beginning of May, was in the ear on the 13th July, and ripe on the 20th August. The wheat sown on the 29th April was ripe on the 14th August. The hottest day this last summer was the 20th of July. Barley harvest commenced in July; finished cutting wheat on the 28th August; slight frost on the 30th of the same month; potatoes taken up first week of October.

6th September:—Flocks of grey geese flying to the south. *Prunus Americana* ripe and very plentiful in the first part of this month, or rather before this month. Flocks of passenger pigeons are in from the north, and leave from the 20th to the last of the month. On the night of the 7th whip-poor-will gave us his parting song. *Coregonus lucidus* enter the river to spawn. The *Coregonus albus* in lake Winnipeg commences spawning about the 19th of October, and ends about the 1st November.

This Register was kept by Mr. Donald Gunn, of the Lower Settlement, Red River. For the details of the Register see the Red River Report for 1857.

GENERAL CHARACTER OF THE REGION WEST OF THE 98TH MERIDIAN IN THE UNITED STATES.

Very great misapprehension has prevailed with regard to the region west of the Mississippi, as well as of the valley drained by the Saskatchewan. Sanguine enthusiasts have laid out new States and Territories on the broad map of the Federation, and peopled them in imagination with bustling, industrious, and wealthy communities. Other visionaries have converted the four hundred thousand square miles drained by the Saskatchewan into a region of unbounded fertility and inexhaustible resources. Whereas, a proper appreciation and use of facts will convince the most sanguine, that the larger portion of this area is, in its present state, unfit for the permanent habitation of man both on account of climate, soil* and absence of fuel.

The candid opinion of Prof. Joseph Henry regarding the adaptation of a large portion of the United States for settlement has been already given, it is confirmed and strengthened by the following excellent summary, from the pen of Major Emory of the United States and Mexican Boundary Commission. It will at once occur to the reader that a knowledge of these facts gives great additional value to the truly fertile valleys of Red River, the Assiniboine, part of the Qu'Appelle, and portions of the South and North Branch of the Saskatchewan.† It determines also the direction in which efforts should be made to people this great wilderness, and guide the progress of settlement in such a manner as will render the country available for that grand desideratum, a route across the continent.

"In the fanciful and exaggerated description given by many of the character of the western half of the continent, some have no doubt been influenced by a desire to favor particular routes of travel for the emigrants to follow; others by a desire to commend themselves to the political favor of those interested in the settlement and sale of the lands; but much the greater portion by estimating the soil alone, which is generally good, without giving due weight to the infrequency of rains, or the absence of the necessary humidity in the atmosphere, to produce a profitable vegetation. But be the motive what it may, the influence has been equally unfortunate by directing legislation and the military occupation of the country, as if it were susceptible of continuous settlement from the peaks of the Alleghanies to the shores of the Pacific."

"Hypothetical geography has proceeded far enough in the United States. In no country has it been carried to such an extent, or been attended with more disastrous consequences. This pernicious system was commenced under the eminent auspices of Baron Humboldt, who, from a few excursions into Mexico, attempted to figure the whole North American continent. It has been followed by individuals to carry out objects of their own. In this way it has come to pass, that, with no other evidence than that furnished by a party of persons travelling on mule back, at the top of their speed, across the continent; the opinion of the country has been held in

* See 'Cretaceous Series,' Geological Report.

† See Preliminary-Report for area of cultivable land.

suspense upon the subject of the proper route for a railway, and even a preference created in the public mind in favor of a route which actual survey has demonstrated to be the most impracticable of all the routes between the 49th and 32nd parallels of latitude. On the same kind of unsubstantial information maps of the whole continent have been produced and engraved in the highest style of art, and sent forth to receive the patronage of Congress, and the applause of geographical societies at home and abroad, while the substantial contributors to accurate geography have seen their works pilfered and distorted, and themselves overlooked and forgotten." * * *

"The plains or basins which I have described as occurring in the mountain system are not the Great Plains of North America which are referred to so often in the newspaper literature of the day, in the expressions, "News from the Plains," "Indian Depredations on the Plains," &c.

"The term "Plains" is applied to the extensive inclined surface reaching from the base of the Rocky Mountains to the shores of the Gulf of Mexico and the valley of the Mississippi, and form a feature in the geography of the western country as notable as any other. Except on the borders of the streams which traverse the plains in their course to the valley of the Mississippi, scarcely anything exists deserving the name of vegetation. The soil is composed of disintegrated rocks, covered by a loam an inch or two in thickness, which is composed of the exuviæ of animals and decayed vegetable matter.

"The growth on them is principally a short but nutritious grass, called buffalo grass, (*Sysleria dyctaloides.*) A narrow strip of alluvial soil, supporting a coarse grass and a few cotton wood trees, marks the line of the water-coarses, which are themselves sufficiently few and far between.

"Whatever may be said to the contrary, these plains west of the 100th meridian are wholly unsusceptible of sustaining an agricultural population, until you reach sufficiently far south to encounter the rains from the tropics.

"The precise limits of these rains I am not prepared to give, but think the Red River (of Louisiana) is, perhaps, as far north as they extend. South of that river the plains are covered with grass of larger and more vigorous growth. That which is most widely spread over the face of the country is the grama or mezquite grass, of which there are many varieties. This is incomparably the most nutritious grass known."

AURORAS.

On the night of October 2nd, when camped on Water-hen river, an Aurora of unusual brilliancy and character, even in these regions, surprised us with the varied magnificence of its display of light and colour. A broad ring of strong auroral light nearly encircled the Pole Star. It possessed an undulatory motion and continually shot forth, towards and beyond the Zenith, vast waves of faint light. They followed one another like huge pulsations—wave after wave—expanding towards the south with undiminished strength and continuing many minutes at a time. Suddenly the waves ceased, the luminous belt or ring increased in brilliancy, lost its regular form, and here and there broke into faint streamers, of a pale yellow colour. The streamers rapidly increasing soon reached the Zenith, and finally meeting beyond it, shot forth from the luminous arc with swift motion and in rapid succession. Their colour varied from straw to pink. The display of streamers is quite common in this part of the continent. The waves are also not unfrequently seen; but none of the Half-breeds or the Indians, whom we saw a few days afterwards, had ever witnessed such a brilliant spectacle as the heavens presented during the the early part of the night, when the immense pulsations, 14 deg. to 20 deg. in breadth, and expanding in their apparent ascent from east to west, rolled in tranquil, noiseless beauty, through the heavens overhead.

At ten p.m., on the 27th of October, when camped on the shores of Lake Manitobah, near Oak Point, a Half-breed awoke me to witness a crimson Aurora of surprising magnificence. Unfortunately, a few clouds were flitting athwart the sky, which prevented the centre arc from being visible, but perhaps they increased the depth of the colour. The light was generally steady at the edges of the clouds. The appearance of streamers was recognized only in the clear portions of the sky and above the clouds, where the rose or crimson tints were much fainter. It reminded me of the reflection of a vast prairie on fire; the deep rose and crimson tints lasted for half an hour; then gave way to white and straw-coloured streamers, occasionally tinged with pale emerald green.

Coloured Auroras are not unfrequently seen during the summer months, but they rarely possess the extraordinary beauty of those which have just been described. These beautiful "dancing spirits of the dead," impart a solemnity and charm to the still night, which must ever remain one of its most delightful characteristics in these regions.

Lake Huron, always attractive in calm summer weather, was peculiarly beautiful on the evening and night of the 25th of July, 1857, during our first voyage to Red River, when lighted up by a magnificent Aurora, as we neared the small Manitoulin Island. The auroral streamers converged beyond the zenith. Its base was marked by a very abrupt and well-defined sheet of light, from which waves and streamers rose from time to time. Masses of light moved continually from west to east, with an undulatory motion, occasionally folding and unfolding, with great regularity and distinctness of outline. A few minutes after ten o'clock, the base of the moving folds was tinted with delicate rose colour, passing, by imperceptible gradations, into faint emerald green above. The calm surface of the lake reflected these delicate colours, and the ever-varying motions of the auroral streamers and waves. The afternoon had been warm, with a fresh south-west breeze, and a thin haze in the same direction overspreading the high shores of the Grand Manitoulin Island.

The beautiful spectacle presented by this Aurora led to the description, hitherto unpublished as far as the narrator was aware, of a spectacle of extraordinary magnificence which had been witnessed by one of our fellow-travellers, a Post-Captain in the English Navy, who was making the tour of the Grand Lakes. This gentleman described his ascent to the summit of the Peak of Teneriffe for the purpose of seeing the sun rise above the waters of the Atlantic from that imposing elevation. At the moment when the red light of the sun began to flash above the unruffled outline of the horizon, overcome with emotion at the splendour of the scene, he turned away to seek a momentary relief in the grey of the west; but unbounded astonishment and admiration seized him, on beholding, instead of a grey blank, a gigantic image of the

* Report on the United States and Mexican Boundary Survey, made under the direction of the Secretary of the Interior, by William H. Emory, Major First Cavalry and United States Commissioner. Washington, 1846, pp. 43-47.

Peak projected on the sky to the full height of forty degrees, and swiftly sinking into the ocean as the sun rose above its eastern outline.

Colonel Lefroy, in 1843 and 1844, enjoyed many excellent opportunities of witnessing Auroras in Rupert's Land, at Fort Chipewyan, Lake Athabasca, lat. 58° 43′ north; long. 105° 35′ 15″ west: and Fort Simpson, lat. 61° 51′ 7″ north; long. 120° 5′ 20″ west.

The following extracts from the "Magnetical and Meteorological Observations"* at those places, contain the results of much valuable experience on points of great interest connected with the display of this beautiful phenomenon. They are followed by some extracts from Sir John Richardson's "Meteorological Observations at Fort Confidence, on Great Bear Lake," lat. 65° 54′ north, and long. 118° 49′ west, relating to the same subject. The extracts have reference to the supposed altitude of Auroras, their connection with the atmosphere, the sound produced by them, and the connexion of Aurora with magnetic disturbance.

Extracts from Colonel Lefroy's Meteorological Observations.

ALTITUDE.

"For want of corresponding observations elsewhere, there are no data for computing the height of any of the displays, but I avail myself of this opportunity of stating, that the impression conveyed to the senses upon many occasions was altogether opposed to the idea of the seat of the display being so distant as it seems to be in lower latitudes." *Captain, (now Colonel) Lefroy's Magnetical and Meteorological observations at Lake Athabasca, and Fort Simpson, page* 141.

CONNECTION WITH THE ATMOSPHERE.

"If the region in which the auroral development takes place be entirely beyond the limits of the atmosphere, as is commonly supposed, it is difficult to conceive any direct connexion between the aurora and the state of that medium, but this question may perhaps be regarded as not finally settled, and it may be worth while to examine the accompanying meteorological features. The first which will be noticed on referring to the meteorological register, is the apparent connection between the occurrence of aurora and a state of calm.—*Ibid*, page 146.

SOUND.

"With regard to the much disputed question of sound, neither the writer, nor his assistant, were ever positive of hearing any, but the latter thought that he did so, on one or two occasions. The result of inquiries upon the subject was, that opinions were nearly equally divided among the educated residents of the country; a small majority of those the writer consulted, agreed that a sound sometimes accompanied the phenomenon, but, among the uneducated and native inhabitants, whose acuteness of sense is probably much superior to that of the other class, a belief in the sound is almost universal, and many individuals assured the writer that they had heard it. Similar testimony has been borne very positively by the assistant at the Observatory of Toronto, upon one or two occasions of great display."—*Ibid*, page 151.

CONNEXION OF AURORA WITH MAGNETIC DISTURBANCES.

"A little experience in North America, whether in Canada, or in the more Northern regions, suffices to correct the impression that every display of aurora, however inconsiderable or distant, is attended by sensible magnetic disturbance." * * * * On the other hand it is unquestionable that the more brilliant displays are almost always attended by magnetic disturbances, as are indeed many of the more moderate ones; exceptions in the first class are very rare, but the writer believes that some can be established, the general conclusion must, however, be that an intimate relation exists between these distinct phenomena, although not that of cause and effect."—*Ibid*, page 151.

* Magnetical and Meteorological Observations at Lake Athabasca and Fort Simpson, by Captain [now Colonel] J. H. Lefroy, R.A.; and at Fort Confidence on Great Bear Lake, by Sir John Richardson, C.B., M.D. Printed by order of Her Majesty's Government. London: Longman.

Extracts from Sir John Richardson's Meteorological Observations.

ALTITUDE AND DISTANCE.

"Several times during the winter the auroral light was seen, both by myself and Dr. Rae, to pass us in front of a mass of cloud. As we were both aware of the ease with which the eye may be deceived in such observations, we watched the displays of the phenomenon with sufficient scepticism to keep the attention on the alert, and no doubt remained on our minds of the reality of the fact. In former years I had seen similar occurrences more frequently and even more manifestly. Thirty years previously I had entertained the belief that the aurora was connected with the formation of cloud, and other changes in the constitution of the atmosphere, and the nightly observations of this winter, all tended to strengthen that opinion."—Page 329.

SOUNDS.

"With respect to sounds of the aurora, the belief prevails in the Arctic regions, that it is occasionally audible, when very bright and active, at which times it is believed by the natives to be near the earth. Having witnessed the phenomena, some thousands of times, without hearing it, I have become sceptical of it ever producing sounds audible on the surface of the earth.—Page 380.

CONNECTION OF THE AURORA WITH MAGNETIC DISTURBANCES.

"On a review of the observations made during the seven months, many instances of the simultaneous occurrence of the fluctuations of the needle with movements in the auroral light were noticed; but there were also examples of fluctuations of the needle in the absence of the aurora, and very numerous ones of brilliant auroras accompanied by a stationary or sluggish needle. I cannot therefore venture to ascribe the movements of the needle in any case, to those of the aurora, or to any particular directions of the beams and arches. I think however, that the needle varied more frequently during the sudden formation of clouds than at other times; and I am also inclined to say, that the formation of clouds often followed brilliant and active auroras. It is a popular belief in the fur districts, that very fine displays of the aurora presage windy weather."—Page 350.

THE TWILIGHT BOW.

One of the most beautiful celestial phenomena visible after sunset and before sunrise from the north-western prairies is the Twilight Bow. The extraordinary clearness of the nights during summer in this region offers a very favourable opportunity for witnessing the delicate colouring which is communicated to the lower atmosphere by the reflected light from the upper illuminated portions. As the appearance of the Twilight Bow is dependant upon the serenity of the atmosphere to a great degree, its occurrence is not frequently observed or recorded in this country.

The Twilight Bow and the causes which produce it are thus described by M. Bravais:* "Immediately after the setting of the sun the curve which forms the separation between the atmospheric zone directly illuminated by the sun, and that which is only illuminated secondarily, or by reflection, receives the name of the *Crepuscular curve* or *Twilight Bow*. Some time after sunset, this bow, in traversing the heavens from east to west, passes the zenith; this epoch forms the end of civil twilight, and is the moment when planets and stars of the first magnitude begin to be visible. The eastern half of the heavens being then removed beyond solar illumination, night commences to all persons in apartments whose windows open to the east. Still later the Twilight Bow itself disappears in the western horizon; it is then the end of the Astronomic Twilight; it is closed night. We may estimate that civil twilight ends when the sun has declined 6° below the horizon, and that a decline of 16° is necessary to terminate the Astronomic Twilight."

* Annuaire Meteorologique de la France for 1850; quoted by L. W. Meek in the Smithsonian Report for 1856.

I often observed the Twilight Bow to be tinged with a delicate rose colour, passing into straw colour, and then into faint emerald green. The line of demarkation between the bow and the illuminated portion of the atmosphere was often very well defined, quite as clearly as in a secondary rainbow. It appeared most brilliant at an altitude of 60 or 70 degrees above the horizon. It descended slowly towards the boundless level, preserving apparently with considerable exactness the form of a parabola. When the Twilight Bow is best developed, the aspect of the prairie is very singular. Towards the east it is cold, cheerless and gloomy; towards the west it is warm, inspirating, and suggestive of pleasant thoughts and cheerful anticipations. No wonder the prairie Indians associate delightful dreams of happy hunting grounds with the setting sun and the beautiful west. They delight to sit silent and thoughtful "in the glory of the sunset," and allow themselves to be transported in imagination

> "To the Islands of the Blessed,
> To the Kingdom of Ponemah,
> To the Land of the hereafter."

ITINERARY.

(I.)

FROM FORT GARRY SOUTHWESTWARD TO THE 49TH PARALLEL, VIA THE ASSINIBOINE AND THE LITTLE SOURIS.

CAMP.		MAIN TRACK distance from—	
		Preceding camp.	Fort Garry.
		St. miles.	St. miles.
No. 1.	*June* 14, 1858.—Commenced exploratory survey. Encamped on the prairie. Good pasturage	0.00	0.61
2.	*June* 15.—*Lane's Post.*—Pursued a good trail through a fertile country, partially settled. Fine prairies adapted for grazing and agriculture. Clumps of poplar. Heavy timber in the bays of the river. A detachment branched off at St. James' church to make a reconnaissance of the Big Ridge, from Stony Mountain to Prairie Portage...	24.00	24.61
3.	*June* 16.—An attempt to survey the Assiniboine up-stream in canoe had to be relinquished, in consequence of the swiftness of the current. Replaced canoe on a cart, and proceeded 15 miles further. Camped at a stagnant pool in the shelter of a bluff of poplar. Good grass. Heavy timber skirting the river	16.00	40.61
4.	*June* 17.—*Prairie Portage.*—By making an early start Prairie Portage was reached at one, P. M. Crossed a level prairie with rich soil and herbage, but nearly destitute of trees. The detachment from Stony Mountain arrived in the afternoon. Good grazing	19.50	60,11
"	*June* 18.—*Prairie Portage.*—Occupied in repairing carts, completing equipment and making preparations to enter the Sioux country. Made a transverse section of the river, and levelled to determine its fall. Heavy thunder showers during the day.		
5.	*June* 19.—*The Bad Woods.*—Being unable to keep pace with the train after entering the Bad Woods, observations with the micrometer had to be suspended and the survey continued with the ordinary instruments for the trail and for reconnoitring. Correcting—by frequent observations—the main track distances determined from the verified mean rate of the wheeled vehicles. The position of prominent points established by cross bearings. Plenty of wood. Animals watered in the Assiniboine. Pasturage light and scanty..........	15.00	75.11
6.	*June* 20.—At the Half-way Bank, overlooking the valley of the Assiniboine, 7½ miles from last camp, the latitude of 49° 46′ was observed. Height of bank, 150 feet above river. Breadth of valley, one mile. Magnetic variation 13° E. Camped at seven P. M., among sand dunes, from the summit of which Pembina Mountain near St. Joseph's was seen. Terrific thunderstorm after sunset. Water in ponds. Herbage short and stinted. Light sandy soil ...	15.00	90.11
7.	*June* 21.—Trail continues among sand dunes, ponds, scattered poplars and willows. Reached Bear's Head Hill, the highest peak of the Sand Hills, about noon, and halted to allow the animals to graze. Before resuming journey a thunder and hail-storm came on. The hailstones (1-1½ inches in diameter) cracked the bark of the canoes on the carts. After proceeding a few miles another violent thunderstorm compelled a camp at Sunset Lake. Good grazing only in detached areas	15.80	105.91.
8.	*June* 22.—Trail still winds around sand hills and between ponds varying from two to thirty chains in diameter. Smokes have now to be made for the animals at every camp. Mosquitoes and bulldogs so annoying as to prevent them from feeding. The great heat of the weather during the day exhausts the animals and retards progress. A terrific thunderstorm, lasting from ¼ to 4 to 6, P. M., rendered an early camp necessary. Lightning very near and vivid. Incessant roar of thunder for an hour and a half. Plenty of water in lakelets. Grass light. Spruce and aspen on the sand hills	13.22	119.13.
9.	*June* 23.—Observed for latitude, etc., at Pine Creek crossing, 130 miles from Fort Garry. A division followed Pine Creek from the cart trail to the Assiniboine, returning by the Devil's Hills (dunes of drifted sand). Still traversing sand dunes, with occasional intervals of light prairie: and grassy areas, between clusters and ranges of sand hills from 30 to 70 feet high, dotted with stunted oaks, and thinly clothed with small balsam spruce and poplar on their flanks. Country improves and passes gradually into rolling prairie after leaving the old Brandon trail. Grazing improved	25.75	144.88.
10.	*June* 24.—Directed course towards the Assiniboine and Souris Forks, reaching the Assiniboine opposite the mouth of the Little Souris, 146 miles from Fort Garry at 5.40, A. M. Halted to make observations, graze the animals and breakfast. Warned this morning to prepare for an attack by the Sioux. The smoke of two fires in the valley of the river indicating their presence. Grasshoppers very numerous and destructive to baggage and harness. Effected the crossing of the Assiniboine, after completing observations; swimming the horses, ferrying		

ITINERARY.—(*Continued.*)

CAMP.		MAIN TRACK distance from— Preceding camp.	Fort Garry.
		St. miles.	St. miles.
	the baggage in canoes and towing the carts and waggon over. Proceeded up the left bank of the Souris; camping four miles from its mouth. Mounted guard during the night to avoid a surprise by the Sioux.	5.00	149.88
No. 11	*June* 25.—Heavy showers of rain early this morning prevented the carts from advancing at the usual hour. Observed for latitude at a small affluent of the Souris. Camped at sunset on the banks of the Souris valley, between the Blue Hills of Brandon and the Blue Hills of the Souris. Valley very deep and broad. Scenery wild and picturesque. Good track over a rolling prairie. Soil sandy loam. Precautionary measures continued. Herbage rich in the valley and in most of the hollows	15.50	165.38
12	*June* 26.—Terrific thunderstorm last night accompanied by boisterous wind and heavy rain. Remained at preceding camp the greater part of the day in order to refresh the horses, and make geological examinations and sketches in the valley. Resuming march at 4 P. M., travelled over two hours and camped at a beautiful point in the Souris valley, opposite Back Fat Creek, a tributary rising in the Back Fat Lakes from which the North Branch of Pembina River also issues. Crossed a rolling prairie of light sandy loam with occasional stony ridges and small lakes. Obtained a magnificent view of the boundless, southwestern prairies, with Turtle Mountain in the distance, before descending into the valley. Men and animals suffer much from the attacks of mosquitoes surrounding them in clouds ...	6.75	172.13
13	*June* 27.—Struck camp at noon, having made the requisite observations and levelled across the valley. Traversed an undulating prairie with gravelly knolls crested with erratics at intervals. A few hummocks of small poplar along margin of valley. Herbage short and scanty on the high ground, rich and exuberant on the low ground and in the alluvial bottoms	8.25	180.38
14	*June* 28.—Striking camp and advancing at daylight a halt was made at 8 A. M. to breakfast and to examine the shales exposed in the valley. Proceeded down river a short distance in canoe. The Souris is here 1-1½ chains broad and 2½-3 feet deep, with a swift current. Camped at 8 P. M after journeying along the crest of the valley, over a light prairie with occasional areas of rich dark soil. Cold and stormy day. Strong north wind. Rain. Grazing good	12.50	192.88
15	*June* 29.—After crossing Plum Brook or Snake Creek and halting to graze the animals at Snake Hill, layers of drift tertiary coal or lignite were discovered in the bank of the Souris. Engaged during the remainder of the day in sinking shafts and exploring for lignite in this locality. Made camp-fires of lignite. Wood and water abundant..	15.05	207.93
"	*June* 30.—*Snake Hill.*—Still occupied in excavating for lignite, making sections and observations. Three men despatched to Oak Lake to hunt with a view to save provisions, returned in the evening with a number of ducks and pelicans. Grazing tolerably good. Plenty of wood and water.		
16	*July* 1.—Struck camp and started train at daylight. Halted for dinner at an old log-house on the banks of the Souris, a winter Trading Post of the Hon Hudson's Bay Company. Crossed the "Round Plain" in afternoon, a beautiful grassy area about four miles in diameter, level as a bowling green, and surrounded by thinly wooded sand hills. Camped on a level plain supporting luxuriant grass. This plain was flooded in 1852 to a considerable depth, and occupies an area of about a mile in width between the Souris and a range of low sand hills	23.37	231.30
17	*July* 2.—Tents struck and brigade equipped for the march at 4 A. M. Traversed an undulating treeless prairie extending to Turtle Mountain on the left. Crossing Half-way Creek and several deep gullies carrying the prairie drainage into the Souris, the train halted at Mandan Creek, another small affluent so called from the numerous mounds or tumuli, near its mouth, said to have been underground houses of the Mandan Indians. A careful examination of the tumuli was made by digging into them, but no vestiges of Indian remains were found. Camped on the banks of Red Deer's Head River, near its confluence with the Souris. Two sets of astronomical observations determined the latitude of this station to be 49° 1′ 44″, or a fraction over two statute miles north of the international boundary, and in about 100° 55′ west longitude. Magnetic deviation, 14° E. Good grazing, wood and water in the valley. Track of Sioux observed	26.25	257.55
18	*July* 3.—Remaining encamped till afternoon to make observations as well as to repair the carts and travelling gear, a detachment with an escort was enabled to make a reconnaissance of Red Deer's Head River to its mouth. All having returned to camp, the horses were harnessed and the journey resumed by the train at 5 P. M. Striking in a S. W. direction, across an undulating prairie strewed with buffalo dung and scored with their tracks, a distant point of Red Deer's Head River within United States Territory was reaced about 8.30 P. M., and a camp formed near a clump of trees growing on the margin of the river, with a view of taking in a supply of wood to be used as fuel in crossing the great treeless prairie lying between the boundary line and Fort Ellice. Some hostile Sioux in ambuscade in the vicinity of the encampment, attempting to stampede the hobbled horses after dark, shewed the necessity of increased precaution and vigilance. The animals were accordingly picketed within the camp ring, and the number of watchers increased to eight. Traversed before camping a vast sandy plain with short and scrubby grass, burnt last year	10.25	267.80

ITINERARY.—(*Continued.*)

(II.)

FROM RED DEER'S HEAD RIVER—A FEW MILES SOUTH OF THE INTERSECTION OF THE BOUNDARY LINE AND THE LITTLE SOURIS—NORTHWARD TO FORT ELLICE.

CAMP.		MAIN TRACK distance from— Preceding camp.	R. Deer's Hd. Riv.
		St. miles.	St. miles.
No. 19	*July* 4.—Sioux heard by the watch during the night, and the tracks of their scouts observed in close proximity to the encampment this morning. Sufficient wood being distributed among the vehicles to last during a passage of five days across the great treeless prairie between this station and Fort Ellice, and camp being broken up at 10 A. M., the train wended its way in a northerly direction for about three hours across a light sandy prairie, dotted everywhere with bleached buffalo bones; and halted about three hours at a small pond with a margin of marsh. The animals being much fatigued by the excessive heat of the weather, only six miles farther were accomplished	13.95	13.95
20	*July* 5.—Breaking up camp at daylight, the train was in travelling order and advanced at 4.30 A. M. Traversed a level plain with small gravelly knolls and low ridges at intervals. Soil generally light sandy loam. Grass short and scanty. Plenty of water in marshes, ponds and stagnant creeks. No wood of any kind as far as the eye can reach. *Bois de vache* rarely seen. Saw several antelopes and shot a female to-day. Camped at sunset	23.00	36.95
21	*July* 6.—Up at dawn. Train in motion about 4, A. M. Halted at Pipestone Creek for breakfast, having accomplished 13.78 miles, after five hours travelling across a light sandy prairie with low knolls and ridges of gravel and boulders. Remaining here to determine the latitude and being delayed some time in fording the stream, owing to the steepness and miryness of its banks, the train did not get under-way again till 2, P. M. Traversed a rolling woodless prairie with hard gravelly soil, supporting a scanty growth of grass, and camped at Boss Creek, a small affluent of the Assiniboine, flowing in a broad valley among low hills and knolls with gentle slopes. Standing Stone Mountain, Boss Hill and Oak Lake were seen from a conical hill near the encampment	23.40	60.35
22	*July* 7.—Horses caught, unhobbled and ready for the march early. Train left camp site at 5, A. M.. Crossing a level plain and fording Boss Hill Creek, a halt was made at a stagnant brook, after traversing a light sandy and gravelly prairie with short herbage. Thence journeying over a rolling prairie with very light soil, in many places covered with boulders and supporting occasional hummocks of poplar and willow, partially burnt, the Assiniboine was reached, a camp pitched at a small affluent, and the animals turned loose to graze a little after 4, P. M.	19.70	80.05
23	*July* 8.—Tents struck at 3, and the train advancing at 4, A. M. Halted to skin and cut up a buffalo bull "run" and shot this morning. Resuming march at 11, A. M., and crossing a rough prairie with hard gravelly soil covered with erratics, the Two Creeks (tributaries of the Assiniboine) were reached about 3, P. M. Having forded the creeks and camped, the remainder of the day was occupied in examining and searching for fossils in the shales exposed in the valley	10.87	90.92
24	*July* 9.—Striking tents early, the horses were caught, and the train, with the exception of one cart, equipped for travel at 2·45, A. M. The ox, although hobbled, could not be found, and three of the party, accompanied by the waggon and driver, hastened on to Fort Ellice, leaving the train to follow after recovering the ox, which was accomplished after a search of seven hours. Crossed an undulating prairie extending to the Assiniboine, with light sandy soil, except in the hollows, where a thin coating of vegetable mould is found. Halted to camp at a gully with stagnant water in the bottom, leading to the Assiniboine. Grass abundant and rich in the depressions. Plenty of water and wood. Buffalo seen again to-day	15.88	106.80
25	*July* 10.—*Fort Ellice.*—Starting at 3, A. M., Fort Ellice was reached early. Before fording Beaver Creek, a level prairie, with soil of light, sandy loam was crossed. The monotony of the plain relieved here and there with clumps of light poplar and low sandy hills. Between Beaver Creek crossing and the fort, a well-beaten track passing over a sandy plain and hills or dunes of white sand were traversed	10.90	117.70
"	*July* 11.—*Fort Ellice.*—Encamped within half a mile of the fort. Remained quietly in camp to-day, (Sunday,) much to the advantage of the wearied animals. Engaged at noon and at night in determining astronomically the position of this station by different sets of observations. Thunderstorm and rain in the afternoon.		

ITINERARY.—(*Continued.*)

(III.)

FROM FORT ELLICE WESTWARD TO THE QU'APPELLE MISSION.

CAMP.		MAIN TRACK distance from— Preceding camp.	Fort Ellice.
		St. miles.	St. miles.
26	*July* 12.—Having completed observations, reconnoitred Beaver Creek to its junction with the Assiniboine and made a traverse of about 4 miles north of the fort to obtain a section of the Qu'Appelle River at its mouth, the Fort Ellice encampment was broken up and the train proceeded westward about 5 P. M. Went into camp at 7.30 P. M. after traversing a light sandy prairie with occasional clumps of small poplar, and several marshes and ponds......	7.50	7.50
27	*July* 13.—Camp struck at daylight and train advancing at the usual hour. Traversed an undulating prairie of light sandy loam with scattered clumps of poplar and willow. Halted to feed after travelling nine miles. Thence journeyed over a rolling prairie of rich sandy loam, clothed with an exhuberant growth of excellent grass. Another halt was made at the Cross Woods, an open belt of light aspen reported to extend from Qu'Appelle River to Pipestone Creek. Camped at sunset in a region of marshy ponds surrounded by light prairie. Terrific thunderstorm just before dark—continued some hours	25.50	33.00
28	*July* 14.—Started at 4 A. M., and traversed a light rolling prairie with gravelly ridges thinly wooded with scattered aspens, succeeded by a wide treeless plain of rich sandy loam. A halt was made for breakfast at a bluff of poplar after accomplishing a distance of 12 miles. Continued the journey across an undulating prairie of light sandy loam with occasional clumps of small poplar and many ponds. Camped late at the beginning of a vast treeless prairie stretching north to the Qu'Appelle. A cold, windy, disagreeable day. Peals of thunder heard overhead in the morning. Detained some time by rain	23.00	56.00
29	*July* 15.—Rose at 3 A. M. and resumed the journey westward across a light undulating open prairie, succeeded by a treeless rolling prairie, in the middle of which, finding some sticks of wood dropped by Indian hunters, the train stopped to graze the animals and breakfast. Dined at the Weed or Bear Berry Ridge, and camped at sunset on an undulating prairie, with clumps of poplar and willows. Soil of prairies traversed to-day generally light with gravelly ridges. Areas of rich loam with good grass in the depressions. Abundance of water in numerous ponds dotting the plain. Wood scarce. Trail runs parallel to the Qu'Appelle at a distance of 12–16 miles. Cold and cloudy in morning. Strong N. W. wind...	26.55	82.55
30	*July* 16.—Camp broken up at 3.15 A. M., and train *en route* before 4 A. M. Halted after 12 miles travel over a vast treeless rolling prairie, with soil and herbage as before. From this station on an open plain, the woods of the Qu'Appelle 12–18 miles off could be seen. Proceeding westward over a sandy prairie, among clumps of poplar and willow, the "Indian Head Hills" were reached and crossed about noon. Finding on the west side of the hills an encampment of Half-breeds journeying from Qu'Appelle Mission to Fort Ellice, an opportunity was afforded for exchanging two horses almost worn-out, for fresh ones. After passing the Qu'Appelle Fort trail, the course lies over a light treeless undulating prairie, sloping gently towards the Qu'Appelle and intersected by several creeks flowing in deep valleys. Camped at sunset. Plenty of wood, water and grass. Mosquitoes in clouds and unusually troublesome in the evening	26.62	109.17
31	*July* 17.—Horses driven into camp, unhobbled and equipped for travelling at 3, A. M. Train departed from camp at 3.40, A. M., and journeyed till 7, A. M., stopping at a point where the party met yesterday left some wood, on a vast level prairie, with dark, rich soil, bearing luxuriant grass. Met a half-breed and Indian family on the plains, gathering and drying 'wild turnips' for winter provisions. Crossed a light, undulating prairie with many knolls, ridges, and marshy ponds, and reached the Church of England Missionary Post at the Fishing Lakes, Qu'Appelle Valley, just before sunset. Forded the Qu'Appelle, and camped on north side of valley	26.40	135.57
"	*July* 18.—*Qu' Appelle Mission.*—To-day being Sunday, all except those left in charge of the encampment, attended Divine service, conducted by the missionary (the Rev. James Settee,) in his log-house. The tired animals much benefitted by the rest and the good grazing in the valley. Thunderstorm at night, with violent showers of heavy rain		
"	*July* 19.—*Qu' Appelle Mission Encampment.* Engaged last night and to-day in making the usual observations for latitude and variation, triangulating to establish the position of prominent points, making a section of valley, levelling river, taking photographs, and preparing for canoe voyages up and down the Qu'Appelle.		

ITINERARY.—(*Continued.*)

(IV.)

FROM THE QU'APPELLE MISSION WESTWARD TO THE ELBOW OF THE SOUTH BRANCH OF THE SASKATCHEWAN VIA THE QU'APPELLE RIVER AND VALLEY.

CAMP.		MAIN TRACK distance from—	
		Preced-ing camp.	Qu'App. Mission.
		St. miles.	St. miles.
No. 32	*July* 20.—Broke up Qu'Appelle Mission encampment at 3, A. M., and commenced the ascent of the Qu'Appelle River in canoe after the following divisions were *en route:* A detachment with three carts, to proceed to Fort Pelly after making a detour to Long or Last Mountain Lake, another with three carts to proceed along the south side of the Qu'Appelle Valley, to meet the canoe division at the Grand Forks of the river, and a third to proceed down the Qu'Appelle in canoe to Fort Ellice, thence by land to Fort Pelly. Embarked in a three fathom birch-bark canoe (brought from Red River,) manned by two voyageurs, and passed through the 3rd and 4th Fishing Lakes. Camped after 18 miles paddling, five of which were against the current of the river meandering through a marsh before debouching into lake No. 4.	18.00	18.00
33	*July* 21.—Little rest obtained in camp last night, the swarms of mosquitoes and sand-flies being annoying beyond measure. Rising at 3 and embarking at 4, A. M., the journey was resumed. The rate of progress against the current of this tortuous river being slow, one of the party proceeded to make a reconnaissance of the valley on foot, whilst the other continued the survey of the river in canoe, ascertaining the rate of current and canoe by log. Delayed by a heavy rain for three hours after halting at 2, P. M. The hill sides of valley, which are upwards of 300 feet high, were ascended at intervals. Sailed and tracked up the river till 8.40, P. M., some time after dark and camped	19.16	37.16
34	*July* 22.—*En route* early. Tracking and paddling alternately against a tortuous muddy stream. Having halted to dine and ascend the sides of valley, an altitude of 3–400 feet, to examine the prairie beyond, a violent thunderstorm coming on prevented re-embarkation for a considerable time. Tracked and paddled till sunset. Camped on the south side of the valley surrounded by clouds of mosquitoes. The river is so tortuous that to-day no fewer than 200 courses and distances were recorded in canoe	14.00	51.16
35	*July* 23.— The Qu'Appelle still meanders through rich alluvial flats, clothed with long rank grass. Its serpentine course from side to side of the broad valley is clearly marked by a close margin of tall willows. Made an early start, proceeding up the river and valley as heretofore. Joined the carts that were in waiting at the appointed *rendezvous*, and encamped not far from the Forks in company with a band of Indians—"Bungays"*	14.32	65.48
No. 36	*July* 24.—Thunderstorm early this morning. The tortuosity of the river and the strength of its current retarded progress so much that it was decided to continue the exploration by land. The canoe was accordingly replaced on a cart and the course of the train directed up the valley. Ascended to the crest of the valley on the north side, and continued the journey for several hours along its margin over a light open prairie. Pitched camp at sunset on a grassy plateau on the hill side of valley. Cree Indians seen......	13.43	78.91
37	*July* 25.—The morning occupied in ascertaining the dimensions of the valley. The measurement from bank to bank ($1\frac{1}{4}$ miles) show that the width of this great excavation is well maintained and the height of the bank computed from observations with the sextant indicate but little diminution in depth. Visited by Crees. Proceeded on the brink of the valley, over a slightly undulating prairie of light sandy and gravelly soil with poor, short grass. Halted at a deep ravine affording wood and water. The trail of the train leading occasionally some distance into the prairie in order to head the deep gorges and ravines ramifying from the valley, rendered it necessary to make detours from the main track at intervals to obtain a more complete reconnaissance. Camped late at a small pond a mile west of the Round Hill, after crossing a rolling prairie of light sandy soil. Unable to make fire; there being neither wood nor *bois de vache* in the vicinity. Herbage scrubby and scant	21.85	100.76
38	*July* 26.—Started at daybreak. Came suddenly upon an encampment of Crees numbering five skin tents on the edge of the valley, near the eastern extremity of Buffalo Pound Hill Lake. The inmates were not up, but the baying of their dogs at our abrupt appearance soon aroused them. Pursued the trail over a light undulating prairie along the margin of the valley. Stopped for a short time without unyoking at Buffalo Pound Hill, a conical elevation at the west end of the lake of the same name, from which the Touchwood Hills, the "Outlook" and the Grand Côteau de Missouri can be seen. These and other prominent points were connected at intervals by numerous cross-bearings. Halted near the Outlook Hills on a rolling gravelly plain with stunted grass. Wood, water and some good grass obtained in the gullies. Passed another Cree encampment after some hours travel over an undulating sandy prairie. The tents (nine in number) were at the head of a broad ravine filled with poplars. After a little delay in distributing tobacco, powder		

* Crees and Ojibways of mixed origin.

ITINERARY.—(*Continued.*)

CAMP.		MAIN TRACK distance from— Preceding camp.	Qu'App. Mission.
		St. miles.	St. miles.
	and ball, and bartering for *Mesaskatomina* berries, the journey was continued and camp pitched a little after sunset. Watch appointed to prevent the Crees from stealing the horses. Camp fires of buffalo dung.	24.52	125.28
No. 39	*July* 27.—Camp struck at daylight and train *en route* at 4, A. M. The numerous deep gorges and ravines breaking the continuity of the valley side, rendered many deviations from a direct course necessary. Traversed a very light sandy prairie strewn with boulders and halted on the brink of the valley at the east end of Sand Hill Lake. Soon surrounded by Crees who had galloped across the valley from their encampment on the opposite side, on getting the first glimpse of the train. Whilst a 'talk' was going on between our guide and the chief of this band, the height of the prairie plateau above Sand Hill Lake was ascertained by levelling, and the width of the valley by measurement. Gave the chief some tobacco and ammunition to distribute, and procured his son as a guide to the "River that turns," and the Saskatchewan. Crossed the valley, which is dry here at this season, and camped not far from the Indian lodges. Grazing poor. Wood and good water very scarce. Buffalo seen twice to-day	20.15	145.43
40	*July* 28.—Left camp site at the usual hour. The train pursued a course over a sterile, stony buffalo plain, thickly dotted with *bois de vache,* and halted early to wait for the return of a division that had branched off to make a reconnaissance of the Eyebrow Hill and Ridge. Indifferent grass and no water, but a supply was obtained for cooking by catching the rain which fell in torrents at noon. A traverse was made on horseback to the west end of Sand Hill Lake, and the exploration continued along the brink of the valley, whilst the carts followed the trail heading the ravines, till reaching a tributary rising in Eyebrow Hill ridge. Camp pitched in the valley at the confluence of this affluent and the Qu'Appelle. The usual observations and levelling conducted here to obtain a section of the valley. Good grass on the flats. Water and a limited supply of wood. Flies tormenting as usual	15.25	160.68
No. 41	*July* 29.—Struck camp and started early. The train recrossed the Qu'Appelle and proceeded along the foot of the northern slope until the mud flats became too wet and springy for the animals. Ascended to the crest of the valley and pursued a circuitous course along its brink among hills of white and yellow sand, quite loose and destitute of vegetation. Having crossed some feeble brooks rising in ponds among the Sand Hills, (feeders on this side, of the great marsh filling the Qu'Appelle valley at the summit-level, and sending its waters to the Assiniboine and the Saskatchewan,) a halt was made to determine the position and make a thorough examination of the height of land. Being soon surrounded by mounted Crees, the train journeyed on to negotiate and parley with their chief Shortstick, who was impounding buffalo among the sand hills farther west, whilst a detachment retraced their steps to the height of land to determine by levelling, the elevation of the feeding marshes and ponds in the valley, above the Saskatchewan. Encamped near the buffalo pound, surrounded by clusters of skin tents. Grazing very poor. Water scarce. Scrub poplar between the sand hills	8.40	169.08
42	*July* 30.—Commencing operations at daylight this morning, the levelling was resumed and the survey of the valley continued without intermission until closing upon the South branch of the Saskatchewan, at 5, P. M. Distance levelled, 11.87 miles. Altitude of the summit-level pond (one of the sources of the Qu'Appelle and the 'River that turns') above the Saskatchewan, 85.89 feet. The train reached the Saskatchewan early in the day, and all preparations were completed for a re-division of the party; the canoe gummed and equipped for a voyage down the South Branch, and the carts repaired for a journey to Fort à la Corne. The train having departed on its way northward, the canoe division embarked about sunset, and after 2.70 miles paddling camped on the right bank of the river......	10.35	179.43 2.70
	To mouth of the "River that turns"...	....	176.73

ITINERARY.—(*Continued.*)

(V.)

FROM THE QU'APPELLE MISSION EASTWARD TO FORT ELLICE VIA THE QU'APPELLE RIVER.

CAMP.		MAIN TRCK distance from—	
		Preced-ing camp.	Qu'App. Mission.
		St. miles.	St. miles.
No. 43	*July* 20.—Embarked this morning in a 2½ fathom canoe with two voyageurs, and commenced the descent of the Qu'Appelle from the beginning of the portion of the river issuing from Fishing Lake No. 3. Obtained the dimensions of the connecting river at its mouth, and steered down the centre of Fishing Lakes 2 and 1, sounding at intervals with the hand lead. Measured the volume of water in the river at its exit from the east end of Lake No. 1. Pitched camp at sunset at the foot of the southern slope, 350 feet below the prairie level...........	....	25.26
44	*July* 21.—Started at daybreak. Paddled till 3 P. M. when it became necessary to camp in consequence of a thunderstorm. The Qu'Appelle continues wonderfully winding and meanders from side to side of its broad valley so often, that the distance made by the river is far greater than that actually accomplished in a direct line................	29.65	54.91
45	*July* 22.—*En route* early. Moored canoe and took breakfast at an area of burnt grass revealing land of good quality. Halted at intervals to determine the dimensions of the river and valley by the usual series of observations. A thunderstorm in the evening occasioned a detention of an hour and three quarters. Camped late	43.15	98.06
46	*July* 23.—Struck camp and embarked at the usual hour, soon passing Pheasant Creek (called by the Crees, *A-kis-koo-wi se-pi-sis*) a small tributary rising in the Pheasant Hills some distance to the north. Entered Crooked Lake, *Ka-wa-wa-ki ka-mac* of the Crees, at noon. Sounded through the lake and left it with some difficulty, its outlet being concealed by rushes. Continued paddling down the river, which maintains its uniform width of about 70 feet and average current of 1¼ miles an hour. Camped at sunset at the mouth of an affluent from the south, called *Ne-pi-me-na-ne Se-pi-sis;* interpreted, Pembina or Summerberry creek	27.78	125.84
No. 47	*July* 24.—Wet morning. Rain increasing; after three hours paddling it compelled a halt of seven hours. Reached *Ka-wa-wi-ya Ka-mac* or Round Lake in the afternoon. Carried a line of soundings through it, as on the other lakes, until arriving at the re-commencement of the river. Thence glided down the river a distance of 2.43 miles, by its serpentine course, and camped at *Assini-pichi-pu-yakan*—the Stony Barrier.	25.93	151.77
48	*July* 25.—Embarked after the heavy rain ceased. Passed in a short time the mouth of a creek, falling in from the south, named *Isquao-wis-te-quau-na-ka-us-ta-ki,* or the creek where the Cree women's skulls lie. Camped late, after passing Little Cut-Arm Creek on the north, and Scissors Creek on the south,—small affluents with very long names in Cree. Swarms of insatiable mosquitoes and other venomous insects as usual	33.13	184.90
49	*July* 26.—Resumed the voyage at dawn. Passed, after two hours travel, Great Cut-arm Creek, another tributary from the north. Halted frequently as before to obtain the position of prominent points in the valley by intersecting bearings and to examine the character of the prairies above. River extremely serpentine. Fine meadow grass on the flats. Flanks of valley and ravines timbered. Pitched camp at the usual hour..................................	33.81	218.71
50	*July* 27.—Left camp at daylight. Passed some places where the whole valley is filled with trees, chiefly poplar, ash, elm, maple and oak. Arrived at the Assiniboine River at sunset, and after making a section of the mouth of the Qu'Appelle, proceeded to Fort Ellice and camped.		
	To mouth of the Qu'Appelle.....................	37.88	256.59

ITINERARY.—(*Continued.*)

(VI.)

FROM FORT ELLICE, NORTH WESTWARD, TO FORT PELLY AND SWAN RIVER, VIA THE WEST SIDE OF THE ASSINNIBOINE.

CAMP.		MAIN TRACK distance from— Preceding camp.	MAIN TRACK distance from— Fort Ellice.
		St. miles.	St. miles.
No. 51	*July* 28.—Occupied the greater part of the day in making preparations for a reconnaissance of the country between the Qu'Appelle and Swan River. Started in the evening from Fort Ellice with a light equipment. Forded the Qu'Appelle three chains from its mouth, and ascending to the upper plateau pursued the trail skirting the Qu'Appelle Valley till dark. Camped on a sandy area covered with creeping juniper. Grass scanty. Small poplar..	4.41	4.41
52	*July* 29.—Equipped for the trail and in motion at daylight. Followed the Qu'Appelle Valley half a mile farther, then struck north-westerly through a woodland district with prairie intervals. Passed a large sandy knoll called Red Deer's Horn Hill. Halted after fording a creek of the same name. Traversed a rough and partially wooded prairie of light sandy soil before crossing Wolverine Creek. Rested as usual at noon, near a conical hill, named *Ne-pa-guy-we-nis*. Forded the Big Valley Creek, and camped about eight miles west of the Assinniboine. Wood and water in abundance. Luxuriant grass. Good land.	28.21	32.62
53	*July* 30.—Left camp early and followed the trail winding between clusters of ponds scattered over a level prairie supporting straggling hummocks of poplar. Crossed another small tributary of the Assinniboine in the evening, and pitched camp beside a marsh Wood and good water. Exuberant growth of willows and grass. Soil, sandy loam ...	24.12	56.74
54	*July* 31.—Struck camp and *en route* at the usual hour. Traversed a fine country with open groves of sapling poplar and most luxuriant vegetation before fording the Two Creeks. Upon fording Stony Creek and re-ascending to the prairie level, a halt was made for the noon-day feed. Crossed an undulating country succeeded by a flat tract abounding in ponds and marshes, some of which impeded progress. In fording the Steep Creek a cart was upset, the crossing place being bad. Encamped among the Beaver Hills, three quarters of a mile beyond the Steep Creek	27.35	84.09
55	*August* 1.—Broke up camp and started in a very heavy rain. Followed a course through an entangled mass of vegetation skirting the Beaver Hills. Breakfasted in a pouring rain after fording White Mud River, a rapid stream seventy feet wide and four feet deep. This crossing occupied some time, being very bad, and the banks of the river steep and slippery. Traversed a very level country with surface soil of rich sandy loam supporting clumps of small poplar, osiers, and a luxuriant growth of various plants. Reached the Assinniboine after passing through some beautiful open woodlands. Forded the Assinniboine and encamped at Fort Pelly. Found the carts from the Mission here.................................. ..	19.45	103.54
"	*August* 2.—*Fort Pelly Encampment.*—Engaged during the day and at night in making the usual observations to determine the latitude and magnetic variation. Preparing for a traverse to Swan River.		
"	*August* 3.—Started for Swan River a little after noon. Crossed Miry Creek and continued down the valley of Snake Creek to its confluence with Swan River. The trail led across La Course Creek and three others falling into Snake Creek. Stopped some time to ascertain the dimensions of Swan River and its fertile valley. Returned over a level country with a surface soil of sandy loam sprinkled with erratics. Hummocks of small poplar and spruce. A few tamarack.		
	To Swan River	9.41	112.95

ITINERARY.—(*Continued.*)

(VII.)

FROM "THE RIVER THAT TURNS," NORTH EASTWARD, TO FORT A LA CORNE, VIA THE SOUTH BRANCH OF THE SASKATCHEWAN.

CAMP.		MAIN TRACK distance from— Preceding camp.	Riv. that Turns.
		St. miles.	St. miles.
(42)	*July* 30.—Launched canoe at the mouth of "The River that Turns," and commenced the track survey of the Saskatchewan (S. B.) Descended the river for half an hour and pitched camp in the vicinity of a rock exposure on the right bank......................	....	2.70
No. 56	*July* 31.—Embarked early. The swift current of the river (three miles an hour) accelerated progress. River half a mile to three quarters of a mile wide, but shallow where broad and much interrupted by sand-bars and mud-flats. Halted at noon to examine an exposure of sandstone. Passed large Cree encampments on both sides of the river, at an Indian crossing place. The Crees "pitching" eastward to avoid the Blackfeet. Mesaskatomina berries in great profusion. Camped at dusk....................................	42.68	45.38
57	*August* 1.—Left camp at the usual hour. Made a transverse section of the river upon halting to breakfast. Camped at sunset in the Moose Woods after some hours paddling through intricate channels between large alluvial islands and flats. Good land on the flats wooded with ash, elm, and aspen. Buffalo seen floating in the river......	34.52	79.90
58	*August* 2.—Passed, soon after embarking this morning, some old shanties of the half-breeds who come to the Moose Woods to barter with the Indians in winter. Halted occasionally and ascended to the brink of the valley to examine the country beyond, which generally consists of a rolling sandy prairie dotted with clumps of poplar. Recorded many sandbars, snags and sawyers to day, and one or two small rippling rapids. Had to moor canoe at 4 P.M. and seek shelter in consequence of a thunderstorm. Camped at 7 P.M. on a low stony point covered with driftwood.....	43.00	122.90
59	*August* 3.—Struck camp and embarked at daylight. Anchored once or twice to measure the rate of current. Found it to maintain an average velocity of three and a quarter miles an hour. In the narrow places it is much swifter. Halted at noon to level along the brink of the river to determine the extent of its fall. Passed some precipitous bluffs of yellow clay in the bays of the river and camped at a small rapid. This rapid offers no impediment to navigation, as its fall is not more than nine inches, and the ruffled water is only on one side of the river, on the other, the channel is smooth and deep..	47.10	169.00
60	*August* 4.—Left camp at sunrise and did not stop for breakfast till 11.30 A.M. Resumed the voyage at 1.30 P.M. River filled in some places with well-wooded alluvial islands and mud-flats in course of formation. The banks are now lined with poplar. A thunderstorm with very heavy rain at half-past 4 P.M., compelled a halt of three quarters of an hour. Stopped to camp at 7.20 P.M...	50.90	219.90
61	*August* 5.—Started at 6.30 A.M. A drizzling rain, that had been falling all the morning, began to pour very heavily about 11 o'clock, rendering it necessary to halt and seek the shelter of some large white spruce trees' which grew at the river side. Continued the journey after the rain had ceased, and, being aided by a very swift current, swept round the great bends of the river with considerable velocity. Current much increased in swiftness, being in many places upwards of four miles an hour. Several portions of the river descended to-day might be termed rapids, the water being quite rough with a heavy groundswell. Arrived at the Grand Forks at 2 20 P.M., and commenced the ascent of the Coal Falls on the North Branch to search for Lignite. Tracking up this impetuous torrent was a slow process, and camp was pitched at a point about two miles from the Forks, only reached at sunset by the canoe. Found Cretaceous fossils.		
	To the Grand Forks	29.83	249.73
62	*August* 6.—Left the tent standing over the baggage and proceeded up the left bank of the river on foot, leaving the voyageurs to follow with the lightened canoe The rapids retarded their progress very much. About five miles from the Forks a mass of the so-called coal of the voyageurs was observed in the drift banks, but none in *situ*, it holds *inoceramus*. Collected a number of specimens and glided swiftly back to the Forks. Saw a half-breed family with a bark canoe at the Forks, preparing to ascend the South Branch to gather Mesaskatomina berries. They had set out from the Nepowewin. Left the Forks at three P.M., and proceeded down the Main Saskatchewan. Arrived at Fort à la Corné a little after sunset, and pitched tent within the Fort enclosure.		
	From the Forks	20.15	
"	*August* 7.—*Fort a la Corne.*—Triangulating to establish the position of prominent points in the valley and to ascertain its dimensions, in the vicinity of the Fort. Sketching the Fort, the Mission, etc. The guide in charge of the train of carts journeying to this rendezvous from the Elbow arrived in the evening, he had left the carts in the morning and pushed on in advance. Determined the magnetic variation.		
"	*August* 8.—*Fort a la Corne.*—Wet all the morning. The carts arrived in the forenoon. This being Sunday some of the party attended service, conducted on the opposite side of the river by the Rev. Henry Budd, a native Missionary. Making preparations for an overland journey to Fort Ellice, and for a continuation of the canoe voyage to Red River, *viâ* the Main Saskatchewan and Lake Winnipeg.		

ITINERARY.—(*Continued.*)

(VIII.)

FROM FORT PELLY, SOUTH WESTWARD, TO THE LITTLE SASKATCHEWAN OR RAPID RIVER, VIA THE FLANKS OF THE DUCK AND RIDING MOUNTAINS—THENCE ALONG THE RAPID RIVER FROM THE SUMMIT OF THE RIDING MOUNTAIN TO THE ASSINIBOINE — THENCE BACK TO THE INTERSECTION OF THE LOWER TRAIL AND THE LITTLE SASKATCHEWAN.

CAMP.		MAIN TRACK distance from—	
		Preceding camp.	Fort Pelly.
		St. miles.	St. miles.
63	*August* 4.—Started from Fort Pelly encampment this morning, equipped for an exploration of the country lying between the Assinniboine River and the Duck and Riding Mountain Ranges. Pursued a S. E. course along the left bank of the Assinniboine over the gentle western slope of the Duck Mountain. Crossed several small brooks, the largest called Sandy Knolls Creek, and rested near the Two Creeks at a point about two miles from the base of the Duck Mountain, and not far from the Assinniboine. Wood, water, and excellent grass in abundance. Camped at 7 P. M. in a region of luxuriant vegetation. Good land............	....	17.72
64	*August* 5.—The train left camp a little before sunrise. Traversed an undulating country with numerous clumps of poplar and low willows. Land good but rather marshy in many places. Forded Pine Creek and rested for two hours at noon near Swampy Creek. Camped at sunset after crossing a fine tract of country with a most exuberant growth of grass and various plants between open groves of young aspen..................	27.45	45.17
65	*August* 6.—Struck camp and started at the usual early hour. Upon the train halting to rest and graze the draught animals a few miles beyond the Swampy River crossing, a detachment set out on horseback at right angles to the trail to make a reconnaisance of the Duck Mountain. Being unable to ford Swamp River at the point where they struck it, they proceeded up its valley until a dense poplar forest, filled with fallen logs occupying the slope of the mountain, compelled them to retrace their steps. Continued over a country almost unchanged in topographical character, except that for the last mile before camping the trees and vegetation bore evidence of having been prostrated and torn by a most violent storm from the west. Met a train of carts in the afternoon, journeying to Fort Pelly...	17.17	62.34
66	*August* 7.—Reached Shell River early. Upon fording this river which separates the Duck and Riding Mountains, a mounted party set off to trace it to its confluence with the Assinniboine, whilst the train went into camp to await their return. Returning late, camp was not moved. Good grazing................	12.32	74.66
"	*August* 8.—Made a detour up the Shell River to day, ascending the valley as far as it was possible to penetrate with horses through the forest of balsam-poplar and whitewood, rendered almost impassable by fallen logs and underbrush. Returned to camp, and determined the latitude of this station (50° 58′ 12″.)	....	....
67	*August* 9.—Resumed the trail before sunrise. After two hours travel the river trail diverging to Fort Ellice was passed, the train taking the mountain trail. Traversed an undulating district with open woodlands. Crossed several brooks, some with bad crossings. Halted for the noon rest on a rolling area with good meadow grass. Forded a creek and camped at 5 P.M., amidst clumps of poplar and willows. Excellent pasturage. Soil rich sandy loam	24.00	98.66
68	*August* 10.—Started at 4.29 A.M. Traversed a marshy undulating tract of country. Vegetation very luxuriant. Ponds very numerous. Country beautiful. Forded and rested at Birdstail Creek; dimensions of creek 40 feet wide and three feet deep. Strong current; canoes might descend but they would find it difficult to return. Valley broad and deep, and partially wooded. Camped at 6.49 at Small Creek. Country fine..........	22.52	121.18
69	*August* 11.—*En route* at 4.21 A.M. Swampy Creek with a bad crossing. Trail followed the flank of the Riding Mountain, skirting an impenetrable forest of aspen stretching to the summit of the mountain. Dined at a rapid brook ten feet wide and three deep. Country beautiful; poplars and willows fringing ponds and lakelets are characteristic of this part of the country. Young maple numerous on the left of the trail. Camped at sunset on the banks of the Little Saskatchewan (Rapid River)	26.10	147.28
"	*August* 12.—Remained in camp and observed for latitude and variation (Lat. 50° 33′ 15″.) (Variation 15° 30 E′.)		
"	*August* 13.—Left carts at camp (69) and proceeded on horseback up the valley of the Little Saskatchewan, for a distance of 15.15 miles on the south-western slope of the Riding Mountain, when a dense barrier of woods opposing further progress and affording no feeding for the animals, compelled the party to return to camp (69).—The country passed over in making this side traverse is of the finest description, and well adapted for farming. The soil is a very rich loam, supporting alternate open woodlands and rich meadows. (Distance travelled in making this side trip to-day, 30.30 miles.)		

ITINERARY.—(*Continued.*)

CAMP.		MAIN TRACK distance from— Preceding camp.	Fort Pelly.
		St. miles.	St. miles.
No. 70	*August* 14.—Striking camp (69) at an early hour, the train ascended to the edge of the valley and journeyed along its margin in a southerly direction, for the purpose of pursuing the river to its junction with the Assiniboine. Crossed an uneven country with rich soil, supporting clumps of poplar and willow. Halted at noon near a lakelet fringed with osiers. Camped in a district of ponds, some of them a quarter to half a mile in diameter. Excellent grass	26.40	
71	*August* 15.—Got off at sunrise. Continued as close to the valley as possible in order to make a topographical delineation of the river. Three hours travel brought the train upon the White Mud River trail leading from Fort Garry to Fort Ellice. Pursued the trail for 0.12 miles, and then diverged to the south-west, continuing along the river. Camped at 3.40 P. M., to make new axle-trees for two of the carts, the old ones being nearly worn out, and there being a supply of post oak in the valley. Good pasturage. Wood. Plenty of water in creeks and lakelets.................................	13.70	
72	*August* 16.—Axle-trees finished and course resumed at 7 A. M. Crossed the lower trail to Fort Ellice after journeying one mile. Traversed a gently undulating prairie with a multitude of ponds and lakelets in the lowlands. Rested near the river at a point where the valley becomes so broad and shallow that it is lost in the surrounding plain. Crossed several tributary creeks and camped on a rolling prairie strewn with boulders..	23.95	

CAMP.		MAIN TRACK, distance from— Preceding camp.	Fort Pelly.
		St. miles.	St. miles.
No. 73	*August* 17.—Started at 6.20 A. M., over an open level prairie, through which the Little Saskatchewan now meanders. Halted for three hours at a point where the slopes of the valley resume their abrupt character, the river cutting through another plateau of light rolling prairie with short and scanty herbage. Crossed some dry coulés intersecting an inclined undulating prairie, and camped in the valley of the Assinniboine at its confluence with the Little Saskatchewan. Valley about a mile wide, filled with most luxuriant grass. North slope treeless, southern slope clothed with poplars	15.67	
	Explored north of the mountain trail crossing	15.15	
	Length of track along the Little Saskatchewan	94.87	
74	*August* 18.—Struck camp upon completing a reconnaissance of the junction of the two valleys, and taking intersecting courses to distant points. Turned north-westerly from camp over a light rolling prairie overspread with an accumulation of boulders. Soon entered and began to retrace the trail of yesterday. Nooned where the yesterday's morning halt was made.		
75	*August* 19.—On the trail at sunrise. Continuing up the Little Saskatchewan Valley. Reached the teamsters' camp (71) on the lower trail to Fort Ellice at 2.40 P.M. Commenced the trail survey of the lower track to Fort Ellice, from the brink of the Little Saskatchewan Valley, one mile from the crossing place, at 5.40 P.M. Camped at 6.20 near the point where men and carts were left on the 16th.		

ITINERARY.—(*Continued.*)

(IX.)

FROM THE LITTLE SASKATCHEWAN FORD WESTWARD TO FORT ELLICE VIA THE LOWER TRAIL.

CAMP.		MAIN TRACK distance from— Preceding camp.	MAIN TRACK distance from— L. Saskatchewan.
		St. miles.	St. miles.
No. 76	*August* 20.—Struck camp (75) 2.15 miles west of the crossing place, and pursued the trail a little before sunrise; winding westerly amidst a labyrinth of ponds and lakelets scattered over a rich undulating prairie supporting a most rank vegetation, gaily tinted with brilliant flowers. Rested for about two hours near the junction of this trail and the White Mud River trail from Prairie Portage. Camped three quarters of a mile from Lac Salé on a gently undulating prairie. Good grass. Straggling clumps of underwood, poplar, and low willow bushes. Soil sandy loam	23.40	25.55
77	*August* 21.—Passed close to Lac Salé, a fine expanse of water one mile in length and half a mile in breadth. 10.70 miles travel brought the train to the southern extremity of Shoal Lake, where camp was pitched in order to enable a division to make a survey of this oblong expanse of water lying nearly at right angles to the trail. The division followed up the eastern shore of the Lake to its northern extremity, and returned to camp. Shoal Lake is 5.65 miles long and 0.25 to 0.45 miles wide, and lies in a broad shallow basin. In the vicinity of Shoal Lake the ground is much covered with a white efflorescence. Good grazing. Wood and water in abundance	10.70	36.25
78	*August* 22.—Crossed the outlet of Shoal Lake connecting it with another lake three fourths of a mile in diameter, and proceeded over an undulating prairie, in several places flat and marshy. Rested for two hours amidst a group of ponds. Crossed a brook soon succeeded by a valley ten chains wide and thirty feet deep, filled with stagnant pools. Rested for two hours in a good pasturing district, and then journeyed across a country unchanged in characteristic features. Forded Birdstail Creek, (a stream at this point twenty-five feet wide and two feet deep, meandering in a valley eighty feet deep and one mile broad,) and camped two miles west of it. Luxuriant herbage. Thickets of young poplar..............................	23.80	60.05
79	*August* 23.—Started at 4 A. M. Crossed a small creek and followed the trail over a beautiful prairie with frequent clumps of poplar Roses, dogwood, and willows very numerous. A few small oaks. Grass and plants luxuriant. Halted two hours at a tributary of Bird's tail creek flowing in a deep valley. Reached the Assinniboine about noon, and pitched camp in the valley near the mouth of Beaver Creek. Met here the train which had just arrived from Fort à la Corne on the Saskatchewan. Animals turned loose to graze on the flats, and the remainder of the day occupied in triangulating in the valley of the Assinniboine, between Qu'Appelle River and Beaver Creek. Made a transverse section of the Assinniboine and Valley..............................	10.80	70.85

ITINERARY.—(*Continued.*)

(X.)

FROM FORT A LA CORNE SOUTHWESTWARD TO THE "LUMPY HILL OF THE WOODS"—THENCE SOUTHEASTWARD TO TOUCHWOOD HILLS AND FORT ELLICE VIA THE CARLTON TRAIL.

CAMP.		MAIN TRACK distance from— Preceding camp.	MAIN TRACK distance from— Fort à la Corne.
		St. miles.	St. miles.
No. 80	*August* 9.—The morning occupied in completing the equipment for a reconnaissance of the country lying between the Grand Forks of the Saskatchewan and the confluence of the Qu'Appelle and Assinniboine Rivers. New axle-trees having been made and fitted to the carts, the train left Fort à la Corne and pursued the trail at 1 P. M. (about an hour and a half after the canoe division had embarked for the voyage down the Saskatchewan.) Ascended the hill-sides of the valley through thickets of aspen, and crossed a belt of B. pine from a quarter to three quarters of a mile wide growing along the margin of the summit plateau. Traversed an undulating country with good soil, bearing open groves of aspen, Banksian pine, and spruce. Forded Long Creek, (a small tributary of the Main Saskatchewan,) and camped on its banks at 7 P. M. Vegetation very luxuriant. Raspberries in profusion. Plenty of wood and water. Good farming country.	....	10.00
81	*August* 10.—Struck camp and commenced taking the trail courses at 4.45 A. M. Ascended the shallow valley of Long Creek, traversing a tract of excellent undulating land. Remains of an ancient aspen forest frequently observed. The huge trunks of burnt trees lying hidden in the long herbage occasion much trouble in traversing this district with carts. Killed a bear, and halted two hours beside a pond 250 yards long, to graze the animals and take breakfast. Nooned at a dilatation of Long Creek, one mile long and 200 yards wide. Re-crossed Long Creek near a point where it issues from a series of lakelets extending westerly for about ten miles. Crossed a hill range running at right angles to the trail, and camped at 7.25 P. M. in view of the Birch Hills. Splendid soil. Fine sloping woodlands interspersed with beautiful meadows. Vegetation everywhere most luxuriant.—Flowers innumerable. Abundance of water in brooks and lakelets. Rain and thunder at night	28.17	38.17
82	*August* 11.—Train in motion at 4.30 A.M. Traversed a fine valley from four to six miles wide, with gentle slopes clothed with very long grass. The trail in this valley follows the windings of a shallow and sometimes dry creek, flowing into the South Branch. Halted for two hours to allow the horses to feed in an extensive wet meadow dotted with ponds—the sources of some feeble streamlets meandering to the Saskatchewan. Continuing parallel to the northern slope of the Birch Hills—a thinly wooded range in which Root River rises,—the trail crosses a very fine grazing or farming district. Rested for three hours at noon in a broad rich valley bounded by gentle hill ranges about five miles west of the Saskatchewan (S. B.) A winding course amidst numerous ponds and streamlets of various sizes, brought the train to Lumpy Hill Creek, a brook with many stagnant dilatations, issuing from the hill of the same name. Pursued this stream for two hours, and pitched camp beside it a little after sunset. Ascended the Lumpy Hill of the Woods through open aspen groves, and obtained from its summit a view of the Bloody, Woody and Birch Hill ranges. Rich soil. Good pasturage. Plenty of wood and water..	30.10	68.27
No. 83	*August* 12.—Started before sunrise across an undulating prairie sprinkled with lakelets, and came upon the Carlton track after two hours travel. After following this leading trail eastward for two hours a halt was made in a region of lakes and ponds lying between low spurs from the Lumpy Hill. In descending from this south-eastern extension of the Lumpy Hill range, the course passes over a succession of hills and dales wooded with aspen clumps, until a level and partially wooded prairie is reached. Crossed several brooks—feeders and outlets of many beautiful lakes,—and camped on a rolling prairie whilst the sun was just sinking below the horizon. Good grass. Soil light gravelly clay on the summit of hills, very rich in low places. Lake water a little brackish. Clumps of aspen. Grasshoppers seen......	29.40	97.67
84	*August* 13.—Left camp at sunrise and journeyed three and a half miles in a fine dry valley surrounded by wooded hills enclosing several beautiful lakes. Then traversed a range of hills and mounds, and passed five miles to the west of the lake "where the Moose died." Whilst the carts pursued the trail several side trips were made on horseback to the more prominent hills and lakes on either hand. Skirted some conical hills rising through an undulating prairie, and entered a very hilly country abounding in lakelets. Boulders on the hills. Rested for three hours at the base of the Big Hill and leaving the boundary of the so-called wooded country, entered upon a treeless undulating prairie. From the summit of Big Hill was seen "Buffalo Cart Plain," lying five miles to the north-east. Followed a sinuous course amidst a labyrinth of dome shaped hills, and camped on a gently rising prairie, at the beginning of the "Woody Range." Soil light and gravelly. Many marshy lakes. Small aspen and willow bluffs. A little rain. Gorgeous sunset...	23.66	121.33
85	*August* 14.—Started train at daylight across a beautiful undulating country, but still the same light soil and short herbage. Rested for two and a half hours at noon on a grassy area surrounded by lakelets and open aspen groves. One and a quarter miles S. E. of a		

ITINERARY.—(*Continued.*)

CAMP.		MAIN TRACK distance from— Preceding camp.	Fort à la Corne.
		St. miles.	St. miles.
	brook flowing into Ashes Lake the Carlton track is joined by the trail from the Moose Woods. Camped at 7.35 P.M. on a vast undulating treeless prairie, called the "Carry-wood Plain." Knolls, hillocks, and lakelets as heretofore. Soil light and herbage scanty. Long Lake seen to the south-west.	33.48	154.81
No. 86	*August* 15.—Raised camp a little after sunrise and proceeded across a beautiful prairie studded at intervals with clusters of conical knolls. Traversed several areas of salt prairie, in many places wet in spring, and skirted the shores of a saline lake with water of a bitter taste. Rested upon a patch of salt ground surrounded by wet prairie and an extensive range of ponds and marshy lakes. Vast numbers of aquatic birds seen in the salt marshes and lakes. A few grasshoppers observed. Mosquitoes and bulldogs still very numerous and tormenting. Crossed a fine woodless prairie, separated from a beautiful undulating ascent by a running stream of cool good water ten feet broad. Reached the summit plateau and journeyed over an excellent tract of country with many beautiful lakes, until reaching Touchwood Hill Fort, where camp was pitched at 8 P. M. Land of the best quality. Small aspen groves. Hill and dale. The richest profusion of vegetation. Soil very superior. Lakes in vast numbers.....................	30.10	184.91
"	*August* 16.—*Touchwood Hill Fort.*—Same camp. Whilst the train remained in camp to-day in order to repair the travelling equipage, as well as to rest the horses and take advantage of the good grazing in this locality, an opportunity was afforded for examining and making a reconnaissance of the Touchwood Hill range and surrounding country. Connected several of the more prominent hills, (by intersecting bearings,) with Last Mountain, Long Lake, *Ka-ou-ta-at-tin-ak* (Heart Hill), Little Touchwood Hills, &c. Much conjuring going on in an Indian (Cree) encampment here. The conjuring drum is loud and incessant to night as well as last night.		
87	*August* 17.—Lovely morning. Made an early start, winding around lakes embosomed in a well wooded chain of hills extending from the Great to the Little Touchwood range, succeeded by a beautiful level country embracing much good land. Crossed many intervals of prairie and hill country blending alternately into each other, and rested betwixt two lakes, one of them one mile in diameter. Good land. No sandy soil seen yet, east of the Heart Hill. Grasshoppers flying. Traversed a fine undulating country and pitched camp upon crossing the flank of the Little Touchwood Hills. Excellent pasturage, wood and water......................	24.50	209.41
88	*August* 18.—Resumed the survey at sunrise. Crossed a very beautiful undulating country. Many small lakes. Aspens on the ridges 6–9 inches in diameter. Killed a badger at the noon halt. Crossed a stream with a rapid current connecting two lakelets, and camped in a treeless valley filled with long rich grass. Fine pasturage country. Picturesque scenery. Soil light on the ridges. Rich vegetable mould in the flats. Herbage very rank. Numerous badger holes........	25.50	234.91
No. 89	*August* 19.—Set out at 5.15 A. M. Thermometer 46°. Heavy westerly wind. Crossed many grassy streamlets connecting chains of ponds, and entered upon a beautiful woodless country with an undulating surface covered with long waving grass. Halted for two hours amidst innumerable ponds and lakelets. Much beautiful meadow land. Good grass, fine soil. Grasshoppers flying. Intersected, at noon, the trail from Fort Pelly to the Qu'Appelle Mission. Traversed in the afternoon a treeless rolling prairie strewn here and there with small boulders, and extending to the horizon in every direction, except on the south and south-west, where the wooded ranges of File Mountain and Pheasant Hill interrupt the uniformity of the outline. Camped on a fine meadow encompassed by round hillocks and lakelets.	24.95	259.86
90	*August* 20.—Heavy dew last night. Before making the noon halt to day, four creeks and several dry valleys were crossed. These streams run with a swift current and unwater a very fine grazing country with much arable land. Camped at sunset in a beautiful hilly district, studded with little aspen bluffs. Good grass. Water in marshy hollows	27.00	286.86
91	*August* 21.—On the trail at sunrise. Traversed a high rolling country until reaching Little Cutarm Creek, an affluent of the Qu'Appelle, 20 feet broad and 1½ feet deep at the ford. Continued across a beautiful open prairie surrounded by aspen groves and grassy knolls, and dined upon fording Big Cutarm Creek, another tributary of the Qu'Appelle flowing in a deep valley. Good grass and aspen clumps at camp. Water in small swamps. Soil light and poor on the knolls and ridges. Some excellent land in the hollows and flats.........	24.40	311.26
92	*August* 22.—Hoar frost this morning. Left camp early, pursuing an easterly course towards a "pilot hill," called "Some one Knocked," or the Spy Hill. This conical elevation of gravel and sand attains an altitude of 120 feet above the prairie, and is the site of old Fort Cutarm. East of the Spy Hill the trail runs nearly parallel to, and not far from the Qu'Appelle valley, and the country deteriorates in character, the soil being sandy, and the herbage indifferent. After the noon halt, four hours travel over a sandy prairie skirted by sand dunes brought the train to the Assinniboine near its confluence with the Qu'Appelle. Forded the Assinniboine and pitched camp on its rich alluvial flats	22.52	333.78
93	*August* 23—Moved camp to the mouth of Beaver Creek. The division returning from a survey of the Riding Mountain, the Little Saskatchewan, &c., soon made its appearance and joined this train after an absence of five weeks. The greater part of to day occupied in making observations.	3.00	336.78

ITINERARY.—(*Continued.*)

(XI.)

FROM FORT ELLICE EASTWARD TO PRAIRIE PORTAGE AND SELKIRK SETTLEMENT VIA THE WHITE MUD RIVER TRAIL.

CAMP.		MAIN TRACK distance from— Preceding camp.	Fort Ellice.
		St. miles.	St. miles.
No. 94	*August* 24.—Struck tents early. The train—which is now a combination of the divisions that separated at the Qu'Appelle Mission on 20th July—started from the Assinniboine and Beaver Creek Forks and commenced the journey eastward for Red River, over the trail which was surveyed westward from its junction with the White Mud River trail on 20th to 23rd August. Camped at sunset. Fertile country. Thickets of aspen. Good pasturage. Plenty of fresh water in ponds.		
95	*August* 25.—Took the trail at sunrise. The courses and distances were not repeated on this portion of the track, having been already recorded between camps 76 and 79. Encamped at the usual hour.		
96	*August* 26.—Underway before sunrise. Crossed alternate open woodlands and prairies studded with beautiful lakes, the haunts of vast flocks of waterfowl. Good soil. Exuberant vegetation. Fine grazing or farming country.		
97	*August* 27.—Morning cold and frosty. Ice on water. Tents frozen. Reached the trail forks (50.80 miles from Fort Ellice) at 7.15 A.M., and commenced the survey of the Upper or White Mud River trail, pursuing a northeasterly course over an open undulating prairie. Rested for two hours at a marshy pond in a district of good pasturage. Crossed a tract of fine rolling land with a profusion of fresh water ponds, and pitched camp at sunset upon a level area wooded with large detached clumps of poplar. Soil, rich sandy loam. Subsoil everywhere gravelly clay. Good wood, water and grass	25.00	75.80
98	*August* 28.—Started at 4 A. M. After 2.15 miles travel the train crossed the trail of the division which passed over this district on 15th August, whilst surveying the Little Saskatchewan. Threaded through a dense jungle of poplars, willows, roses and twining plants until reaching the Little Saskatchewan where a halt was called for breakfast, upon fording the stream at 7 A. M. River 68 feet wide, 3 feet deep, current $3\frac{1}{2}$ miles an hour. Valley 100 feet deep, and $\frac{1}{4}$–$\frac{1}{2}$ a mile broad. Long waving grass in valley with a most luxuriant underwood of cherry, maple and hazel. After leaving the valley the trail leads through a dense forest of poplar and whitewood—the trees generally tall and straight, and averaging 18 inches to 2 feet in diameter. Halted at noon upon a small open space affording good grazing and water. Crossed five streamlets issuing from the Riding Mountain, along the southern flank of which the trail now lies. Camped upon a tract of burnt land thickly covered with oak stumps sprouting again. Fine farming land. Grass good, but in small areas	21.00	96.80
No. 99	*August* 29.—This day being Sunday, the train did not leave camp till 8.30 A. M. Traversed a gently undulating country covered with low willows and burnt oaks. Soil rich sandy loam. Halted at noon upon a nearly level prairie covered with rich grass and brilliant flowers, encompassed by light aspen groves on the south, and a close forest of poplar extending to the summit of the Riding Mountain on the north. Crossed a sluggish brook and after traversing a fine grazing country came upon the White Mud River. Continued along the north bank of this river until sunset. Beautiful country. Excellent wood and pasture land. The valley timbered with balsam, poplar, aspen, oak, maple and ash.	23.00	119.80
100	*August* 30.—Raising camp early, the trail was resumed at daylight. Wended through a close wood of poplar with intervals of wet prairie and good hay ground. Made the morning halt of two hours near a rich wet meadow upon which the animals fed greedily. Continued the journey across a fine country densely covered with sapling poplar, except along the track. Saw two jumping deer. Much white efflorescence along the path. Forded White Mud River at noon and journeyed down its southern bank for some miles through a fine farming country. Sugar maple forests alternating with large areas of unusually long and luxuriant grass. Recrossed White Mud River, kept along and camped on its north bank, $1\frac{1}{2}$ miles from the ford. Level land. Rich soil. Fine grass and wood. The River 55 feet wide, 4 feet deep.............	25.00	144.80
101	*August* 31.—Took an early start and proceeded along the north bank of the White Mud River as before. Traversed a very fine agricultural country diversified with beautiful woodlands and extensive open meadows. Grass and many varieties of plants wonderfully luxuriant. After travelling 9.25 miles the train recrossed the river, whilst a division embarked in canoe to make a track survey of the stream to its mouth. The train journeyed 5.40 miles farther and camped to await the return of the canoe party from Lake Manitobah. In crossing Rat Creek just before camping, all the horses stuck in its deep miry bottom. The canoe detachment in descending White Mud River to Lake Manitobah, a distance of 15.80 miles by its meanderings, startled vast numbers of ducks and other waterfowl. The portion of the river examined in canoe is a fine large stream flowing in a broad level valley. The width of the river		

ITINERARY.—(*Continued.*)

CAMP.		MAIN TRACK distance from— Preceding camp.	Fort Ellice.
		St. miles.	St. miles.
	increases from 70 feet, to 150 feet at its mouth. Depth from 5 to 7 feet. Before debouching into Lake Manitobah it receives Rat Creek, a tributary rising in the Sand Hills on the Assinniboine a little west of Prairie Portage. Another division made a detour from camp to Lake Manitobah and brought back the canoe on a cart. Traversed in returning a fine, level, hay country with occasional cranberry marshes and "oak openings." Passed two log shanties. Plenty of long rich grass and sunflowers, but a scarcity of fire-wood at camp	14.65	159.45
No 102	*September* 1.—Moved camp early and continued up Rat Creek about two miles, thence proceeded southeasterly across an open level prairie with long rank grass growing on a surface soil of rich clayey loam. Thence continued for a considerable distance alongside of a dry valley five chains wide, intersecting a level treeless plain, and leading towards Lake Manitobah. Crossed several other dry valleys ramifying from this ancient watercourse into a vast woodless prairie, and struck the Red River trail at noon a quarter of a mile west of the site of camp 4 of June 17 and 18—Prairie Portage.		
	To old camp (4,) Prairie Portage	16.55	176.00
No.103	*September* 2.—Left Prairie Portage for Selkirk Settlement *viâ* the Lane's Post and White Horse Plain trail running north of the Assinniboine.		
104	*September* 3.—Upon the trail early. As this trail is that which was traversed by the train whilst outward bound in June, the return track survey was discontinued at Prairie Portage and each days travel from thence to Red River is not recorded.		
	September 4.—Arrived at Selkirk Settlement. To Fort Garry	....	236.11

(XII.)

FROM FORT A LA CORNE TO SELKIRK SETTLEMENT VIA THE MAIN SASKATCHEWAN AND THE WEST COAST OF LAKE WINNIPEG.

CAMP.		MAIN TRACK distance from— Preceding camp.	Fort à la Corne.
		St. miles.	St. miles.
No 105	*August* 9.—Completed a series of observations and—embarking in a small birch rind canoe with two voyageurs at 11.30 A. M.—resumed the track survey of the Saskatchewan. Paddled steadily and maintained an average velocity of $5\frac{1}{2}$ miles an hour as heretofore, being aided by the uniformly swift current of the river. In sweeping round the gigantic bays on alternate sides of the river, many turbulent rills were observed emptying the drainage of the upper plateaux down the face of the high clay bluffs into the Saskatchewan. Passed through a tumultous current in rounding some of the points, caused by stony shallows extending into the river at these places. Camped on a low stony point, an hour after passing the last of the clay cliffs coming out upon the river. Fine country for farming; well wooded with large aspen, balsam-spruce and poplar. Plenty of dry driftwood lining the brink of the river..	....	23.06
106	*August* 10.—Descended, soon after embarking to-day, several small rapids, at the points. Although these so-called rapids flow very swiftly and with considerable turbulence, they would not obstruct the navigation of the river as they are only on one side of, and do not affect the main channel which is quite tranquil and deep. Passed numerous islands and camped at sunset on a low and rich alluvial flat, thickly wooded.		
No 107	Country well adapted for agriculture. Mud flats in the river..	53.10	76.16
	August 11.—Started at 4.10 A. M. River sometimes spreads to a width of upwards of $\frac{1}{4}$ of a mile and meanders by several channels among large wooded islands. Current swift. Frequent sand bars and snags, water mark 2 feet higher than present level. Trees along the bank much scratched by ice. Very rich alluvial land, bordering the river, well timbered with poplar, birch, aspen spruce, &c. Many islands. Water very muddy. Violent thunderstorm and rain at night	46.97	123.13
108	*August* 12.—Resumed the voyage at daybreak. The current is becoming slacker; being now $2\frac{1}{2}$ miles an hour. Banks lower and more lightly timbered. Many large snags and sawyers. Reached Big Stone River at noon. Proceeded to Pemican Portage 1.40 miles farther down, and camped. Flats covered with willows and alders. Poplar, ash, elm, sugar maple and spruce along the margin of the river............................	29.35	152.48
109	*Augnst* 13.—Crossed from the Saskatchewan to Cumberland House, *viâ* Pemican Portage, a distance of 1.86 miles. Had to pass through a great reedy marsh $1\frac{1}{4}$ miles across. Water 2-3 feet deep. Some good land in the immediate vicinity of Cumberland.		

ITINERARY.—(*Continued.*)

CAMP.		MAIN TRCK distance from— Preceding camp.	Fort à la Corne.
		St. miles.	St. miles.
No 109	*August* 14.—*Cumberland House.*—Remained here to-day in order to procure a new 2½ fm. canoe which was in process of construction. Saw Messrs. Stewart and Anderson, gentlemen in the service of the Hon. Hudson's Bay Company, who went in search of Sir John Franklin in 1855, and descended Back's Great Fish River to the Artic Sea in bark canoes.		
"	*August* 15.—*Cumberland House.*—To-day being Sunday, although all preparations were completed for starting, the journey was not resumed. Fine weather. Cold at night. Mosquitoes becoming less numerous.		
110	*August* 16.—Left Cumberland at 9 A. M. and reached the Saskatchewan after 6.25 miles paddling through Big Stone River. Passed the beginning of Pemican Portage (Camp 108) at noon. Made several observations to ascertain the volume of water and fall of the Saskatchewan, near camp, 4.50 miles below Tearing River or 19.38 from Pemican Portage. Banks very low and flat, covered with willows and scrub poplar River frequently impeded by sand-bars, mud-flats and shoals. Mean current two miles an hour. Rain during the night ...	19.38	171.86
111	*August* 17.—Embarked at 4 A. M. and passed Fishing Weir Creek after 1½ hours paddling. Glided amidst a number of beautiful islands before passing the Rat Root carrying place, an Indian pitching trail leading to lakes north of the Saskatchewan. Swept swiftly round the Big Bend and rested for an hour after passing a portion of the river which bears a strong resemblance to Rainy River, only the banks are much lower and not so well wooded. Drifted past White Fish Creek and arrived at the Pas at sunset. Camped near Christ Church	46.10	217.96
112	*August* 18.—Left the Pas this morning. After travelling a short distance, came to a channel forking off from the main river and forming a chord to one of its great bends. Whilst at the noon rest, near a branch leading to Moose Lake, a strong south wind arose accompanied by rain. Passed Muskrat Island, a very large island abounding iu muskrats and consequently much resorted to by Indians. After landing to camp, a thunderstorm and heavy rain came on. Much hay ground on the flats in rear of a light belt of brushwood lining the river, but country is now altogether too low and swampy for agricultural purposes. Saw beaver, muskrat and black fox to-day	25.25	243.21
113	*August* 19.—Course now lies through the great alluvial delta of the Saskatchewan. Embarked early and soon passing Marshy Lake, entered a labyrinth of intricate ramifications of the main river reticulating amidst vast muddy flats and shallow marshes. Camped a few miles below Muddy Lake, on the last spot of dry ground to be found before entering Cedar Lake. Willows and grass for fuel. Many sunken shoals and snags. Channels very shallow..............	35.50	276.71

CAMP.		MAIN TRACK distance from— Preceding camp.	Fort à la Corne.
		St. miles.	St. miles.
No 114	*August* 20.—Started from camp at the usual hour and entered Cedar Lake after 2.70 miles travel. Coasted along the north shore sometimes betwixt islands and sometimes making long traverses across deep bays. Encamped at 6 P. M. on a narrow point from whence a contrary wind prevented farther progress	21.60	298.31
115	*August* 21.—Left camp at daybreak. Entered the re-commencement of the Saskatchewan east of Cedar Lake at noon. Saw some buildings just erected by the Hon. Hudson's Bay Company on the banks of the river, for a trading post (Cedar Lake House). Soon arrived at Cross Lake Rapid, and after an hour's work in levelling and measuring the rapid and portaging, made the traverse of Cross Lake. Proceeded down the river till reaching the foot of the second rapid east of Cross Lake at dusk. Met a brigade of boats bound up. Good timber and some good clay land along the margin of river. Horizontal limestone frequently exposed...	27.00	325.31
116	*August* 22.—Reached the summit of the Grand Rapid at 9 A. M. Occupied seven hours in levelling and making a survey of the rapid, as well as portaging, examining the rock formation, sketching and making a general reconnaissance. Ran the lower portion of the rapid and arrived at the mouth of the Saskatchewan at 6 P. M. Continued 2.80 miles farther along the coast of Lake Winnipeg and camped at dark	11.61	336.92
117	*August* 23.—Embarked and resumed the track survey at day light. Crossed a succession of deep funnel-shaped bays, branching into a flat swampy country, and halted to cook dinner at the neck of Cape Kitchinashi. Tracked from thence to the extremity of the cape and camped late. Observed the magnetic deviation. Coast composed of open marshes in front of a vast tamerac and spruce swamp	31.28	368.20
118	*August* 24.—*En route* at sunrise. Coasted along until reaching the Gull Islands at 4 P.M. when a violent easterly wind arose and prevented farther progress. Camped on one of the islands, three miles from main land. No wood	23.10	391.30
119	*August* 25.—Detained by contrary wind until 2 P. M. Pushed on to the next island and thence to a point on the mainland, and camped on a narrow beach, separated, as usual, from the great tamerac swamp by a wide marsh Violent thunderstorm at noon, with high wind and heavy rain. Uncommonly brilliant aurora at night.............	7.50	398.80
"	*August* 26.—Same camp. Unable to stir from camp to-day in consequence of a violent and very cold N. W. wind which arose in the night and continued for 24 hours, raising a wonderfully tempestuous sea upon the lake.		
120	*August* 27.—Got off at daylight. Touched at an exposure of limestone and collected		

ITINERARY.—(*Continued.*)

CAMP.		MAIN TRACK distance from— Preceding camp.	Fort à la Corne.
		St. miles.	St. miles.
	some organic remains. Had to put into the estuary of a rivulet, affording shelter for canoes or boats, for four hours, in consequence of a brisk opposing wind which sprang up. The wind increasing, compelled a night camp after creeping along the shore a few miles farther. Lake separated as before, from a boundless swamp, by a narrow sand beach strewn with driftwood......	17.33	416.13
No 121	*August* 28.—Left last night's bivouac at the usual hour and made the noon halt at Warpath River. After verifying the rate of canoe by a standard measured along the beach, the course was resumed. Made great progress by tracking along the coast and camped opposite Caribou Island............	33.50	449.63
122	*August* 29.—Started at dawn. Rested at Limestone Point after making some wide traverses against a strong contrary wind. This highly fossiliferous exposure afforded some good specimens. After rounding the point, had to contend against a stronger wind and heavier sea than before. After a struggle of two hours in an angry sea, reached a small sand island and camped..	15.97	465.60
123	*August* 30.—Succeeded in reaching the Little Saskatchewan at 11 A. M., although the wind continued all last night and this morning. Ascended the river to the rapids and found there an encampment of Swampys engaged in fishing. A portion of this Indian band had just returned from the Grand Rapid. Returned to the mouth of the river and camped after measuring its volume of water and penetrating into the great muskeg through which it has excavated its way	6.45	472.05
124	*August* 31.—Left the mouth of the river at daybreak. Continued paddling steadily until reaching the commencement of a broad traverse when a very strong head wind compelled a halt. A vast wilderness of swamps and marshes as heretofore......	7.40	479.45
125	*September* 1.—Embarked early and crossed Mantagao Seebe Bay undersail, in a high rolling sea. The wind became so violent and opposing that it occasioned a detention of six hours after making this traverse. Still the same interminable muskeg and marsh. By forcing five hours against the wind, Point Wigwam was reached at 6 P. M., where a camp was formed in the lee of a few stinted willows growing in a patch of sand surrounded by a vast marsh.........	18.73	498.18
"	*September*, 2, 3, *and* 4.—Same camp. Windbound three days and nights by a violent and continuous hurricane from the N. N W. which raised a most tempestuous sea upon the lake. Pemican almost exhausted—have to live on short allowance. Sustained much cold and rain : having no tent and no wood.		
126	*September* 5.—Got off at last. Wind more moderate but still contrary. Passed the Bushkega Islands and contended with the wind until an attempt to round Point Turnagain, compelled rather an abrupt landing on a lee shore in a high surf. Collected some specimens here, from a thin exposure of limestone............	7.70	505.88
No 127	*September* 6.—Up at daybreak but the wind did not moderate sufficiently to permit re-embarkation till 9 A. M. Coasted and tracked against the wind, round Lynx Bay; and made a meal upon sand cherries at noon. Rounded the Cat Head at two o'clock and continued on until a high contrary wind blowing across Kinwow Bay, compelled a sudden camp upon Macbeth's Point, a narrow boulder-promontary so called from a "lopstick" made by that person.	15.26	521.14
"	*September* 7.—Same camp. The gale continued all night and increased to a perfect hurricane during the day raising a sea upon the lake in which no canoe could live. Set snares for rabbits but caught none. Men dispirited by want of food.		
128	*September* 8.—Pushed off in a heavy swell to make the long traverse across Kinwow Bay. Had some tough wet work in the middle of the traverse in consequence of a sudden squall that blew up. Reached the Wicked Point at noon when a strong easterly wind arose and stopped further progress. Sand dunes. Cranberry marshes and swamps. Clear and sunny during the day. Aurora at night. Clouds rising	10.13	531.27
129	*September* 9.—The wind of yesterday having fallen considerably; by starting early and pushing along shore in the lee, the Pike Head was reached at 10.30 A. M. Ascended the Jack Fish River and spent the day in repairing the weir across the river in order to catch a supply of fish, if possible. Rain in the evening...	9.53	540.80
"	*September* 10.—Caught a supply of fish last night, but did not resume the voyage in consequence of unfavourable wind. Rained heavily all day.		
130	*September* 11.—A wet and stormy morning —wind blowing hard from the N. W. Set out upon the lake early—the wind being favourable and having moderated a little. Ran on under a blanket-sail till near sundown when the wind fell altogether. Continued paddling until night set in, and camped on Louis Island in the mouth of Fisher Bay	11.40	252.20
131	*September* 12.—Up long before daylight preparing to start, but greatly dissappointed to find a smart head wind blowing from the south. Embarked, however, at daybreak and worked steadily although making little headway. Crossed to Great Moose Island and from thence by several other wide traverses to Whiteway's Point and camped at dark	20.20	572.40
132	*September* 13.—Crossed from Whiteway's Point to the Dog's Head and ran along the east coast of the lake until passing Loon's Straits. Re-crossed the lake from		

ITINERARY.—(*Continued.*)

CAMP.		MAIN TRACK distance from— Preceding camp.	Fort à la Corne.
		St. miles.	St. miles.
	thence to Grindstone Point. After examining and drawing the rock exposure here, continued on to the Little Grindstone Point and camped very late	35.10	607.50
No 133	*September* 14.—Started at 5 A. M., and crept along shore, in the lee and contending with head wind alternately. Passed the Grassy Narrows and reached the Sandy Bar at nightfall. Good boat harbour within the Sandy Bar and some good land reported in the vicinity..........	28.80	636.30
No 134	*September* 15.—Moved off before daylight. Cold morning. Passed Drunken River and ran along a coast revealing fine clay banks well wooded with aspen. Pushed on, after a short rest at the Willow Islands, and arrived at the mouth of Red River at dark...	42.50	678.80
135	*September* 16.—Left Lake Winnipeg at 6 A. M., and arrived at the Middle Settlement, Red River, at 11 P. M..........................	33.00	711.80

(XIII.)

FROM SELKIRK SETTLEMENT SOUTHEASTWARD TOWARDS LAKE OF THE WOODS AND BACK VIA LA RIVIÈRE SEINE OR GERMAN CREEK.

CAMP.		MAIN TRACK distance from— Preceding camp.	Fort Garry.
		St. miles.	St. miles.
No 135	*September* 18.—All the morning preparing for an exploration of the country east of Red River. Set out from the Middle Settlement with a small equipment, and having procured some additional supplies at Fort Garry, crossed the Assinniboine and Red River, and camped a mile from a bridge over la Rivière Seine. Fine night. Very warm.		
"	*September* 19.—Same camp, (Sunday.) The horses having strayed during the night all hands were occupied to day in searching for them but without success. Extraordinary hot day. A reddish thick haze, like smoke in the atmosphere. Large flocks of geese flying to the south. Immense flocks of black-birds (the crow) flying to the south also.		
136	*September* 20.—Very cold morning. The horses and mule were brought into camp early. Started at 11 A. M. to follow the picket line run for the purpose of locating a road last year. Camped after accomplishing 14 miles about ⅔ ds. of which lies under water averaging 18 inches deep. Small islands or low ridges bearing young osiers and aspen, scattered here and there through extensive wet prairies Fascines and side ditching would be requisite on the greater portion of the located line traversed to day.		
137	*September* 21.—Very cold last night and wet this morning. Unable to start until 8 o'clock in consequence of the heavy rain. Many detours from the picket line had to be made to day in order to get the animals across the marshes and quagmires through which it passes. Proceed to an island opposite the 22nd mile post, but a quagmire beyond, impassable for either men or horses, prevented farther advancement Turned back and camped near same place as last night. Very bad ground for horses. Soft and full of fallen logs. A horse under a cart fell four times in about 50 yards.		
No 138	*September* 22.—Cold morning—ground quite white. Very cold last night. Started early and reached the banks of Red River at noon. Pitched camp beside the old track and made preparations to set out again.		
139	*September* 23.—Dark cloudy morning. Drove the animals into camp at daybreak. Started at 6.20 A. M. and pursued a S. E. course along a good trail leading over fine rich land. Stopped two hours to feed at Legemenaire's Mill on the banks of la Rivière Seine (or German Creek.) Traversed a straight dry track running by the side of German Creek for some miles and nearly the whole distance through a country fit for settlement particularly at Oak Creek (where camp is pitched to night) there being plenty of firewood and oaks sufficiently large for building purposes. Rich loamy soil. Has been an oppressively hot day.......	18.00	20.00
140	*September* 24.—Left Oak Creek and continued southerly across an extensive tract of prairie land with occasional wet places, but upon the whole, well adapted for a road and for settlement. Land excellent. Vegetation luxuriant. Plenty of woods. Fine hay and pasture meadows. Good water. Camped one mile south of German Creek...........	20.00	40.00
141	*September* 25.—Cloudy. Thunder in the distance. Wended through a tangled jungle of osiers and red willow concealing burnt logs of aspen until reaching an extensive *bois brulé* through which the carts could not be hauled on account of the windfalls and the great quantity of prostrated burnt timber. Left the carts, and taking a fortnight's provisions continued the survey with pack-horses. Made slow progress along the valley of German Creek through a close forest of burnt timber. Soil lighter but still good and dry. The bill-hook and axe is all that is required for making a road here. Observed for latitude and variation		51.00

ITINERARY.—(*Continued.*)

CAMP.		MAIN TRACK distance from— Preceding camp.	MAIN TRACK distance from— Fort Garry.
		St. miles.	St. miles.
No 142	*September* 26.—Resumed the journey up German Creek, traversing a country with lighter soil and timber, but still supporting luxuriant vegetation, and well adapted for a road. Bridged a creek and crossed one or two marshes on account of the thickness of the forest, but a dry road could be located in the *bois brulé*. Thick groves of cypress, spruce, young aspen and willow. Camped at a place where the wood became so amazingly dense and so strewn with fallen logs that pack-horses could not force their way through. Cloudy and rainy,..................	14.00	65.00
143	*September* 27.—Slept beneath some large balsam-spruce, and poplar last night, and rising this morning before sunrise, started on foot to make an attempt to reach the Lake of the Woods. Reached la Rivière Seine after much toilsome climbing and scrambling over high heaps of fallen trees lying in every direction. Here the Indian guide (Penisi "the little bird") came to a halt and, although tempting offers were made to him, he could not be persuaded to go farther, having reached the boundary of the Lake of the Woods Indian's Country. On this account as well as owing to the nature of the country—there being a tamarack and cedar swamp from hence 15 miles wide which would take three days to cross on foot before arriving at the next dry ground—it was deemed expedient to return	4.00	69.00
No 144	*September* 28.—Retraversed the tract of country examined yesterday and the day before and camped near Morin's house, the nucleus of a new settlement beginning, at the crossing place of la Rivière Seine.		
145	*September* 29.—Forded la Rivière Seine and followed a good trail leading a considerable distance north of the river. Crossed a dry level prairie with much good land. Rested 1½ hours at a portion of the river rising in a marsh and flowing by the side of a tamarack and spruce swamp. Continued along a good track passing occasionally through willow marshes and wet meadows. Camped near the site of camp 135. Good pasturage and hay ground.		

(XIV.)

FROM RED RIVER WESTWARD ALONG LA RIVIERE SALÉ—THENCE SOUTHWESTWARD TO PEMBINA MOUNTAIN—THENCE ACROSS THE BLUE HILLS OF THE SOURIS—THENCE ACROSS THE ASSINIBOINE TO PRAIRIE PORTAGE.

CAMP.		MAIN TRACK distance from— Preceding camp.	MAIN TRACK distance from— Fort Garry.
		St. miles.	St. miles.
No 146	*September* 30.—Very cold last night. Struck camp (145) and moved the train across to the west side of the river, to commence an exploration of the country between the Assinniboine and the United States frontier. Proceeded along the Pembina trail for about five miles, and camped. Fine evening. Settlers drawing their nets in the river......	....	5.00
147	*October* 1.—Heavy rain towards morning. Left the Pembina trail where it intersects La Rivière Salé, 9.15 miles from Fort Garry. Proceeded up the south bank of La Rivière Salé, across a beautiful fertile prairie, and rested at a bend of the river where there are numerous salt springs. Camped twelve miles from the mouth of the river, on a magnificent, treeless level prairie. Fine farming and pasture land. The river yields plenty of fish, and swarms with duck and geese at certain seasons. Has been an unpleasantly hot day. A smoky appearance all around..	16.25	21.25
148	*October* 2.—Much rain last night. A dark, foggy, warm morning. Journeyed on from point to point of the river, which is now larger, and meanders, with a margin of large elm and oak, through a boundless open prairie. Traversed to-day as well as yesterday a very beautiful and fertile district, most suitable for a settlement, especially along the river, where the timber is very good, consisting chiefly of oak, but there is also much elm and ash. A steady pouring rain came on at 10 A.M., and compelled an early camp near Landrie's house and farm, a pioneer settlement on the banks of La Rivière Salé, about twenty miles from its mouth..	8.00	29.25
No.149	*October* 3.—Great quantity of rain fell last night and this morning. Wind high and very cold. Left camp and continued close along the river until reaching a point where it is crossed by the buffalo hunters trail, leading from the White Horse Plains to Turtle Mountain. There being numerous muskegs (marshes) beyond this point, the carts were left here and a reconnaissance of the river made on horseback until reaching a fork which could not be forded, and above which La Rivière Salé becomes very small. Returned and camped at the crossing place, as the hunters' trail, which has to be pursued now to Pembina Mountain, leads across a boundless prairie upon which wood and water cannot be reached to night............	11.00	40.25

ITINERARY.—(*Continued.*)

CAMP.		MAIN TRACK distance from— Preceding camp.	Fort Garry.
		St. miles.	St. miles.
No 150	*October* 4.—Took the hunters' trail at daybreak and continued across a most truly magnificent treeless prairie, luxuriantly clothed with long waving grass, wonderfully thick. Rested for two hours at noon beside a clump of willows, breaking the monotony of an unbounded oceanlike plain. Crossed afterwards alternate wet and dry prairies, until reaching La Rivière d'Isle des Bois. Camped at sunset on the banks of this affluent of Scratching River. Observed for latitude and variation. Some wet land here, but much that is dry and excellent. Plenty of good oak timber along the river. Has been a lovely day, bright and sunny, but rather cold	27.00	67.25
151	*October* 5.—Raised camp and crossed La Rivière d'Isle des Bois, which is fifteen feet wide and two feet deep. Traversed a fine fertile country sprinkled here and there with clumps of young aspens and a few oaks. Dined at La Rivière Tabac, another small tributary of Scratching River. Continued across a fine open prairie, and passed over the bed of an ancient lake three-fourths of a mile in diameter. Crossed some feeble streamlets and the dry beds of ancient water-courses before camping at Little Bridge Creek, a partially dry stream with many stagnant dilatations. Very stormy and rainy at night	19.00	86.25
152	*October* 6.—Morning cloudy—cold N. W. wind. Reached the base of Pembina Mountain or Ridge after two miles travel from Little Bridge Creek, past open woods of oak, the commencement of "the (so-called) forest," which stretches hence to Prairie Portage. Ascended Pembina Mountain, which is here nothing but a long gradual ascent, or rather a succession of easy steps rising from a lower prairie plateau to a more elevated table land The flank of the "Mountain," from the base to the summit, is clothed with groves of oak and aspen, and strewn with innumerable boulders. Entered upon the "round prairie," after gaining the crest of the mountain. Here a solitary half-breed, who had deserted from a band of buffalo hunters, came out of a clump of willows and looked with astonishment upon the train. He was at first thought to be a Sioux spy. Crossed the "round prairie," which separates Pembina Mountain from the Blue Hills, and halted at noon beside a clump of oak separated by a lakelet from a high conical knoll called the Calf's Tent. Left the hunters' trail to Pembina, and commenced a westerly course amidst thick clumps of poplar and willow. Poor grass and no water at night camp. Has been a miserable cold day—frequent showers of sleet with high wind	19.50	105.75
153	*October* 7.—Cold and stormy last night. Ground this morning covered thickly with snow. Vigilant precautions observed at camp as usual, and a sharp look out kept for Sioux upon the trail. Crossed an undulating prairie, and entered a very hilly district without water, said to be the commencement of the Blue Hills. Skirted 'Le Grand Coulé,' a dry valley ten chains wide, and camped upon crossing one of its ramifying branches, having found in it a scanty supply of water at last. Undulating country. Light sandy loam. Clumps of willows and aspens. Short scanty grass. Has been one of the coldest days this season. Freezing, and a high westerly wind the whole day.....	13.50	119.25
No.154	*October* 8.—Hard frost last night. Morning calm and cloudy. Turned back a short distance upon the course of yesterday, and then struck northerly in order to make a close examination of the poplar forest, "Le Grand Bois" of the buffalo hunters, which is always in view on the right, but still a considerable distance off (5–6 miles). Halted to observe for latitude, and search for water for the animals, soon after entering the so-called forest, which is nothing more than a succession of close clumps of young aspen, poplar, and willow, partially clothing a high upland country. The men despatched in quest of water had to go about four miles before finding any. Camped near a small lake encompassed by burnt timber on the north, which prevented a passage that way. Grazing very poor and scanty. Has been a bright and warm day. Light wind. Quite calm in evening	9.50	128.75
155	*October* 9.—Left camp and went round by the south end of the lake, which is about a mile long. Pursued a winding course, over a hilly district, amidst dense groves of poplars, almost a forest. Soil sandy and clayey loam. Fragments of shale appear wherever the ground is turned up by badgers. Got out of the thickest part of the forest after much wandering to and fro in search of a good track for the carts. Filled the water casks at a lake two miles long and crossed an undulating tract densely covered with willow bushes. At 4 P. M. came upon the trail which was left yesterday morning. Camped after some miles travel westward across an undulating district, partially clothed with brushwood. Has been a beautiful day, quite mild	14.50	143.25
156	*October* 10.—Froze very hard last night. Very cold this morning. Started from camp at daybreak. Forded a swampy brook (half a mile from camp) before crossing a ridge or narrow chain of stony hills, 30–90 feet high, running N. W. and S. E. Crossed several streamlets and rested at noon on an undulating prairie surrounded by round hills. A very heavy rain came on in the afternoon and compelled an early camp. More hilly than heretofore. Round long hills, like ridges; and conical hills. Clumps of poplars here and there, and willows spread all over, where formerly there was a dense poplar forest. A few oaks struggling for existence, but many prostrated. The main woods seen on the right, five to seven miles away	18.00	161.25
157	*October* 11.—Rained till dawn, then snowed till 6 A. M. A gloomy, cold morning. A flock of geese flying to the south was brought down this morning, by imitating their cry,		

ITINERARY.—(*Continued.*)

CAMP.		MAIN TRACK distance from— Preceding camp.	Fort Garry.
		St. miles.	St. miles.
	and one of their number killed. Resumed the journey along the edge of "Le Grand Coulé de la Grosse Butte," a very deep dry valley. Crossed this great unwatered valley, which derives its name from a very prominent object, La Grosse Butte, a solitary conical hill 200 feet high, two and half miles to the south. Continued along the south side of Le Grand Coulé over an undulating country. In crossing a small creek, flowing in a deep valley, before entering a very hilly district, some of the horses got mired. Crossed several ranges of hills and dales, and camped on the margin of a small lake in the centre of the Blue Hills..............	11.00	172.25
No 158	*October* 12.—Pursued a winding course over the Blue Hill range, and forded Cypress River, (a tributary of the Assinniboine,) after passing several beautiful lakes embosomed in wooded dells. To-days' journey was rather heavy for the animals, being so much up hill and down dale. Camped upon crossing a deep, broad, dry valley, as large as that of the Assinniboine. The Blue Hills terminate at this valley, being all on the east side of it. Some of the hills near it are 2–300 feet high, and many of them are crowned with oaks and poplars. Thick fog in the afternoon, and heavy rain commenced at 7 P. M..	21.50	193.75
159	*October* 13.—Rain alternately with snow lasted the whole night. Raining and sleeting all the morning. Took a northerly course towards the Assinniboine. Traversed a level soft prairie and found some difficulty in crossing a swampy creek. Crossed a plateau covered with young oaks, succeeded by sand hills extending to the Assinniboine. Forded the Assinniboine and struck north-easterly, across a region of sand hills sparsely covered with 'creeping juniper,' stunted aspens and oaks. Camped at 5 P. M. Scarcely any grazing. Water in ponds	13.00	206.75
No 160	*October* 14.—Resumed the journey across sand hills covered with ground juniper and "Kini-kinik." At 10.20 A. M. came upon the trail which was pursued by the train whilst *en route* to the Little Souris in June. Recognized it by a collection of small skulls of rabbits hanging on trees,—Indian offerings to Manitou Camped after seven hours travel from this point..................	22.50	229.25
161	*October* 15.—Started early and arrived at Prairie Portage at noon. Pitched camp near the site of camp (4) of June 17. Rain set in. Distant thunder	13.50	242.75
"	*October* 16.—*Prairie Portage.*—Turned the animals into the glebe to graze, through the kindness of Archdeacon Cochrane. Most of the day occupied in endeavouring to procure a guide or packman, and preparing for a traverse on foot into the forest on the south side of the river. Set in very wet in the evening.		
"	*October* 17, 18, 19.—*Prairie Portage.*—Occupied three days in making explorations of the poplar forest, and in traversing the belt of heavy hardwood lining the valley of the Assinniboine,—for description of which see Reports of Progress, page 18.		
	October 20.—Started with the train at sunrise from Prairie Portage for Selkirk Settlement, taking the inner trail leading close along the Assinniboine, there being more wood and water that way. Much snow in places.		

(XV.)

FROM SELKIRK SETTLEMENT TO THE SALT SPRINGS ON WINNIPEGO-SIS LAKE AND THE SUMMIT OF THE RIDING MOUNTAIN, VIA THE WEST COAST OF LAKE WINNIPEG, THE LITTLE SASKATCHEWAN OR DAUPHIN RIVER, ST. MARTIN'S LAKE, PARTRIDGE CROP RIVER, MANITOBAH LAKE, WATER-HEN RIVER AND LAKE, WINNIPEGO-SIS LAKE, MOSS RIVER AND DAUPHIN LAKE.

CAMP.		MAIN TRACK distance from— Preceding camp.	Middle Settlem't.
		St. miles.	St. miles.
No 162	*September* 18.—Embarked in a freighter's boat equipped for a lake voyage with a crew of seven voyageurs. Hoisted sail and got underway from the Middle Settlement at 10 A. M. Ran at a good rate before a light southerly breeze down the Red River. Passed the Stone Fort and landed at Sugar Point at noon to cook dinner. Camped below the Indian Village at sunset. Weather warm and mosquitoes troublesome in evening. Comet visible	....	26.00
No 163	*September* 19.—Pushed off at daybreak, sailing and rowing alternately. Reached the mouth of Red River ("north branch") at 10 A. M. and pulled 2½ miles northward into Lake Winnipeg with the intention of sailing to a point south of the Willow Islands, but a heavy fog coming on with a head wind from the north, the boat had to be put about and steered back to the mouth of the river through the mist. The fog cleared off at noon, but the unfavourable wind freshened up and prevented farther progress ...	7.00	33.00

ITINERARY.—(*Continued.*)

CAMP.		MAIL TRACK distance from— Preceding camp.	Middle Settlem't.
		St. miles.	St. miles.
No 163	*September* 20.—*Mouth of Red River.*—Same camp. Windbound. Wind continued high during the night and blew hard all day from the north. A very stormy sea on the lake and the bar covered with huge breakers. Cold and cloudy all day. Duck and geese southward-bound flying very high and swift before the gale. Examined the coast.		
164	*September* 21.—Wind went down during the night. Started from the mouth of the river before daylight. Sounded across the bar and pulled from point to point along the coast; taking the courses and computing the intermediate distances by dead-reckoning. Although a track survey of this coast had just been completed in canoe, it became necessary to delineate the boat's track along the coast in order to plot, upon the chart, the soundings which were made with the hand lead at intervals of 10 minutes or oftener if necessary, whilst the boat was in motion—commencing at the mouth of Red River Observed frequently with an improved log-line to obtain the rate of the boat; making the requisite allowances and corrections. Cooked dinner at tbe first point beyond the Willow Islands. Met here an Indian, in canoe, from whom a moose nose was procured. Rowed on till a little after dark and camped near Drunken River. Has been a beautiful day. Cold in the morning	26.50	59.50
165	*September* 22.—Left Drunken River at 4.50 A. M. Spread sail and ran swiftly before a fair wind past the Sandy Bar and through the Grassy Narrows. Stopped to cook breakfast upon Guano Island and set sail again, taking a straight course past the Greater and Lesser Black Islands, to Deer Island to examine a very fine exposure of limestone and sandstone upon it. Saw Magnus' brigade of boats from York Factory scudding along the main shore some distance off. Remained 3¾ hours upon the island collecting specimens, &c. Resumed the voyage at 4.40 and sailed to Grindstone Point where the boat was hauled up and camp pitched at 7 P. M. On this last traverse the lead shewed a depth of 8–9 fathoms—the deepest yet recorded	41.00	100.50
166	*September* 23.—The morning occupied in examining the rock and obtaining specimens. Embarked at 8.30 A. M. to make the traverse to the N. E. shore of the lake. Sailed with a "crimp" wind until making the little Granite Islands when the wind chopped round and blew hard from the north. The boat being very leewardly on account of the flatness of her floor and the want of keel, it was found necessary to put about and seek shelter. Found a harbour at Punk Island after driving before the gale, upon a retrograde course. Rain set in and wind continued high all day compelling a camp, but the time was occupied in exploring the island, and the rock escarpments exposed upon it..	3.50	97.00
167	*September* 24.—Aroused the crew at 2 and got underway at 3 A. M. (long before daylight). Pulled out a short distance and raised sail. An E. S. E. wind pushed the boat on at a good speed across Great Washow Bay and past the Bull's Head. Stopped at 8.20 A. M. at Limestone Cave Point. Embarked again after examining the outcrop of rock, and collecting fossils. Sailed through the Dog's Head Straits, thence across the mouth of Fisher Bay past Black Bear, Great Moose and Juniper Islands to the Pike Head. Cooked dinner at the Pike Head River and started again. Continued sailing on past Wicked Point, Cat Head, &c., and stopped to camp at 11 P. M. at Point Turnagain after 20 hours' travel. Lay down to rest on the beach at midnight.	81.50	178.50
No 168	*September* 25.—Raised camp at 3 and started at 4 A. M. recording the courses and soundings and registering the boat's rate and time, by lamp light. Ran with a light breeze past Bushkega Islands and, after daylight came in, took a straight course from the Sturgeon Isles to the Little Saskatchewan. Made close soundings over the bar and entered the mouth of the river at 9.30 A. M. Pulled and tracked until reaching an Indian encampment some distance up the river. The Indians on seeing the boat approaching commenced a brisk fusilade with their flint guns. This welcome salute caused numbers of Indians to congregate here from all points	28.00	206.50
169	*September* 26.—Tracking up this rapid river is slow and laborious work—the tracking ground is bad and it requires four men attached to the towing line to haul the boat. Reached the end of the tracking ground, or where the river flows through a vast marsh, at 3.20 P.M. and camped, (there being a strong head wind from the south) as the oars could give the boat no headway against the strong wind and current. Plenty of cranberries near camp. Indians have followed the boat all day in their little canoes, the squaws dragging them with lines of twisted bark passing round their bodies....	15.00	221·50
170	*September* 27.—Sailed in many of the reaches of the river with rather a squally wind. Entered St. Martin's Lake at noon and stopped at a boulder-point three miles from the commencement of the Little Saskatchewan. Could not proceed farther than this point in consequence of a high adverse wind blowing. Shot a laughing goose. The marsh here is encompassed by a semi-circular barrier of boulders	6.00	227.50
171	*September* 28.—Lake calm. Started a little before day-light. Rowed through the narrows, sounding every minute and stopped for breakfast at 7.40 on Sugar Island. Collected some specimens of the rock formation and steered for three small gneiss islands, (St. Martin's Rocks) pulled thence to Thunder Island and procured there a number of fossils from an exposure of limestone. Started again, after a heavy thunderstorm had passed over, and pulled against a head wind to Fisher Island which was reached at dark. Moored the boat and slept in her, there being no camping ground on shore	22.00	249.50

ITINERARY.—(*Continued.*)

CAMP.		MAIN TRACK distance from— Preceding camp.	MAIN TRACK distance from— Middle Settlem't.
		St. miles.	St. miles.
No 172	*September* 29.—Much rain last night. Entered the mouth of Partridge Crop River at 10 A.M. Proceeded up this stream, meandering by many channels through tall reeds and rushes, and arrived at Fairford at 3 P. M. The Indians had arrived some hours in advance and they became very noisy in the evening after receiving their annual supply of liquor at the Hon. Hudson Bay Company's Post here. Frost at night	15.00	264.50
173	*September* 30.—Entered Lake Manitobah at noon. Coasted along the east shore and camped at 6 P. M. at Flat Rock Bay in order to examine a highly fossiliferous exposure of limestone. Some stunted poplar, birch, and oak along the coast. Tamarac swamp in the rear	12.00	276.50
174	*October* 1.—Took in a number of specimens and pulled to Steep Rock Point to examine another outcrop 20 feet high, thence made a straight course across the lake to Point Pao-nan sounding, &c., as usual. Passed between the point and Cherry Island at noon; thence ran on with a fair wind till 7.30 P. M. (after dark) and camped at Sandy Point on the west side of the lake...	27.00	303.50
175	*October* 2.—Hoisted sail and started at 6 A. M. A heavy rain commenced at 8 A. M. and continued all day. Took breakfast at 10.25 A. M. on an island off the mouth of Water Hen River. Struck sail and pulled up Water Hen River through a great marsh. Camped at 4 P. M. on the first wooded dry ground reached	30.50	333.50
176	*October* 3.—Continued tracking and rowing up Water Hen River alternately. Reached the "Turning Point" at sunset and camped near some Indian and Half-breed saltmakers who were proceeding in a boat to Oak Point with a cargoe of salt from the Salt Springs. Flat swampy country, poorly timbered. River shallow in some places	14.25	347.75
177	*October* 4.—Clear and frosty last night with a strong N. W. wind. Passed the southern extremity of Water Hen Lake (apparently a dilatation of the river) and sailed through the remainder of the river into Winnipegosis Lake. Stopped to cook dinner at Point Ermine and sailed upon a straight course thence to Snake Island. Slept in the boat.	25.00	372.75
178	*October* 5.—Cold morning. Collected some very fine specimens and fossils from the limestone exposed on Snake Island. Saw vast numbers of "scarfs" (crow ducks) flying. Embarked at 10 A. M. and ran at a high speed under reefed canvas to the Salt Springs. Had to discharge cargoe rapidly and haul the boat up on the beach, having landed on a lee shore. Engaged during the rest of the day in examining and surveying the Salt Works, and measuring the height of the Springs above the lake. Wet weather. Shot a number of ducks ...	5.50	378.25
No 179	*October* 6.—Left the Salt Springs at 10 and reached the mouth of Moss River at 11 A. M. Passed a good log-house built and inhabited by Indians on the banks of the river. Upon halting to examine a rock exposure half a mile from the lake, the Indians came up requesting a "smoke." Continued upstream and camped after ascending the second rapid. The first rapid falls 2½ feet, and is very shoal and full of boulders. The boat had to be lightened and poled up. The second rapid is 10 chains long and has a fall of 2½ feet. In order to ascend it the boat had to be emptied and dragged up—all hands wading in the water except the steersman. Some good land on the immediate banks of the river but it soon passes into muskeg	10.25	388.50
180	*October* 7.—Started at 7 and reached the third rapid at 8 A. M. Poled the boat up. Fall 18 inches. Length, three chains. Entered Dauphin Lake at 4 P. M and continued along the west coast till 6 P. M. Had a magnificent view of the Riding Mountain upon entering the lake. Very cold and raw during the day	16.75	405.25
181	*October* 8.—Coasted five miles farther and landed at a point wooded with oaks, near a great marsh in which were vast flocks of ducks and geese. Levelled to obtain a profile of the country surrounding the lake. Hauled up the boat and made preparations to start on foot for the summit of the Riding Mountain to-morrow. Fine country for grazing. Has been a fine day	5.00	410.25
182	*October* 9.—Left the boat in charge of three men and started with the remainder of the party to make the ascent of the Riding Mountain. Pursued a straight southerly course to the highest or nearest peak of the mountain, measuring the distances by pacing and by rate. Crossed some fine meadow land, then entered upon a very wet marshy country. Open marsh and savannah between dry gravelly strips covered with scrub poplar alternating with quaking bogs and alder and tamarack swamps. Rested for the night on a scrub oak ridge after a cold wet fatiguing march	11.00	421.25
183	*October* 10.—Commenced ascending the slope of the mountain this morning. Found it rather toilsome work, tearing through tangled brushwood in a thick forest and crawling up the steep acclivities. Whilst taking dinner upon a high rounded peak within two miles of the summit, a brown bear made his appearance. A well directed shot brought him down as he was walking quietly off. Ascended to the summit and made a camp of brush to keep off a heavy snow that came on. Supper of bear's meat.	7.50	428.75

ITINERARY.—(*Continued.*)

(XVI.)

FROM THE SUMMIT OF THE RIDING MOUNTAIN TO MANITOBAH HOUSE AND ISLAND.

CAMP.		MAIN TRACK distance from— Preced-ing camp.	Riding Mount'n.
		St. miles.	St. miles.
No 184	*October* 11.—The whole face of the country covered with snow, which fell to the depth of six inches last night. Commenced the descent of the Mountain after making several traverses into the heavy forest which grows on the fine table land on the summit. The descent was rather difficult, the steep slopes being rendered very slippery by the snow. Trudged on through melting snow and slush until reaching the end of the solid ground at the foot of the lowest slope, and camped very wet and cold. Rain and sleet.	....	7.50
185	*October* 12.—Commenced the march early this morning across quaking marshes and muskegs which occupy the region between the base of the mountain and Dauphin Lake. Reached the boat encampment at 2.30 P. M. and spent the remainder of the day in drying wet clothes, &c.	11.00	18.50
186	*October* 13.—Launched the boat and coasted round to a point near the mouth of Turtle River from whence the exploration across the country to Manitobah House commences. Camped here in order to procure Tâwápit or one of his sons as Indian guide.	7.60	26.10
187	*October* 14.—Started with Tâwápit's son at dawn, for Turtle River, leaving Dauphin Lake on our left. Crossed Turtle River at 10 and entered a region of bog, marsh and aspen ridge. The abrupt flanks of Riding Mountain continued visible for many miles. Camped at night on a ridge. Bogs very bad, fully three quarters to four-fifths of the country is bog and marsh. Night cold. Hard frost.	21.00	47.1
188	*October* 15.—Arrived at the Ridge Pitching Track at 9.30 A. M. Pursued the excellent road it offered for 3¼ miles—then struck into swamps and bogs again. Horses mired. Were compelled to carry food and blankets and force the horses through the bogs—at 3 P. M., reached Crow Creek, and in half an hour Sucker Creek. Arrived at nightfall at Ebb and Flow Lake much fatigued	27.00	74.10
189	*October* 16.—Slept in Ojibway birch bark tent. Excellent breakfast of white fish, potatoes and rabbits Indian boy brought in a mink he had trapped. The Indian to whom the tent belonged has already set seventy traps, and the hunting season for most fur bearing animals is begun. Gallopped on an excellent Buffalo runner to Manitobah House passing through a low, wet, but good grazing country—arrived at Manitobah House, at noon.	12 75	86.85
No 190	*October* 17.—*Manitobah House.* — Sunday. Stormy and cold. Stayed at Manitobah House enjoying the hospitality of Mr. and Mrs. Mackenzie. Messenger arrived from Fairford. Snow storm began in the afternoon—continued all night.		
"	*October* 18 *and* 19.—*Manitobah House.*—Snow covers the ground nine inches deep. Snowbirds in flocks—ducks flying south. Day passed in writing letters and journal. Wrote Report. South wind. No boat.		
"	*October* 20 *and* 21.—South wind. Smoke from the burning prairies. Warm days, snow melting fast. Men's allowance at this Post three white fish per day. Walked through surrounding country; visited the Freeman's House and the Fish Stages. Found limestone exposure with glacial groves; also a former Lake Ridge, 14 feet above present altitude. Drift clay, 4 feet deep. Gneissoid and limestone boulders.		
"	*October* 22. — *Manitobah House.* — Visited and explored the Sugar Island. The mouth of Ebb and Flow Lake. The Narrows. Shot 'Stock' ducks. Immense accumulations of reeds about the Islands and Ebb and Flow Lake. Prepared for voyage to Manitobah Island.		
191	*October* 23.—Started with Whiteway at 10 A. M. in company with the carpenter (halfbreed) who built Mr. Mackenzie's house. Reached Island at noon. Explored and made plan of Island and surrounding country. Collected fossils. Camped on Manitobah Island	9.50	96.35
"	*October* 24.— *Manitobah Island.* — No boat or any sign of division from Dauphin Lake. Saw Indians, but they would not approach the Island. Made a collection of the different strata of rocks, plants, shrubs, &c. South wind. Beautiful weather, being the warm, genial period called Indian Summer. Whiteway hunted—killed duck and mink. Indians hunting near but would not approach the Island.		
"	*October* 28. — *Manitobah Island.* — Lovely day. Canoed round the coast. Prepared a large beacon fire. At 10 P. M. boat arrived, and the whole party camped on the Island.		

ITINERARY.—(*Continued.*)

(XVII.)

FROM DAUPHIN LAKE TO OAK POINT ON LAKE MANITOBAH, THENCE TO SELKIRK SETTLEMENT.

CAMP.		MAIN TRACK distance from— Preceding camp.	MAIN TRACK distance from— Dauphin Lake.
		St. miles.	St. miles.
No 192	*October* 13.—Started at 3 P. M. from a point near the mouth of Turtle River after disembarking the division which was to make an overland journey to Manitobah House. Ran along the east coast of Dauphin Lake, sounding as heretofore, as far as the wind would permit, and camped at sunset. Clear and frosty	....	8.00
193	*October* 14.—Unable to advance this morning in consequence of a strong head wind blowing from the N. W. Got underway in the evening, the wind having fallen. Rowed steadily along the N. E. shore until 10.20 P. M. when the boat was moored for the night, as the moon became obscured with clouds and haze	10.00	18.00
194	*October* 15. — Entered Moss River after an easy run with a light wind. Arrived at the Indian house, near the mouth of the river at 6 P.M., and stopped for the night. The Indians gave us a feast of moose nose in their warm and comfortable log shanty. Very cold and cloudy	17.50	35.50
195	*October* 16.—Strong head wind blowing from the east across Lake Winipego-sis this morning. Had some heavy pulling from the time of entering the lake until getting to leeward of a point about two miles from the mouth of Moss River. Wind blew from same quarter till evening when it veered round to the northwest causing the surf to beat upon the beach with great violence. Hauled up the boat high and dry after discharging the heavier part of her lading. Examined the coast and collected fossils from rock in position. Much rain during the day	2.00	37.50
"	*October* 17.—Same camp. Aroused the crew at 3 A. M. as the wind had increased in violence and the water had risen so much (2½ feet) that the breakers threatened to knock the boat to pieces. Discharged everything from the boat and spent most of the night in dragging her over over the beach to save her from the fury of the waves. Gale from the N. W. blew hard all day. Cold and snowing.		
196	*October* 18. — Four inches of snow on the ground this morning. Wind blew from the same quarter till noon when it turned a little more to the West—sufficiently favorable to take the boat across the lake. Had some difficulty in launching the boat on account of the heavy surf. Pulled against the wind to the point and hoisted sail. Ran under close reefed canvas, with a side wind, to Salt Point, thence pulled along the west shore of the inlet of Water-hen River, and camped on a point where an old Half-breed man and his Indian wife were "tenting"	12.75	50.25
No 197	*October* 19.—Got underway before daylight, a fair wind from the south having sprung up. Took the eastern branch of the Water-hen River running from the inlet to the great bend. Course lay against the wind beyond Water-hen Lake. Camped at a ¼ to 7 P. M near the islands or narrow part of the river after heavy pulling all afternoon. Observed the magnetic variation of 16 15′ E	25.00	75.25
198	*October* 20.—Reached the mouth of the river and entered Lake Manitobah at noon. Met four boats bound to the Salt Springs for cargoes of salt. Sailed, with the wind on starboard quarter, to Basin and Elm Islands. Had to stop on the latter in consequence of the wind becoming contrary. Has been a beautiful day—the beginning of Indian summer	15.00	90.25
199	*October* 21.—Embarked at 8.30 A. M. The lake nearly calm. A light wind from the south. Rowed to a point on the mainland and collected a number of geological specimens at an outcrop of horizontal limestone. Pulled on, over the calm surface of the lake and halted to cook supper opposite the "Point without Poles." Embarked again at 8 P. M. and ran with a light breeze, on the course to Point Pao-nan. A fog arose at 10 P. M. and the shore was made with difficulty some distance short of the point. Has been a beautiful Indian summer-day. Warm and hazy. Sounds audible a great distance	36.00	126.25
200	*October* 22.—Set sail at daylight. Ran with a "crimp" wind past the Pao-nan and across to the other side of the lake to a point which the boat was unable to weather and beyond which the course along the coast lay against the wind. Hauled the boat up on the gravel beach enclosing a marsh which lines the coast everywhere. Fine Indian summer-like day	10.00	136.25
"	*October* 23–24. — Same camp. Wind still contrary. Blew hard from the south these two days and nights. Shot a number of prairie hen on the point. Observed the magnetic variation of 15° E. on the 23rd. Cold, cloudy and raining at intervals on the 24th. No wood on the point. Unable to keep up a fire.		
201	*October* 25.—The wind moderated at noon. Shoved off the boat and started at 2 P. M., but had some heavy pulling and made little headway against the wind until sunset, when it became quite calm. Stopped at the narrows at 8 P. M. and sent men ashore to cook supper, the water being too shoal to allow the boat to get near the land. Started again and plied the oars until 10.15 P. M. when a beacon fire and some signal shots		

ITINERARY.—(*Continued.*)

CAMP.		MAIN TRACK distance from— Preceding camp.	MAIN TRACK distance from— Dauphin Lake.
		St. miles.	St. miles.
	on Manitobah Island revealed the camp of the division which had come round from Riding Mountain to this point	11.00	147.25
No 202	*October* 26.—Embarked at 7 A. M. and pulled through the narrows against a light southerly wind until reaching Manitobah House at 1.30 P. M. Started again at 3 rowing against the same wind till 5 P. M. and camped upon Pelican Island	13.60	160.85
203	*October* 27.—Started before daylight. Pushed on along the N. E. shore of the lake until arriving at Monkman's Point, near Swan Creek. Found Monkman and some others from Red River fishing here. They had a large number of white fish drying and smoking for winter use. Beautiful aurora at night	27.50	188.35
204	*October* 28.—Sharp frost last night. Ran along shore with a light N. E. wind and turned into the channel which leads across Marshy Point through a vast marsh. Got out of marsh at 1 P. M. after much difficulty—having to drag the boat in many places through mud where the water is shallow. Arrived at Oak Point at 2 P. M. A number of Red River settlers encamped here in Indian wigwams, carrying on their annual autumn fishing. Hauled up the boat on rollers and loaded three ox carts with the cargo of baggage and fossils to be transported to Selkirk Settlement. Encamped near John Monkman's house, half a mile from the lake	10.00	198.35
No 205	*October* 29.—Very sharp frost last night. Procured three horses from John Monkman and started with the train of ox carts for Red River. Crossed a rich and fertile prairie with scattered groves of scrub oak, poplar and willow. Camped at sunset beside a clump of poplar saplings on an area of dark rich soil with gravelly subsoil	13.00	211.35
206	*October* 30.—Hard frost last night. Skirted the southwestern shore of Shoal Lake past a fine locality for a settlement—the land being rich with beautiful grassy lawns and hay meadows between oak orchards and belts of poplar near the margin of the lake. Camped at "Bell's Hummock" a clump of fair sized poplar, enclosing a pond of good water. A favorite camping place............	19.00	230.35
	October 31.—Froze very hard last night. Took an early start and reached Stony Mountain at noon after traversing several low ridges intersecting beautiful prairies. Spent 2½ hours in quest of fossils at the mountain and pushed on to the settlement. Reached the Scotch church at 6.30 P. M................	37.00	267.35

DISTANCES FROM FORT GARRY TO IMPORTANT POINTS IN RUPERT'S LAND.

Description of Route.	Localities.	Intermediate Distances.	Distance from Fort Garry.	Remarks.
		Statute miles.	Statute miles.	
Cart Trail..................	Fort Ellice	....	236.11	Viâ the White Mud River trail.
" "	Qu'Appelle Mission..................................	136.57	371.68	" the trail south of the Qu'Appelle.
" "	Elbow of the South Branch of the Saskatchewan	176.73	548.41	" the Qu'Appelle Valley.
Cart Trail......	Fort Pelly..	....	339.65	Viâ Fort Ellice and the trail on the West side of the Assiniboine.
Cart Trail	Touchwood Hills	....	387.98	Viâ Fort Ellice and the Carlton trail.
" "	Grand Forks of the Saskatchewan...............	184.91	572.89	" the trail.
Boat Navigation.........	Little Saskatchewan or Dauphin River.........	....	206.50	Viâ the west coast of Lake Winnipeg.
" "	Main Saskatchewan................................	137.93	344.43	" " " "
" "	Grand Forks of the Saskatchewan..............	354.27	698.70	" the Saskatchewan River.
" "	Elbow of the South Branch of the Saskatchewan	249.73	948.43	" " "

GEOLOGICAL REPORT.

GEOLOGICAL REPORT.

CHAPTER XVI.

SURFACE GEOLOGY OF A PART OF THE VALLEY OF LAKE WINNIPEG.

Abraded, Polished, and Grooved Rocks on Baril Portage—Sturgeon Lake—On the Winnipeg—Lakes Manitobah and Winnepego-sis—Polished Pavement on South Branch—Erratics on the Qu'Appelle, at the Moose Woods, on Cut-Arm Creek, Assiniboine, West of Mississippi, on Souris—Beaches between Lakes Superior and Winnipeg—Great Dog Portage—Character of—Sand Bank—Section of—Dr. Hitchcock's views—Beach at Prairie Portage—Portage de Millieu—The Big Ridge on Red River—On the Assiniboine—Near Dauphin Lake—Pembina Mountain—Lines of Boulders—On South Branch—On St. Martin's and Manitobah Lake—Character of Pembina Mountain—Dr. Owen's description—At the Bad Woods—At the Grand Forks—Ridges on the Riding and Duck Mountains—Correspond with Ridges on the Great Dog Portage—Probable former connection of Grand Coteau de Missouri, Turtle, Riding, Duck, Thunder, Porcupine, and Pasquia Mountains—Ancient River Valleys The Qu'Appelle—The Little Souris—Sand Hills and Dunes—Their Distribution— Circular Depressions — Effects of Denudation—The Valley proper of Lake Winnipeg denuded—Outcrop of Formations—Conform to the general trend of the Laurentian Series.

The surface of the country between Lake Superior and the South Branch of the Saskatchewan exhibits the following phenomena at different localities:

1. GROOVED, SCRATCHED, POLISHED AND ABRADED ROCKS.
2. ERRATICS.
3. ANCIENT SEA AND LAKE BEACHES AND TERRACES.
4. ANCIENT RIVER VALLEYS.
5. SAND HILLS AND DUNES.
6. CIRCULAR DEPRESSIONS.
7. REMARKABLE EFFECTS OF DENUDATION.

1. *Grooved, Scratched, Polished and Abraded Rocks.*

Instances of the action of ice in abrading and polishing extensive surfaces of rock are very numerous on the canoe route from Lake Superior to Lake Winnipeg. The first wide expanse noticed on the west side of the watershed is at Baril Portage, 143 miles from Lake Superior, and 1500 feet above the sea. Where Mille Lacs becomes narrow on approaching Baril Portage, gneissoid hills and islands about 100 feet high show a well defined stratification dipping north, at an angle of about 15°, and on that side smooth, and sometimes roughly polished; on the south side, they are precipitous and abrupt. The same character was noticed at the Baril Portage. The north-eastern exposure of the rocks there was smooth, the southern rugged, and often precipitous.

On Sturgeon Lake, 208 miles from Lake Superior, and 1156 feet above the sea, the north-eastern extremities of hill ranges slope to the water's edge, and when bare are always found to be evenly smoothed and ground down. The aspect of the south and south-western exposures is that of precipitous escarpments.

When on the Winnipeg in 1857, I ascended an abraded granite hill about 250 feet high, and obtained from its summit a very extensive view of the surrounding country. The broad river, with its numerous deep bays, was seen stretching far to the north, and all around smooth dome-shaped hills, similar to the one on which I stood, showed their bare and scantily wooded summits in every direction. The general surface was either bare, and so smooth and polished as to make walking dangerous, or else thickly covered with cariboo moss and tripe de roche.

This description applies to a vast area drained by the Winnipeg. In 1858, we frequently ascended the smoothed and polished rocks, on which glacial grooves were easily traced for long distances; sometimes, but not often, boulders were found resting upon the polished surfaces. On one occasion I attempted to ascend a round dome-shaped mound forming the summit of a granite hill, but its beautifully polished surface prevented me from obtaining a footing. The action of atmospheric agents had only succeeded in dimming its beauty, but had not destroyed its smoothness.

Grooves and scratches occur on the limestones of Lakes Winnipeg and Manitobah, where the surface has been preserved from atmospheric agencies, but whether they were of recent origin or connected with the drift, is not certain.

By far the most curious instance of modern ice action occurs in the valley of the South Branch, already described (see Chapter V.) The polished pavement on the edges of that river is a curious and instructive illustration of the manner in which boulders and ice may leave behind them lasting memorials, graven on stone, of their long continued action, even on the banks of a river.

2. *Erratics.*

The distribution of boulders or erratics in the area explored, may be traced, as in Canada, to at least two epochs: 1st. The Drift and Boulder period, during which by far the larger number were torn from the parent rock and carried by ice to their present resting places. 2nd. The Recent period, including the re-arrangement of ancient boulders and the distribution of fresh supplies by the action of ice. Where erratics are distributed in unusual quantities, their position is marked on the large map. The largest boulder was seen in the valley of the Qu'Appelle, its position is shown on the 'Track Survey of the Qu'Appelle.' The circumference of this enormous erratic is 78 ft., and it exposes a portion above ground at least 14 ft. in altitude. The next largest, one of limestone, was seen on the prairies below the Moose Woods; it is about 16 feet high, and at least sixty in circumference, is very jagged, and consists of immense slabs, whose edges project two and three feet. Near it are many others of the same kind, but of smaller dimensions. Near Little Cut Arm Creek, an affluent of the Qu'Appelle, large unfossiliferous boulders are very numerous. One of gneiss measured thirteen feet in diameter. North of the Assiniboine, near the Big Ridge, large boulders are also abundant, and when magnified by refraction look like tents in the level prairies. Twice we were deceived by this appearance and led several miles from our course by their resemblance to a cluster of tents.

In speaking of the boulders in the Western Prairies, Dr. Owen says: "On the west side of the Mississippi, in the vast prairie region of Iowa, the attention of the geologist is frequently arrested by erratic blocks of enormous dimensions, scattered here and there, and half sunk in the ground. Unlike the boulders we have just been considering, they are far from their original situation. As they rise amid the ocean of grass they may be seen for miles; and in the absence of more conspicuous objects they form the principal landmarks of the traveller. The largest of them might, in an inhabited country, very well be mistaken for cabins in the distance. The one here represented was measured and found to be fifty feet in circumference, and twelve feet high. It is probable that at least one half of the rock is buried in the ground. Hence may be gathered some idea of its huge dimensions."*

The drift on the Blue Hills of the Souris is of local origin, and consists almost exclusively of the shales which form the outcrop of the Cretaceous rocks whose limit is defined by the Pembina Mountain. Its age is consequently posterior to that of the true boulder drift, which is so generally distributed over the high prairies to the west.

In Lake Winnipeg, ice every year brings vast boulders and fragments of rock of the Laurentian series which occupy its eastern shores. Many of these are distributed in the shallows and on the beaches of the western side; these phenomena resemble in miniature the stupendous operations described by Arctic travellers as continually occurring on the shores of the Arctic Ocean.

In Lake Manitobah long lines of boulders are accumulating in shallows and forming extensive reefs; the same operation is going on in all the lakes of this region, and is instrumental in diminishing the area of the lake in one direction, which is probably compensated by a wearing away of the coast in other places. Several of these modern accumulations formed by a re-arrangement of the boulders of the older drift are noticed in preceding chapters. Taken as a whole, and in connection with the destruction of the coasts, they afford a striking illustration of the changes now taking place in the relations of land and water throughout the Lake Region.

3. *Beaches and Terraces.*

The most remarkable beach and terrace, showing an ancient coast line between Lake Superior and Lake Winnipeg, is undoubtedly that which separates Great Dog from Little Dog Lake on the Kaministiquia canoe route. I have thus described it in my report on the Red River Expedition of 1857.

The Great Dog Portage, 55 miles from Lake Superior by the canoe route, rises 490 feet above the level of the Little Dog Lake, and the greatest elevation of the ridge cannot be less than 500 feet above it. The difference between the levels of Little and Great Dog Lakes is 347·81 feet, and the length of the portage between them one mile and fifty-three chains.

The base of the Great Dog Mountain consists of a gneissoid rock supporting numerous boulders and fragments of the same material. A level plateau of clay then occurs for about a quarter of a mile, at an altitude of 283 feet above Little Dog Lake, from which rises, at a very acute angle an immense bank or ridge of stratified sand, holding small water worn pebbles. The bank of sand continues to the summit of the portage, or 185 feet above the clay plateau. The portage path does not pass over the highest part of the sand ridge. East of the path it is probable that its summit is 500 feet above the Little Dog Lake.

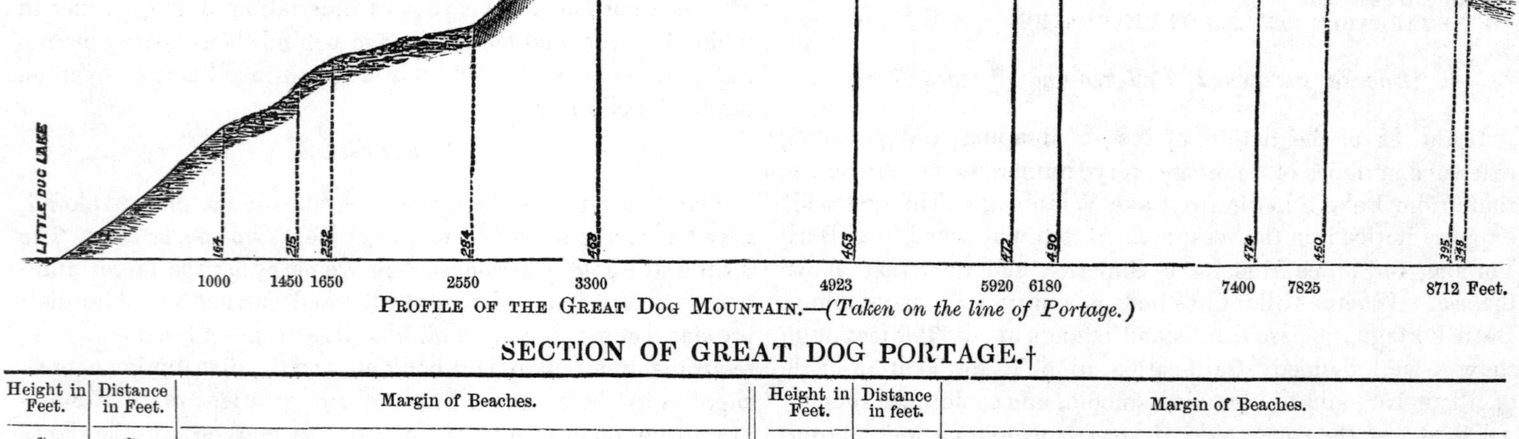

PROFILE OF THE GREAT DOG MOUNTAIN.—*(Taken on the line of Portage.)*

SECTION OF GREAT DOG PORTAGE.†

Height in Feet.	Distance in Feet.	Margin of Beaches.	Height in Feet.	Distance in feet.	Margin of Beaches.
ft.	ft.		ft.	ft.	
163.53	1000	Beginning of 1st Plateau.	472.60	5920	End of 3rd Plateau.
215.00	1450	Termination of do.	490.00	6180	Summit of level and commencement of 4th Plateau.
251.74	1650	Beginning of 2nd do.	474.00	7400	End of 4th Plateau, and commencement of descent to edge of cliff.
283.78	2550	End of 2nd Plateau, and commencement of Sand Bank.	395.00	8640	End of descent.
468.19	3300	Commencement of 3rd Plateau.	348.00	8712	Bottom of Cliff, and level of Great Dog Lake.

* Owen's Geological Survey of Wisconsin, Iowa and Minnesota, page 144.

† This Section was kindly furnished me by Mr. Napier, Engineer to the Red River Exploring Expedition of 1857.

Here then we have a terrace 472 feet above Little Dog Lake, or 835 feet above Lake Superior, or 1435 feet above the sea.

This ancient beach furnishes an admirable proof of Dr. Hitchcock's expectation that higher beaches than those measured by Sir Wm. Logan on the shores of Lake Superior would be found in that region. Dr. Hitchcock says in his Surface Geology, page 63, (Smithsonian Contributions,) "I will only add, that if it be admitted that the facts adduced in this paper prove the presence, since the Drift period, of the Ocean at a height of 2000 or even 1200 feet, above its present level, then it must have extended over nearly all of our western country; and unless Professor Agassiz says that he had his eye upon this matter along the shores of Superior, I cannot avoid entertaining the expectation, that what I call beaches will yet be found at a much higher level there, than the 331 feet terrace, measured by Mr. (now Sir William) Logan."

I am inclined to think that another beach and terrace can be recognized at Prairie Portage, one hundred and four miles by the canoe route from Lake Superior; its altitude would correspond with that on the Great Dog. Prairie Portage passes over the height of land, but not the highest land on the route, and its course lies first south west up a steep wooded hill, without rock exposure, but composed of drift clays, sand, and numerous boulders; it then enters a narrow valley, which terminates in a small lake, about five acres in area and 20 feet deep, occupying a hollow among the hills on the height of land. The portage path continues on in the same direction until the Height of Land Lake is reached, a small sheet of water, about a square mile in area, and 157 feet above Cold Water Lake. The utmost elevation reached on the Prairie Portage is probably 190 feet above Cold Water Lake or nearly 900 feet above Lake Superior. Portage du Milieu, one hundred and five miles from Lake Superior passes over a low sandy ridge. It is 869 feet above Lake Superior, or 1469 feet above the sea.

In the valley of Lake Winnipeg the first prominent beach is the Big Ridge. This has been partially described in my Report on the Red River Expedition of 1857. Last year I had an opportunity of tracing it for a very great distance near the shores of Lake Manitobah.

Commencing east of Red River, a few miles from Lake Winnipeg, this ridge pursues a south-westerly course until it approaches Red River, within four miles of the Middle Settlement; here it was ascertained by levelling to be 67½ ft. above the Prairie; on the opposite side of the river, a beach on Stony Mountain corresponds with the Big Ridge, and beyond that curious Island in the Prairie, it is observed forming the limit of a former extension of the valley of Lake Winnipeg. On the east side of Red River the Big Ridge is traced nearly due south from the Middle Settlement to where it crosses the Roseau, forty-six miles from the mouth of that stream, and on or near the 49th parallel. It is next met with at Pine Creek, in the State of Minnesota, and from this point it may be said to form a continuous level gravel road, beautifully arched, and about one hundred feet broad, the whole distance to the shores of Lake Winnipeg, more than 120 miles.

On the west side of Red River, north of the 49th parallel and north of the Assiniboine, we followed the Big Ridge from a point about three miles west of Stony Mountain to near Prairie Portage. Here it appears to have been removed by the agency of the Prairie Portage River and the waters of the Assiniboine, which during very high floods pass from the valley of that river into Lake Manitobah. It, or one of a few feet higher elevation, was again observed on White Mud River, about 20 miles west of Lake Manitobah. Here it resembled in every particular the ridge on the east of side Red River, being about 100 to 120 feet broad, and 25 feet above the level of the Prairie. It was again noticed in the rear of Manitobah House, where the same characteristics were preserved. It probably crosses the Assiniboine three or four miles west of Prairie Portage. The general contour of this Ridge is shown on the map.

In the rear of Dauphin Lake, the next ridge in the ascending series occurs, it forms an excellent pitching track for Indians on the east flank of the Riding Mountain. Probably these ridges are found close together, at the foot of the Pembina Mountain, where no less than four distinct steps occur, as shown on the map. The summit of these steps may be the plateau whose altitude was ascertained by Dr. Owen to be 210 feet above the prairie level, and the first and second steps may be a part of the Big Ridge, limiting the lowest level prairies of Red River and the Assiniboine.

The lower prairies enclosed by the Big Ridge are everywhere intersected by small subordinate ridges, which often die out, and are evidently the remains of shoals formed in the shallow bed of Lake Winnipeg when its waters were limited by the Big Ridge. Many opportunities for observing the present formation of similar shoals occurred in Lake Manitobah, St. Martin's Lake, Lake Winnipeg and Dauphin Lake. These, when the lakes become drained, will have the form of ridges in the level country then exposed. Indeed it may be said that the region between Dauphin Mountain and Lake Manitobah in the direction of Ebb and Flow Lake and south of that body of water, is but recently drained, or still in process of draining, being removed from the surface of Ebb and Flow Lake by a very few feet and covered with water to a large extent in the spring. At present it consists of marsh, bog and ridge, in continued succession. When completely drained, the country will resemble the present prairies of the Assiniboine, with the gentle rich depressions, and the low, dry, gravelly ridges.

The long lines of boulders exposed in two parallel, horizontal rows, about twenty feet apart, in the drift of the South Branch, are the records of former shallow lakes or seas in that region. They may represent a coast line, but more probably low ridges formed under water, upon which the boulders were stranded. In Lake Manitobah and St. Martin's Lake modern instances now in process of arrangement, are visible for many miles in length. In these shallow lakes, the boulders brought year by year by ice from the neighbouring shore accumulate upon long, narrow spits, and ultimately form breakwaters or islands. The same process may have occurred with the boulders on the South Branch. The fine layers of stratified mud, easily split into thin leaves, which lie just above them, show conclusively that they were deposited in quiet water; their horizontality proves that they occupied an ancient coast or ridge below the comparatively tranquil water of a lake of limited extent. The vast accumulations of sand and clay above them establish the antiquity of the arrangement, and the occurrence of two such layers parallel to one another, and separated by a considerable accumulation of clay and sand, leads to the inference that the conditions which established the existence of one layer also prevailed during the arrangement of the other. It may be that these are boulders distributed over the level floor of a former lake or sea, and they may cover a vast area; if so, it only proves that the agents which brought them, operated a second time after a long interval, and with similar results.

The Pembina Mountain is *par excellence* the ancient beach in the valley of Lake Winnipeg. Dr. Owen described it as it occurs a few miles south of the 49th parallel: "After a hot and fatiguing ride over the plains, we arrived an hour after sunset at the foot of the Pembina Mountain. In the twilight as we stood at our encampment on the plain, it looked as if it might be three hundred feet or more in height; but in the morning, by broad daylight, it seemed less. When I came to measure it, I was somewhat surprised that it did not exceed 210 feet. I observed on this as on many other occasions that a hill rising out of a level plain, appears higher than it really is, especially when, as in this case, the trees on its flank and summit are of small growth. Pembina Mountain is in fact, no mountain at all, nor yet a hill. It is a terrace of table-land, the ancient shore of a great body of water, that once filled the whole of the Red River valley. On its summit it is quite level and extends so, for about five miles westward, to another terrace, the summit of which I was told is level with the great Buffalo Plains, that stretch away toward the Missouri, the hunting grounds of the Sioux and the half-breed population of Red River.

Instead of being composed of ledges of rock, as I was led to suppose, it is a mass of incoherent sand, gravel and shingle so entirely destitute of cement, that with the hand alone a hole several feet deep may be excavated in a few minutes. The Pembina River has cut through this material a deep, narrow valley, but little elevated above the adjacent plain. Along its banks are precipices of sand, surmounted by gravel and a few boulders. I was told that it was impossible to ascend these banks. So loose is the deposit, that, no sooner is an ascent attempted, than the stones fifty or a hundred feet above, are detached, and come tumbling down at such an alarming rate that the climber is glad to make his escape."*

An inspection of the map will show the contour of the Pembina Mountain as far as ascertained. It will be observed that where Mr. Dickenson ascended it, 15 miles north of the 49th parallel it occurs in four distinct terraces. It crosses the Assiniboine near the Bad Woods, blends with the Riding and Duck Mountains, and probably appears again on the Main Saskatchewan, twenty-two miles from the Grand Forks. The elevation of the entire country east of this long ancient coast line is about 700 feet above the level of the ocean, and it forms the boundary of a distinct tract of lowland, in part surpassingly rich, as over the Red River and Assiniboine prairies, and the region on the Main Saskatchewan slightly elevated above the area subjected to annual overflow; part covered with swamp, marsh, or level limestone rock, on which a few inches of soil affords nourishment to small spruce, tamarac and aspen; and finally, by a shallow water area extending over 13,100 square miles, and embracing lakes which rank with the first class in point of superficies on this continent.

High above the Pembina Mountain the steps and plateaux of the Riding and Duck Mountains arise in well-defined succession. On the southern and south-western slopes of these ranges the terraces are distinctly defined, on the north-east and north sides the Riding and Duck Mountains present a precipitous escarpment which is elevated fully one thousand feet above Lake Winnipeg, or more than 1600 feet above the sea.

Standing on the edge of the escarpment of the Riding Mountain and looking in the direction of Dauphin Lake, a gulf, some two hundred and fifty feet deep is succeeded by two ranges, one lower than the other, of cone-shaped hills covered with boulders. The hills are parallel to the general trend of the escarpment, sometimes they are lost on the plateaux on which they rest. In other places they stand out as bold eminences, shewing the extent of the denudation which gave rise to them. These ranges of conical hills correspond with terraces on the west side of the mountain. They are the result of the same denuding forces which have left their impress upon the west flank, and were formed by the unequal wearing away of the east flank, at the time when the terraces on the opposite side were in process of arrangement.

I estimated the summit of Bear Hill, one of the most prominent of the conical hills separated from the edge of the escarpment by a deep valley, at 800 feet above Lake Winnipeg; if to this altitude we add 628 feet, the height of Lake Winnipeg above the sea, the elevation of the first terrace below the summit of the mountain will be about 1428 feet. This altitude corresponds in a remarkable manner with the sand bank on the Great Dog Portage, which has been found to have an elevation of 1438 feet above the Ocean. Great Dog Portage is 500 miles distant in an air line from Bear Hill, on the Riding Mountain. The second tier of conical hills stands upon the second plateau from the summit, and very probably corresponds with the Pembina Mountain; the altitude of the summit of Pembina Mountain above the sea is about 950 feet, and that of the second plateau, according to our estimate, nearly the same.

The denudation which has taken place in the valley of Lake Winnipeg is enormous. Five hundred feet above Dauphin Lake the Cretaceous shales crop out on the north-eastern flank; their position is nearly horizontal, and their thickness very great; they must have extended very far to the north-east, probably to the north shore of Lake Winnipeg, covering the horizontal limetones which occur at the Dog's Head and elsewhere on the western coast of that lake. It is not unlikely that future observations will establish a former connection between the Grand Coteau de Missouri, the Turtle, Riding, Duck, Thunder, Porcupine and Pasquia Mountains. It seems to me that they were formerly all part of one grand table land consisting of Cretaceous and Tertiary formations, which have been subjected to enormous denudation, and covered to a large extent with drift clays and sands, and with boulders of the unfossiliferous rocks.

4. *Ancient River Valleys.*

These records of former water-courses have been noticed in a preceding chapter (XV). Next to the valley of the Qu'Appelle, the old course of the Little Souris through the depression now occupied by the Back-fat Lakes is the most curious and imposing. Standing upon one of the most prominent of the Blue Hills of the Souris, near their southern extremity, the ancient valley can be traced as far as the first lake, which is distinctly seen by the unassisted eye, and with a good marine telescope its outline is plainly visible. Back-fat Creek flows with a sluggish current to join the Souris from these lakes in a westerly direction, while an arm of the Pembina River issues from their eastern extremity and flows into Red River. The Little Souris here pursues a course at right angles to its former

* Page 179. Geological Survey of Iowa, Wisconsin and Minnesota.

valley, and has excavated a channel from three to four hundred feet deep through the loose drift of the Blue Hills, and the Cretaceous rocks which underlie it.

5. *Sand Hills and Dunes.*

The most extensive of these unstable ranges are shown on the large map, and the position of those of smaller dimensions is indicated by a note.

It is needless to remark that the region they occupy is almost absolutely barren. Many of the hills and dunes are continually exposing fresh surfaces, sometimes beautifully ripple marked. The probability of their being the remains of Tertiary deposits, is noticed in a subsequent chapter. The following are the most extensive ranges:

1. Sand hills and dunes of the Assiniboine, extending from the Bad Woods to a short distance beyond Pine Creek, forty miles.
2. Sand hills of the Souris.
3. Sand hills and dunes of the Qu'Appelle.
4. Sand hills and dunes of the South Branch.
5. Sand and gravel ridges north of the Touchwood Hills.

6. *Circular Depressions.*

This curious disposition of the drift, probably due to a rearrangement of its materials, is of not uncommon occurrence south-east of the Touchwood Hills. Circular depressions varying from 100 yards to half a mile in diameter, appear in the prairies, generally surrounded by a ridge of sand or gravel. Many of them are quite dry, others hold water, often but not always brackish. The deepest and largest depression noticed was about 600 yards across and 40 feet below the general level.

7. *Effects of Denudation.*

An adequate conception of the effects of denudation in the valley of Lake Winnipeg can be best attained if we revert to the period when the Cretaceous shales now forming the flanks of the Turtle, Riding, Duck, Porcupine and Pasquia Mountains, resting probably upon Devonian Rocks, occupied the basins of Lakes Manitobah and Winnipeg, and found their eastern limits near the present outcrop of the Laurentian series. In order to complete our view of the extent of this great physical movement, we must conceive the same shales and sandstones in part overlaid by Tertiaries, filling the depressions or valleys in the Cretaceous rocks, (the result of denudation) and forming with that elevated tract an extensive, wide-spread table-land. These relations become more evident upon an inspection of the sections. The great gulf, nearly one thousand feet deep, between the summit of the Duck and Riding Mountains and the Laurentides has been in great part excavated by denuding forces during and since the Tertiary period.

In the section and on the map the Thunder, Porcupine and Pasquia Mountains are represented as being capped by Cretaceous rocks, but it is not improbable from the circumstance that Lignite has been found in the drift of the valley of Swan River, and that Indians who hunt in this region speak confidently of the occurrence of Lignite near the summit of Thunder and Porcupine Mountains, that patches of Tertiary formations which have escaped denudation may still exist there. Thus much appears certain, that the denudation of the valley of Lake Winnipeg belongs part to the Tertiary and part to the Post-Tertiary epochs. The great valleys leading to the Post-Tertiary sea, which was the main agent in effecting the denudation, were excavated posterior to the boulder drift period. These are the Main Saskatchewan, Red Deer River, Swan River, Valley River, and the Assinniboine, all of which cut the Cretaceous shales at right angles to the denuded face of the series of escarpments which these rocks in great part form.

The outcrops of the different formations as far as they are known, follow the general direction of the rim of the basin of unfossiliferous rocks in which they are deposited with remark able uniformity. Conforming to the direction of the Laurentian system exposed on the east side of Lake Winnipeg, the Silurian series stretches from Pembina on the 49th parallel, to the Saskatchewan on the 54th, and thence towards the Arctic Sea.* Following its outcrop the Devonian series is symmetrically developed between the same distant boundaries; but the most singular feature of this region is that the soft Cretaceous shales should also conform with tolerable exactness to the exposed edges of the unfossiliferous rim of the great basin in which they lie. The occurrence of Cretaceous forms in this valley of the Mackenzie, is a remarkable proof of the extension of this series in that direction. The present nucleus of the fossiliferous basin is occupied by the great Lignite formation of the Tertiaries of the Grand Coteau de Missouri; and so symmetrical is this arrangement, that a line drawn through any part of the country from that part of the Grand Coteau de Missouri which lies within British Territory, to any point between Pembina and the Grand Forks of the Saskatchewan, would pass over proportionally extensive areas of the Tertiary, Cretaceous, Devonian, Silurian and Laurentian series.

* See Mr. Isbister's Map,—Proceedings of the Royal Geological Society.

CHAPTER XVII.

THE LAURENTIAN SERIES.—THE SILURIAN SERIES.—THE DEVONIAN SERIES.

Distribution of Formations.—The LAURENTIAN SERIES*—The Laurentides—The Laurentian System described—Economic Materials in—Distribution of the Laurentian Series in the Basin of Lake Winnipeg.—The* SILURIAN SERIES*—The Chazy Formation—Deer Island—Grindstone Point—The Potsdam Sandstone—Probable Fossils in the Laurentian Series—Potsdam Sandstone on the South Shore of Lake Superior—The Bird's Eye Limestone—The Hudson River Group.—The* DEVONIAN SERIES*—Salt Springs—List of Salt Springs where Salt is gathered and manufactured—Mode of extracting Salt by solar evaporation—Formations superior to the Devonian—Western Limit of the Devonian Series—The Riding Mountain—Absence of Drift proofs—Limit of area in which formations between the Devonian and Cretaceous may be found—Probable absence of the Carboniferous Series—The Nebraska Series—Kansas Rock—Permian Series—Jurassic or Triassic Series probable in Kansas—Cretaceous Rocks repose on Jurassic in Nebraska—Probability of the occurrence of the Coal measures in the Basin of Lake Winnipeg.*

DISTRIBUTION OF FORMATIONS.

The distribution of Series of Formations in the order of their occurrence in the Valley of Lake Winnipeg and the Saskatchewan is as follows:

1. LAURENTIAN SERIES.
2. SILURIAN "
3. DEVONIAN "
4. CRETACEOUS "
5. TERTIARY "

THE LAURENTIAN SERIES.

The whole eastern coast of Lake Winnipeg and the adjacent islands, are Laurentian. Sir John Richardson, who voyaged along this shore in his journey to the Arctic sea remarks that "along the whole eastern shore the granite, gneiss, and trap rocks are every where exposed, the first named rock being the most extensive; and nowhere do these masses rise to the altitude of Hills."* The origin of the name Laurentian and the character of the rock series which compose this system is described by Sir William Logan and Mr. Hunt in the following extract from a 'Sketch of the Geology of Canada.'

THE LAURENTIDES.†

"The province of Canada is traversed, through its whole length, by a mountainous region dividing it into two basins, which may be distinguished as the Northern and the Southern basins. These mountains, which have been named the Laurentides, form the North shore of the St. Lawrence, from the Gulf as far as Cape Tourmente, near Quebec; from which point they leave the river, and while they follow its general direction become more and more remote, until near Montreal, they are at a distance of ten leagues from the St. Lawrence. Going further Westward, this mountainous region follows the line of the Ottawa, and crosses this river near the *Lac des Chats*, fifty leagues from Montreal. Thence taking a southward direction, it reaches the St. Lawrence near the outlet of Lake Ontario, and from this point running north-westward, the southern limit of this formation, reaches the south-eastern extremity of Lake Huron, at Matchedash Bay, and forms the Eastern shore of the Lake as far as the 47th degree of latitude, where quitting this lake, the formation gains Lake Superior, and extends in a north-west direction to the Arctic Sea.

"To the south of the St. Lawrence, this same region covers a considerable space between the Lakes Ontario and Champlain, and constitutes the Adirondack mountains. With this exception and perhaps also a small exposure in Arkansas and another near the sources of the Mississippi, this formation is not found to the south of the St. Lawrence, and as it belongs especially to the valley of this river and constitutes the Laurentide Mountains, the Geological Commission of Canada has distinguished it by the name of the *Laurentian system.*"

THE LAURENTIAN SYSTEM.

"The rocks of this system are, almost without exception, ancient sedimentary strata, which have become highly crystalline. They have been very much disturbed and form ranges of hills, having a direction nearly north-east and south-west, rising to the height of 2,000 or 3,000 feet, and even higher. The rocks of this formation are the most ancient known on the American continent, and correspond probably to the oldest gneiss of Finland and Scandinavia and to some similar rocks in the North of Scotland.

"The rocks of the Laurentian formation are in great part crystalline Schists, for the most part gneissoid or hornblendic. Associated with these Schists, are found large stratified masses of a crystalline rock, which is composed almost entirely of a lime and soda Felspar. This rock is sometimes fine grained, but more often porphyritic, and contains cleavable masses of Felspar, sometimes several inches in diameter; these Felspars are triclinic, and have ordinarily the composition of Andesine, Labradorite, Anorthite, or of intermediate varieties. Their colours are various, but the cleavable Felspars are generally bluish or reddish, and often give colored reflections. Hypersthene is very generally disseminated in these felspathic rocks, but always in small quantity. Titanic iron-ore is also found

* *Arctic Searching Expedition*, page 360. Am. Ed.

† A Sketch of the Geology of Canada serving to explain the geological map and collection of Economic Materials sent to the Universal Exhibition at Paris, 1855, by W. E. Logan, F.R.S., and T. Sterry Hunt, A.M.

in them, in a great number of places, sometimes in small grains, but often in considerable masses.

"With Schists and Felspars are found strata of Quartzite, associated with crystalline Limestones, which occupy an important place in this formation. These limestones occur in beds of from a few feet to three hundred feet in thickness, and often present a succession of thin beds intercalated with beds of gneiss or quartzite; these latter are sometimes quartzite conglomerates, and have in certain cases a base of dolomite. Associated with these limestones, are sometimes found beds composed in great part of Wollastonite and of Pyroxene, species which evidently owe their origin to the metamorphism of silicious limestones. Beds of Dolomite and Limestone more or less magnesian, are often interstratified with the pure limestones of this formation.

"The limestones of this system are rarely compact, and most frequently are coarsely granulated. They are white or reddish, bluish or grayish, and these colours are often arranged in bands which coincide with the stratification. The principal mineral species met with in these limestones, are Apatite, Fluor, Serpentine, Phlogopite, Scapolite, Orthoclase, Pyroxene, Hornblende, Wollastonite, Quartz, Idocrase, Garnet, brown Tourmaline, Chondrodite, Spinel, Corundum, Zircon, Sphene, Magnetic and Specular Iron, and Graphite. The Chondrodite and Graphite are often arranged in bands parallel with the stratification. Beds of a mixture of Wollastonite and Pyroxene are sometimes met with, which are very rich in Zircon, Sphene, Garnet and Idocrase. The most crystalline varieties of these limestones often exhale a very fetid odour when bruised. The limestones of this formation do not yield everywhere well crystallized minerals; near the Bay of Quinté there are beds met with which still preserve the sedimentary character, and show only the commencement of metamorphism.

"The conditions in which they are sometimes found, indicate that the agents which have rendered these limestones crystalline, have been such as to render the carbonate of lime almost liquid, and that, while in that state, it has undergone great pressure. As evidence of this opinion, we find that the limestone often fills fissures in the adjacent silicious strata, and envelopes the detached, and often, folded fragments of these less fusible beds precisely like an igneous rock.

"The crystalline Schists, Felspars, Quartzites and Felspars which we have described, make up the stratified portion of the Laurentian system, but there are besides, intrusive Granites, Syenites and Diorites, which form important masses; the Granites are sometimes Albitic, and often contain black Tourmaline mica in large plates, Zircon and Sulphuret of Molybdenum.

"Among the economic minerals of this formation, the ores of iron are the most important, and are generally found associated with the limestones."

The Laurentian rocks which form the east coast of Lake Winnipeg strike off at its north-east corner, and, passing to the north of Moose Lake, go on to Beaver Lake.*

The only exposure of Laurentian rocks seen within the area explored west of Lake Winnipeg were observed in St. Martin Lake; they have been described in Chapter IX, page 91.

* Sir John Richardson's Journal of a Boat Voyage, &c., page 49. Am. Ed.

THE SILURIAN SERIES.

Nearly the whole length of the western coast of Lake Winnipeg is composed of Limestones, Sandstones, and Shales of Silurian age. From Big Black Island to the Rapids on Red River, the formations are concealed by quaternary deposits. On the south-east coast limestone is occasionally seen in position, but its junction with the Laurentian series near the mouth of the Winnipeg is concealed by drift.

The formations which have been recognized on Lake Winnipeg, and in the valley of Red River, are

1. The Chazy Formation.
2. The Bird's-eye "
3. The Trenton "
4. The Hudson River Group.

Chazy Formation.

The following section occurs on Deer Island, and for the sake of convenience this and other sections are introduced in the order of their occurrence.

No. 1. Four feet of dark green argillo-arenaceous shale, with thin layers of sandstone of uneven thickness—Fucoids very abundant in the sandstone. The weathered sandstone is reddish brown; fresh surfaces are white or gray. White Iron Pyrites, assimilating the forms of disks, spheroids and shells, occurs in the sandstone. A *Modiolopsis* is common in the Shale.

No. 2. In many respects like the former; the sandstone layers are from one to four inches in thickness, and predominate over the shaly portions. Its thickness is six feet. The character of these formations (1 and 2) is very variable; the green argillaceous portion sometimes predominates, and occasionally the sandstone.

No. 3. Ten feet of sandstone with green bands of a soft argillaceous rock, from one quarter to four inches in thickness. The sandstone often white, but generally red. A persistent green band, a few inches thick, filled with obscure forms, resembling fucoids, is very characteristic. The red coloured sandstone is often soft and friable, the white frequently embodied in the red. Both red and white contain obscure organic forms. The green patches which are found throughout the sandstone contain impressions of fucoids; an *Orthoceras* was found in the sandstone. In some parts of the exposure on Deer Island the sandstone layers are much harder, although partaking of the characters already described. When thus hard, the white portion is extremely brilliant, of a pure white, and very silicious; it would form an excellent material for the manufacture of glass. Forms coloured brown, often pervade the white sandstone and appear to resemble fucoids and corals replaced by brown ochreous sand.

No. 4. Eighteen feet of limestone, perfectly horizontal, very hard, and breaking off the cliff where the soft sandstone has been weathered away in huge rhomboidal slabs, eight to twenty-five feet in diameter, and four to ten inches thick.

The surface of the limestone shows silicified shells and corals, among the shells an *Orthoceras* nine inches in diameter was seen, with others belonging to the genus *Rhynchonella*. (Page 86.)

The rocks at Grindstone Point, about six miles north of Deer Island, are similar to those described in the foregoing section. Being further north, the exposure is higher, and the sandstone bands more fully shown. Beneath No. 1 of Deer Island, a hard, yellow, compact sandstone is exposed for a space of four feet above the level of the water. Strata No. 1 and No. 2 of

Cliffs of Chazy Limestone at Grindstone Point, Lake Winnipeg, shewing the General Character of the Coast.

Deer Island appear in a slightly different form here: the sandstone bands are thicker; the green shaly portion more distinct as a separate band, and two feet thick; while above the hard yellow sandstone, the base of No. 1 appears in the form of a purple band of very soft sandstone, about one foot in thickness, containing a vast number of *stains*, which seem to have been occasioned by fucoids.

The lithological character of the hard, yellow, compact sandstone beneath No. 1 of Deer Island, when compared with the sandstones, shales, and limestones which lie above it, suggests the idea that it may belong to the Potsdam sandstone formation. The occurrence of well known chazy forms in the superior strata, remove all doubt as to their age; but further investigation might establish the existence of the formation which lies at the base of the fossiliferous rocks, as far as these are known, in this remote region.*

The lithological character of the Potsdam sandstone on and near the south shore of Lake Superior resembles the soft and friable Chazy sandstone of Lake Winnipeg in a very remarkable manner. It is not without interest that rocks belonging to formations possessing so close a vertical relationship should exhibit lithological characters almost identical in localities fully six hundred miles apart. Neither will it be thought improbable that more extended investigations may establish a still closer connection. Messrs. Foster and Whitney thus describe the Lake Superior sandstone in their Report on the Geology of the Lake Superior Land District:

"The Potsdam sandstone of New York is a quartzose rock, whose particles are firmly aggregated, while the same rock, on the northern slope of Lake Michigan, is so slightly coherent that it may be crushed in the hand. The calciferous sandstone of New York, when traced west, passes into a magnesian limestone. Even in that State, according to Hall, groups which, at one extremity, are of great importance, and well characterized by fossils, cannot be identified at the other." (p. 114.) * * * * "In descending the river, (the Menomonee,) it is first observed near the foot of the Chippewa Island. The subjacent rocks in this vicinity consist of talcose slates, in nearly vertical beds, intermingled with dark, compact, igneous rocks and crystalline greenstone. Their contour is very irregular, as though they had been abraded before the deposition of the arenaceous beds which occupy the inequalities in the surface of the more ancient rocks, in horizontal layers. The greatest inclination observed in the superior rocks was 3° to the south-east. *The sandstone consists of alternating bands of red and white, and is so friable, when first removed, that it may be crushed in the hand.* The grains are coarse and silicious, adhering together without any visible cement. After having parted with the water disseminated through the pores, it acquires a considerable degree of consistency, and is little acted on by the weather." (Page 132.) * * * * * * * *

"In the neighbourhood of Pleasant Valley, about twelve miles west of Strong's Landing, on the Fox River, it is exposed in several low escarpments, succeeded by the calciferous sandstone which here presents its usual characters. From this re-

* Until lately the Potsdam Sandstone has been supposed to represent the epoch when organic life was first introduced by the Creator on the surface of our globe. Recent discoveries tend to throw back the first peopling of the world into a past so indefinitely remote, that all preconceived ideas of the organic history of the world become unsettled and at fault. At the late meeting of the American Scientific Association, Sir William Logan exhibited a map illustrating the distribution of some of the bands of chrystalline limestone interstratified with the gneiss of the Laurentian Series—the Azoic Series of some geologists. The following notice is from *The Canadian Naturalist and Geologist*, page 300:—

"Although the Laurentian Series has hitherto been considered azoic, a search for fossils in them has not been neglected. Such search is naturally conducted with great difficulties. Any organic remains which may have been entombed in these limestones, would, if they retained their calcareous character, be almost certainly obliterated by crystalization, and it would be only through their replacement by a different mineral substance that there would be a chance of some of the forms being preserved. No such instances had been observed on the investigation of the Rouge and its vicinity, but from another locality in the Laurentian formation, Mr. John McMullen, one of the explorers of the Geological Survey, had obtained specimens well worthy of attention. They consisted of parallel or apparently concentric layers resembling those of the coral Stromatocerium, except that they anastomoze at various parts; these layers consist of crystalline pyroxene, while the interstices are filled with crystalized carbonate of lime. These specimens had recalled to recollection others which had been obtained from Dr. Wilson of Perth some years ago, and had not then been regarded with sufficient attention. In these similar forms are composed of green serpentine, concretionary, while the interstices are filled with white dolomite. If it be supposed that both are the result of mere unaided mineral arrangement, it would seem strange that identical forms should result from such different minerals in places so far apart. If the specimens had been obtained from the altered rocks of the Lower Silurian series there would have been little hesitation in pronouncing them to be fossils. The resemblance of these forms to Stromatocerium from the Bird's-eye limestone, when the coral has been replaced by concretionary silica, is very striking. In the pyroxenic specimens, the pyroxene and the carbonate of lime being both white, the forms, although weathered into strong relief on the surface, are not perceptible in fresh fractures until the fragments are subjected to an acid, the application of which shows the structure running throughout the mass. Several specimens of these supposed fossils were exhibited to the section."

gion, its southern limit stretches to the west and north-west. The country here presents a feature which continues to the Mississippi River. The hills appear to be outliers, capped by the calciferous sandstone, or succeeding limestones, while the valleys and the lower part of the escarpments are composed of the Potsdam. The rock is fine-grained, of a light yellow color and very friable. Some of the superior beds, which are thin, have been wrought for grindstones. The friable character of this sandstone is one of its most prominent features, and, owing to this circumstance, the escarpments are not usually high, or abrupt, unless it has been protected by the overlying rock. In its want of cohesion, it differs, in a very marked degree, from the prevailing character of this rock, as developed in New York and Canada, where it is usually, though not always, compact. It is not, however, unlike the sandstone of the Pictured Rocks, and is less friable than that of the Mississippi and St. Croix region. The almost uninterrupted continuity with which this rock can be traced, even from its eastern extension through Canada, and along the northern shore of Lake Huron to the St. Mary's River, and thence westerly, leaves no doubt as to its true position and identity in age with the Potsdam sandstone of New York. If we were at a loss in thus tracing it continuously, we have still the evidence of the succeeding fossiliferous strata, which show, conclusively, the same relations to this sandstone as they do to its equivalent in New York. With both these evidences combined, we cannot hesitate for a moment in our conclusion regarding its age and place in the series." (Page 133.)

Fine exposures of the chazy formation occur on Punk Island (see page 87); and along the west coast north of Big Grindstone part as far as the Cat Head. They appear in the form of cliffs, varying from 25 to 45 feet in altitude at nearly all points and promontories. The character of the rock is described in Chapter VII. At the narrows the three limestone promontories, the Bull's Head, Limestone Cave Point, and Whiteway's Point, approach within a few miles of the Laurentian series on the east coast. The strait from Whiteway's Point to the Dog's Head is not more than three miles across. Before this narrow channel was excavated, Lake Winnipeg must have been divided into two parts, like Lakes Manitobah and Winnipego-sis, and it is not improbable that near the Dog's Head a rapid river or falls once existed. The relation of the two lakes would then resemble the present relation of Lake Manitobah, the Little Saskatchewan and Lake Winnipeg.

The Birdseye and Trenton Limestone.

The whole of the coast on the north-west side of Lake Winnipeg is represented by Sir John Richardson to be occupied by the Birdseye Limestone. Near the First and Second Rocky Points the strata contain many gigantic orthoceratites which have been described by Mr. Stokes in the Geological Transactions.*

In Pine Island Lake, there are exposures thirty feet in altitude, containing *Orthocerata* and *Receptaculites neptunii.*† The strike is south-west by west and north-east by east, being at right angles to the general direction of the Laurentides.

The Hudson River Group.

This formation appears in cliffs five and twenty feet high at the Stone Fort, Red River. It is also exposed near the rapids. Most of the forts and churches in the Settlements are constructed of stone from this rock. The color of its weathered surface is a pale yellowish gray, but of fresh surfaces, a white gray.* Dr. Owen visited Red River Settlements in 1848† and described the fossils he found near the Stone Fort in his Report published in 1852. Dr. Owen says:

"About twenty miles below the mouth of the Assinniboine, near lower Fort Garry, solid ledges of limestone are exposed of a light buff colour, sometimes mottled, spotted, or banded with light brown. Immediately opposite the Fort, a considerable amount of rock has been quarried, and used in the construction of the building. In these beds, I succeeded in finding several well-defined and characteristic fossils, sufficient to establish, without the least doubt, the age of the Red River limestones.

They are: *Favosites basaltica; Coscinopora Sulcata;* hemispherical masses of *Syringopora; Chætetes lycoperdon;* a *Conularia;* a small, beautiful undetermined species of *Pleurorhynchus; Ormoceros Brongniarti; Pleurotomaria lenticularis* (?); *Leptæna alternata; Leptæna plano-convexa* (?); *Calymene senaria;* and several specimens of the shield of *Illænus crassicauda.*

Many of these are identically the same fossils which occur in the lower part of F. 3, in Wisconsin and Iowa, in the blue limestones of Indiana, Ohio, Kentucky, and Tennessee, and also in the lower Silurian of Europe.

The Coscinopora is precisely the same as the coral, which is particularly characteristic of the lower beds of the upper Magnesian limestone of Wisconsin. The specimens of Favosites Basaltica cannot be distinguished from those which abound in the upper Magnesian limestones of Wisconsin and Iowa, and the lower Coralline beds of the Falls of the Ohio. It is also worthy of note that these limestones of Red River, like their equivalent in Iowa and Wisconsin, are highly magnesian, containing from seventeen to forty per cent. of the carbonate of that alkaline earth.

Beyond the settlements of Red River, no opportunity is afforded on that stream for making further observations on the rock formations of the country.

A mile or two below the Cree Village, the river enters a tract of low land, and then meanders for more than twenty miles through a morass, before it finally disembogues into Lake Winnipeg.

On the south shore of that lake, however, I again had an opportunity of inspecting fossiliferous limestones in situ. At the two localities where I succeeded in obtaining a view of them, they were much disturbed, dipping either at a high angle, or, standing vertically. On Poplar Point, they are quite thin-bedded, and contain besides small *Entrochites*, large varieties of *Endoceras*. In a small bay, near Big Swamp Point, the limestone is seen jutting out beneath heavy, loose masses of crystalline rocks, some of which weigh hundreds of tons. The surfaces of many of the limestone slabs at this locality are crowded with well-preserved specimens of the characteristic fossil *Leptæna alternata*."

THE DEVONIAN SERIES.

In consequence of the extreme flatness of the country the junction of the Silurian and Devonian Series has been only ap-

* Journal of a Boat Voyage through Rupert's Land, p. 49, Am. Ed.

† *Ibid*, page 54.

* See Red River Report, page 294.

† Geological Report on Wisconsin, Iowa and Minnesota, page 181.

proximately determined, chiefly by the occurrence of the Saline springs which distinguish the Devonian Series in this region. In all cases where Saline springs were seen issuing from rock in position, Devonian formations were recognized by characteristic fossils. Several of these localities have been described in Chapter XI.

In 1823, Mr. Keating* noticed the Salt springs in Minnesota State and Dacotah Territory, far south of the boundary line. Even at that early period in the history of the Settlements on Red River, five hundred dollars were cleared by one individual during one winter from the sale of the salt he had manufactured from springs near Pembina. The price of salt in the Settlement was then six dollars per barrel weighing eighty pounds. At a spring on Saline River, south of the boundary line, Major Long's party found the *Salicornia herbacea* growing very abundantly around it. "Mr. Schweinitz states, on the authority of Mr. Nuttall, that this is the only inland locality of this plant, besides the Onondaga Salt Springs in the State of New York."

In the valley of la Rivière Salé, Salt springs are very numerous, and the ground in their vicinity is frequently covered with a thick incrustation. Many years since the half-breeds of the settlement used to collect salt from this valley for domestic purposes. The names Saline creeks and Salt points on Red River, north of the 49th parallel, were given in consequence of springs strongly impregnated with salt occurring there, but south and west of Stony Mountain no rocks in position have been observed east of Pembina Mountain. The whole country is nearly horizontal, having a mean elevation of about 130 feet above Lake Winnipeg.

Subjoined is a table shewing the localities, north of the 49th parallel, where Salt springs occur, distinguishing between springs from which salt has been and has not been manufactured or collected as a crust on the surface of the ground :

1. Salt Brook	Red River.
2. Salt Point	"
3. La Rivière Salé	Collected from incrustations by the side of the springs. These incrustations are often two inches in thickness.
4. Salt Point	Winnepego-sis Lake.
5. Turtle River	Dauphin Lake—collected by Indians.
6. Crane River	Manitobah Lake — collected by Indians.
7. Monkman's Salt Works.	Winnepego-sis Lake—manufactured by John Monkman, Chapter X.
8. Swan River	Manufactured for H. B. Co.
9. West Coast of Winnipego-sis Lake in many places.	
10. West Coast of Lake Manitobah in many places.	
12. The Pas Mountain.	

It has been already stated (Chapter X.) that the processes employed in the manufacture of salt in Rupert's Land are of the rudest description. By the employment of simple artifices the yield might be greatly increased, and its market value reduced to one fourth the price it brings at the Settlements. In the valley of La Riviere Salé, about twenty-six miles from Fort Garry, springs issue from the sides of the hills in positions very favourable for the employment of solar evaporation in shallow basins, which might be excavated at a lower level than the spring, and salt extracted without the employment of artificial heat; an immense advantage in a country where fuel is scarce and labour dear.

* Major Long's Expedition to the Sources of St. Peter's River.

In the State of New York between 500,000 and 600,000 bushels of salt are now made annually by solar evaporation. Wooden vats are employed, with moveable roofs, so that the brine may be protected at the approach of unfavourable weather. The average daily supply of brine at these works during six months of the year is 2,000,000 gallons, and the cost per barrel of 300 lbs. is one dollar. Salt made by the boiling process weighs 56 pounds to the bushel, solar made salt 75 pounds. By the boiling process at Onondaga the cast iron kettles, holding from 50 to 70 gallons each, are disposed in double rows above suitable furnaces technically called 'blocks.' Each block contains from 50 to 70 kettles, and manufactures during eight months of the year from 20,000 to 25,000 bushels of salt.

In 1800 the number of bushels of salt made at the Onondaga Salt Works was 50,000; in 1810, 450,000 bushels; in 1830, 1,435,446 bushels; in 1840, 2,622,305 bushels; in 1850, 4,268,919 bushels; and in 1857, 4,300,000 bushels.

The strength of the brine is measured by a 'Salometer,' whose zero is distilled water, and maximum, represented by 100, is water saturated with common salt. The brines of Onondaga vary from 76° to 44°. Wells which do not furnish brine above 50° are not considered worth working.

The sea-water at Nantucket gives a bushel of salt to every 380 gallons; at the salt springs of Zanesville, Ohio, 95 gallons furnish the same quantity of salt, while the old wells of Onondaga yield one bushel from 40 to 45 gallons, and the new wells at Syracuse the same quantity from 30 to 35 gallons of brine.*

The value of the salt trade in the United States may be inferred from the following statistics :—

	Bushels
In 1840 the quantity of foreign salt imported was	8,183,203
In 1850 " " " " "	11,224,185
In 1857 " " " " "	17,165,704

The value of the foreign salt consumed in 1857 amounted to nearly 2,000,000 dollars, and the value of foreign and domestic salt exported from the States during the same year was 230,000 dollars.

In Mr. Sterry Hunt's Report† for 1855 the excellent method pursued in France for the manufacture of salt from sea water, is described at length, and many features of this process might be very profitably employed in Rupert's land.

The most eastern exposure of the Devonian Series, recognized by fossils of that age occurs on Thunder Island, St. Martin's Lake; the most westerly exposure is seen on Moss River, and it is between these two points that, as far as known, brine springs are most numerous. Barren areas surrounding

* The History, Commerce, Sources, Manufacture, and economical value of Salt consumed in and exported from the United States; by William C. Dennis, of Key West, Florida.—Patent Office Report, 1857.

† Report for the year 1855 of T. Sterry Hunt, Esq., Chemist and Mineralogist to the Geological Survey, addressed to Sir William Edmond Logan, F.R.S., Director of the Geological Survey of Canada.

brine springs are of frequent occurrence at the foot of the range of hills from the Riding Mountain to the Pas. In a country nearly horizontal, where the attitude of the rocks conforms to the general surface, it will be at all times very difficult to determine the precise line of junction between succeeding series, and fortunately in the present instance the brine springs which undoubtedly have their source in Devonian rocks, afford an excellent guide in determining the outcrop and extent of the series.

As far as my observations enabled me to judge there is no difference in the general aspect of the country occupied by the Silurian and Devonian Series in this region. The rock of either age almost everywhere approaches the surface and is covered with a few inches of vegetable mould. Where fires have occured the soil is burned away and the bare surface exposed. Very few areas of drift were seen; the most imposing being some low hills on St. Martin's Lake. Denuding forces appear to have cut down the surface of the country to one nearly uniform level from the Riding Mountain ranges to the Laurentides. The upper extremity only of this excavated valley being covered many feet deep with quaternary deposits through which Red River, the Assiniboine and White Mud River have cut their channels.

The western limits of the Devonian Series are shown on the map to follow the boundary of the Great Cretaceous Table Land so well defined by Pembina Mountain, Riding Mountain, Duck Mountain, Porcupine Hill, the Pas Mountain, and the high plateau similar to Pembina Mountain which stretches from the Pas to the Main Saskatchewan, near and below Fort à la Corne. The country as the base of this continuous boundary is uniformly horizontal, and while Devonian rocks in position were seen within thirty miles, and brine springs within ten miles of Cretaceous Shales on the precipitous flanks of the Riding Hill range, yet no evidence of any intermediate formation was visible.

During the ascent of the Riding Mountain, a very careful search was made for traces in the drift of the higher series, in the hope of obtaining evidence of the existence of Carboniferous rocks, but without success. The boulders so numerous on the ridges and the successive terraces, were carefully examined but they were found to be derived altogether from the Laurentian Series, or the limestone of Lake Winnipeg or the superior Cretaceous Shales.

The presence of fragments of any particular rock in the drift of Canada affords presumptive evidence of the existence of the parent rock in position some distance to the north of the place where the detritus is found.

If rocks occupying a position between the Devonian and Cretaceous Series exist on the flanks of the Riding Mountain, it is probable that traces would have been discovered in the drift. The space in which members of the Carboniferous Series or superior formations might occur, is narrowed down to a strip ten miles in breadth between the Salt Springs south of Dauphin Lake and the outcrop of the Cretaceous Shales on the flanks of the Mountain. (See Chapter X. for a description of the ascent of the Riding Mountain.) At least seven miles of this distance is so nearly horizontal that it does not rise twenty feet above Dauphin Lake, and the dip of the Devonian Strata is uniformly at a very small angle to the south-west, where exposures were seen on Manitobah Lake. (Small local deviations from a uniform dip on Snake Island and Moss River are noticed in Chapter X. and XI.) The Cretaceous Shales were found exposed on the flanks of the mountain, about 400 feet above Dauphin Lake, and the rise from the level country at the foot of the mountain to that altitude is embraced within two and a half or three miles; yet within this narrow limit the drift on the slopes between each terrace, on the terraces themselves, or in the bottom of gullies excavated by mountain streams, gave no evidence of other rocks than those already named. It must be admitted that the time I could devote to an examination of the boulders was short, and a more minute search might give other results.

With this negative evidence in view, it appears tolerably certain that the Carboniferous Series is not represented in the only locality where it may be looked for with much chance of success. Nevertheless, between the Devonian and Cretaceous Series in the basin of Lake Winnipeg there is still a vertical section fully four hundred feet in altitude, which is concealed by drift on the flanks of the Riding Mountain, covering a horizontal area two and a half to three miles broad. It is possible that within this narrow limit, or further to the north where the area may be broader, rocks of Carboniferous, Permian, Triassic, or Jurassic age, may be yet found. With a view to show the relation which the Cretaceous and Carboniferous Series have to one another in lower latitudes, the following brief notice of their occurrence in Nebraska and Kansas is introduced.

In Nebraska the Carboniferous Series, or the coal measures, are exposed at the mouth of the Platte,* and extend up the river about 50 miles, when they dip beneath the water level of the Missouri. They are overlaid by No. 1 of the Nebraska section of the Cretaceous Series in latitude 41.5°, long. 96° Cretaceous and Tertiary formations then occupy the valley of the Missouri as far as Fort Benton, lat, 47.54°, long. 110°, and extend into British America, as shown on the map which accompanies this report. Hence it appears that ten degrees of latitude south of the Riding Mountain, the Cretaceous Series repose on the Carboniferous without the intervention of Permian, Triassic or Jurassic Rocks.

In Kansas Territory, on the Kansas and Smoky Hill Rivers, an elaborate section has been made by Messrs. F. B. Meek and F. V. Hayden,† commencing with the Cretaceous Sandstones on the summit of the Smoky Hills, lat. 38° 30′ N., long, 98° W., and descending through the various intermediate formations seen along the Smoky Hill and Kansas Rivers to the mouth of the Big Blue River on the Kansas. This section, over one thousand feet vertically, passes from the Cretaceous to the upper coal measures, and includes rocks of Permian age. Messrs. Meek and Hayden remark, in relation to this section, "It will be observed we have in this general section, without attempting to draw lines between the systems or great primary divisions, presented in regular succession the various beds with the fossils found in each, from the Cretaceous Sandstone on the summits of the Smoky Hills, down through several hundred feet of intermediate doubtful strata, so as to include the beds containing Permian types of fossils, and a considerable thickness of rocks, in which we find great numbers of upper coal measures forms. We have preferred to give the section in this form, because, in the first place, the upper coal

* Notes explanatory of a map and section illustrating the Geological Structure of the country bordering on the Missouri River, &c., by F. V. Hayden, M.D.

† Geological Explorations in Kansas Territory, by F. B. Meek and F. V. Hayden, published in the proceedings of the Academy of Natural Sciences at Philadelphia.

measures of this region pass by such imperceptible gradations into the Permian above, that it is very difficult to determine, with our present information, at what particular horizon we should draw the line between them, while on the other hand it is equally difficult to define the limits between the Permian and beds above, in which we found no fossils."(1)

Jurassic or Triassic formations may occur above the Permian in the section just referred to. Messrs. Meek and Hayden state that "between No. 5 (of the Section) and the Cretaceous above, there is still a rather extensive series of beds in which we found no organic remains; these may be Jurassic or Triassic or both, though as we have elsewhere suggested, we rather incline to the opinion that they may prove to belong to the former." (2)

Formation No. 1 of the Nebraska series of the Cretaceous rocks has not yet been recognized in Rupert's Land. This formation reposes on Jurassic rocks in Nebraska Territory at the Black Hills (3). It rests, as before stated, upon the limestones of the coal measures on the Missouri, near the 42nd parallel.

"There is at the base of the Cretaceous System, at distantly separated localities in Nebraska, Kansas, Arkansas, Texas, New Mexico, Alabama and New Jersey, if not indeed everywhere in North America where that System is well developed, (at any rate east of the Rocky Mountains,) a series of various colored clays and sandstones, and beds of sand, often of great thickness, in which organic remains, excepting leaves of apparently dicotyledonous plants, fossil wood, and obscure casts of shells, are very rarely found, but which everywhere preserves a uniformity of lithological and other characters, pointing unmistakeably to a similarity of physical conditions during their deposition, over immense areas.

"Although the weight of evidence thus far favors the conclusion that this lower series is of the age of the Lower Green Sand, or Neocomien, of the old world, we yet want *positive* evidence that portions of it may not be older than any part of the Cretaceous System." (4)

Judging therefore, solely from the relation which the Cretaceous Series bears to formations beneath them in their development through Rupert's Land, Nebraska and Kansas, we might expect to find on the Riding Mountain in the vertical section (400 feet), concealed by drift, beneath formation No. 4, (see succeeding chapter,) either formation No. 1, 2 and 3 of the Nebraska section, or members of the Jurassic and Permian as well as the Carboniferous Series.

The prospect of any member of the true Coal Measures being found on the flanks of the Riding, Duck, Porcupine or Pas Mountains, becomes in consequence of the ascertained existence of other series beneath the Cretaceous in the same geological basin, rather unfavorable, but is certainly far from being without hope.

It is very gratifying to know that on the western side of the great Basin between the Laurentides and the Rocky Mountains, within the limit of the Saskatchewan Valley, the Carboniferous Series are represented. Sir Roderick Murchison, in his address at the Anniversary Meeting of the Royal Geographical Society, in referring to the splendid results of the Palliser Expedition, says, "Thus in addition to the determination of latitude, longitude, and the altitude of the mountains and two of their passes, Dr. Hector presents us with a sketch of the physical and geological structure of the chain, with its axis of slaty sub-crystalline rocks, overlaid by limestones of Devonian and Carboniferous age, and flanked on the eastern face by Carboniferous Sandstone, representing, probably, our own coal fields, the whole followed by those Cretaceous and Tertiary deposits which constitute the subsoil of the vast and rich prairies watered by the North and South Saskatchewan, and their affluents." *

(1) Page 19, Geological Explorations in Kansas.

(2) Page 21, *ibid.*

(3) On the Lower Cretaceous beds of Kansas and Nebraska, by F. B. Meek and F. W. Hayden.—Proceedings Acad. Nat. Sci. Phil, Dec. 1858,—published in Am. Jour. Sci. page 219, 1859.

(4) Remarks on the Tertiary and Cretaceous formations of Nebraska, &c. &c., by F. B. Meek and F. V. Hayden, M.D.

* Page 318, Proceedings of the Royal Geological Society, Vol. III., No. 4.

CHAPTER XVIII.

THE CRETACEOUS SERIES.—THE TERTIARY SERIES.

Great Extent of the Cretaceous Series in Rupert's Land,—Cretaceous Series in the United States—Vertical Section in Nebraska Territory—Formation No. 1—Formation No. 2—Probable distribution on the North Branch of the Saskatchewan—Formation No. 3—Formation No. 4—Distribution on the Little Souris—The Assiniboine—The Qu'Appelle—Formation No. 5—Distribution on the Qu'Appelle—The South Branch of the Saskatchewan—The Tertiary Series—Sand Dunes probably derived from Tertiary Rocks—Importance of—Lignite—Distribution in America—Distribution and importance of in Europe.

THE CRETACEOUS SERIES.

By far the greater portion of the country explored in 1858 is underlaid by the different formations of the Cretaceous Series. They were seen in position on the Little Souris in longitude 100° 30 W, and on the South Branch in longitude 106° 35. Between these widely separated points they were noticed in many places on the Assiniboine, the Qu'Appelle and their affluents. This important series, as it occurs in Nebraska, has been carefully studied and admirably described by Messrs. Meek and Hayden. In the notes explanatory of a Map and Section illustrating the geological structure of the country bordering on the Missouri River, Dr. Hayden has described the rocks of Nebraska Territory* where the Cretaceous series is best developed, and as this division, styled the NEBRASKA SECTION forms the standard to which the Cretaceous rocks of

* The first reliable accounts we have of the general physical characters of the Upper Missouri country, were given to the world in the report of Lewis and Clark's expedition to the Columbia in 1804–5–6. The exploration of these gentlemen, in addition to bringing out a large amount of information of a different character, established the fact of the occurrence of Cretaceous rocks at the Great Bend of the Missouri below Fort Pierre, and of the existence of what was supposed to be "stone coal" (Lignite) in the Mandan country. Various beds of clay, sand, sandstone, &c., were mentioned in their report, but without any suggestions respecting their age.

In 1832, the Prince of Wied and party also ascended the Missouri to its sources; and the result of his explorations, embodying a great amount of highly interesting information respecting the geography, natural history, &c., of the country explored, have been published in the form of a large quarto volume, accompanied by a magnificent folio atlas of plates, illustrating the scenery of the country, and the manners and customs of its native tribes, in a style of art rarely equalled on this side of the Atlantic. Respecting the geology of the country, however, the Prince's expedition added little of importance beyond the discovery of *Mosasaurus Missouriensis*, to the results of Lewis and Clark's Expedition.*

Mr. Nicollet, the well known Geographer, visited this country in 1839, ascending the Missouri to Fort Pierre, and making on his way up a fine collection of Cretaceous fossils at the Great Bend.† Although passing rapidly through the country, he formed a tolerably correct idea of its geology, and gave in his report a vertical section of the Cretaceous rock seen below Fort Pierre which is correct, excepting that he seems to have had no knowledge of No. 2, and as we think, without sufficient reason, represented two of the subdivisions of No. 3 as distinct formations; No. 1 he appears to have referred to the carboniferous system. As he did not go above Fort Pierre, he probably saw nothing of No. 5. though some of its characteristic fossils were presented to him by gentlemen connected with the Fur Company.

In 1843, Mr. Edward Harris, who accompanied the celebrated Ornithologist Audubon to the mouth of Yellowstone River, brought back specimens from various localities along the Missouri River, some of which verified the statements of former explorers, while others gave evidence of the existence of a fresh-water formation near Fort Union.

At various times after this, specimens of mammalian remains were brought in by gentlemen connected with the American Fur Company, indicating the existence of an interesting tertiary deposit on White River; the first account of which was published by Dr. H. A. Prout, of St. Louis, in the American Journal of Science, 1847.

In 1847, Dr. John Evans, one of Dr. Owen's assistants in the geological survey of the Chippeway Land District, was sent by that gentleman on an expedition to the Mauvaises Terres of White River, and brought back a fine collection of Mammalian and Chelonian remains, which were investigated by Prof. Leidy, of Philadelphia. He also collected at the Great Bend, Sage Creek and Fox Hills, many interesting Cretaceous fossils, which were investigated by Dr. D. D. Owen, and published in his final report in 1852. Dr. Evans' observations, embracing a section of the Bad Lands, together with a description of their physical features, were also published in this report.

In the following year Mr. Thaddeus A. Culbertson, visited the Upper Missouri country under the auspices of the Smithsonian Institution, during which expedition he collected some interesting vertebrate remains from the White River formations. He also ascended the Missouri on the Fur Company's boat to a point above Fort Union, noting the character of the face of the country, and the occurrence of lignite beds at various localities.

In the spring of 1853, Dr. Evans again visited this country incidentally, while on his way to Oregon Territory, in the geological survey of which he was engaged, under the patronage of the general government. During this expedition he made another extensive collection of vertebrate remains, and some fresh-water mollusca at the Bad Lands of White River, as well as some interesting Cretaceous fossils from Sage Creek. The mammalian remains of this expedition were studied by Prof. Leidy, and the other fossils by Dr. Evans and Dr. Shumard, and published in the Proceedings of the Acad. Nat. Sc. at Philadelphia, and the Acad. Sciences of St. Louis.

At the same time (1853) the writers of this paper were employed by Prof. Jas. Hall, of Albany, N. Y., to visit the Bad Lands of White River, for the purpose of making a collection of the Tertiary and Cretaceous fossils of that region. This expedition brought back an extensive and interesting collection of vertebrate remains from the Bad Lands. and of Cretaceous fossils from Sage Creek, as well as from Great Bend and other localities along the Missouri below Fort Pierre. The first were investigated by Prof. Leidy, and published in the Proceedings of the Acad. Nat. Sc. at Philadelphia; and the latter by Prof. Hall and one of the writers,* and published in the Transactions Acad. Arts and Sciences, Boston.

In this latter paper a brief vertical section of the rocks seen during the expedition, and a complete list of all the mollusca then known from the Cretaceous and Tertiary rocks of that country, were given. The fact that the fossils characterizing the Cretaceous formations of Texas and New Mexico belong to different types from those occurring in the northwest, was also in this paper made known for the first time, in the following words: "Among all the collections made in Texas by Dr. Rœmer and others, and of all those brought by the Boundary Survey Expedition, and other surveying and exploring parties, which we have seen, there is but a single species which we regard as doubtfully identical with one from Nebraska. This is *Inoceramus Barabini*, Morton (I. Crispii, Mantell.) (?)"

A summary of the leading results of this expedition, throwing light upon the general geology of the country, its soil, scenery, &c., was likewise given to the

* The Prince lost nearly all his geological specimens by the burning of the Fur Company's steamboat.

† These, together with others given to him at Fort Pierce, were investigated by Dr. Morton, and published in the Jour. Acad. Nat. Sc. Philada., Vol. 8, p. 207.

* Mr. Meek.

the North-west are referred, the following notice of the series is abbreviated from their explanatory notes and remarks.*

The history on the preceding page, of the discoveries in Nebraska Territory, is contained in the introduction to Messrs. Meek and Hayden's "Remarks on the Tertiary and Cretaceous Formations of Nebraska, and the parallelism of the latter with those of other portions of the United States and Territories."

Subjoined is the vertical section of the geological formations of Nebraska Territory, with their extension into Rupert's Land, as far as determined :—

VERTICAL SECTION OF THE GEOLOGICAL FORMATIONS OF NEBRASKA TERRITORY, AS FAR AS DETERMINED, WITH THEIR EXTENSION INTO RUPERT'S LAND.

			Subdivisions.	Localities.	Estimated thickness.	Localities in Rupert's Land.
TERTIARY SYSTEM.	Miocene.		Light colored indurated clays, with occasional beds of sandstone, conglomerate and whitish limestone. Great numbers of *mammalian* and *chelonian* remains with a few fresh-water and land shells.—(Bad Lands of White River.) Beds of clay, sand, sandstone, and lignite, containing great numbers of fresh-water and land *mollusca*, with a few marine or estuary shells; remains of plants, *Saurians*, *Trionyx*, &c.—(Great Lignite Basin.) Sand, sandstone, clays, and very impure lignite, with remains of fresh-water, land, and a few estuary shells, *Saurians*, fishes, *Trionyx*, &c.—(Bad Lands of Judith.)	*Mauvaises Terres* of White River. Great extent of country on both sides of the Missouri between Heart and Milk Rivers; on the Yellow Stone. Bad Land at the mouth of Judith River, &c.	About 900 ? feet.	Grand Côteau de Missouri.
CRETACEOUS SYSTEM.	No. 5.		Gray and yellowish arenaceous clays, and sandstones, sometimes weathering to pink color; containing *Belemnitella bulbosa*, *Nautilus Dekayi*, *Ammonites placenta*, *A. lobatus*, *Scaphites Conradi*, *Baculites ovatus*, and great numbers of other marine mollusca.	Moreau trading post, and under the Tertiary of Sage and Bear Creeks. Fox Hills.	100 to 150 feet.	South branch of the Saskatchewan—*Scaphites Conradi*, *Nautilus Dekayi*, *Avicula Linguæformis*, *Avicula Nebrascana*.
	No. 4.		Bluish and dark gray plastic clays, containing *Nautilus Dekayi*, *Ammonites placenta*, *Baculites ovatus*, and *B. compressus*, with numerous other marine mollusca,—remains of *Mosasaurus*.	Great area about Fort Pierre and along the Missouri below there. Under No. 5, at Sage and Bear Creeks. Great Bend of the Missouri. Near Milk and Muscle Shell Rivers.	350 feet.	Little Souris River—containing *Anomia Flemingi*, *Inoceramus Canadensis*, *Leda Hindi*. Two Creeks, Assiniboine River, *Natica obliquata*, *Avellana Concinna*; *Ammonites*—South branch of the Saskatchewan—*Leda Evansi*, *Ammonites placenta*, *Scaphites Nodosus*.
	No. 3.	20 feet.	Dark, very fine unctuous clay, containing much carbonaceous matter, with veins and seams of selenite, sulphuret of iron, fish and scales, (local.)	Bluffs along the Missouri below the Great Bend. Extends to Big Sioux River, and occurs along the latter stream.	150 feet. New Jersey and Alabama.	North Branch of the Saskatchewan at the Coal Falls (?) Sharks teeth—Scales of Fish—Inoceramus.*
		100 feet.	Lead gray calcareous marl, weathering above to a yellowish tint. Scales and other remains of fishes—*Ostrea congesta* — passing downwards into			
		30 feet.	Light gray or yellowish limestone, containing great numbers of *Inoceramus problematicus*, fish scales, and *Ostrea congesta*.			

public by Prof. Hall in an interesting paper read before the American Association for the Advancement of Science, at the Providence meeting.

Subsequent to all these expeditions, one of the writers† again visited Nebraska, and spent two years in traversing various portions of that country; part of which time he was aided by Col. A. J. Vaughan, Indian agent, and afterwards by Mr. Alexander Culbertson, and other gentlemen of the American Fur Company. During this expedition he explored the Missouri to the vicinity of Fort Benton and the Yellow Stone to the mouth of Big Horn River. Also considerable portions of the Bad Lands of White River, and other districts not immediately bordering on the Missouri. The vertebrate remains collected by him, as may be seen by reference to the various papers by Prof. Leidy in the Proceedings of the Academy, embrace a larger number of species than all those previously known from that country, many of which belong to new and remarkable genera. Large collections of mollusca were also obtained from the Cretaceous and Tertiary formations, and have since been published by us, together with remarks on the general geology of the country, in a series of papers in the Proceedings of the Academy Nat. Sc. Phila. Vol. viii.

Again, in 1856, the same one of the writers returned to that country in connection with a government expedition under the direction of Lieut. G. K. Warren. The new Cretaceous and Tertiary invertebrate remains, together with the new facts in regard to the geology of the country, collected by this expedition, form the basis of this paper.

Up to the publication of our first paper, about fifty-six new species of Cretaceous and Tertiary mollusca had been published from Nebraska, by Drs. Mortons Owen, Evans and Shumard, and by Prof. Hall and one of the writers. Since that ime, sixteen additional new species have been published by Drs. Evans and Shumard, making in all seventy-two species hitherto published by others from that country. Our own investigations (including those here described) have made known one hundred and fifty new species, and two new genera, many of the former of which also belong to types not hitherto recognized in this country. Of these one huudred and fifty species, fifty-four (if we include the Judith River, fresh-water and estuary species) belong to the Tertiary system, and ninety-six to the Cretaceous. Fifty of the Tertiary species belong to fresh-water and land types, and four to genera inhabiting salt and brackish waters; being about four-fifths of all the land and fresh-water Tertiary species hitherto made known from American formations. The geological position, and vertical range of all our new species, and several of those published by others from the north west as well as a number of the well known and widely distributed forms such as *Scaphites Conradi*, *Ammonites lobatus*, *A. placenta*, *Nautilus Dekayi*, *Inoceramus problematicus*, *Mosasaurus Missouriensi*,‡ *&c.*, have been determined with considerable accuracy; so that we have now the means of tracing out the parallelism between these deposits and their equivalents in other countries.

* Notes explanatory of a Map and Section illustrating the geological structure of the country on the Missouri River from the mouth of the Platte River to Fort Benton, in lat. 47° 30′ N., long. 110° 30′ W., by F. V. Hayden, M. D. Proc. Acad Nat. Sci. Phil., May, 1857.

† Dr. Hayden.

* Mr. Meek thinks the specimens may belong to either No. 2 or No. 3.

‡ In a section of the Nebraska formations accompanying the last paper communicated by us to the Academy, we gave as the position in which the remains of *Mosasaurus* occur in the north-west, the upper part of No. 5. This was in consequence of erroneous information in regard to the locality from which the specimens given to the Prince of Neu Wied were obtained. The locality (at the Great Bend of Missouri) has since been visited by one of us, and many specimens obtained; they occupy a horizon about the middle of No. 4 of the section.

VERTICAL SECTION OF NEBRASKA—(*Continued.*)

		Subdivisions.	Localities.	Estimated thickness.	Localities in Rupert's Land.
CRETACEOUS SYSTEM.	No. 2.	Dark gray laminated clay; scales and other remains of fishes, small Ammonites, *Inoceramus problematicus? Serpula*, small oyster-like *O. congesta*, &c.	Along the Missouri Bluffs, from ten miles above James River to Big Sioux River.	90 feet. Wanting in	Assiniboine—Scales of Fish.* North Branch of the Saskatchewan at the Coal Falls (?)
	No. 1.	Yellowish and reddish friable sandstone, with alternations of dark and whitish clays. Seams and beds of impure lignite, fossil wood, impressions of dicotyledonous leaves; *Solen, Pectunculus, Cyprina*, &c. this bed is not positively known to belong to the Cretaceous system.	Near the mouth of Big Sioux River, and between there and Council Bluffs. Near Judith River?	90 to 100 feet, or more.	Not recognized in Area Explored.
CARBONIFEROUS SYSTEM.		Yellow limestone, containing *Fusulina cylindrica, Terebratula subtilita, Spirifer Meusebachanus Allorisma regularis*, and other fossils of the coal measures.	Forms shoals in the Missouri River at De Soto; 15 to 20 feet exposed at Council Bluffs, at low stages of the river.	Unknown.	Not recognized in Area Explored.

Formation No. 1, of Vertical Section.

The following excellent descriptions of the formations constituting the Nebraska Section are from the clear and concise "Notes Explanatory of a Map, &c," by F. B. Meek and F. V. Hayden, M.D. They will serve as an admirable guide for the study of the development of the Cretaceous series in the part of Rupert's Land referred to in this Report.

In the order of superposition, Formation No. 1 rests directly upon the true limestones of the Coal Measures. Its first exposure seen along the Missouri is at Wood's Bluffs, right bank, about eighty miles above the mouth of the Platte, and it dips beneath the water level of the Missouri, a few miles below the mouth of the Vermillion. Its general character is a coarse grained friable sandstone, very ferruginous, of a yellow or reddish yellow color, with thin beds of impure lignite and various colored clays. It contains very few fossils, mostly of the genera *Solen, Cyprina* and *Pectenculus*, also fossil wood, and numerous impressions of dicotyledonous leaves, similar to the common willow. Its entire thickness is estimated at ninety to one hundred feet, but it may be more.

This formation has not yet been recognized in Rupert's Land. In Nebraska it reposes upon the upper members of the Carboniferous series near the mouth of the Platte (lat. 41° 40'), and it overlies Jurassic rocks at the Black Hills.†

Formation No. 2 of Vertical Section.

This Formation is first revealed in thin outliers below the mouth of Big Sioux River, and on that stream six miles above its mouth it caps the Bluffs, apparently mingling to some extent with the succeeding bed, and containing at this locality large numbers of *Inoceramus problematicus* and fragments of fishes. Near the mouth of Iowa Creek and above, it shows itself worthy of a separate position in the series. It is composed of a dark leaden gray laminated plastic clay, containing few fossils, but great quantities of the sulphate of lime in crystals, assuming a variety of beautiful forms. Its greatest thickness is seen five miles below the mouth of James River. At Dorion's Hills it is seen at low water mark. Entire thickness estimated at ninety feet. Fossils, *Ammonites, Inoceramus, Cytheria, Serpula, Ostrea* and abundant fish remains.

This formation has been recognized on the Assiniboine.

On the North Branch of the Saskatchewan, a few miles above the Grand Forks, huge masses of a dark coloured, almost black shale, with sharp, well preserved edges jut out of the banks, and are exposed whenever portions of the face of the clay cliffs fall into the river. Their appearance is such as to justify the expectation that rock in position from which they originated is close at hand. Some specimens which I procured and sent to Mr. Meek, contain, according to that gentleman, fish scales, sharks' teeth and *Inoceramus*, which renders it almost certain that the masses were detached from rocks belonging to formation 2, or 3, of the vertical section. I have therefore assigned in the foregoing table, the locality Coal Falls, North Branch of the Saskatchewan, with a note of interrogation, as the probable outcrop of one or both of these divisions of the Vertical Section.

Formation No. 3 of Vertical Section.

The geographical distribution of this formation and its influence on the scenery render it one of the most interesting on the Missouri. It is first seen in thin outliers near the mouth of Big Sioux River, and becomes quite conspicuous on the summits of the Bluffs ten miles above the Iowa Creek. At Dorion's Hills it reaches to the water's edge and is the prevailing formation from thence to the foot of the Great Bend, where it passes by a gentle dip beneath the water level of the Missouri. At Dorion's Hills there is a fine section of this bed about eighty feet exposed above the water's edge, containing its most abundant and characteristic fossil, *Ostrea congesta*. In many places as opposite the mouth of Running Water, it assumes the form of a long series of precipitous bluffs, giving a pleasing variety to the general monotony of the scenery. This is one of the principal characteristic external features of this formation.

The upper portion of this rock is a yellowish and gray calcareous marl, very soft and yielding so that it is easily cut up into numerous ravines by the temporary streams, and thus the bluffs along this part of the Missouri often present the appearance of a series of cones. The lower stratum, however, is more compact and forms a soft bluish gray limestone.

Though so well developed and covering so wide an area, the middle and upper portions, at least, of this rock can never be made useful for building purposes. Quite soft and friable in places, when detached, it absorbs moisture rapidly and crumbles in pieces. Being a rich calcareous marl, it may be used at some future time as a fertilizer.

The fossils of this formation, though belonging to few species, as far as is yet known, are numerous in individuals. A species of oyster (*O. congesta*) is found in great quantities throughout the bed, and in localities *Inoceramus problematicus* is abundant. Fish remains, though consisting mostly of scales and obscure fragments, are disseminated throughout the deposit, several species of which have already been identified and described by Dr. Leidy. Entire thickness of this bed about one hundred and fifty feet.

Near First Cedar Island, a very singular bed makes its appearance superimposed on No. 3, which we shall consider as probably forming a local upper member of that formation. It extends up the Missouri River to a point near the Great Bend, a distance of about eighty miles. Lithologically it is a dull black, unctuous clay, destitute of any grit, and does not effervesce with an acid. It contains some carbonaceous matter and great quantities of Selenite in crystals.

This formation as stated above, probably occurs on the North Branch of the Saskatchewan, at the Coal Ealls.

* Beds Nos. II and III as well as No. IV and V may in some localities merge into one another. See foot-note, page 130, of the Geology and Palæontology—Mexican Boundary Survey.

† Descriptions of the Species and Genera of Fossils collected by Dr. F. V. Hayden in Nebraska Territory, under the direction of Lieut. C. K. Warren, U. S. Topographical Engineer, with some Remarks on the Tertiary and Cretaceous Formations of Nebraska, and the parallelism of the latter with those of other portions of the United States and Territories, by F. B. M., and F. V. Hayden, M.D.

Formation No. 4 of Vertical Section.

This formation is the most important one in the Cretaceous System of the north-west, not only in regard to its thickness and its geographical distribution, but also in its influence on the agricultural capabilities of the country. It is only second in interest to the succeeding bed in the number, beauty and variety of its organic remains. Commencing about ten miles above the mouth of James River, where it is seen only in thin outliers capping the distant hills or bluffs, it continues gradually assuming a greater thickness as we ascend the Missouri until reaching the Great Bend, where it monopolizes the whole region, giving to the country underlaid by it a most gloomy and sterile aspect. At the Great Bend it attains a thickness of two hundred feet, and continues to occupy the country bordering on the Missouri, to the mouth of Grand River, where in consequence of the dip of the strata, it passes gradually beneath the level of the river.

After dipping beneath the water level between Grand and Cannon Ball Rivers, this formation again rises to the surface about thirty miles below the mouth of Milk River, (far up towards the sources of the Missouri), by a reversed dip of the strata, from beneath the northern portion of the Great Lignite Basin, as will be seen by reference to the section on the map. Near the mouth of the Muscle Shell River it occupies the whole country for a distance of about eighty miles, and thins out upon the tops of the hills near the mouth of the Judith River.

In summing up the extent of country underlaid by this great formation, we find that south of the Lignite Basin, it occupies an area of two hundred miles in length and one hundred in breath, or twenty thousand square miles. North of the Great Lignite Basin, commencing at its first appearance near Milk River, we find it covering an area of two hundred miles in length and sixty in breath, or about twelve thousand square miles. I have been thus particular in estimating its approximate limits and extent of surface on account of its influence on the future destiny of that region. Wherever this deposit prevails it renders the country more completely sterile than any other geological formation I have seen in the north-west. We see from the above estimate that it renders barren over thirty thousand square miles of the valley of the Missouri.

The fossils of this formation are too numerous to mention in detail. The upper and lower members appear to be exceedingly fossilliferous, while the intervening portions of considerable thickness contain only a few imperfect specimens of *Cephalopoda* and the bones of *Mosasaurus Missouriensis*. The entire thickness of this formation may be estimated at about three hundred and fifty feet.

The formation is probably more extensively developed in Rupert's Land than any other member of the Cretaceous Series.

The most easterly exposure, where it holds characteristic fossils, is on the Little Souris. Fifteen miles from the mouth of that river it consists of a very fissile, dark-blue argillaceous shale, holding numerous concretions containing a large percentage of iron. Some very obscure fossils were found in it, with fragments of *Inoceramus Canadensis*.* The shale weathers ash-white; and the exposure on the Little Souris is seventy feet thick in horizontal layers.

Where the river has excavated a passage through the Blue Hills of the Souris, the rock frequently occurs in cliffs, the dip being 3° south. Fragments and perfect forms, but very fragile, of *Inoceramus Canadensis*, (Meek,) are very common. The ferruginous concretions are disposed in regular layers, and constitute a marked feature of the rocks of this valley.

A few miles west of the Blue Hills the dip of a very remarkable exposure of shale, with bands of ferruginous concretions, facing the south, was levelled with the utmost care, and found to be perfectly horizontal. At the base of the exposure, and on a level with the water's edge, a layer occurs full of gigantic *Inoceramus*, probably the same species as those before mentioned. One specimen measured eight inches and a half in diameter, it was very fragile; but the peculiar prismatic structure of the shell was remarkably well preserved. On attempting to raise it, it separated into thousands of minute prisms.

A search for fossils here was more successful, and resulted in the discovery of several new species, which are named and described in Chapter XIX, by Mr. Meek. Among the fossils were *Anomia Flemingi*, (Meek,) N. sp.; *Inoceramus Canadensis*, (Meek,) N. sp.; *Leda Hindi*, (Meek,) N. sp.

* See Chapter XIX, by Mr. F. B. Meek.

On the Two Creeks, an affluent of the Assiniboine, the same formation exist. Among the fossils collected there were *Natica obliquata, Anvellana concinna, Ammonites* (sp. undt), &c. &c.

On the Qu'Appelle River this rock is again seen, below the Big Cut-arm Creek, and also near the Scissors Creek. Although no organic forms were procured, yet the lithological aspect of the rock is the same as on the Little Souris. The same remark applies to the outcrop on the Riding Mountain.

An exposure, a few miles below the mouth of the "River that Turns" on the South Branch of the Saskatchewan, contains at its base a hard Calcareous Sandstone, containing *Avicula Linguæformis*, below it is a soft sandstone destitute of fossils. This section is described on page 62. It is not improbable that the strata above the second concretionary layer pass into formation five of the Nebraska section, and represent the upper Cretaceous in this region.

Formation No. 5 of Vertical Section.

This very interesting bed, though differing lithologically from the preceding one, contains many of the same species of fossils. It is worthy, however, of a distinct position in the series, not only from its extent, thickness and difference of composition, but also from the more favorable influence that it exerts upon the country underlaid by it. In ascending the Missouri River it first makes its appearance near the mouth of Grand River, about one hundred and fifty miles above Fort Pierre. Near *Butte aux Grès* it becomes quite conspicuous, acquiring a thickness of eighty or one hundred feet, and containing great quantities of organic remains. Here it forms an extension of what is called Fox Ridge, a series of high hills having a north-west and south-west course, crossing the Missouri River into Minnesota at this point. Its north eastern limits I have not ascertained. In its south western extension it continues for a considerable distance nearly parallel with the Missouri, crosses the Moreau River about thirty miles above its mouth, then forms a high dividing ridge betwen the Moreau and Sheyenne Rivers, at which locality it first took its name. Continuing thence its south westerly course, it crosses the Sheyenne, and is seen again in its full thickness at the heads of Opening Creek and Teton River, forming a high ridge from which tributaries of the Sheyenne and Teton take their rise. The little streams flowing into the Sheyenne have a north westerly course, while those emptying into the Teton take a south easterly direction. We thus find that this bed underlies an area of about two hundred miles in length and fifty miles in breath, or about ten thousand square miles.

The general character of Formation No 5 is a yellow arenaceous and argillaceous grit, containing much ferruginous matter, and in localities a profusion of Molluscous fossils. It forms a much more fertile soil, more hearty and luxuriant vegetation, sustains a finer growth of timber than Formation No. 4, and abounds in springs of good water.

Like No. 4, this bed yields a great abundance of quite perfect and well preserved organic remains. Many of the species approximate so closely to Tertiary forms, that did we not find them everywhere associated with Ammonites, Scaphites and other genera which are not known to have existed later than the Cretaceous epoch, we should at once pronounce the formation in which they occur Tertiary. The whole thickness of this bed is estimated at one hundred to one hundred and fifty feet.

The first exposure of this formation is probably found on the Eyebrow Hill stream, where it joins the Qu'Appelle Valley. A ferruginous clay in yellow and red layers reposes on a hard greenish coloured sandstone, seamed with veins of Selenite, and containing huge concretions. No fossils were found in the rock.

The upper part of the section on the South Branch containing concretions full of *Avicula Nebrascana* is doubtless the representation of No. 5 in this region. A description of this section is given on page 62, and of another, fifty miles from the Qu'Appelle on the South Branch, on page 63. Among the specimens procured from the South Branch belonging to this formation were *Scaphites Conradi*, *Nautilus Dekayi*, *Avicula linguæformis*, *Avicula Nebrascana*, *Rostellaria Americana*.*

* See Chapter XIX., by Mr. Meek.

THE TERTIARY SERIES.

No evidence of Tertiary rocks in position east of the South Branch of the Saskatchewan was obtained during the exploration. On an island in the prairie called the Wood Hills, referred to in chapter II, Lignite is reported to exist in position, and the fragments showed me by Charles Pratt were similar to those obtained from the boulder Lignite on the Little Souris. On the crest and abrupt sides of the Riding, Porcupine and Thunder Mountains, the Indians affirm that beds of Lignite exist, a statement rendered probable by the occurrence of worn fragments in the drift of the valleys of the rivers flowing from those eminences.

The sand dunes which form so distinguishing a feature near the Elbow of the South Branch may have been derived from Tertiary sandstones formerly overlying the upper Cretaceous rocks in that vicinity. West of the South Branch, sand hills, quite bare, and certainly not less than 100 feet high were seen at a considerable distance, and also numerous sand hills were observed south of the Qu'Appelle, east of the Elbow of the South Branch. In a foot note, on page 139 of the Geology and Palæontology of the Mexican Boundary Line, Prof. James Hall says that the drifting sands of the south-west, like those of the north appear to be derived from the sandstones of the Tertiary period.

No rock was seen in position on the Eyebrow Hill Range, although, from the circumstance that upper Cretaceous rocks occur *in situ* in the Qu'Appelle Valley, five miles north-west and 300 feet below the summit of the Ridge, there is little reason to doubt, that as on the Grand Coteau de Missouri, of which the Eyebrow Hill Range is a northerly extension, Tertiary rocks in position do exist there.

Sand hills and dunes form an important physical feature in the surface Geology of the part of Rupert's Land under consideration. In a former chapter a short notice is given of their distribution, and reference is here made to it in view of the probable relationship which may ultimately be established between sand dunes and hills and the remains of former Tertiaries. If future investigations should establish the origin of these sand dunes and hills, and show that they are the widely distributed remains of Tertiary rocks, the antiquity of the valley of the Qu'Appelle will be cleared of much doubt.

LIGNITE.

Although the Lignites are not generally available for economic purposes, yet some seams sufficiently pure for use are known to exist in the great Lignite basin of the Upper Missouri. A brief notice of the character of this important material as it occurs in the Tertiary rocks of the north-west, will enable a tolerably accurate judgment to be formed of its probable value as a source of fuel in Rupert's Land.

The great Lignite Basin of the Missouri extends from the 100th to the 108th degree of west longitude, and from the 45th degree of north latitude to an undescribed limit, probably through the valley of the Saskatchewan to the valley of the Mackenzie.

Dr. Hayden, who traced the great Missouri formation up that river for a distance of six hundred miles, and up the Yellowstone for three hundred miles, considers that the fossils obtained from it show conclusively that it possesses the mixed character of a fresh water and estuary deposit, and that it cannot be older than the Miocene period. It is composed of clays, sands, sandstones and Lignites. The extent of country known to be occupied by this basin, as it occurs on the Missouri and its tributaries exceeds sixty thousand miles. The beds of Lignite in this extensive formation vary in thickness as well as in purity at different localities. On the Yellowstone they are found seven feet in thickness. At Fort Berthold on the Missouri a two-foot bed is pure enough to be used as fuel.*

Governor Stevens, in his Report of the Exploration of a route for the Pacific Railway, says that Lignite has been traced from the Coulées of the Mouse River to the head waters of Milk River, a distance of five hundred miles, *apparently* underlying the whole of that extensive district of country, with a thickness of bed varying from a few inches to six feet; he regards it as a source of fuel not to be overlooked.†

I do not enumerate the Lignites described by Sir John Richardson and others as occurring at Edmonton, and various places on the North and South Branches of the Saskatchewan, for the obvious reason that no doubt by this time a full and complete description of their value as a source of fuel on the North Branch, has been already prepared by Dr. Hector, who would enjoy unusual facilities when at Edmonton for studying their development and economic value. On the South Branch they are said to exist, by Sir Alexander Mackenzie, in long. 116° W; but as the country between the Elbow and the mouth of Bow River is still a *terra incognita*, it is not improbable that important Lignite beds may be found much further east than the longitude specified by that illustrious traveller.‡

At Nanino, Vancouver's Island, Lignite beds, long conjectured to be of Tertiary age, have been worked to some extent for the San Francisco market, and to supply steamers which touch there.‖ The doubts which have existed respecting the age of the Vancouver Coal have recently been set at rest by Mr. Bauerman, who in a geological description of a part of Vancouver's Island, transmitted to Sir Roderick Murchison, confirms the opinion that the Coal of Vancouver is of Tertiary age.§

Lignite exists in abundance on the Rio del Norte, the river forming part of the boundary line between the United States and Mexico. Some specimens are so bituminous as to be of no use in the blacksmith's forge, where it runs together and becomes baked into a solid mass. Seams of Lignite 3 to 4 feet thick are exposed on Elm Creek, a tributary of the Del Norte, and have been used and found valuable in a blacksmith's forge. This Lignite occurs in Cretaceous formations.

In Europe, Tertiary Lignite deposits possess considerable economic value. They are worked in France, Germany and Switzerland. In England, the Lignites of Devonshire, associated with beds of clay, are about seventy feet thick. The strata of Lignite coal near the surface vary from eighteen inches to four feet in thickness, separated by beds of brownish clay of about the same dimensions. The lowermost stratum of Lignite coal is sixteen feet thick.¶

* Page 9. Remarks on the Tertiary and Cretaceous Formations of Nebraska, &c., by F. B. Meek and F. V. Hayden, M.D.

† Pacific Railway Reports. Vol. I, page 95.

‡ Foot Note, page 110, Am. Ed. Sir John Richardson's Arctic Searching Expedition.

‖ Pacific Railway Report, Vol. VI, Geological Report.

§ Sir Roderick Murchison's Address—at the Anniversary Meeting of the Royal Geographical Society.

¶ Phillips.

CHAPTER XIX.

REMARKS ON THE CRETACEOUS FOSSILS COLLECTED BY PROFESSOR HENRY Y. HIND, ON THE ASSINIBOINE AND SASKATCHEWAN EXPLORING EXPEDITION, WITH DESCRIPTIONS OF SOME NEW SPECIES.

BY F. B. MEEK.

Remarks — List of Fossils collected — Plants — Mollusca — Anomia Flemingi—Inoceramus Canadensis—Avicula linguæformis — Avicula Nebrascana — Leda Hindi — Leda Evansi—Rostellaria Americana—Natica obliquata—Avellana concinna—Ammonites Placenta—Ammonites, sp. undt.— Ammonites Barnstoni — Ammonites Billingsi — Scaphites nodosus—Scaphites Conradi—Nautilus Dekayi.

The specimens submitted by Professor Hind from the Assiniboine and Saskatchewan country, together with a portion of the same collection previously sent by Mr. Billings to Dr. Hayden and the writer, establish the fact of the existence in that region, of three of the five subdivisions into which the Cretaceous rocks of Nebraska are separable.* Some of those from a locality on the Assiniboine, one hundred and fifty miles west of Fort Garry, present exactly the lithological characters of Formation No. 2 of the Nebraska section, and contain small scales of fishes undistinguishable from specimens collected in that formation by Dr. Hayden on the Missouri above the mouth of Big Sioux River, and near the Black Hills.

Others more recently sent by Professor Hind, collected on Little Souris River, and near the mouth of the Two Creeks on the Assiniboine, evidently belong to a higher position in the series. Amongst these I recognize *Leda Evansi*, *Natica obliquata* and *Avellana concinna*, all of which occur in the upper part of No. 4 and in No. 5 of the Nebraska section, but are more common in the former. As the matrix in which they occur presents exactly the lithological characters of No. 4, and is quite unlike any part of No. 5 of the Nebraska section, there is little room to doubt that the bed in which they were found, represents the former of these rocks.

Several of the specimens obtained near Sand Hill Lake on Qu'Appelle River, and the South Branch of the Saskatchewan, are from a green sandstone, which is more indurated but in other respects more like the green sands of New Jersey than any I have before seen from north-western localities. In some of these, there are great numbers of *Avicula linguæformis* and *A. Nebrascana*, the first of which occurs in both Nos. 4 and 5 of the Nebraska section, but is more abundant in the latter; and the other is nearly or quite restricted to No. 5, where the two formations are not blended as is sometimes the case. As this rock differs entirely in its lithological characters from Formation No. 4,—while No. 5 is often highly arenaceous, and sometimes assumes a slight greenish tinge, at the higher northern localities in the Upper Missouri country,—the probability is that it represents No. 5, or the most recent member of the Cretaceous series of the north-west.

Amongst the specimens collected on the Saskatchewan are *Ammonites placenta*,* *Nautilus Dekayi*, and apparently a variety of *Scaphites nodosus*, all of which are generally characteristic of the upper part of Formation No. 4, but probably sometimes pass up into No. 5. Others from the same localities contain *Rostellaria Americana* and fragments of *Scaphites Conradi*, which are restricted to No. 5 where these two upper formations are not blended.

Amongst all the collections from this region, I see nothing indicating the existence of Formations Nos. 1 and 3 of the Nebraska series, though they may occur there.

The two *Ammonites* from McKenzie's River, are not alone sufficient to determine the age of the rock from which they were obtained; the larger one bears considerable resemblance in form and general appearance to several Jurassic species, though they may belong to the Cretaceous epoch. It is very desirable that a good series of specimens should be obtained from this remote northern locality, not only for the purpose of determining the age of the formation, but for the light they might throw upon far more interesting questions respecting the probable climatic conditions in these high northern latitudes during the Secondary Period.

LIST OF THE CRETACEOUS FOSSILS COLLECTED, WITH DESCRIPTIONS OF THE NEW SPECIES.

PLANTS.

No. 1.—Several impressions apparently of the stems of marine plants occur in the specimens from the locality on the Assiniboine, near the mouth of the Two Creeks.

No. 2.—Along with the above there are also specimens of a very curious spiral body, differing from any fossil I ever before met with. It is a long, slender, slightly flattened, or subcylindrical body, measuring in every part of its length about 0.18 inch in its greatest diameter, and very regularly

* The Cretaceous series of Nebraska consists of five distinct subdivisions, which have, for convenience, been numbered 1, 2, 3, &c., from the lowest upwards.

* When this specimen was first sent to Dr. Hayden and the writer, we were not aware of the fact that any other Cretaceous fossils had been found in that region, and suggested that it might possibly have been carried north by the Indians from some of the Upper Missouri localities. The other specimens, however, obtained from there, remove all doubts in regard to the existence of Cretaceous rocks on the Saskatchewan.

coiled into a spiral form, the turns being widely disconnected like those of a cork screw. Each turn measures about 0.58 inch across, and there are five turns in a length of 2.15 inches. It is smooth, and shows no organic structure under a common pocket lens, the organic matter having been replaced by the fine sediment of which the matrix is composed. Unless these are the tendrils or root-like appendages by which some floating plant clung to marine bodies, I can form no conjecture in regard to their nature.

MOLLUSCA.

LAMELLIBRANCHIATA.

Gen. ANOMIA.—*Lin.*

No. 3.—Anomia Flemingi, *N. sp.*

Plate 1, Figs. 2 and 3.

Shell oval or sub-circular, compressed plano-convex, extremely thin and fragile. Lower valve flat and apparently more nearly circular than the other. Upper valve depressed convex, rounded in front, and more broadly and less regularly rounded on the ventral side; posterior margin obliquely subtruncate from the dorsal side, rather abruptly rounded, and waved so as to form a broad very obscure fold at its connection with the ventral margin; beak small, compressed, located near the middle of the cardinal edge, but not projecting beyond it. Surface marked by small obscure lines of growth. Length 1.10 inches; breadth from beak to opposite side, one inch.

In formation No. 4 of the Cretaceous beds in Nebraska, there is a species something like this, which Dr. Hayden and I have described (but not yet published) under the name of *A. subtrigonalis*. The species now before me, however, is much more compressed, and more rounded in outline. It differs from *A. tellinoides* of Morton (Synop. Org. Rem., p. 61, pl. 5, fig. 11,) in being straighter on the cardinal side, and in having the umbo of the upper valve much less prominent and gibbous. Named after Mr. John Fleming, one of the gentlemen connected with the Saskatchewan expedition.

Locality and position.—Little Souris River, in soft lead gray argillaceous rock, or indurated clay, probably of the age of the fourth division of the Cretaceous series in Nebraska.

Gen. INOCERAMUS.—*Sowerby.*

No. 4.—Inoceramus Canadensis, *N. sp.*

Plate 1, Figs. 4 and 5.

Shell broad oblong-oval, compressed, apparently very nearly equi-valve; anterior side rounded; posterior side longer and more broadly rounded or sub-truncate; base forming a semi-oval curve; hinge straight, of medium length; beaks small, compressed, scarcely rising above the hinge line, located near the anterior side, not very oblique; surface ornamented by small obscure irregular concentric undulations, and fine closely arranged rather indistinct lines of growth, which are generally only seen on the outer fibrous layer. Length of largest specimen about 3.35 inches; height near 2.80 inches.

The specimens of this species in the collection are imperfect, but retain enough of its characters to show it is distinct from any of the known species in the Nebraska formations. It resembles somewhat *I. Sagensis*, Owen, (Report, Wisconsin, Iowa and Minnesota, Tab. VII. fig. 3,) but is much more compressed, and longer in proportion to its height.

It also bears some resemblance to *I. regularis*, D'Obigny, (Pal. Franc, T. 3, pl. 410,) but is not near so deeply rounded on the ventral border, and is more compressed.

Locality and position, same as last.

Gen. AVICULA, *Klein.*

No. 5.—Avicula linguæformis.

Plate 1, Fig. 6.

Avicula linguœformis, Evans and Shumard, Proceed. Acad. Nat. Sci. Phila. Vol. VII., p. 163.

Locality and position.—Sandy Hills, South Branch Saskatchewan. Height of Land in the Qu'Appelle Valley, near the Elbow of South Branch of the Saskatchewan,—Upper Cretaceous.

No. 6.—Avicula Nebrascana.

Plate 1, Fig. 7.

Avicula Nebrascana, Evans and Shumard, Trans. Acad. Sci. St. Louis. Vol. I., p. 38.

Locality and position.—South Branch of the Saskatchewan.—Upper Cretaceous.

Gen. LEDA, *Schumacher.*

No. 7.—Leda Hindi, *N. sp.*

Plate 1, Figs. 8 and 9.

Shell small, sub-ovate, compressed; anterior side narrowly rounded; pallial border forming a broad semi-oval or semi-ovate curve, not crenulate within; posterior side a little longer than the other, much compressed, distinctly sinuous below, and provided with a narrow, short, obtusely pointed rostriform extension above; umbones depressed, located a little in advance of the middle; hinge having about twelve teeth in front of the beaks, and probably more behind; surface ornamented by distinct, regularly arranged, rather strong concentric lines. Length 0.35 inch; height 0.18 inch.

This is a very neat little shell, which will be readily distinguished from any of the species yet known in the Nebraska Cretaceous rocks, by the distinct sinus in its postero-ventral margin. Even where the border is broken away the curve of the concentric lines will always show that the sinus did exist in its margin.

The specimen does not show the pallial line, but in form and general appearance the shell is more like *Leda* than *Nucula*, it may, however, possibly belong to the latter genus.

The specific name is given in honor of Prof. Henry Y. Hind of Trinity College, Toronto, in Charge of the Assiniboine and Saskatchewan Exploring Expedition, to whose zeal and industry we are indebted for much interesting information respecting the geology and topography of the country explored.

Locality and position.—Little Souris River, from an equivalent to No. 4 of the Nebraska section..

No. 8.—Leda Evansi.

Leda Evansi, Meek and Hayden, Proceed. Acad. Nat. Sci. Phila., Ap. 1856, p. 84.

Locality and position.—South Branch of the Saskatchewan; same geological position as last.

GASTEROPODA.

Gen. ROSTELLARIA.—*Lamk.*

No. 9.—Rostellaria Americana.

Rostellaria Americana.—Evans and Shumard, Trans. St. Louis Acad. Sci., Vol. I. p. 42.

Locality and position.—South Branch of the Saskatchewan, upper Cretaceous.

Gen. NATICA.—*Adanson.*

No. 10.—Natica obliquata.

Natica obliquata.—Hall and Meek, Mem. Acad. Arts and Sci., Boston, Vol. V. n. s. p. 384, pl. 3, fig. 1.

Locality and position.—Two Creeks, on the Assiniboine; in bed representing Formation No. 4 of the Nebraska Cretaceous.

Gen. AVELLANA.—*D'Obigny.*

No. 11.—Avellana concinna.

Acteon concinna.—Hall and Meek, Mem. Am. Acad. Arts and Scien., Boston, Vol. V. n. s. p. 388, pl. 2, fig. 6.

The specimen of this species, first figured in the paper above cited, is either a young individual, or the outer lip was broken away; for that now before me, which is evidently the same species, has a strong thickened outer lip; consequently, it cannot be a true *Acteon*, but agrees more nearly with the characters of the genus *Avellana*.

Locality and position.—Same as last

CEPHALOPODA.

Gen. AMMONITES.—*Bruguiere.*

No. 12.—Ammonites placenta.

Ammonites placenta.—Dekay, N. Y. Lyc. Nat. Hist., Vol. II. pl. 5, fig. 2; Jour. Acad. Nat. Sci., Phila., Vol. VI. p. 88, &c.; Morton, Synop. Org. Rem., p. 36, pl. 2, figs. 1 and 2.

Locality and position.—South Branch of the Saskatchewan, from an equivalent of Formation No. 4 of the Nebraska Cretaceous series.

No. 13.—Ammonites.—Sp. undt. (fragments.)

Locality and position.—Two Creeks, Èq. No. 4 of Nebraska Cretaceous.

No. 14.—Ammonites Barnstoni, *n. sp.*

Plate 2, Figs. 1–3.

Shell compressed-subglabose, broadly rounded on the dorsum, and prominent or subangular around the umbilicus which is deep, conical, and nearly as broad as the outer whorl.' Volutions having their greater diameter at right angles to that of the shell; each of the inner ones about three-fourths hidden in the profound ventral groove of the succeeding turn. Surface ornamented by distinct regular costæ, which are sharply elevated around the umbilicus, into small elongated subnodose prominences; and at less than half the distance across the sides of the whorl, their number is increased nearly threefold by division and implantation; after which they become of uniform size, and arch gently forward in passing over the dorsum.

The septa are deeply divided into five principal lobes and six saddles, which are crowded together, and variously branched and subdivided. The dorsal lobe is a little longer than wide, and has three branches on each side—the two terminal of which are nearly straight and parallel; the first two lateral branches above these, are nearly of the same size, but more diverging; while the third pair are much smaller, —and all sharply digitate, and more or less subdivided. The dorsal saddle is longer than wide, contracted in the middle, and irregularly divided into four unequal branches, the two terminal of which are subdivided into two branchlets each, and all obtusely digitate, and variously sinuous in the margins.

The superior lateral lobe is longer than the dorsal lobe, but very irregularly branched, and, like it, provided with numerous sharp digitations on all its divisions; at the extremity it has three very unequal branches, the middle one of which is much longer than the others, and very slender; the other two are small, unequal, opposite, and diverging;—that on the right being subdivided nearly to its base: above these there are several other unequal alternating lateral branches, one of which on the right side is much larger than the others. The lateral saddle is rather smaller than the dorsal, and divided at the extremity into two very unequal branches, of which the one on the left is larger than the other, and again deeply divided into two bifid and deeply sinuous brachlets. The inferior lateral lobe is much smaller than the superior, and very irregularly divided into two or three alternating unequal lateral branches on each side, and one terminal branch, with numerous sinuosities. The ventral lobe is very small, and simply digitate.

This species bears considerable resemblance in form, and, in the size and character of its umbilicus, to the Jurassic species *A. irens* D'Obigny (Pal. Franc., Tome I. p. 562, pl. 222), but differs in having the costæ pinched up into little subnodose prominences around the umbilicus, and bifurcating on the sides; they are also much more arched in passing over the dorsum. It is quite different from any of the described species from the Nebraska rocks, though I think I have seen some fragments of it in Lieut. Warren's collections from No. 4 of the Nebraska Cretaceous subdivisions.

The specific name is given in honour of Mr. Geo. Barnston, chief factor of the Hudson's Bay Company, who discovered it in the valley of Mackenzie's River. It is probably a Cretaceous species, but may be of Jurassic age.

No. 15.—Ammonites Billingsi, *n. sp.*

Plate 2, Figs. 4, 5 and 6.

Shell moderately compressed, or subdiscoidal; dorsum rounded; umbilicus very small; volutions having their greater breadth at right angles to the shorter diameter of the shell, increasing rather rapidly in size, or more than doubling their diameter each turn; inner ones entirely embraced, and hidden in the ventral groove of the last turn; surface apparently smooth, but showing very faint traces of radiating costæ, which arch a little in crossing the dorsum.

Dorsal lobe longer than wide, provided with three branches on each side, the two terminal of which are much longer than the others, and each subdivided,—the subdivisions being short, and each having two or three small digitations; the first two lateral branches above these are small, opposite, very diverging, and bifid or digitate; and the third pair very small, and apparently simple. The dorsal saddle is as long as the dorsal lobe, but narrower, and has three or four short obtusely rounded branches on each side. The superior lateral lobe is nearly as large as the dorsal saddle, and has three subequal branches at the extremity,—that on the dorsal side being bifurcate, with digitate divisions; and the middle, and other lateral divisions, are provided with three or more small digitations each. The inferior lateral lobe is much smaller than the superior lateral, and has much the same form, excepting that its terminal division is proportionally larger, and the principal lateral division on the dorsal side is not so deeply divided. The ventral lobe is a little smaller, but in other respects very similar to the inferior lateral lobe; between it and the umbilicus there appears to be one or two smaller auxiliary ventral lobes, which seem to show a tendency to branch in the same way as the principal ventral lobe.

The specimen from which the foregoing description was made out, is evidently a young shell; consequently, adult individuals of the same species may be expected to possess much more distinct costæ. The lobes and saddles of the septa, in old shells, will also be found much more deeply divided and more complex, but the mode of branching probably remains the same from the time the principal divisions are formed.

As the specimen described was found in the matrix filling the umbilicus of *A. Barnstoni* (being only 0.67 inch in its greatest diameter), it might be supposed by those who know how widely the *Ammonites* sometimes vary at different ages, that it may be the young of that species. It presents fundamental differences, however, in the mode of branching of the lobes and saddles of its septa, that cannot be due to different stages of development. In addition to this, I found along with it a much smaller specimen, evidently the young of *A. Barnstoni*, which shows that the young of that species did not vary in form materially from the adult, and is quite different from the species now under consideration.

It has much the form of *A. Halli*, Meek and Hayden (Proceed. Acad. Nat. Sci. Phil., Vol. VIII. p. 70), and there are no differences in the structure of the dorsal lobes of the two, that might not be due to different degrees of development. Their superior lateral lobes and dorsal saddles, however, present radical differences, such as we never see in the same species, however widely they may differ in size or age.

I have named this species in honour of Mr. E. Billings, the accomplished Palæontologist of the Canadian Geological Survey.

Gen. SCAPHITES.—*Parkinson.*

No. 16.—Scaphites nodosus? *Var.*

Plate 2, Figs. 7 and 8.

Scaphitus nodosus [?]—Owen, 1852. Rept. Iowa Wiscn. and Min., p. , pl. , fig.

Locality.—South Branch of the Saskatchewan, from an equivalent of Formation No. 4 of Nebraska Sec.

No. 17.—Scaphites Conradi.

Ammonites Conradi.—Morton, 1834. Synop. Org. Rem., p. 39, pl. 19, fig. 4.

Scaphites Conradi.—D'Obigny, 1850. Prodromus, p. 214.

Ammonites Nebrascensis, &c.—Owen, 1852. Rep. Iowa, &c.

Scaphites Conradi.—Meek and Hayden, 1856. Acad. Nat. Sci. Phila., p. 281.

Locality and Position.—South Branch of the Saskatchewan; No. 5, Nebraska Section, or most recent Cretaceous.

Gen. NAUTILUS.—*Bruguiere.*

No. 18.—Nautilus Dekayi.

Plate 2, Figs. 9 and 10.

Nautilus Dekayi.—Morton, 1834. Synop. Org. Rem., pl. viii. fig. 4, and pl. xiii. fig. 4.

Locality and Position.—South Branch of the Saskatchewan;—Upper Cretaceous.

CHAPTER XX.

ON SOME OF THE SILURIAN AND DEVONIAN FOSSILS COLLECTED BY PROFESSOR HENRY Y. HIND, ON THE ASSINIBOINE AND SASKATCHEWAN EXPLORING EXPEDITION.

BY E. BILLINGS, F.G.S.

OFFICE OF THE GEOLOGICAL SURVEY OF CANADA,
Montreal, 15th Nov., 1859.

The Silurian fossils from Lake Winnipeg and the Saskatchewan are interesting, but unfortunately, many of the specimens are in such a bad state of preservation that little can be said about them, except to indicate the species to which they appear to belong. The following constitute the principal part of the collection:

PLANTÆ.

Two species of Fucoids from Punk Island in Lake Winnipeg resembling forms which occur in the Chazy sandstone.

ZOOPHYTA.

The only coral is a species allied to *Columnaria alveolata.* It is from Grindstone Point, Lake Winnipeg.

ECHINODERMATA.

Columns of a large *Glyptocrinus* allied to *G. ramulosus* occur at Punk Island and Grindstone Point, and besides these at the latter locality were found several plates of a *Glyptocystites* closely allied to *G. multiporus.*

BRACHIOPODA.

Two specimens of a plaited Rhynconella a little smaller than *R. plena* were found at Punk Island.

LAMELLIBRANCHIATA.

MODIOLOPSIS PARVIUSCULA (N. S.)

This species closely resembles *M. modiolaris* (Conrad) but is always much smaller. It is transversely elongate, anterior extremity small, rounded half the width of the posterior; the latter obliquely truncated and somewhat straight from the end of the hinge line for rather more than half the width, then rounded at the lower posterior angle. Hinge line straight or a little arched, full three-fourths the whole length of the shell. The umbones are less than one-fifth the length from the anterior extremity. The valves are moderately convex, obscurely and obliquely carinate from the umbones towards the lower posterior angle. In many specimens the ventral margin is concave near the anterior extremity, as if for the purpose of a byssus. Surface with obscure concentric undulations of growth. Length of large specimen, one and a half inch. In general they are a good deal smaller.

This shell so much resembles *M. modiolaris* that I have long hesitated as to the propriety of giving it a separate name. It is very widely distributed, since we have specimens from Lake Winnipeg at Punk Island, from the Pallideau Islands in Lake Huron where it occurs in strata which hold fossils of the Chazy, Black River and Trenton limestones, and from near Cornwall and the Island of Montreal in the Chazy.

Besides the above there are several small nearly circular fossils from Punk Island, which appear to be casts of some lamellibranchiate shell.

GASTEROPODA.

Trochonema umbilicata (Hall, Sp.) This species occurs at Lake Winnipeg and at the Little Saskatchewan in considerable numbers. A species allied to *Pleurotomaria rotuloides* (Hall) is common at Punk Island, and a *Maclurea* allied to *M. Logani* (Salter), but with more slender whorls was found at Punk Island and the Little Saskatchewan. One of the specimens has the operculum in place, but is destitute of the shell and somewhat distorted. None of the Gasteropoda have the shell preserved.

CEPHALOPODA.

ORTHOCERAS SIMPSONI (N. S.)

Plate 1, Fig. 1.

The specimen is a portion of the siphuncle, nine inches and one-fourth in length, eleven lines in diameter at the larger extremity, and ten at the smaller. It is nearly cylindrical with a broad, shallow constriction above and below each of the narrow annulations which mark the attachment of the septa. There are eight of those septal rings at the following distances from each other, commencing at the smaller extremity. Between the 1st and 2nd, fourteen lines; 2nd and 3rd, twelve lines; 3rd and 4th, ten and a half lines; 4th and 5th, thirteen and a half lines; 5th and 6th, fifteen lines; 6th and 7th, thirteen and a half lines; 7th and 8th, twelve and a half lines. The annulations are nearly at right angles to the length, and we must infer from this fact either that the septa are scarcely at all concave, or that the siphuncle must be central, or very nearly so. If in an orthoceratite the septa are flat then no matter whether the siphuncle be central or not, the septal annulations must be at right angles, but if the septa are concave then the annulations will be oblique if the siphuncle be at all removed from the centre. My impression is, that this is a large orthoceratite with distant septa and a nearly central siphuncle since the annulations have a scarcely perceptible obliquity.

It is one of those species in which the siphuncle became gradually filled with a solid calcareous animal secretion, with the exception of a narrow cylindrical channel along the centre. This central canal is clearly indicated in the specimen, and has a diameter of nearly two lines.

Dedicated to Sir George Simpson, Governor of the Hudson's Bay Company.

Locality and Formation.—Cat Head, Lake Winnipeg, supposed to be Silurian.

Besides the above, there are several other cephalopods, all of which are in a bad state of preservation, and cannot be determined without much study and comparison.

A small serpulites appears to be common at Punk Island; it much resembles the large species of the Chazy limestone.

The occurrence of *M. parviuscula*, *H. umbilicata*, the *Maclurea*, and *Glyptocystites* are quite sufficient to show that the localities where they have been collected are Lower Silurian, and most probably about the age of the Black River and Chazy limestones.

DEVONIAN.

The following are the fossils from Snake Island in Lake Winnipego-sis.

Atrypa reticularis (Linne,) in abundance, both the common form with moderately coarse ribs and the more finely striated varieties, *Atrypa aspera* (Schlotheim). The specimens very closely resemble those figured by Professor Hall, in his new work, the "Geology of Iowa," plate 6, figs. 3, *a*, *b*, *c*, *d*, but are a little more pointed in front. A fine *Orthis* agrees well with the figures and descriptions of *O. iowensis*, (Hall,) Geology of Iowa, plate 2, fig. 4, but is a little longer. The proportions are the same, but the length, breadth and depth are each two lines greater than the figures. Besides these there are fragments of several other Brachiopods, among which are two small species of *Productus*.

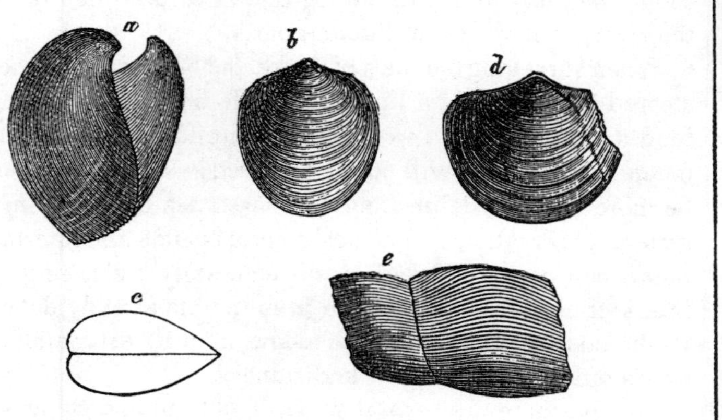

FIGURE 1, FOSSILS FROM SNAKE ISLAND.

Fig. 1, *a*, *Orthis Iowensis*, (Hall) side view.
b, *Lucina occidentalis* (Billings.)
c, *do* *do* Outlines of same, side view.
d, *Lucina elliptica* (Conrad.)
e, *Loxonema nexilis?*

The lamellibranchiate shells are *Lucina elliptica* (Conrad) a species of the corniferous limestone and Hamilton groups of Canada and New York, and a new species of the same genus, which I propose to call *L. occidentalis*.

Of gasteropoda there are two species of *Euomphalus*, and a fragment of a *Loxonema*, most probably, *L. nexilis*.

The Cephalopoda consist of fragments of *Orthoceras*, *Gomphoceras* and a species of *Nautilus* or *Gyroceras*.

Although we have none of the characteristic spirifers corals or trilobites to guide us, yet I think that upon the evidence of the above fossils we can safely say that this locality is Devonian, and most probably about the age of the Hamilton group.

The fossils from the Manitoba Islands are mostly the same as those of Snake Island, with the addition of two species of Chonetes and fragments of a large fish. There is also here a large *Stromatopora*, probably *S. concentrica*.

At Thunder Island, St. Martin's Lake, the Stromatopora occurs, with abundance of a small *Strophomena* and some corals, not determinable.

Lucina Occidentalis. N. S.

Oval, length about one ninth greater than the width, hinge line gently convex, cardinal extremities obtusely rounded, anterior and posterior margins gently convex, sub-parallel ventral margin rounded or a little pointed in the centre; beaks central, small, pointed, incurved, nearly in contact with each other, and turned a little towards the anterior extremity; both valves moderately convex and marked with concentric undulations of growth.

Length of specimen, nine lines, width eight lines, depth of both valves, five lines. The greatest width is at about one-fourth the length below the beaks, from which level the margins converge but little, until within two-fifths of the length of the front, when they become more strongly curved.

Locality and *Formation*, Snake Island, Lake Winnipego-sis. Devonian.

(PRODUCTUS——?)

Supposed to be from Carboniferous Limestone.

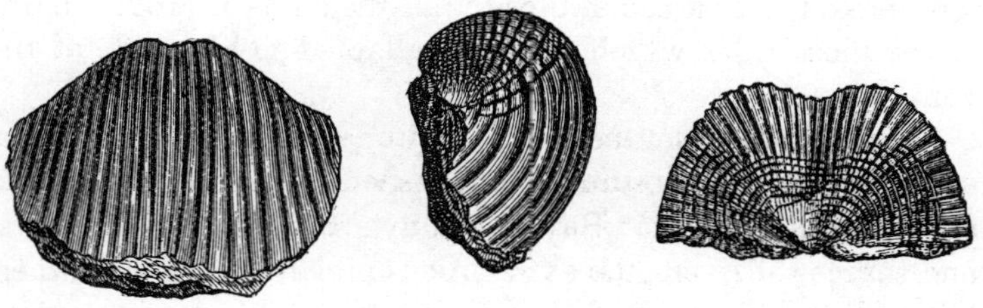

FIG. 1. FIG. 2. FIG. 3.

Fig. 1, *Productus*, From Red River.
Fig. 2, " Side view.
Fig. 3, " Rostral extremity.

There is some evidence of the existence of at least a portion of the carboniferous system in this region. The fossil procured from the half-breed, who said he collected it from "the solid" rock, at some place on the Red River is a *Productus* of the group *Semireticulati*, all of which appear to be confined to the carboniferous series. The specimen is not worn and presents all the appearance of having been freshly broken from the rock. If it were procured from a boulder, then there must be carboniferous limestone north of the locality, as no boulders have travelled from the south.*

E. BILLINGS.

* The following label was attached to this specimen, "Given to me by a Half-breed, who declared he picked it from 'solid rock' in Red River."
(Doubtful,) H. Y. H.

The importance of any evidence of the Carboniferous Series in the Valley of Lake Winnipeg, can not be too highly rated, although I do not think that much reliance is to be placed upon the statement in the foregoing label. 'Solid rock' occurs, as far as known in the Valley of Red River, in two places only, at the Stone Fort, and above, at and below the Rapids, from which places both Dr. Owen and I obtained Lower Silurian fossils from rock in position. It is probable that the specimen was procured from a boulder; but boulders are brought *north* each year from Minnesota by the ice of Red River; it is therefore quite possible that the specimen figured above was brought by ice from the south. Under any circumstances, its presence within 30 miles of the mouth of Red River is an important fact, and affords good ground for hope, that if the Carboniferous Series are not represented on the flanks of the Riding, Duck, and Porcupine Mountains, they will be found in the State of Minnesota, or Dacotah, on the north side of the Height of Land, and in the Valley of Red River. H. Y. H.

APPENDIX.

I.

METHODS TO BE PURSUED IN DETERMINING THE DATA FOR THE BASIS OF THE MAPS AND REPORTS OF THIS EXPLORATION.

In order to determine, within the limited period allotted for field operations, the topographical and geological character of the region indicated for exploration, and to describe faithfully and in detail, its characteristic features and adaptability for settlement, it is necessary that the most expeditous method of conducting the exploratory survey be adopted, combined at the same time with every possible accuracy. As it may become advisable during the progress of the exploration to form different divisions, the following rules and suggestions are designed for general guidance, in order that the explorations and surveys may be made on a uniform system. An extensive equipment of instruments may not be supplied to each observer, he must therefore make the best use of those with which he is provided, and follow those rules which are best adapted to his mode of travelling.

Observations for latitude and longitude should be made whenever there is an opportunity, and especially at such places as the Honorable Hudson Bay Company's Forts, the mouths, forks and sources of rivers, the extremities of lakes, and at prominent hills. The magnetic variation should, if possible, be determined at every convenient camp. The delineation of the topography of the country between established positions is to be accomplished by track-survey. The courses, and cross-bearings to all conspicuous points, are to be taken by magnetic compass, and the intermediate itinerary distances to be ascertained by micrometer, or viameter, or by the measured and corrected velocity of the carts, canoes, or boats. With a view to make a complete reconnaissance of a considerable breadth of country, lateral traverses should be made at stated intervals, on either side of the main lines of exploration.

When surveying rivers or lakes in a boat or canoe, the instruments essentially required for the track, are a watch, a magnetic compass, a log-line and a sounding-line. At every bend of a river the direction of the reach in front is to be taken with the compass, and when the reach is very long the boat must be stopped in order that the course may be taken more accurately. The times of arriving at, and departing from, each bend, or the vertex of two courses, and the length of any halt upon a reach or course, are to be carefully noted. The velocity of the boat is to be determined by the log-line, with which frequent observations are to be made, particularly when any change in the rate is supposed to occur. In rivers it is first necessary to measure the velocity of the current, as it has to be added to, or subtracted from, the *apparent* rate of the boat, indicated by log-line before the true rate is ascertained. The depth, particularly of large rivers and lakes, is to be taken at close intervals, and the height of any water-mark above the present level. The width of the rivers is to be recorded (from measurement when possible) whenever it seems to vary. The height of the banks and flood-marks are also to be noted. The position and dimensions of islands, tributary streams, sand-bars, boulders, &c., are to be ascertained. It being very difficult to estimate correctly the fall or length of swift rapids, it will be necessary to make instrumental observations for this purpose, at least whenever it is possible to do so, and when they occur on large rivers, very particular descriptions of them, and their portages, if there are any, should be given. Accurate cross-sections of rivers, with the mean rate of current at each place, should be made as frequently as possible. Whenever it can be done, it would be most desirable in addition to taking cross-sections and rate of current, to ascertain by levelling, the fall of the river in some *measured* distance as a quarter, or half a mile. These observations and measurements will be of the greatest use in determining the descent in rivers whose general dimensions and rate of current are known, thereby enabling sections or profiles to be made of them hereafter. In ascertaining the rate of current, it should be measured with the log-line at certain intervals *across* the river, as it varies in different parts.

When surveying the coast of a lake, the boat or canoe should be steered in as straight a line as possible from one point or headland to another, and propelled at a uniform rate, so that the compass or log-line will not be required so often, and there will be more time for delineating the coast, taking soundings, and general observations. The positions of islands and intermediate points can be established more accurately by taking several intersecting bearings to them from points already determined on the course, which is the base-line, than by estimation, as the eye is oftentimes deceived in distances.

On land there are several ways of obtaining distances expeditiously, differing in accuracy according to the nature of the ground. In an *open hilly* country, Rochon's micrometer-telescope is the best, but it may be found to retard progress. On *level* ground a viameter gives very accurate results; there are many occasions, however, when it cannot be used. Determining the track distances by the time and rate of travelling, will probably be the method most used on this survey. The rate therefore at which the carts travel should be known as near as can be, and should be adhered to as much as possible. Three miles an hour is the average rate at which horses walk, but it can be tried occasionally by timing them on a *measured* distance. Due allowances must of course be made for undulations in the ground and the windings of the track. The position of distant hills or other conspicuous objects, and the width of valleys, should be determined by triangulation when the ground is suitable for measuring a base-line. The heights of hills or mountains, and the depths of valleys, should be computed trigonometrically when the level or barometer is not used. The names of all rivers, lakes, etc., should be ascertained from the Indians or Half-breeds, and information procured from them relative to those parts not explored. The approximate positions and dimensions of lakes, rivers, hills, etc., according to the Indians and others, may be made use of in constructing a map of the country, but it should be strictly mentioned, and nothing should be laid down as a fact, which has not been surveyed and examined.

In addition to the topographical, geological and general character of the region to be explored (the nature of the soil, timber, vegetation, economic materials, etc., etc., specified in the general instructions, and of which *exact* descriptions should be given) it is unnecessary to state in detail what should be observed in the country, as everything should be noted. The field-books, of which different kinds are provided for the several methods of surveying, must be kept in such a clear manner that the notes recorded can be understood and plotted by other persons than the observer if necessary.

II.

TABLE OF THE PORTAGES, DÉCHARGES, RAPIDS, LAKES, LAKE STRAITS AND NAVIGABLE CHANNELS ON THE PIGEON RIVER ROUTE (THE OLD NORTHWEST COMPANY'S ROUTE) FROM LAKE SUPERIOR TO RAINY LAKE, SHEWING THEIR LENGTHS AND DISTANCE FROM LAKE SUPERIOR.*

Portages.	Length in St. miles.	Décharges.	Length in St. miles.	Rapids.	Length in St. miles.	Lakes.	Length in St. miles.	Lake Straits and Navigable Channels.	Length in St. miles.	Distance from Lake Superior. Statute miles.	Remarks.
No. 1. Grand Portage	8.16									0.00	
								1. Pigeon R.	1.50	8.16	
2. Partridge Portage.	0.25									9.66	
								2. Pigeon R.	4.25	9.91	Strong current. Canoes poled up part of the way.
		No. 1. Décharge	0.25							14.16	
								3. Pigeon R.	1.75	14.41	
		2. Décharge	0.35							16.16	
								4. Pigeon R.	3.13	16.51	
3. Fowl Portage	1.13									19.64	
						No. 1. Fowl Lake	4.55			20.77	
4. Moose Portage	0.41									25.32	
						2. Moose Lake	4.25			25.73	
5. Great Cherry Por.	0.48									29.98	
						3. Lake	0.25			30.46	
6. Mud Portage	0.15									30.71	
						4. Lake	0.30			30.86	
7. Lesser Cherry P.	0.13									31.16	
						5. Mountain L.	7.81			31.29	
8. Watap Portage	0.30									39.10	
						6. Watap Lake	3.70			39.40	
9. Great New Por.	1.46									43.10	
						7. Rose Lake	3.00			44.56	
10. Portage	0.01									47.56	4 feet fall.
						8. Mud Lake	2.62			47.57	
11. Portage	0.21									50.19	
						9. South Lake	2.84			50.40	
12. Height of Land P.	0.26									53.24	
						10. Gun Flint L. E	2.77			53.50	
		3. Décharge	0.01							56.27	4 feet fall.
						11. Gun Flint L. W.	8.92			56.28	
				No. 1. Rapid	0.01					65.20	2 feet fall. North canoes let down by line.
								5. Lake Strait	0.25	65.21	
13. Little Rock Por.	0.02									65.46	
								6. Lake Strait	0.75	65.48	
				2. Rapid	0.01					66.23	
								7. Lake Strait	0.25	66.24	
14. Mill Fall Portage	0.06									66.49	
								8. Lake Strait	0.25	66.55	
15. Island Portage	0.29									66.80	
						12. Lake	1.35			67.09	
				3. Rapid	0.02					68.44	3 ft. fall, 2 ft. deep. North canoes let down by line.
						13. Lake	0.85			68.46	
				4. Rapid	0.07				0.75	69.31	5 ft fall, 2 chs. wide. North canoes let down by line.
								9. Lake Strait		69.38	
				5. Rapid	0.05				0.50	70.13	4 ft. fall, 25 ft. wide North canoes let down by line.
								10. Lake Str.		70.18	
				6. Rapid	0.04				0.12	70.68	3 ft. fall. North canoes let down by line.
								11. Lake Str.		70.72	
				7. Rapid	0.01				0.20	70.84	1 ft. fall. Run by loaded North canoes.
								12. Lake Str.		70.85	
				8. Rapid	0.01				0.45	71.05	1 ft. fall. Run by loaded North canoes.
								13. Lake Str.		71.06	
				9. Rapid	0.02					71.51	2½ feet fall.
						14. Lake	5.35		0.70	71.53	
		4. Décharge	0.05							76.88	5 feet fall.
								14. Lake Str.		76.93	

* The distances are from the International Boundary Survey—made according to the 7th article of the Treaty of Ghent.

TABLE OF PORTAGES, &c.—(*Continued.*)

Portages.	Length in St. miles.	Decharges.	Length in St. miles.	Rapids.	Length in St. miles.	Lakes.	Length in St. miles.	Lake Straits and Navigable Channels.	Length in St. miles.	Distances from Lake Superior. Statute miles.	Remarks.
No. 16. Portage	0.01									77.63	
						No. 15. L. Seiganagah	10.07			77.64	
17. Portage	0.01									87.71	
						16. Swamp Lake.	0.82			87.72	
18. Swamp Portage..	0.24									88.54	
						17. Cypress Lake	5.35			88.78	
19. Portage	0.02									94.13	
						18. Knife Lake..	10.70		0.20	94.15	
20. Portage	0.04									104.85	
								15. Lake Str.		104.89	
21. Portage	0.09									105.09	
						19. Lake	0.60			105.18	
22. Carp Portage ...	0.15									105.78	
						20. Birch Lake.	4.00			105.93	
23. Portage	0.11									109.03	
						21. Basswood L.	16.00			110.04	
24. Portage	0.10									126.04	
								16. Lake Str.	0.25	126.14	
25. Fir Portage	0.20									126.39	
								17. Lake Str.	0.22	126.59	
				No. 10. Rapid	0.04					126.81	3 feet fall.
								18. Lake Str.	0.35	126.85	
				11. Rapid	0.02					127.20	2 feet fall.
								19. Lake Str.	1.50	127.22	
26. Portage	0.07									128.72	
						22. Lake	0.65			128.79	
27. Portage	0.09									129.44	
						23. Crooked L..	16.80			129.53	
28. Curtain Fall Por.	0.10									146.33	
				12. Rapid	0.03					146.43	3 feet fall.
						24. Iron Lake...	4.50			146.46	
29. Bottle Portage...	0.25									150.96	
						25. Nequaquon L	22.10			151.21	
30. Portage	0.12									173.31	
						26. Lake	5.60			173.43	
31. Portage	0.15									179.03	
								20. Lake Str.	0.50	179.18	
32. Portage	0.04									179.68	
								21. Loon's Narrows ...	7.60	179.72	
						27. Sand Point L	9.67			187.32	
						28. Nameukan L	5.20			196.99	
33. Nu Portage	0.08									202.19	
								22. Lakelet...	0.25	202.27	
34. Portage	0.14									202.52	
								23. Lake Str.	5.20	202.66	
Total	15.33	Total	0.66	Total...	0.33	Total	160.62	Total..	30.92	207.86	Statute miles to Rainy Lake.

SYNOPSIS OF THE FOREGOING TABLE OF THE PIGEON RIVER ROUTE.

	Statute miles.	
Land Carriage	15.33	16.32 stat. miles Interrupted Navigation.
Décharges	0.66	
Rapids	0.33	
Lakes	160.62	191.64 stat. miles Navigable Water.
Lake Straits	30.92	

Aggregate distance... 207.86 from Lake Superior to Rainy Lake.

Distance from Lake Superior to Rainy Lake, *viâ* the Kaministiquia Route, = 263.34 statute miles.

III.

INDIAN SUMMER.

Indian summer is a phenomenon of constant yearly occurrence and marked characteristics in the North West. The following table, kindly furnished from the private memoranda of Mr. James Walker, Assistant at the Provincial Observatory, establishes the fact that the hazy, warm, mellow weather we term Indian Summer is a periodical phenomenon in Canada, but the cause does not appear to be quite understood. The characters of Indian Summer are more decided in the North West than in the neighbourhood of Lake Ontario. Sounds are distinctly audible at great distances; objects are difficult to discern unless close at hand; the weather is warm and oppressive, the atmosphere hazy and calm, and every object appears to wear a tranquil and drowsy aspect.

INDIAN SUMMER AT TORONTO.

1840 to 1859 inclusive.
[20 years.]

Year.	Commencement.	Termination.	No. of days.	Remarks.
1840......	1st November ...	5th November ...	5	
1841......	29th October	2nd November ...	5	
1842......	28th October	4th November ...	8	
1843......	23rd October	25th October	3	
1844......	22nd October	26th October	5	And 2nd to 7th Nov. (6 days.)
1845......	24th October	29th October	6	
1846......	4th November ...	7th November ...	4	
1847......	28th October	31st October	4	
1848......	20th November ...	23rd November ...	4	
1849......	13th November ...	18th November ...	6	
1850......	7th November ...	13th November ...	7	
1851......	6th October	11th October	6	
1852......	16th November ...	21st November ...	6	
1853......	12th October	20th October	9	Well marked.
1854......	24th October	28th October	5	Not well marked.
1855......	16th October	26th October	11	Not well marked.
1856......	19th October	22nd October......	4	Very dense fog.
1857......	5th October	12th October	8	And 2nd to 8th Nov. (7 days.)
1858......	18th October	28th October	11	
1859......	2nd November ...	8th November ...	7	Well marked.
Mean result ...	27th October	2nd November ...	6 days.	J. W.

IV.

I.—TABLE OF MAGNETIC VARIATIONS.

Locality.	N. Latitude.	W. Long.	Variation.	Date.
Toronto	43° 39′ 24″ ...	75° 17′ 33″...	2° 06′ W....	1859.
Drummond's Island (Lake Huron)	46° 00′ 00″ ...	84° 00′ 00″...	00° 00′	Line of no variation, 1851.*
Fort William (Lake Superior)	48° 23′ 30″ ...	89° 27′ 10″...	8° 45′ E.†..	
Dog River (Foot of Dog Lake)	....	....	7° 1′ E.‡..	
Kaministiquia (Height of Land)	48° 56′ 00″ ...	....	7° 26′ E.‡..	
Rainy Lake	....	....	10° to 12° E.§	
Assiniboine River	49° 46′ 19″ ...	98° 20′ 00″...	13° 00′ E. ...	June 20, 1858.
Little Souris River	49° 41′ 00″ ...	99° 35′ 00″...	15° 00′ E. ...	June 25, "
Camp 117. Cape Kitchinashi, Lake Winnipeg	53° 8′ 00″ ...	97° 28′ 00″...	15° 00′ E. ...	August 23, "
Camp 125. Point Wigwam, Lake Winnipeg	52° 10′ 00″ ...	97° 39′ 00″...	15° 00′ E. ...	September 2, "
Camp 200. Lake Manitobah..................................	51° 17′ 00″ ...	98° 54′ 00″...	15° 00′ E. ...	October 23, "
Camp 69. Little Saskatchewan	50° 33′ 15″ ...	100° 6′ 00″...	15° 30′ E. ...	August 12, "
Camp 197. Waterhen River	51° 54′ 00″ ...	99° 55′ 00″...	16° 15′ E. ...	October 19, "
Camp 17. Red Deer's Head River...........................	49° 1′ 44″ ...	100° 55′ 00″...	16° 53′ E. ...	July 2, "
Camp 25. Fort Ellice	50° 23′ 39″ ...	101° 15′ 00″...	17° 30′ E. ...	July 11, "
Camp 31. Qu'Appelle Mission	50° 49′ 40″ ...	103° 27′ 00″...	18° 00′ E. ...	July 19, "
Camp 55. Fort Pelly	51° 47′ 22″ ...	101° 56′ 00″...	19° 30′ E. ...	August 2, "
Camp 62. Fort à la Corne	53° 30′ 00″ ...	104° 30′ 00″...	22° 30′ E. ...	August 7, "

* J. W. Foster and J. D. Whitney, 1851.

† Bayfield, 1824.

‡ Murray, 1846.

§ Thompson, 1826.

II.—MEMORANDA RELATIVE TO THE ABSOLUTE MAGNETIC DECLINATION AT TORONTO, FROM 1840 TO 1859.

MONTH.	1840.	1841.	1842.	1843.	1844.	1845.	1846.	1847.	1848.	1849.	1850.	1851.	1852.	1853.	1854.	1855.	1856.	1857.	1858.	1859.
		W.	W.			° ′	° ′	° ′	° ′	° ′	° ′	° ′	° ′	° ′	° ′	° ′	° ′	° ′		
January	....			....	....	1.26.0	1.31.5	1.32.2	1.34.5	1.38.4	1.36.5	1.39.5	....	....	1.50.9	....	1.54.5	1.58.6		
February	....	1.12.0	1.16.6	....	....	1.24.4	1.29.4	1.33.1	1.35.0	1.35.2	1.37.5	1.41.4	....	....	1.45.0	....	1.55.2	1.59.1		
March	....			....	....	1.28.5	1.29.1	1.32.0	1.34.6	1.36.8	1.38.5	1.39.6	....	....	1.48.6	....	1.55.3	2.00.9		
April	....			....	....	1.29.8	1.30.0	1.33.2	1.35.6	1.35.4	1.37.5	1.40.5	....	....	1.47.3	....	1.56.1	1.59.4		
May	....	1.14.9	1.19.4	....	....	1.30.3	1.29.9	1.32.2	1.34.6	1.37.0	1.37.1	1.40.9	....	....	1.47.5	....	1.55.8	1.58.4		
June	....			....	...	1.29.2	1.28.8	1.32.3	1.35.2	1.36.1	1.38.1	1.41.0	....	W.	....	....	1.55.9	1.58.7		
July	....			....	....	1.29.1	1.31.6	1.32.3	1.34.6	1.36.1	1.36.2	1.40.0	....	1.43.8	....	W.	1.55.3	1.59.3		
August	....	1.15.6	1.18.4	....	W.	1.28.6	1.30.5	1.32.7	1.36.4	1.35.7	1.39.9	1.41.7	....	1.47.3	....	1.51.3	1.55.1	2.01.0		
September	....			....	1°.27′.2*	1.30.9	1.32.1	1.34.2	1.35.7	1.37.2	1.40.4	1.42.3	....	....	....	1.51.3	1.57.6	2.00.4		
October	....			....	....	1.22.2	1.31.7	1.34.8	1.37.3	1.37.2	1.41.2	1.41.6	....	....	....	1.53.4	1.57.0	2.01.6		
November	....	1.14.9	1.21.3	....	....	1.29.9	1.32.3	1.34.7	1.36.2	1.38.5	1.40.7	1.40.1	....	....	....	1.55.2	1.58.0	2 02.5app		
December	....			....	....	1.31.3	1.32.1	1.35.1	1.35.1	1.39.1	1.39.5	1.41.3	....	....	....	1.55.3	1.57.8	2.03.9		
		1.14.3	1.18.9	1°.23′.5	1.27.2	1.29.1	1.30.8	1.33.2	1.35.4	1.36.9	1.38.6	1.40.9	1.43.4	1.45.5	1.47.9	1.53.3	1.56.1	2.00.3	2°.03	2°.06.0

* *The result of the Observations between 31st August and 7th Sept.*—Vol. i. p. 9.

	W.
The Mean Westerly Declination at Toronto in 1841, was	1° 14.3′
" " " in 1857. was	2 0.3
The Mean Annual Increase on 16 years	= 2.9
The Westerly Declination at Toronto, August 1859, may be asserted to be	W. 2° 06′.0

REMARKS AND AUTHORITY.

Extract.—"Every one of the Results show that the Westerly Declination is increasing. Their difference in amount arises partly from error, introduced by variations in the direction of the line of detorsion, and partly from actual Magnetic irregularities. The mean is an annual increase of West Declination, amounting to 4′.77, with a probable error of 0′.24."—*Toronto Observations,* Vol. i. pp. 9, 10, and 11. "*Sabine.*"

REMARKS AND AUTHORITY.

Extracts.

" The Declinations at the Mean Epoch, July 1st, 1848, =1° 34′.91 West."—"*Sabine.*"

" The mean annual increase of West Declination, in the years 1845 to 1851 inclusive = + 1°.952."—*Toronto Observations,* Vol. ii. pp. 3, 4. and 5. "*Sabine.*"

SCOBIE'S ALMANAC, 1848.

" The mean variation at Toronto for the year 1846, was 1° 30′ West. It increases about 3′ every year."—"*Lefroy.*"

1849.

"The mean variation for 1847 = 1° 36′ [?] West ; increasing 3′ every year."

1850.

"The mean variation for 1848, was 1° 34′ West ; increasing 2′ every year."

1852.

"The mean variation for 1850, was 1° 39′."—"*Lefroy.*"

REMARKS AND AUTHORITY,

Extract.—"The annual secular change from July, 1851, to April, 1854 (two years and nine months), was an increase of 2′.54 ; and assuming the circumstances of the new series to be strictly comparable with those of the old series, the increase from April, 1854, to October, 1855, is at the annual rate of 3′.4.

"It seems probable, therefore, that the rate of the secular increase of west declination at Toronto is augmenting."—Vol. iii. page 126. "*Sabine.*"

III.—MAGNETIC VARIATIONS AT TORONTO, FROM 1841 TO 1859 INCLUSIVE.

Year.	Declination.	Annual Difference.	Year.	Declination.	Annual Difference.
	° ′	′		° ′ W.	′
1840			1850	1.38.6	+1.7
1841	1.14.3		1851	1.40.9	+2.3
1842	1.18.9	+4.6	1852	+1.43.4*	+2.5*
1843	1.23 5*	+4.6*	1853	+1.45.5*	+2.1*
1844	1·27.2*	+3.7*	1854	+1.47.9*	+2.4*
1845	1.29.1	+1.9*	1855	+1.53.3*	+5.4*
1846	1.30.8	+1.7	1856	1.56.1	+2.8*
1847	1.33.2	+2.4	1857	2.00.3	+4.2
1848	1.35.4	+2.2	1858	+2.03.0*	+2.7*
1849	1.36.9	+1.5	1859	+2.06.0*	+3.0*

Mean declination in 1841 was	1°.14′.3
Do do 1859 is	2°.06′.0
Increase in 18 years	51′.7
Mean annual increase	+2′.9

* The entries marked thus * are the nearest approximation deduced from the previous annual increase; or they are the means of a series of observations taken during the year.

V.

A LIST OF THE WATER-COLOUR DRAWINGS AND PHOTOGRAPHS ACCOMPANYING THIS REPORT

LIST OF WATER-COLOUR DRAWINGS, BY W. HIND, FROM SKETCHES TAKEN BY J. FLEMING, ASSISTANT SURVEYOR TO THE EXPEDITION.

LAKE WINNIPEG.

1. *The Grindstone Point:* shewing exposures of Limestone and Sandstone. A characteristic scene on the west coast.

2. *Deer Island:* shewing escarpments of Limestone reposing on Sandstone. A characteristic scene on the west side of islands and on the west coast.

3. *Coast Scene near the mouth of Red River:* shewing the increase of land by the throwing up of sand beaches, and the formation of marshes in their rear.

4. *The Cat Head:* shewing the precipitous cliffs of limestone at this point and along the coast.

THE SASKATCHEWAN.

5. *The Grand Rapid of the Saskatchewan.*—Shewing the upper and most precipitous portion of the Grand Rapid, with the perpendicular cliffs of limestone on either side.

6. *The Saskatchewan at Fort à la Corne.*—Shewing the Honourable Hudson's Bay Company's Fort (right bank of the River,) and the Nepowewin Mission (Ch. of Eng.) on the left bank.

7. *Cumberland House.*—One of the principal Forts of the Honourable Hudson's Bay Company, situated on Pine Island Lake, a tributary of the Saskatchewan.

8. *The Pas or Cumberland Missionary Station* (Ch. of Eng.) on the Saskatchewan. On the right bank are Christ Church and the Parsonage.

LAKE MANITOBAH AND ST. MARTIN'S LAKE.

9. *Fairford.*—A Missionary Station (Ch. of Eng.) on the Partridge Crop River; a stream flowing from Lake Manitobah into St. Martin's Lake. (View, looking up the river.)

10. *Fairford.*—Second view (looking down.)

11. *Sugar Island, St. Martin's Lake,* shewing its peculiar rock formation.

THE QU'APPELLE OR CALLING RIVER.

12. *The Valley of the Qu'Appelle at the Mission* (Ch. of Eng.); shewing the character of the excavation, and the treeless prairie on the south bank. Dimensions of Valley, 265 feet deep, 1 mile 21 chains wide.

13. *Qu'Appelle Lakes, Fishing Lake No.* 3.—The Qu'Appelle Lakes are 8 in number and vary from 4 to 16 miles in length, from 15 to 66 feet in depth, and from half a mile to 1½ mile in width.

THE LITTLE SOURIS, OR MOUSE RIVER.

14. *View of the Valley near the Blue Hills of the Souris;* shewing the great treeless prairie extending to the Grand Coteau de Missouri.

15. *The partially wooded Valley of the Little Souris,* near Back-Fat Creek, an affluent from the Back-Fat Lakes.

16. *The Valley of the Little Souris,* in its passage through a portion of the Blue Hills; shewing the character of the excavation.

THE ASSINIBOINE RIVER.

17. *View from the Half-way Bank;* shewing the Great Wooded Valley through which the river meanders. In the distance is Pembina Mountain with the partially wooded country intervening.

18. *Fort Ellice;* on Beaver Creek, a small tributary of the Assiniboine flowing through a very deep but short valley. The Fort is one of the chief provision depôts of the Hon. Hudson Bay Company.

19. *Mode of preparing dried Buffalo Meat* on the prairie; Red River Carts.

20. *Stony Mountain.*

LIST OF PHOTOGRAPHS TAKEN BY HUMPHREY L. HIME,

THE RED RIVER.

1. View of Red River from the Stone Fort.

2. View of Red River from St. Andrew's Church, four miles above the Stone Fort.

3. Red River; Middle Settlement, eight miles below Fort Garry.

4. Freighter's Boat on the banks of Red River, seven miles below Fort Garry.

5. Bishop's Court, (the residence of the Bishop of Rupert's Land) on the banks of Red River.

These Photographs exhibit the general character of the river.

CHURCHES OF SELKIRK SETTLEMENT.

6. Cathedral of St. Boniface (Roman Catholic) and Nunnery on the banks of Red River, opposite Fort Garry.
7. St. John's Church, two miles below Fort Garry. (Ch. of Eng.)
8. Presbyterian Church and Parsonage, seven miles below Fort Garry.
9. St. Paul's Church, Parsonage and School House, 8½ miles below Fort Garry. (Ch. of Eng.)
10. St. Andrew's Church, (Rapids Church,) 16 miles below Fort Garry. (Ch. of Eng.)
11. St. Andrew's Parsonage.

HOUSES AND STORES OF THE SETTLERS.

12. Residence of Chief Factor, (the late Mr. Bird,) Middle Settlement.

13. Residence of Mr. Bannatyne, near Fort Garry.

14. Mr. McDermot's store, near Fort Garry.

15. Quarters of the Assiniboine and Saskatchewan Exploring Expedition, Middle Settlement.

16. Farm Houses and Wind-mills, Middle Settlement.

INDIAN TENTS AND GRAVES.

17. Ojibway Tents on the banks of Red River, near the Middle Settlement.

18. Tents in the Prairie, west of the Settlement.

19. Birch Bark Tents, west bank of Red River, Middle Settlement.

20. Indian Graves, covered with split sticks.

21. Indian Graves, covered with birch bark.

THE PRAIRIE.

22. The Prairie, on the Banks of Red River, looking south.

23. The Prairie, looking west.

FORTS AND STORES OF THE HONOURABLE HUDSON'S BAY COMPANY.

24. Fort Garry: at the confluence of Red River and the Assiniboine.

25. Hon. Hudson's Bay Company's Officers' Quarters: Lower or Stone Fort.

26. Fur Store: interior of Lower or Stone Fort.

NATIVE RACES.

27. John McKay: a Cree Half-breed.

28. Letitia: a Cree Half-breed.

29. Susan: a Swampy-Cree Half-breed.

30. Wigwam: an Ojibway Half-breed, Lake Superior.

31. An Ojibway Squaw with Papoose.

32. Red River Freighter's Boat.

33. Dog Carioles; Expedition returning to Crow Wing, by the winter road.

INDEX.

PLATE I.

EXPLANATION OF PLATE I.

ORTHOCERAS SIMPSONI. (Page 186.)

Figure 1.—A fragment of the siphuncle of this species.

ANOMIA FLEMINGI. (Page 183.)

Figures 2 *and* 3.—Two different specimens of this species. The fine concentric lines represent shading.

INOCERAMUS CANADENSIS. (Page 183.)

Figure 4.—Left valve.

Figure 5.—Right valve. The finer concentric lines represent shading. The small figure ×4 shows the fine striæ seen on the outer fibrous layer of the shell, magnified four times.

AVICULA LINGUÆFORMIS. (Page 183.)

Figure 6.—An imperfect specimen of this species. The lines represent shading.

AVICULA NEBRASCANA.

Figure 7.—Two specimens of this species.

LEDA HINDI. (Page 183.)

Figure 8.—Natural size.

" 9.—The same enlarged.

For Figure 10, see Mr. Meek's remarks, Plants, No. 2, page 182.

PLATE I.

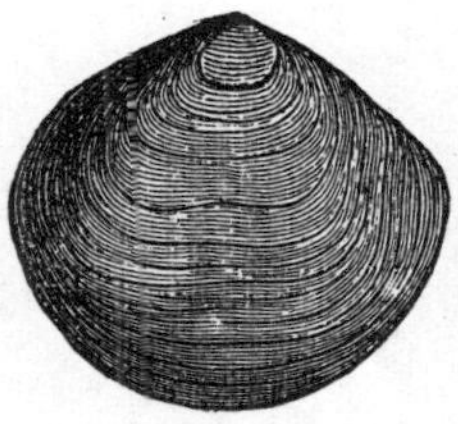

Fig. 2.

Fig. 3.

Fig. 8.

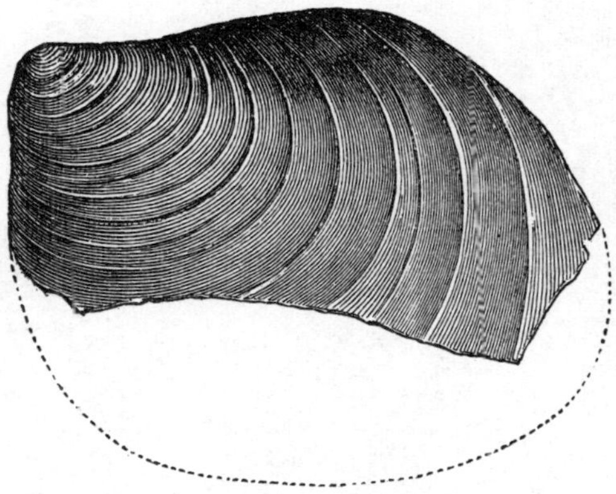

Fig. 4

Fig. 9.

Fig. 6.

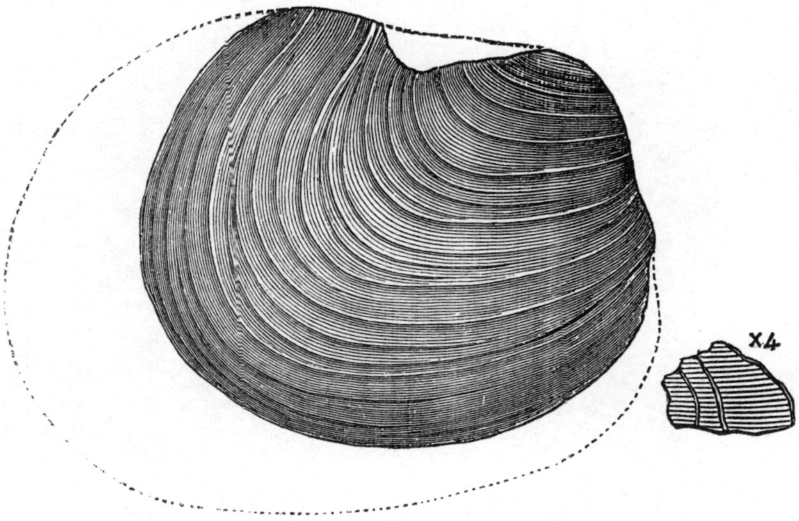

Fig. 5.

Fig. 7.

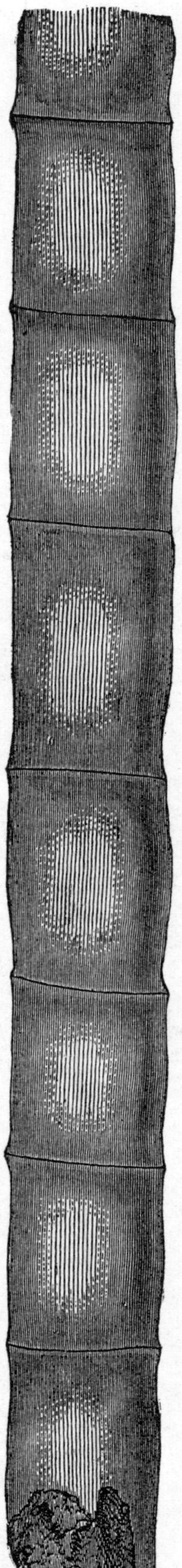

Fig. 1.

Fig. 10.

H. Smith, del., Montreal.

G. G. Vasey, Engraver.

PLATE II.

EXPLANATION OF PLATE II.

AMMONITES BARNSTONI. (Page 184.)

Figure 1.—Side view, shewing the deep umbilicus. | *Figure* 2.—Front view of same specimen.

Figure 3.—Diagram of one of the septa.

AMMONITES BILLINGSI. (Page 184.)

Figure 4.—Side view. | *Figure* 5.—One of the septa enlarged.

Figure 6.—Front view of same specimen.

SCAPHITES NODOSUS [?] *Var.* (Page 185.)

Figure 7.—Side view.

Figure 8.—Front view. The fine lines represent shading.

The specimen figured shows the remains of two rows of tubercles on the dorsum—one on each side. They are much worn, and have been unfortunately omitted altogether by the artist.

NAUTILUS DEKAYI. (Page 185.)

Figure 9.—Front view. | *Figure* 10.—Side view.

PLATE II.

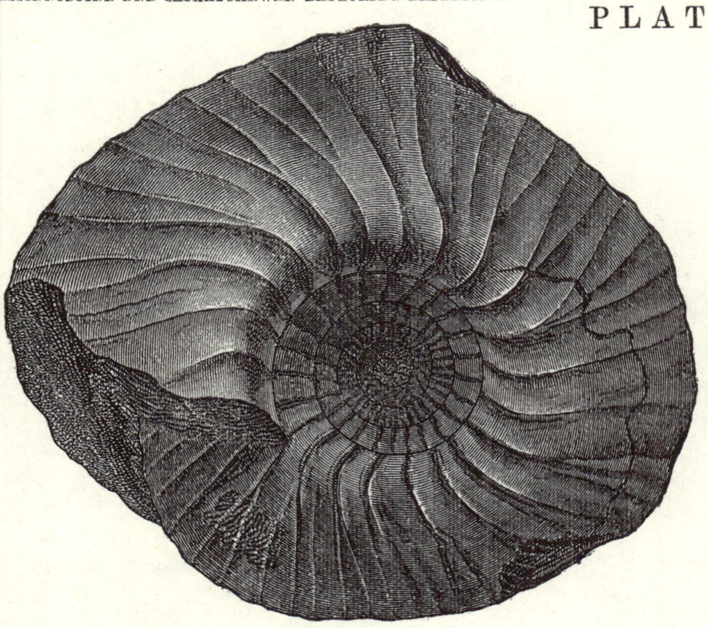

Fig. 1.

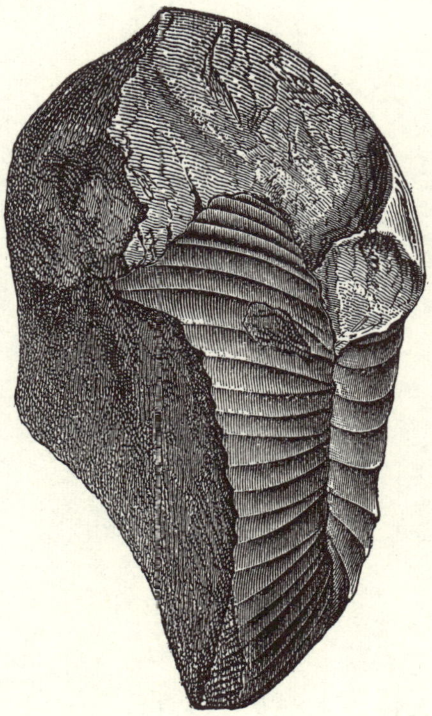

Fig. 2.

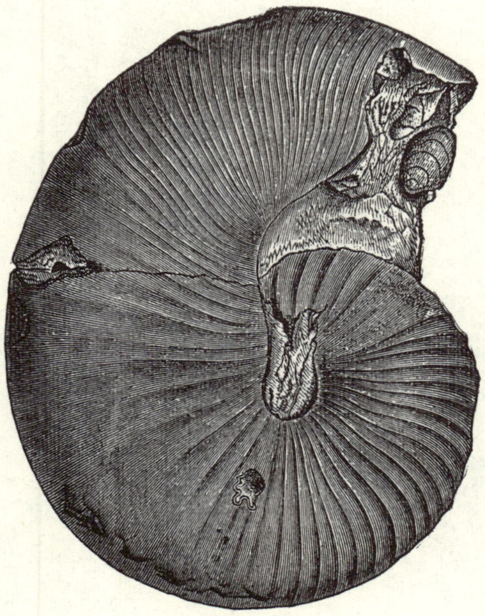

Fig. 7.

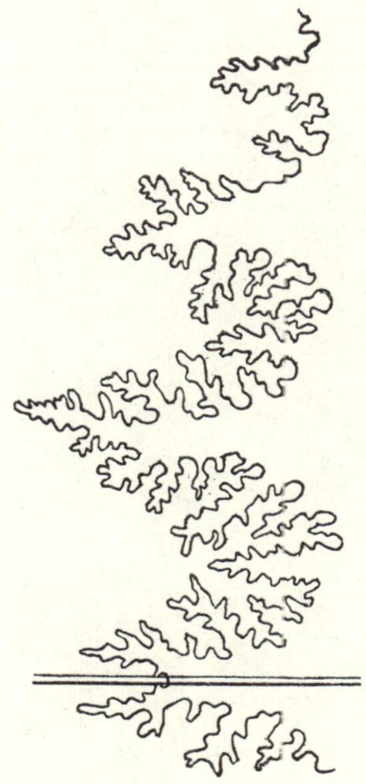

Fig. 3.

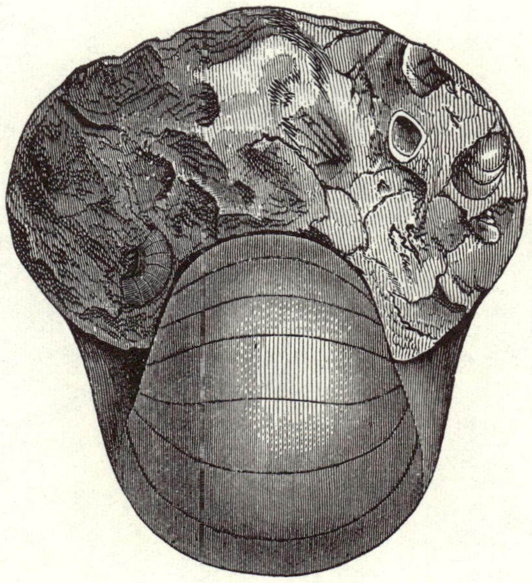

Fig. 9.

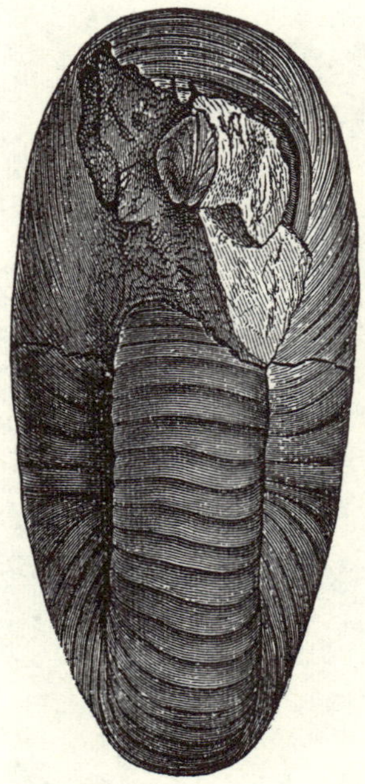

Fig. 4.

Fig. 5.

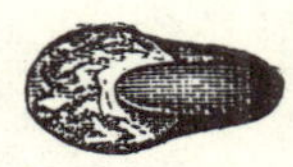

Fig. 6.

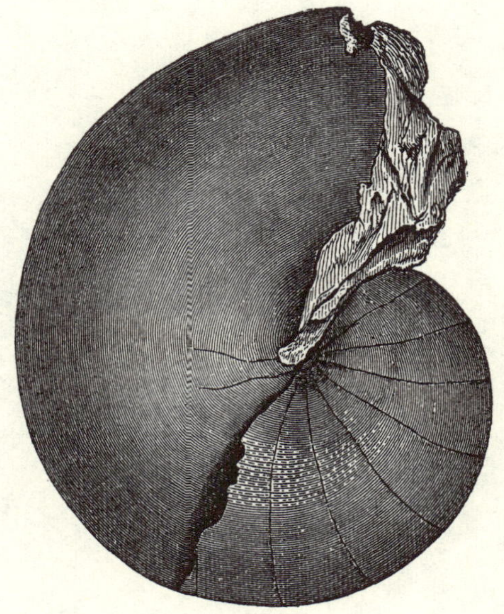

Fig. 10.

H. Smith, del., Montreal.] Fig. 8. [G. G Vasey, Engraver.